ROUTLEDGE HANDBOOK OF PLANNING AND MANAGEMENT OF GLOBAL STRATEGIC INFRASTRUCTURE PROJECTS

This book examines complex challenges in managing major strategic economic and social infrastructure projects. It is divided into four primary themes: value-based approach to infrastructure systems appraisal, enabling planning and execution, financing and contracting strategies for infrastructure systems and digitising major infrastructure delivery.

Within these four themes, the chapters of the book cover:

- the value and benefits of infrastructure projects
- planning for resilient major infrastructure projects
- sustainable major infrastructure development and management, including during mega events
- improving infrastructure project financing
- stakeholder engagement and multi-partner collaborations
- delivering major infrastructure projects effectively and efficiently
- whole-life-cycle performance, operations and maintenance
- relationship risks on major infrastructure projects
- public-private partnerships, design thinking principles, and innovation and technology.

By drawing on insights from their research, the editors and contributors bring a fresh perspective to the transformation of major strategic infrastructure projects. This text is designed to help policymakers and investors select and prioritise their infrastructure needs beyond the constraining logic of political cycles. It offers a practical set of recommendations for governments on attracting private capital for infrastructure projects while creating clear social and economic value for their citizens. Through theoretical underpinning, empirical data and in-depth informative global case studies, the book presents an essential resource for students, researchers, practitioners and policymakers interested in all aspects of strategic infrastructure planning, project management, construction management, engineering and business management.

Edward Ochieng is Professor of Project Management at the British University in Dubai, Dubai.

Tarila Zuofa is a Senior Lecturer in Project Management at Manchester Metropolitan University, UK.

Sulafa Badi is an Associate Professor of Management and Organisational Behaviour at the British University in Dubai, Dubai.

ROUTLEDGE HANDBOOK OF PLANNING AND MANAGEMENT OF GLOBAL STRATEGIC INFRASTRUCTURE PROJECTS

Edited by
Edward Ochieng, Tarila Zuofa and Sulafa Badi

LONDON AND NEW YORK

First published 2021
by Routledge
2 Park Square, Milton Park, Abingdon, Oxon OX14 4RN

and by Routledge
52 Vanderbilt Avenue, New York, NY 10017

Routledge is an imprint of the Taylor & Francis Group, an informa business

© 2021 selection and editorial matter, Edward Ochieng, Tarila Zuofa and Sulafa Badi;
individual chapters, the contributors

The right of Edward Ochieng, Tarila Zuofa and Sulafa Badi to be identified as the authors
of the editorial material, and of the authors for their individual chapters,
has been asserted in accordance with sections 77 and 78 of the Copyright,
Designs and Patents Act 1988.

All rights reserved. No part of this book may be reprinted or reproduced
or utilised in any form or by any electronic, mechanical, or other means, now
known or hereafter invented, including photocopying and recording, or in
any information storage or retrieval system, without permission in writing from the publishers.

Trademark notice: Product or corporate names may be trademarks or registered trademarks,
and are used only for identification and explanation without intent to infringe.

British Library Cataloguing-in-Publication Data
A catalogue record for this book is available from the British Library

Library of Congress Cataloging-in-Publication Data
Names: Ochieng, Edward, editor. | Zuofa, Tarila, editor. | Badi, Sulafa, editor.
Title: Routledge handbook of planning and management of global strategic infrastructure projects / edited by
Edward Ochieng, Tarila Zuofa and Sulafa Badi.
Description: Milton Park, Abingdon, Oxon ; New York, NY : Routledge, 2021. | Includes bibliographical
references and index.
Subjects: LCSH: Infrastructure (Economics)--Planning. | Infrastructure (Economics)--Management.
Classification: LCC HC79.C3 R685 2021 (print) | LCC HC79.C3 (ebook) | DDC 363.6068/4--dc23
LC record available at https://lccn.loc.gov/2020037308
LC ebook record available at https://lccn.loc.gov/2020037309

ISBN: 978-0-367-47748-6 (hbk)
ISBN: 978-1-003-03638-8 (ebk)

Typeset in Bembo
by KnowledgeWorks Global Ltd

CONTENTS

List of figures	xvii
List of tables	xix
List of vignettes	xxi
Notes on contributors	xxiii
Preface	xxxiii
List of abbreviations	xxxvii

1 Introduction   1

Edward Ochieng, Tarila Zuofa and Sulafa Badi

1.1   *Learning outcomes of this chapter 1*
1.2   *Primary aims and objectives of the book 1*
1.3   *The structure of the book 2*
1.4   *Definition and scope of strategic infrastructure systems 3*
1.5   *Global infrastructure outlook 6*
1.6   *Trends in global infrastructure spending 7*
1.7   *Covid-19: the impact on construction and infrastructure 9*
1.8   *The impact of fiscal and monetary policies on infrastructure spending 10*
1.9   *Need for sustainable infrastructure 12*
1.10   *Chapter summary 14*

PART I
Value–based approach to infrastructure systems appraisal   17

2 Significance: The need for better benefits realisation in megaprojects   19

Kate Davis, Jeffrey Pinto and Francesco Di Maddaloni

2.1   *Introduction 19*
      2.1.1   *Chapter aim and objectives 19*
      2.1.2   *Learning outcomes 20*
2.2   *Overview of public infrastructure projects 20*

2.3   *Cui bono? Defining the nature of project value 21*

2.4   *Value as satisfaction versus consumption 22*

2.5   *Elements of project value management 23*

2.6   *Propositional elements in understanding project value management 26*

2.7   *Benefits realisation: resources, competences, and capabilities 28*

2.8   *Management of stakeholders vs. management for stakeholders 29*

2.9   *Benefits realisation and stakeholder perceptions 30*

2.10 *Chapter summary 34*
     2.10.1   *Chapter discussion questions 34*

2.11 *Case study: High Speed 2 (United Kingdom) 35*
     2.11.1   *Actual and forecasted costs 36*
     2.11.2   *Reasons for timescale and cost overruns 36*
     2.11.3   *Delivery of expected benefits 36*
     2.11.4   *Key stakeholders 40*
     2.11.5   *Case study discussion questions 41*

3 Master planning for resilient major infrastructure projects   47
Tarila Zuofa

3.1   *Introduction 47*
     3.1.1   *Chapter aims and objectives 48*
     3.1.2   *Learning outcomes 48*

3.2   *Characterising key resilience concepts 49*
     3.2.1   *Robustness 49*
     3.2.2   *Adaptability 49*
     3.2.3   *Flexibility 49*

3.3   *Resilience 50*
     3.3.1   *A global perspective on resilience 51*

3.4   *Climate resilient infrastructure 52*
     3.4.1   *Planning and designing climate-resilient infrastructure 54*

3.5   *Appraising strategies for strengthening infrastructure resilience: a sectorial approach 55*
     3.5.1   *Cities and urban developments 55*
     3.5.2   *Power sector 57*
     3.5.3   *Transportation sector 58*
     3.5.4   *Water infrastructure 59*

3.6   *Towards resilient infrastructure 61*
     3.6.1   *Any alternative tools for measuring resilience success? 61*
     3.6.2   *Advancing the aspiration for resilient major infrastructure 63*

3.7   *Chapter summary 64*
     3.7.1   *Chapter discussion questions 64*

3.8   *Case study: Remembering the Haitian 2010 earthquake 65*
     3.8.1   *Case discussion questions 65*

Contents

4 Achieving sustainable major infrastructure projects: development and
   management   71
   Tarila Zuofa
   4.1   *Introduction 71*
         4.1.1   *Chapter aims and objectives 72*
         4.1.2   *Learning outcomes 72*
   4.2   *Towards sustainable development and sustainable infrastructure 73*
   4.3   *Global initiatives and the call for sustainable infrastructure 75*
         4.3.1   *An overview of global sustainability initiatives 75*
         4.3.2   *The global call for sustainable infrastructure 76*
   4.4   *Prerequisites for implementing major sustainable infrastructure 79*
   4.5   *The importance of project leadership in sustainable infrastructure development 84*
         4.5.1   *Enhancing sustainability practices through
                 sustainability leadership 85*
   4.6   *Systems and tools for stimulating sustainable infrastructure 87*
         4.6.1   *Sustainability benchmarking and continuous improvement 88*
         4.6.2   *Attaining best practice in benchmarking
                 major infrastructure projects 89*
   4.7   *Achieving sustainable major infrastructure projects
         development and management post-COVID-19 91*
   4.8   *Chapter summary 93*
   4.9   *Chapter discussion questions 93*
   4.10 *Case study: The Eko Atlantic city project 94*
         4.10.1 *Preamble 94*
         4.10.2 *The Eko Atlantic City development project 95*
   4.11 *Case discussion questions 96*

5 Planning for sustainable infrastructure development during mega-events:
   an Expo 2020 case study   103
   Samih Yehia and Ashly Pinnington
   5.1   *Introduction 103*
         5.1.1   *Mega-event and sustainable infrastructure development 103*
         5.1.2   *Chapter aim and objectives 104*
         5.1.3   *Learning outcomes 105*
   5.2   *The core area of mega-event sustainability 105*
         5.2.1   *Transportation 105*
         5.2.2   *Utilities 106*
         5.2.3   *Construction 107*
   5.3   *Building green major infrastructure development 108*
   5.4   *Role of leadership in reducing carbon emissions 108*
   5.5   *Importance of planning for legacy to achieve sustainable infrastructure 109*
   5.6   *Integrating green principles within the awarding
         system for hosting sustainable mega-event 110*

5.7 *Applying green considerations within the design stage 110*

5.8 *Sustainability of the mega-event life cycle 111*

5.9 *Chapter discussion questions for further reflection 112*

5.10 *Expo 2020 case study 114*

    5.10.1 *Sustainable construction 115*

    5.10.2 *Sustainable utilities 117*

    5.10.3 *Sustainable mobility 120*

5.11 *Chapter summary 125*

5.12 *Review and case discussion questions 126*

    5.12.1 *Recommendation 1: hosting a sustainable mega-event requires developing a deep understanding of a mega-event's complex, multiple impacts 126*

    5.12.2 *Recommendation 2: mega-events require intensive planning 127*

    5.12.3 *Recommendation 3: developing sustainable infrastructure for hosting a mega-event requires multi-stakeholder engagement occurring on multiple levels 128*

    5.12.4 *Recommendation 4: leadership engagement on environmental impacts increases attention to its attainment 128*

    5.12.5 *Recommendation 5: sustainable design and planning are a key factor 129*

    5.12.6 *Recommendation 6: legacy consideration should be planned during the design stage 130*

    5.12.7 *Recommendation 7: leadership emphasis on sustainability is mandatory for sustainable infrastructure development of mega-event 130*

6 Stakeholder engagement in major infrastructure projects   135
Hemanta Doloi

6.1 *Introduction 135*

    6.1.1 *Chapter aim and objectives 136*

    6.1.2 *Learning outcomes 137*

6.2 *Defining stakeholders 137*

6.3 *The identification of stakeholders and their interest 138*

6.4 *Tools and techniques for effective stakeholder engagement 139*

6.5 *Formulation of appropriate stakeholder engagement strategies 141*

    6.5.1 *Identification of stakeholders, project contexts and issues 141*

    6.5.2 *Mapping of stakeholders with project issues 141*

    6.5.3 *Stakeholder interest, impact and expectations 142*

    6.5.4 *Evaluation of social performance 142*

    6.5.5 *Monitor, control and maintain 142*

6.6 *Assessment of stakeholders impacts and their inter-relationships 143*

6.7 *Social network analysis and network characteristics 144*

    6.7.1 *Evaluation of social value performance 145*

6.8 *Chapter discussion questions 147*

Contents

6.9   *Case study 148*
     6.9.1   *Case study: A foreshore development project (Project A) 148*
     6.9.2   *Case discussion questions 150*
6.10   *Chapter summary 152*

7   Managing multi-partner collaborations on major infrastructure projects   155
   Diana Ominde and Edward Ochieng
  7.1   *Introduction 155*
     7.1.1   *Chapter aim and objectives 155*
     7.1.2   *Learning outcomes 156*
  7.2   *Planning for a comprehensive multi-partner framework for infrastructure systems delivery 156*
     7.2.1   *Collaboration mediators 157*
     7.2.2   *Knowledge integration capacity 158*
     7.2.3   *Project collaboration quality 159*
  7.3   *Challenges and barriers to multi-partner collaborations 161*
     7.3.1   *Organisational culture 161*
     7.3.2   *Management approaches 162*
     7.3.3   *Relational challenges 162*
     7.3.4   *Communication 164*
     7.3.5   *Benefits of multi-collaborations 165*
     7.3.6   *Faster innovation 165*
     7.3.7   *Efficiency improved 166*
  7.4   *Standardising multi-collaborations best practices 166*
     7.4.1   *Stakeholder engagement 166*
     7.4.2   *Communication standards and tools 167*
  7.5   *Chapter summary 168*
     7.5.1   *Chapter discussion questions 169*
  7.6   *Case study 169*
     7.6.1   *Nairobi-Naivasha section 169*
     7.6.2   *Relationship with the community 170*
     7.6.3   *Case discussion questions 170*

PART II
Enabling planning and execution   **173**

8   Delivering major infrastructure projects effectively and efficiently 175
   Ambisisi Ambituuni
  8.1   *Introduction 175*
  8.2   *Chapter aim and objectives 175*
     8.2.1   *Learning outcomes 176*
  8.3   *Prioritising major infrastructure projects 176*
  8.4   *Major infrastructure prioritisation and cost-benefit analysis 177*
  8.5   *Appraising the current infrastructure situation and government needs 179*

Contents

8.6   *Creating a vision and goals for the future infrastructure assets 181*
       8.6.1   *Creating an SDG-based vision 181*
       8.6.2   *Understanding the tensions of infrastructure vision and goals 182*
       8.6.3   *Drafting an infrastructure vision 184*
8.7   *Ensuring major infrastructure projects are delivere deffectively and efficiently 184*
8.8   *Finalising the plan for major infrastructure projects 186*
8.9   *Moving from planning to delivery 188*
8.10   *Managing delays and cost overrun 188*
8.11   *Chapter summary 191*
8.12   *Chapter discussion questions 191*
8.13   *Case studies 192*
       8.13.1   *Case 1: Country-based comparison of project appraisal processes in decision support 192*
       8.13.2   *Case 2: Delivering an SDG-based major infrastructure projects: The Indonesian case 193*

9   Enhancing the whole life-cycle performance of major infrastructure projects   197
    Edward Ochieng
    9.1   *Introduction 197*
       9.1.1   *Chapter aim and objectives 198*
       9.1.2   *Learning outcomes 198*
    9.2   *Whole life costing of infrastructure investment 198*
       9.2.1   *Life cycle cost analysis 199*
       9.2.2   *Embedding circular economy principles into infrastructure whole life costing 201*
    9.3   *Collaborative procurement models 203*
    9.4   *Fostering early contractor involvement 207*
    9.5   *Mandating the use of building information modelling 210*
    9.6   *Reducing the costs at preliminary design phase and pre-construction phase 214*
    9.7   *Chapter summary 216*
       9.7.1   *Chapter discussion questions 216*
    9.8   *Case study: E39 coastal highway route infrastructure project (Norway) 217*
       9.8.1   *Technical aspects of the project 217*
       9.8.2   *Case discussion questions 218*

10   The role of operations and maintenance in infrastructure management   221
    Edward Ochieng and Diana Ominde
    10.1   *Introduction 221*
       10.1.1   *Chapter aim and objectives 222*
       10.1.2   *Learning outcomes 222*
    10.2   *Definition of operations and maintenance 223*
    10.3   *Infrastructure maintenance and operations research 225*
    10.4   *Implementing operations and maintenance best practices 227*

Contents

        10.4.1 *Maximise asset utilisation 231*
        10.4.2 *Enhance quality for users 231*
  10.5 *Alignment of infrastructure spending with economic growth 238*
  10.6 *Measuring infrastructure spending 239*
        10.6.1 *Measuring performance of infrastructure 241*
        10.6.2 *Performance indicators for transport infrastructure 242*
  10.7 *Operations and maintenance challenges 245*
  10.8 *Chapter summary 245*
        10.8.1 *Chapter discussion questions 246*
  10.9 *Case: Infrastructure systems in the United States 246*
        10.9.1 *How does the United States compare internationally? 247*
        10.9.2 *Case discussion questions 247*

PART III
Financing and contracting strategies for infrastructure systems   **251**

11 Improving the financing and development of major infrastructure projects   253
    Edward Ochieng and Maria Papadaki
  11.1 *Introduction 253*
        11.1.1 *Chapter aim and objectives 254*
        11.1.2 *Learning outcomes 254*
  11.2 *Project identification capacity 254*
        11.2.1 *Project identification risks 257*
        11.2.2 *Project identification transaction costs 258*
        11.2.3 *Project identification market and other disciplines
                 on investment decisions 259*
        11.2.4 *Project identification analytical framework 260*
  11.3 *Sustainable infrastructure governance in developing,
       developed and emerging economies 262*
        11.3.1 *Key constraints of achieving sustainable infrastructure
                 development in developed and developing economies 266*
  11.4 *General principles of infrastructure cost benefit analysis (CBA) 268*
        11.4.1 *Infrastructure project appraisal through CBA 270*
  11.5 *Investment financing methods: assessment options 283*
        11.5.1 *Private activity bonds 284*
        11.5.2 *Public-private partnerships 284*
        11.5.3 *Cross-border public-private partnership 285*
        11.5.4 *Regional infrastructure funds 288*
        11.5.5 *Project bonds 289*
        11.5.6 *Innovative infrastructure financing 291*
  11.6 *Chapter summary 292*
        11.6.1 *Chapter discussion questions 292*
  11.7 *Case: Improving infrastructure financing in Brazil 292*
        11.7.1 *Case discussion questions 293*

Contents

12 Regulatory process for infrastructure systems development 297
Nicholas Chileshe and Neema Kavishe

12.1 *Introduction 297*
    12.1.1 *Infrastructure regulatory frameworks 297*
    12.1.2 *Chapter aim and objectives 298*
    12.1.3 *Learning outcomes 298*

12.2 *Approaches to evaluating regulatory effectiveness 298*
    12.2.1 *Regulatory effectiveness – the case of Tanzania 299*
    12.2.2 *Components of regulatory effectiveness 299*

12.3 *How to recognise good and bad infrastructure regulations 300*
    12.3.1 *Background to Tanzania 300*
    12.3.2 *Good regulations 300*
    12.3.3 *Examples of bad regulations 301*
    12.3.4 *Best practice, recommendations for good governance 302*

12.4 *Managing political and regulatory risk 302*
    12.4.1 *Conceptualisation of risks 303*
    12.4.2 *CSFs for managing RM practices for infrastructure projects 305*

12.5 *Benchmarks for regulatory governance key standards 306*
    12.5.1 *Benchmarking principle 1: Independence 307*
    12.5.2 *Benchmarking principle 2: Accountability 307*
    12.5.3 *Benchmarking principle 3: Transparency and public participation 307*
    12.5.4 *Benchmarking principle 4: Predictability 308*
    12.5.5 *Benchmarking principle 5: Clarify roles 308*
    12.5.6 *Benchmarking principle 6: Completeness and clarity in rules 308*
    12.5.7 *Benchmarking principle 7: Proportionality 309*
    12.5.8 *Benchmarking principle 8: Requisite powers 309*
    12.5.9 *Benchmarking principle 9: Appropriate institutional characteristics 309*
    12.5.10 *Benchmarking principle 10: Integrity 309*

12.6 *Reviewing standards, procedures and tools 309*
12.7 *Infrastructure contracts and achieving international commitments 310*
12.8 *Chapter summary 312*
12.9 *Chapter discussion questions 313*
12.10 *Case study: Insights into Tanzania's delayed infrastructure project planned, delayed and suspended: Bagamoyo Port project 313*
    12.10.1 *The China factor 314*
12.11 *Case discussion questions 316*

13 Managing relationship risks on major infrastructure projects 321
David Bryde, Simon Taylor and Roger Joby

13.1 *Introduction 321*
    13.1.1 *Chapter aim and objectives 321*
    13.1.2 *Learning outcomes 322*

13.2 *Performance and productivity of major infrastructure projects 322*

Contents

13.3 *The nature of relationships in major infrastructure projects 323*

13.4 *Relational risk 324*

13.5 *Relational governance 325*

13.6 *Relational contracting 326*

13.7 *Frameworks for managing relational risk 329*

    13.7.1 *Relational indicators/critical success factors (CSFs) 329*

    13.7.2 *Organisational climate 330*

13.8 *Frameworks for relationship management 331*

13.9 *Chapter summary 334*

    13.9.1 *Chapter discussion questions 334*

13.10 *Case study (for the purpose of confidentiality projectlocation and name withheld) 335*

    13.10.1 *Introduction 335*

    13.10.2 *Overview of the programmes 335*

    13.10.3 *Programme A 336*

    13.10.4 *Programme B 337*

    13.10.5 *What can we learn from these two different approaches? 338*

    13.10.6 *Case discussions questions 339*

14 Empowering public–private partnership in major infrastructure systems   343

Neema Kavishe and Nicholas Chileshe

14.1 *Introduction 343*

    14.1.1 *Empowering infrastructure public-private partnerships 343*

    14.1.2 *Chapter aim and objectives 344*

    14.1.3 *Learning outcomes 344*

14.2 *Preparing PPPs 344*

14.3 *The challenges for PPPs in Tanzania 345*

    14.3.1 *Inadequate PPP skills and knowledge 346*

    14.3.2 *Poor PPP contract and tender documents 346*

    14.3.3 *Inadequate project management and monitoring by the public sector 346*

    14.3.4 *Inadequate legal framework 347*

    14.3.5 *Misinformation on financial capacity of private partners 347*

    14.3.6 *Lack of competition 347*

14.4 *Managing the rigorous PPP process 347*

    14.4.1 *Classify priority project 349*

    14.4.2 *Appraise as PPP 349*

    14.4.3 *Configure as PPP projects 349*

    14.4.4 *Prepare PPP contract 349*

    14.4.5 *Manage PPP process 350*

    14.4.6 *Manage PPP agreement 350*

14.5 *Conducting a bankable feasibility study 351*

14.6 *Structuring a balanced risk allocation 352*

14.7 *Creating a conducive PPP environment 352*

Contents

14.8   *Chapter summary 354*
14.9   *Chapter discussion questions 355*
14.10 *Case study: Lessons from a failed Dege Eco Village PPP housing project 355*
      14.10.1 *PPP preparation 356*
      14.10.2 *Regulatory issues 356*
      14.10.3 *Contractual issues 357*
      14.10.4 *Risks 357*
14.11 *Case discussion questions 357*

PART IV
Digitising major infrastructure delivery   361

15 Applying design thinking principles on major infrastructure projects   363
   Ximing Ruan and Geraldine Hudson
   15.1 *Introduction 363*
       15.1.1 *Chapter aim and objectives 364*
       15.1.2 *Learning outcomes 364*
   15.2 *Design thinking 364*
   15.3 *Design thinking models 369*
   15.4 *Systems thinking 371*
   15.5 *Systems thinking on infrastructure projects 372*
       15.5.1 *Hard systems thinking model 376*
       15.5.2 *Soft systems methodology 377*
       15.5.3 *Tools for systems thinking 377*
   15.6 *Systems thinking models for major infrastructure projects 388*
   15.7 *Chapter summary 393*
       15.7.1 *Chapter discussion questions 394*
   15.8 *Case study: Connecting Bristol strategy (UK) 394*
       15.8.1 *Background 395*
       15.8.2 *Highlights of the SMART city strategy 395*
       15.8.3 *Introduction 395*
       15.8.4 *Being smart and sustainable in response to global and future*
             challenges 396
       15.8.5 *World-class connectivity in Bristol 396*
       15.8.6 *City-wide innovation ecosystem 397*
       15.8.7 *Responsible innovation 397*
       15.8.8 *Public service innovation 397*
       15.8.9 *Explore, enable and lead 398*
       15.8.10 *Bristol design principles 398*
       15.8.11 *Case discussion questions 398*

16 Digital transformation and the cybersecurity of infrastructure systems
in the oil and gas sector   401
Sulafa Badi and Huwida Said
16.1 *Introduction 401*
    16.1.1 *Aim and objectives 402*
    16.1.2 *Learning outcomes 403*
16.2 *The oil and gas sector 403*
16.3 *Digital transformation in the oil and gas sector 404*
16.4 *Cybersecurity risks in the oil and gas sector 404*
    16.4.1 *What is cybersecurity risk? 404*
    16.4.2 *Cybersecurity risks in oil and gas infrastructure systems 412*
16.5 *Cyber-proofing oil and gas infrastructure systems 416*
    16.5.1 *Technological determinants of cybersecurity 416*
    16.5.2 *Organisational determinants of cybersecurity 418*
    16.5.3 *Environmental determinants of cybersecurity 420*
16.6 *Chapter summary 422*
16.7 *Chapter discussion questions 423*
16.8 *Case study: The Baku-Tbilisi-Ceyhan pipeline cyberattack 423*
    16.8.1 *Case discussion questions 424*

17 Infrastructure megaprojects as enablers of digital innovation transitions   431
Eleni Papadonikolaki and Bethan Morgan
17.1 *Introduction 431*
    17.1.1 *Chapter aim and objectives 432*
    17.1.2 *Learning outcomes 432*
17.2 *Theoretical background and knowledge gap 432*
    17.2.1 *Innovation footprint of infrastructure megaprojects 432*
    17.2.2 *Contextual transitions from innovation to digital innovation 433*
    17.2.3 *Research setting of innovation in infrastructure megaprojects 434*
    17.2.4 *Knowledge gap 435*
17.3 *Research method and data 435*
    17.3.1 *Methodological rationale 435*
    17.3.2 *Methods for data collection and analysis 436*
17.4 *Data presentation and findings 436*
    17.4.1 *Framing digital innovation in UK infrastructure megaprojects 436*
    17.4.2 *Pathway 1: Calling for cultural and technological
change – Substitution (up to 1998) 437*
    17.4.3 *Pathway 2: Reorienting towards new digital
technologies – Transformation (1998–2011) 438*
    17.4.4 *Pathway 3: Legitimising digital innovation – Reconfiguration
(2011–2016) 440*
    17.4.5 *Pathway 4: Renewing megaprojects via digital
innovation – Re-alignment (2016–ongoing) 441*

17.4.6 *Mapping multi-level transitions of digital innovationin the UK construction sector* 442

17.5 *Discussion* 444

    17.5.1 *Megaprojects as niches for digital innovation* 444

    17.5.2 *Identifying transition pathways of digital innovation in the UK construction sector* 444

    17.5.3 *Contribution to theory and knowledge* 445

    17.5.4 *Managerial implications* 445

17.6 *Chapter summary* 446

    17.6.1 *Chapter discussion questions* 446

17.7 *Case study: Tideway's digital evolution: how digitalisation drives productivity* 447

    17.7.1 *Case discussion questions* 447

Index   451

FIGURES

1.1	Standard classification of economic and soft infrastructure systems 5
2.1	Timeline of the development of value management 23
2.2	Project value 23
2.3	Chain of benefits 25
2.4	Five project value propositions 27
2.5	Current proposed HS2 route 37
2.6	Actual costs 38
2.7	Available funding and current forecast cost and schedule estimates 39
3.1	Principles for achieving resilience in water infrastructure 62
3.2	Advancing the aspiration for resilient major infrastructure 63
4.1	Four dimensions of infrastructure sustainability 74
4.2	Global infrastructure trends and needs, including investments to meet SDGs 77
4.3	Prerequisites for implementing major sustainable infrastructure 79
4.4	Benchmarking guidelines 90
4.5	Integrated framework for delivery of sustainable infrastructure 92
4.6	Partial pictorial view of Eko Atlantic City 94
4.7	Part view of Eko Atlantic City 95
6.1	Four broad categories of stakeholders and underlying characteristics 140
6.2	Schematic diagram of the stakeholder engagement process in public projects 143
6.3	Conceptual measurement of social value performance against the threshold value over the project life cycle 144
6.4	Two-mode network between the respondents and project issues in Project A (based on degree centrality in the impact network) 149
6.5	Concentric map of the respondents in Project A in the impact network 150
6.6	Social value performance in Project A (design stage) 151
8.1	Cost-benefit analysis used by the UK government 178
8.2	SDGs 183
8.3	Moving from vision to delivery of major infrastructure projects 187
9.1	An illustration of a whole life cycle cost analysis 200
9.2	BIM regularisation manifesto 213
9.3	An illustration of a submerged floating tube bridge 218
10.1	Doctor Yellow 225
10.2	Maintenance and replacement cost of infrastructure evaluated 228

xvii

Figures

10.3	A summary of asset utilisation strategies 232
10.4	A summary of the four maintenance strategies 237
10.5	Life cycle energy framework for road transport infrastructure 240
10.6	Stakeholders in the transport sector and performance measurement 243
11.1	Efficiency considerations of good infrastructure investment decisions 256
11.2	Analytical infrastructure framework 261
11.3	Public investment management assessment (PIMA) framework 264
11.4	An all-inclusive methodical of the strengths and weaknesses of a nation's public infrastructure 265
11.5	Example of cumulated and punctual probability distribution of the ENPV 282
11.6	Public-private partnership model 286
11.7	Cost of PPP financing, relative to government debt finance 287
11.8	Principal stages of a project issuance 290
12.1	Diagnostic infrastructure project risk management 304
14.1	Typical PPP process 348
14.2	Interrelated PPPs' institutional capacity 354
15.1	Phases, activities and deliverables across a construction project life cycle 366
15.2	Three competing forces 368
15.3	Design thinking model 369
15.4	Design thinking model 370
15.5	The potential impact of systems thinking 373
15.6	Transforming to systems thinking 374
15.7	Seven-stage model of SSM 378
15.8	Types of systems and projects 379
15.9	Types of project management methods 381
15.10	Application of systems thinking to system development life cycle 382
15.11	Systems thinking fishbone diagram 383
15.12	Rich picture diagram 384
15.13	Actor map 385
15.14	Concept map 386
15.15	Trend map 387
15.16	Causal loop diagram 388
15.17	An example of an enterprise system 392
15.18	The zones of complexity 393
15.19	Bristol SMART city essential pillars 395
16.1	Oil and gas sector technological development trajectory 405
16.2	Digital transformation in the oil and gas sector 409
16.3	Cybersecurity risk 411
16.4	The oil and gas operation streams 413
17.1	Theoretical setting of the chapter framed around MLP: (1) actors and social groups, (2) rules and institutions, (3) digital technologies and STS (infrastructure megaprojects) based on Geels (2004) 435
17.2	Timeline of digital innovation in UK construction influencing and being influenced by institutions, actors and megaprojects as socio-technical systems (STS) of digital innovation 443

TABLES

2.1 Reasons for cost and time deviations 40
2.2 Key stakeholders 41
3.1 Aspects of resilience 50
3.2 Definitions of resilience 51
3.3 Resilience improvement strategies for selected transportation infrastructure 60
4.1 Guiding principles for dimensions of sustainability 75
4.2 Summary of common leadership theories and their main features 84
4.3 Sustainability rating systems and phase applicability 87
7.1 Elements of collaboration quality modified 160
8.1 Strategic Infrastructure Planner Framework: 14 economic infrastructure readiness parameters 180
8.2 Project performance metrics 194
9.1 An overview of the main collaborative contracting models 205
10.1 Current status of infrastructure in the United States as appraised by the American Society of Civil Engineers 227
10.2 Procurement levels that can be considered by operators 234
10.3 French highways sustainability framework 235
10.4 Key determinants of performance indicators 244
11.1 Investment risk in infrastructure 257
11.2 Credit information reinforces market signals 259
11.3 An example of the choice of the counterfactual scenario 269
11.4 Calculation of the return on investment ($/£/EUR thousands) 274
11.5 Calculation of the return on national capital (£/$/EUR thousands) 274
11.6 Financial sustainability (£/$/EUR thousands) 275
11.7 Sensitivity analysis 279
11.8 Switching values 279
11.9 Results of Monte Carlo simulation 281
11.10 Infrastructure investment by sector 283
11.11 Sources of infrastructure financing 294

xix

Tables

12.1 Summary of eight of the 10 mega-infrastructure projects in Africa funded by China 311
14.1 List of identified risks associated with PPPs in a project life cycle 353
15.1 Top ten design principles 365
15.2 Design principles for national infrastructure 367
15.3 Design thinking models 369
15.4 Systems thinking techniques 380
15.5 The five axioms of systems thinking 389
15.6 Corollaries of the five axioms of systems thinking 390
15.7 Clarifications of systems 394
16.1 Main digital technologies and their applications in oil and gas infrastructure systems 406
16.2 Points of entry for different cyberattacks and vulnerabilities across the oil and gas operation streams 414
16.3 Cybersecurity regulatory frameworks 420
17.1 Transition pathways of digital innovation in UK construction along the threecategories (institutions, actors and technologies), based on frameworkby Geels (2004) 437

VIGNETTES

3.1 The deep-water container terminal (DCT) in Gdansk, Poland 53
3.2 Snapshot of resilient urban and city infrastructure developments in selected areas 56
4.1 A tale of two major infrastructure projects 78
10.1 Maintenance management of Tokaido Shinkansen case study 224

CONTRIBUTORS

Edward G. Ochieng, PhD, PGCertHELT, MSc, BSc(Hons), FHEA, AHEA, is a Professor of Project Management at the British University in Dubai. Prior to joining the university, Edward taught in four different UK institutions: London South Bank University, Cranfield University, Liverpool John Moores University and Robert Gordon University. Edward has extensive experience and knowledge relating to organisational challenges and solution development for managing large capital and heavy engineering projects. Edward's research interest with people and organisational challenges continues but has now been complemented by the need for a wider understanding of infrastructure development, management of project processes, project value creation, capital effectiveness, project complexity and political economy and the management of projects. Edward has edited 31 book chapters and published 4 books, 28 book chapters and over 90 refereed papers in high-ranking journals and conferences. He collaborated with academics, industrialists and the Infrastructure Project Authority (IPA-UK) to examine frameworks that could be used to identify and estimate benefits and value (including capital effectiveness) at the front end of public sector projects. He has secured an estimated £717,825.00 from a range of funders for several projects and a consultancy in project management. In addition, he collaborated with academics and industrialists to investigate the promising new technology cryogenic energy storage (CES) to solve the problem of how to store excess renewable energy; the research team was awarded €7 million. He has supervised nine PhD students to successful completion and over 300 MSc/MBA industrial projects by research. Edward is currently supervising 10 PhD students.

Tarila Zuofa, PhD, PGCertHE, MSc, BTech, FHEA, PRINCE2, is a Senior Lecturer in Project Management at Manchester Metropolitan University. He is also the Programme Leader for the MSc in Project Management. Since joining the university, he has supported the revalidation of MSc Project Management through the development and production of materials for different new modules. Tarila has worked at several other UK HEIs as Teaching Fellow, Lecturer, Senior Lecturer,

Senior Teaching Fellow as well as visiting lecturer in a Middle East HEI. In some of these universities, he was responsible for preparing documentation that resulted in the successful accreditation of their MSc Project Management programmes by leading professional bodies like the Association for Project Management. Over the years, Tarila's main teaching and research have had a consistent emphasis on risk management in the oil and gas sector and project management in general. However, this is now complemented by the need to understand soft issues in project management such as sustainability, governance, and stakeholder integration primarily in developing countries and then other countries. Tarila currently supervises PhD students as well as MSc/MBA projects by research and has acted as external examiner in some UK universities. Tarila has published in and served as reviewer for several high ranking, built environment engineering and business journals. He has been invited to present his research findings in leading international conferences like PMI Research and Education Conference and the Academy of Management Annual Meeting. Prior to joining the academic world, Tarila was an offshore marine engineer and undertook projects mainly in the Gulf of Guinea. He is also a PRINCE2® Practitioner, which allows him to apply his Project Management knowledge in consultancy activities across various sectors and in projects of any scale.

Sulafa Badi, BSc Arch, MSc, PhD, FHEA, is Associate Professor of Management and Organisational Behaviour at the British University in Dubai and Honorary Senior Fellow at the Bartlett School of Construction and Project Management, University College London (UK). She started her career as an architect before joining academia and holds an MSc in Construction Economics and Management and a PhD in Project Management, both from University Collge London. Sulafa's research has explored the management of large infrastructure projects in a wide range of contexts. Her research has had both academic and practical impact and involved studies in the United Kingdom, China, India and the Middle East. She secured research funding from the UK Engineering and Physical Sciences Research Council (EPSRC), the Royal Institute of Chartered Surveyors (RICS), Transport for London (TfL), Innovate UK and the British University in Dubai. Sulafa has presented the findings of her research at many international research conferences, winning several Best Paper awards. She also published her research in leading international journals, including *Industrial Marketing Management*, *Project Management Journal*, *Construction Management and Economics*, and the *International Journal of Managing Projects in Business*. Sulafa is an Associate Fellow of the Higher Education Academy (UK) and has taught at both Master and Doctoral levels in the subjects of project management, supply chain management, social networks in project and enterprise organisations, organisational behaviour, operations management and managing change in organisations.Book Contributors:

Kate Davis is a Senior Lecturer at Kingston University in London specialising in the areas of strategic organisational project management, consultancy and designing and delivering courses that have a real impact. She has particular expertise in the development and implementation of modules with large student numbers, pioneering the development and deployment of technology to improve teaching and learning. She is recognised for innovative approaches to promote interaction with employers resulting in high graduate employment. She completed her PhD in 2016 which examines multiple stakeholder perceptions of project success. She also participates in research projects in the emerging topic of organisational project management. She is currently developing qualifications in Strategic Organisational Project Management, in consultation with a panel of industry experts, to enhance Kingston University's reputation in the field of project management.

Jeffrey K. Pinto is the Andrew Morrow and Elizabeth Lee Black Chair in the Management of Technology in the Sam and Irene Black School of Business at Penn State University, Behrend College. He is the lead faculty member for Penn State's Master of Project Management program and was a visiting scholar at the Kemmy School of Business, University of Limerick. The author or editor of 28 books and over 150 scientific papers that have appeared in a variety of academic and practitioner journals, books, conference proceedings, video lessons, and technical reports, his work has been translated into nine languages. He has served as keynote speaker and as a member of organizing committees for a number of international research conferences. Dr. Pinto served as Editor of the *Project Management Journal* from 1990 to 1996. With over 30 years' experience in the field of project management, Dr. Pinto is a two-time recipient of the Distinguished Contribution Award from the Project Management Institute (1997, 2001) for outstanding service to the project management profession. He received PMI's Research Achievement Award in 2009 for outstanding contributions to project management research. In 2017, he was honored with the International Project Management Association's Research Achievement Award for his research career in project management. He has written a best-selling college textbook, *Project Management: Achieving Competitive Advantage*, published by Pearson and currently in its fifth edition. Dr. Pinto recently recorded a set of project management lectures for Pearson's LiveLessons video series. With Dr. Ray Venkataraman, he is the author of a second textbook, *Operations Management: Managing Global Supply Chains*, published by Sage and currently being revised for a second edition.

Francesco Di Maddaloni is a Lecturer in Project Management at the Kingston Business School, Kingston University of London. His main research interests are in the areas of project governance and strategic planning, with particular attention on the social aspects and sustainability of major infrastructure and construction projects at the local, regional, and national level. On these themes he has presented at numerous conferences and published in the *International Journal of Project Management*, where he is also a member of the International Editorial Board. Francesco is the lead of the Project Management Research Group at Kingston University with the aim of attracting talented researchers, building collaborative links between industry and academia and making a positive impact on businesses, economy, and social environments. He is Teaching Fellow at the Higher Education Academy (FHEA) and PRINCE2 qualified.

Samih Yehia is a part-time Lecturer at the British University in Dubai specialising in the area of sustainable business management. He has particular expertise in the development and implementation of business modules that insure business continuity and sustainable growth. He is recognised for creating and enhancing innovative approaches to promote business models that consider multiple pillars of the sustainability. He completed his PhD in 2020 which examined multiple business sustainability contemplations during the preparation stage of hosting a mega-event. He is currently developing business models for mega-manufacturers in the capacity of senior manager.

Ashly H. Pinnington is Professor in Management at the British University in Dubai; previously for 12 years he was Dean of Faculty and then Dean of Research. Prior to joining the university in 2007, Ashly worked in the UK and Australia for 15 years in research intensive universities and 6 years in post-1992 universities. He holds a PhD in Management (1991) awarded by Henley Business School, University of Reading. His current research interests are in the Anthropocene and Sustainability, Social Responsibility, Talent Management and the Management of Professionals. Ashly has published in numerous journals including *Organization and Environment, Journal of Management Studies, Human Relations, Organization Studies, British Journal of Management, Human Resource Management Journal, International Journal of Human Resource Management, Group and Organisation Management*, and the *International Journal of Project Management*. He has collaborated with academic researchers residing mainly in the UK, Australia and the UAE, and worked on research projects funded by the European Commission, Leverhulme Trust, Australian Research Council, European Social Fund, and the UAE Federal Demographic Council.

Hemanta Doloi is currently an Associate Professor in Construction Management in the Faculty of Architecture, Building and Planning at the University of Melbourne, Australia. He is the director of the Smart Villages Lab (SVL) that focuses on the data-driven research in the area of "rural construction and development" under the auspice of the SmartVillages program. He leads the trans-disciplinary research in smart villages for developing solutions for affordable housing and infrastructure systems, generating new theories for education and governance and empowering rural communities. Hemanta won the Australian Institute of Quantity Surveyor's Infinite Value award in teaching and research, recognising the excellence of scholarships impacting wider community in the profession. He serves as editorial members of a few prestigious journals including the *International Journal of Project Management, Built Environment Project and Asset Management* and *Construction Innovation: Information, Process, Management*. He is a founding Chair of the *International Conference on Smart Villages and Rural Development (COSVARD)*. He has widely published in the areas of project management, infrastructure planning and policy and construction economics and management.

Diana Ominde is a Doctoral Fellow at Strathmore University. She is currently pursuing a PhD in Project Management at Strathmore Business School. Her research predominantly focuses on stakeholder management value creation and optimisation of information technology projects. Her research interests include appraisal of mega projects, mega project complexity, stakeholder integration, optimisation of ICT projects, multi-partner collaboration and project-based organisational development. In 2019, Diana presented at the PMI congress in Delft, The Netherlands, where she shared her knowledge on project complexity, delivery and operational effectiveness of projects. Diana has published one refereed journal paper and two conference papers. She holds a Bachelor's of Commerce degree in Finance and Management Science and Masters of Commerce in Management Science. She has vast experience in project management and organisational development.

Ambisisi Ambituuni is a Senior Lecturer in Project Management at the School of Strategy and Leadership, Coventry University, United Kingdom. He is a Certified Prince2 Project Practitioner with research interests in project and risk management and has published a number of peer-reviewed journal articles in this area. Ambi holds a PhD from the School of Civil Engineering and Geoscience at Newcastle University, UK, for a body of work which developed a risk management framework for the supply and distribution of petroleum products. Ambi is also interested in understanding the interface between engineering projects and knowledge transfer/management. He is also interested in regulatory developments, corporate social responsibilities and supply chain management. Ambi is a recipient of the prestigious Petroleum Technology Development Fund (PTDF) for both his MSc and PhD. He also contributed and won numerous research grants, notable amongst which is his contribution to the NERC funding application whilst at Newcastle which, amongst other factors, earned a consortium of universities including Newcastle Durham, Glasgow, St. Andrews and Stirling Universities the sum of £5 million as research funding from NERC. Ambi is an experienced academic with leadership responsibilities across Coventry University and at course and module levels. He teaches at both UG and PG levels and also supervises a number of PhD students. He holds an MSc in Project Management (with distinction), a BSc (Hons) in Building Engineering, and a NEBOSH General Certificate in Health and Safety. He is a Certified e-Business Professional and a Fellow of the UK Higher Education Academy. His industry experience spans across the construction industry, oil and gas, transport and telecommunications industry.

Maria Papadaki is an Assistant Professor at the British University in Dubai (BUiD) and a Managing Director for the Dubai Centre for Risk and Innovation (DCRI). She has over 10 years of experience in Risk Management from both academia and industry, with numerous years in the implementation, development, improvement and management of risk frameworks, tools and techniques. Maria worked previously at Rolls-Royce plc, leading different roles in the areas of Enteprise Risk Management, Project Management and Supply Chain Management. She was also appointed by the University of Manchester as a Senior Relationship Manager and a Head of the PMO office for BUiD. Maria initiated the idea and led the development of the first innovation Hub (H2B) in Crete, Greece, on behalf of the Heraklion Chamber of Commerce. She joined BUiD in 2016, as a Managing Director for DCRI, and lead the development of digital certificates for BUiD's graduates which made the University first in the Middle East and third in the world implementing blockchain technology in education. In 2018, she was appointed to the Board of Directors for the Institute of Risk Management in London. Under this portfolio she is leading the global education and training standards strategy for the Institute . Maria is a visiting Lecturer at the University of Manchester and her research focuses on innovation, blockchain, artificial intelligence, and enterprise risk management.

Contributors

Nicholas Chileshe is an Associate Professor in Construction and Project Management in the Academic Unit of UniSA STEM, at the University of South Australia. He is also the Deputy Director of Scarce Resources and Circular Economy (ScaRCE) University Research Concentration. Nicholas obtained his MSc in Civil Engineering from Vinnitsa Polytechnic Institute (USSR) in 1988, then an MSc in Construction Management in 1996, and a PhD in Construction Management in 2004 from Sheffield Hallam University, UK. A qualified Bridge Engineer by profession, Nicholas Chileshe is actively engaged in scholarly research work and has authored and co-authored more than 250 refereed journal and conference publications since joining academia in 1999. His current research interests include contemporary issues in developing countries such as public-private partnerships (PPPs), waste management and reverse logistics. He is interested in exploring how some of the concepts – such as quality management, risk management and reverse logistics – can act as catalysts for the evaluation of sustainability issues within the Construction and Project Management disciplines. Dr. Chileshe currently serves on a number of Editorial Boards such as the *International Journal of Project Management* and *International Journal of Building Pathology and Adaptation* and since January 2019 as one of the Deputy Editors of the *Engineering, Construction and Architectural Management Journal*.

Neema Kavishe is a Quantity Surveyor by profession and currently works as a Lecturer in the School of Architecture Construction Economics and Management (SACEM) at Ardhi University (Dar es Salaam) for the past 10 years. She is also an acting Director for the Ardhi University Endowment Fund. Neema completed her PhD in Civil Engineering from the University of Birmingham, UK. Her PhD Thesis was on the subject of Public-Private Partnership (PPP) and she is a Certified Public-Private Partnership (PPP) Foundation. Neema graduated from the University of Dar es Salaam in Tanzania with an honours degree in Building Economics in 2007 and received an MSc degree in Construction Economics and Management from the Ardhi University in Dar es Salaam Tanzania in 2010. She has experience in teaching and supervising dissertations at both undergraduate and postgraduate levels. She is also actively involved in research and has a number of publications in referred journals and conference proceedings.

David J. Bryde is Director of Research and Knowledge Transfer and Professor of Project Management at Liverpool Business School, John Moores University. Professor Bryde is particularly interested in how projects contribute to sustainable development through the design, procurement and delivery of new products, services, systems and infrastructures. This includes topics such as relational/psycho-social, lean/agile project management, sustainability and the operation of networks in project-intensive organisations. His most recently completed research projects which were funded by the Association of Project Management (APM), investigated relational aspects in the management of outsourced projects and coping strategies used by project managers to deal with challenging situations. He is currently the Lead Project Coordinator for the EU Marie Sklodowska Curie Research and Innovation Staff Exchange (RISE) Horizon 2020 Work Programme "Being Lean and Seen: Meeting the challenges of delivering projects successfully in the 21st century." This is a four-year project involving nine academic and non-academic partners from the UK, Germany and Malaysia. See https://www.ljmu.ac.uk/micro sites/being-lean-and-seen for more details. He is also currently involved in an APM funded research project exploring how the social and knowledge networks of directors influence aspects of Corporate Environment Performance in project-extensive organisations. Professor Bryde is widely published, with over 100 journal papers, research monographs, book chapters, conference presentations, invited keynote speeches/guest lectures/presentations, expert interviews and articles. He has supervised in over 20 doctoral students to successful completion and has examined numerous PhD/DBAs as an external and an internal examiner. With Jake Holloway and Roger Joby, he is author of *A Practical Guide to Dealing with Difficult Stakeholders*, published by Gower.

Simon Taylor's career started in the steel manufacturing industry in 1995, learning mechanical design and fabrication. As technology advanced so did his skill set and he moved into product design for the food service sector working for clients such as Nestle, Unilever and Virgin. As part of the end to end product design process, planning and project management became more and more part of his working life and this became something he decided to focus on as he took a job as a project manager working with the London Underground supply chain. Being accustomed to processes and procedures within the LU network he moved to work directly for London Underground during the PPP contract (public-private partnership) working closely with Metronet and their suppliers. It was here that his career as a planner finally took shape working on stations and line upgrade projects. He became a full-time planner working on the Victoria Line Upgrade (VLU) and was then promoted to Programme Planning Manager in 2006. After successful early delivery of the VLU he became head of planning for deep tube line upgrades and subsequently head of planning at Transport for London (TfL) where he was responsible for planning within capital projects across all transport modes. He was also heavily involved in planning and now controls career development, including direct responsibility for the TfL planning apprenticeship. Simon joined the £56 billion High Speed Two programme in March 2015 as Head of Programme Planning where he was responsible for planning and planning capability on the largest infrastructure programme in Europe. He is now a co-founder of th3rd curve ltd, ex board member for the Association for Project, an active member of the APM Planning Monitoring and Control SIG, and co-author of the *APM Guide on Planning, Monitoring and Control*.

Roger Joby is a Visiting Research Fellow at Liverpool Business School, John Moores University and an international project management consultant and educator with over 40 years' experience, principally for Clinical Research Organizations. He was the recipient of the Pharmaceutical Contract Management Group (PCMG) Lifetime Fellowship Award in 2019.

Ximing Ruan is a Senior Lecturer in Strategy and Operations Management at Bristol Business School, University of the West of England and remains active in research. He supervises PhD students and also supervised DBA students for University of Liverpool. He has also worked with colleagues to successfully secure a £0.8 million research fund, which started in April 2016 for 24 months. Dr. Ruan has received British Council-Newton Funds to present research outcomes at six international workshops in 2016 and 2017, which were organised in Thailand (March 2016), Mexico (May 2016), China (August and October 2016), Kazakhstan (November 2016) and Brazil (May 2017). Dr. Ruan presented his research findings as keynote speaker at the International Conference on Management, Economics and Finance in Istanbul in 2016 and 2019.

Geraldine Hudson is a Senior Lecturer in Strategy and Operations Management at Bristol Business School, University of the West of England. Since joining UWE, Geraldine has redeveloped and led the Honours programme in Project Management and has contributed to Project Management teaching at the postgraduate level. She is a dissertation supervisor on the MSc and MBA programmes. Geraldine is completing her Doctorate in Education at this institution on the pedagogy of Project Management. Preceding teaching, Geraldine had fifteen years' experience working in Project Management with a specialism in guiding product development projects to launch.

Huwida Said is an Associate Professor in the College of Technological Innovation at Zayed University, Dubai. Recently he was appointed Assistant Chair of the Computer and Applied Technology department (CAT) in the same college. Dr. Said received her Bachelor of Engineering in 1995 in Electrical and Electronics Engineering from the University of Wales Swansea, UK, and a Ph.D. in 2000 in Computer Sciences from the University of Reading, UK. Her teaching interests are Computer Information Security, Database Security, Information Security Fundamentals, Network Security, Internet and Web Security, Wireless Security, Digital Computer Forensics, and Introduction to Programming. Dr. Said's research interests include Computer Information Security, Computer Networking, Internet Security, Computer Forensics and recently Blockchain and Database Systems. She is the recipient of several prestigious awards and research grants. She has served as a chair and TPC member of several IEEE conferences and as a reviewer for prestigious journals. She is a member of several research labs at Zayed University and professional organisations.

Eleni Papadonikolaki is an Associate Professor in Building Information Modelling (BIM) and Management at the Bartlett School of Construction and Project Management at University College London (UCL). She is a co-partner in a Digital Outlook partnership and a consultant in the area of digital innovation and management. Eleni holds a PhD on the Alignment of Partnering with Construction IT from Delft University of Technology in the Netherlands, a MSc in Digital Technologies, also from Delft and an Engineering Diploma in Architectural Engineering from the National Technical University of Athens (NTUA), Greece. Bringing practical experience of working as an architect engineer and design manager since 2006 on a number of complex and international projects in Europe and the Middle East, she is researching and helping teams manage the interfaces between digital technology and management. Eleni is a Member of the Association for Project Management (MAPM), Member of the Architects Registration Board (ARB) in the UK and a Chartered Engineer of the Technical Chamber of Greece (TEE-TCG). Eleni is the author of over 50 peer reviewed publications, for instance, in the *International Journal of Project Management, Construction Management and Economics*

and others. She has secured and delivered collaborative research projects £5.5 million as Principal and Co-Investigator funded by the European and UK research councils. Eleni is teaching at postgraduate and executive levels and is a steering committee member and researcher in the UCL Construction Blockchain Consortium (CBC), a social enterprise to promote and co-create digital innovations in the built environment. She is the Director of the MSc in Digital Engineering Management at UCL where she develops the new generation of project leaders for digital transformation.

Bethan Morgan, PhD, MBA, BA (Hons), CIM, currently explores the profound opportunities and challenges facing the products and production of the built environment created through digital transformation. Beth combines substantial management skills with expertise in the teaching and research of digitalization in the construction industry. She is building on this experience in her current role as an Honorary Lecturer at UCL and co-founder of Digital Outlook Partnership. Beth's particular research interests lie in the organizational and institutional changes created and necessitated by digitalization, and the evolution of digital capabilities in firms and industries. She holds a PhD titled "Organizing for technology in practice: implementing Building Information Modelling in a design firm" and is the holder of research grants and a EPSRC Fellowship. Beth holds a MBA (Distinction) from the University of Nottingham, UK, and BA (Hons) in Architectural Studies from the University of Newcastle upon Tyne. She has authored a variety of journal and conference papers. Prior to starting her academic career, Beth gained extensive industry and policy experience working for major firms and most recently as Director of Research and Innovation at Constructing Excellence. A qualified teacher in delivering Harvard Teaching Cases, together with Dr. Papadonikolaki she is pioneering the introduction of this pedagogy to the Built Environment sector through Digital Outlook.

PREFACE

Strategic infrastructure projects are the backbone that interconnects our modern economies. The most strategic infrastructure investments are functional and create the greatest impact in terms of economic growth, social uplift and sustainability. It is generally assumed that for every dollar of capital spent on public infrastructure investment the gross domestic product of the country will increase by approximately US $0.05 to US $0.25 trillion. Thus, the competitive economic advantage clearly depends on a country's infrastructure vision and long-term planning. In fact, infrastructure projects will only drive sustained economic growth when it is properly aligned with the country's priorities. In other words, it is vital for policymakers to work at the topmost position of their countries' agendas to successfully convert their strategies into action. Yet, there is no succinct book available to help policymakers and investors in selecting and prioritising their infrastructure needs beyond the constraining logic of political cycles.

As infrastructure contributes to the economic output of a nation, they are an essential measure of productivity. However, developing efficient and cost-effective interdependent or interconnected social and economic infrastructure projects is challenging. When deciding what to build, infrastructure planners and policymakers often need to balance complex political, economic, social, and environmental trade-offs. At the same time, economic and social infrastructure projects require large capital investments, and public finances are increasingly strained. Stakeholders often face a difficult choice; they optimise an existing, flawed asset or they build something new; either they continue with a proven asset or they invest in sustainable technology. Interdependent or interconnected economic and social infrastructure projects that proceed often exceed timelines and budgets.

As shown in this book, governments around the world are facing an acute need for new or modernised infrastructure projects; the estimated shortfall in global infrastructure debt and equity investment is at least US $1 trillion per year. There is no fundamental scarcity of private capital – investors are frequently falling short of their target allocations. Despite infrastructure's attractiveness as an asset class and the reduced role of traditional financing, investors are struggling to find opportunities that are globally competitive on a risk-adjusted return basis. Driven by the need to embrace long-term solutions for increasingly complex and interconnected global challenges and strategies that support a more sustainable and inclusive future, this book offers a practical set of recommendations for governments on attracting private capital for infrastructure projects while creating clear social and economic value for their citizens. Through theoretical underpinning, empirical data and case studies, the book provides support to students, researchers, practitioners and policymakers as they continue the challenging quest of improving and understanding the complex strategic planning and management of infrastructure projects. By drawing on insights from their research, the editors and contributors bring a fresh perspective to the transformation of infrastructure projects. Many of our observations and recommendations can be utilised in developed and developing economies.

Key observations and proposed areas for further research

Within the context of this book, the trends and research themes that will shape the future of infrastructure systems include:

Digital technologies: As established from the reviewed literature, technology is having an unprecedented impact on the delivery of infrastructure projects. There has been an increase of usage of robots – from autonomous rovers that can be used to enhance the efficiency and quality of site inspections to mechanical arms that can automate highly repetitive tasks like brick-laying. The introduction of automation construction sites could significantly enhance productivity while ensuring a safer work environment. For instance, drones have become increasingly common in infrastructure project delivery. More research is required to look into how drones can be used to improve safety and productivity in complex construction sites.

- The rise of artificial intelligence has made a mark on infrastructure projects: however, more research is required to look into digital capabilities surrounding predictive design, digital building twins and the use of augmented reality at planning, design and construction phases. This can lead to the elimination of waste and cost savings. Internationally, infrastructure systems still remain among the least digitally transformed. As shown in this book, many conventional infrastructure systems have been designed, built, operated and maintained in much the same way for decades, despite huge strides made in other sectors. It is worth mentioning that the main barrier to the adoption of innovation has always been lack of an enabling environment from governments.
- More research is required to look into how clients and contractors can create long-term effective partnership models for delivery and funding. The infrastructure research community needs to look into how the use of a life cycle approach to infrastructure delivery works with all phases of an infrastructure project: feasibility, planning, design, construction, maintenance and operation).
- As infrastructure operators, owners and planners need to become more adept at using data analytics; it will be vital for the research community to look into how data analytics can be used to achieve operational efficiency. In addition, more research is needed to look into how the use of data analytics can be used to enhance the overall infrastructure planning process and create a much stronger integration between supply and demand. This could lead to a more accurate evidence-based decision-making process throughout the infrastructure ecosystem.
- Project monitoring is moving beyond documenting cost overruns and construction delays to include more real-time and forward-looking insights. More research is required to look into how the application of digital technology and real-time data can be used by infrastructure project managers to make better long-term informed decisions around scheduling labour and materials for major infrastructure projects. Digitally optimised infrastructure operations can provide infrastructure project managers with actionable data that can be used to put long-term infrastructure projects back on track quickly.

Preface

Intended audience

This book examines some key issues relevant for students, practitioners and policymakers interested in managing interdependent and interconnected infrastructure projects. Academic researchers and PhD students can also use this book to pursue new directions in major strategic economic and social infrastructure project management. It is the first inclusive reference book in this field, incorporating value and benefits of infrastructure projects, planning for resilient major infrastructure projects, sustainable major infrastructure development and management, sustainable infrastructure development during mega-events, stakeholder engagement in large infrastructure projects, the whole-life-cycle performance of major infrastructure projects, operating and maintaining infrastructure projects, improving infrastructure projects financing, the regulatory process for major infrastructure development, relationship risks on major infrastructure projects, a public-private partnership in major infrastructure projects, design thinking principles on major infrastructure projects, innovation and technology for major sustainable infrastructure projects. To ensure the alignment of the above infrastructure themes, the book has been divided into four primary themes: a value-based approach to infrastructure projects appraisal, planning and execution, financing and contracting strategies for infrastructure projects and digitising major infrastructure projects.

ABBREVIATIONS

AAAA	Addis Ababa Action Agenda
ADB	Asian Development Bank
AEC	Architecture, Engineering and Construction
AfDB	African Development Bank Group
AHEL	Azimio Housing Estate Limited
AHS	Affordable Housing Schemes
AI	Artificial Intelligence
AIIB	Asian Infrastructure Investment Bank
AIM	Asset Information Model
APM	Association of Project Management
AR	Augmented Reality
ARP	Address Resolution Protocol
ASDSO	Association of State Dam Safety Officials
ASTM	American Society for Testing and Materials
ASP	Active Server Pages
BAU	Business-as-Usual
BCRs	Benefit-cost Ratios
BIAC	Benchmark Indicative Asset Cost
BIG	Bjarke Ingels Group
BIM	Building Information Modelling
BINAC	Benchmark Indicative Non-Asset Cost
BNEF	Bloomberg New Energy Finance
BOO	Build Own Operate
BOOT	Build, Own, Operate, Transfer
BOT	Build-Operate-Transfer
BRI	Belt and Road Initiative
BS	British Standard
BSI	British Standards Institute
BTC	Baku-Tbilisi-Ceyhan
CAPEX	Capital Expenditure
CASS	Corporate Administrative Support Services
CAV	Connected and Autonomous Vehicles
CBA	Cost-Benefit Analyses

Abbreviations

CBRMS	Collaborative Business Relationship Management System
C and C	Command and Control
CCPPP	Canadian Council for Public Private Partnerships
CDE	Common Data Environment
CE	Constructing Excellence
CEEQUAL	Civil Engineering Environmental Quality Assessment and Awards
CERs	Certified Emission Reductions
CES	Certification Event Sustainability
CERT	Computer Emergency Response Team
CIM	Construction Information Modelling
CIOB	Chartered Institute of Building
CIOs	Chief Information Officers
CO$_2$e	CO$_2$-equivalent
COP21	Conference of Parties 21
CPI	Consumer Price Index
CPSs	Cyber-Physical Systems
CRM	China Railway Materials
CSF	Critical Success Factors
CSIS	Center for Strategic and International Studies
CSR	Corporate Social Responsibility
CTSS	Corporate Technology Support Services
CVF	Competing Values Framework
DB	Design and Build
DBFO	Design-Build-Finance-Operate
DBFOT	Design Build Finance Operate Transfer
DBOM	Design-Build-Operate-Maintain
DBOT	Design Build Operate Transfer
DBS	Development Bank of Singapore
DCT	Deep-water Container Terminal
DC	Degree Centrality
DC	Design Construct
DEA	Data Envelopment Analysis
DEC	Dubai Exhibition Centre
DEFRA	Department for Environment, Food and Rural Affairs
DEWA	Dubai Water and Electricity Authority
DFI	Direct Foreign Investments
DfT	Department for Transport
DIG	Debt-Investment-Growth
DNS	Domain Name Server
DoS	Denial-of-Service
DOT	US Department of Transport
DP	Dubai Port
DSA	Debt Sustainability Assessments
DSCs	Decision Support Centres
DTC	Dubai Taxi Corporation
DWC	Dubai World Central
DWTC	Dubai World Trade Centre

Abbreviations

DXB	Dubai International
EAC	East African Community
EBRD	European Bank for Reconstruction and Development
ECC	Engineering and Construction Contract
ECC	Economic Rate of Return
ECI	Early Contractor Involvement
ECOWAS	Economic Community of West African
EHS	Environmental, Health, and Safety
EIA	Environmental Impact Assessment
EIU	The Economist Intelligence Unit
ELC	Events Life Cycle
EMPOWER	Emirates Central Cooling System Corporation
EMS	Environmental Management System
EN	European Standard
EPEC	European PPP Expertise Centre
EPC	Engineering Procurement and Construction
EPCM	Engineering Procurement Construction Management
EPS	Earnings per Share
ESG	Environmental, Social and Governance Principles
ESIA	Environmental and Social Impact Assessments
ERR	Economic Rate of Return
ETC	Energy Transitions Commission
EU	European Union
EV	Eigenvector Centrality
FCC	Forecast Life Cost
FCMB	First City Monument Bank
FDR	Financial Discount Rate
FNPV	Financial Net Present Value
FRR	Financial Rate of Return
GCC	Gulf Cooperation Council
GHG	Greenhouse Gas
GIB	Global Infrastructure Basel
GICA	Global Infrastructure Connectivity Alliance
GII	Global Infrastructure Initiative
GoH	Government of Haiti
GOT	Government of Tanzania
GREs	Government-related Entities
GSISS	Global State of Information Security Survey
GVA	Gross Value Added
GWh	Gigawatt hours
HFA	Hyogo Framework for Action
HNWI	High-net-worth Individuals
HRF	Haiti Reconstruction Fund
HSAP	Hydropower Sustainability Assessment Protocol
HST	Hard System Thinking
HS1	High Speed One
HS2	High Speed Two

Abbreviations

HTTP	Hypertext Transfer Protocol
HTTPS	Secure Hypertext Transfer Protocols
HVAC	Heating, Ventilation, and Air Conditions System
IA	Infrastructure Australia
IaDB	Inter-American Development Bank
IAI	International Alliance for Interoperability
ICE	Institution of Civil Engineers
ICG	Infrastructure Client Group
ICIF	International Centre for Infrastructure Futures
ICLEI	Local Governments for Sustainability
ICRC	International Committee of the Red Cross
ICS	Industrial Control Systems
ICT	Information and Communication Technologies
IDA	International Development Association
IDS	Intrusion Detection
IFC	International Finance Corporation
IFG	Institute for Government
IISD	International Institute for Sustainable Development
INCOSE	International Council on Systems Engineering
INVEST	Infrastructure Voluntary Evaluation Sustainability Tool
IPA	Infrastructure and Projects Authority
IPD	Integrated Project Delivery
IPCC	Intergovernmental Panel on Climate Change
IPP	Independent Power Producer
IPS	Intrusion Prevention Systems
IRR	Internal Rate of Return
IS	Infrastructure Sustainability Rating Scheme
ISI	Institute for Sustainable Infrastructure
ISO	International Organisation for Standardisation
IoT	Internet of Things
ITS	Intelligent Traffic Solutions
IT	Information Technology
JPCERT/CC	Japan Computer Emergency Response Team Coordination Centre
JS	Javascript
KADCO	Kilimanjaro Airports Development Company
KOC	Kuwait Oil Company
KwIDF	Kuwait Intelligent Digital Field
LA	Licensing Agency
LCCA	Life Cycle Cost Analysis
LCC	Life Cycle Cost
LEED	Leadership in Energy and Environmental Design
LMX	Leader-Member Exchange
LoC	Loss of Control
LoV	Loss of View
MAC	Media Access Control
MDB	Multilateral Development Banks
MED	Multi-effect Desalination Process

Abbreviations

MIGD	Million Imperial Gallons per Day
MITM	Man-in-the-Middle
MLP	Multi-Level Perspective
MSF	Multi-stage Flashing
MTDS	Medium Term Debt Management Strategy
MW	Megawatts
NAIC	National Infrastructure Advisory Council
NAO	National Audit Office
NCE	New Climate Economy
NEC	New Engineering Contract
NGO	Non-governmental Organisations
NHC	National Housing Corporation
NIA	National Infrastructure Assessment
NIB	Nordic Investment Bank
NIC	National Infrastructure Commission
NPFF	National Planning Policy Framework
NPV	Net Present Value
NSIP	Nationally Significant Infrastructure Project
NSSF	National Social Security Fund
OCBC	Oversea-Chinese Banking Corporation
ODD	Organisational Design and Development
OECD	Organisation for Economic Co-operation and Development
OG	Olympic Games
OPEX	Operational Expenditure
OT	Operational Technology
P-A	Principal-Agent
PAB	Private Activity Bonds
PAS	Publicly Available Specifications
PDCA	Plan-Do-Check-Act
PERT	Programme Evaluation and Review Techniques
PHP	Hypertext Preprocessor
PIMA	Public Investment Management Assessment
PIR	Project Initiation Routemap
PLA	Participatory Learning and Action
PLCs	Programmable Logic Controllers
PMBOK	Project Management Body of Knowledge
PMI	Project Management Institute
PPF	Project Preparation Facilities
PPIAF	Public-Private Infrastructure Advisory Facility
P2P	Peer-to-Peer
PPP	Public Private Partnership
PSTM	Project Stakeholder Typology Model
PTA	Public Transport Agency
RA	Rail Agency
RISE	Regulatory Indicators for Sustainable Energy Tool
RLU	Royal Lestari Utama
RMP	Relationship Management Plan

Abbreviations

RO	Reverse Osmosis
ROCE	Return on Capital Employed
RTA	Dubai's Road and Transport Authority
RTUs	Remote Terminal Units
SA	System Analysis
SADC	Southern African Development Community
SAVE	Society of American Value Engineers
SCADA	Supervisory Control and Data Acquisition
SCG	Strategy and Corporate Governance
SDG	Sustainable Development Goals
SDIP	Sustainable Development Investment Partnership
SDR	Social Discount Rate
SDTBL	Sustainable Development Triple Bottom Line
SE	System Engineering
SEPM	Systems Engineering and Project Management
SGR	Standard Gauge Railway
SIS	Safety and Instrument Systems
SNA	Social Network Analysis
SOE	State-owned Enterprises
SOSM	System of Systems Methodologies
SPI	Social Performance Indicators
SSL	Secure Socket Layer
SSM	Soft Systems Methodology
SST	Soft System Thinking
STAR	Southern Tagalog Arterial Road
STAR	Sustainable Transport Appraisal Rating
STS	Socio-Technical Systems
SuRe	Standard for Sustainable and Resilient Infrastructure
T and M	Time and Material
TBL	Triple Bottom Line
TCC	Target-Cost-Contract
TCPS	Traditional Construction Procurement System
TE2100	Thames Estuary 2100 Project
TEU	Twenty-Foot Equivalent Units
TIC	Tanzania Investment Centre
TICTS	Tanzania International Container Terminal Services
TIES	Transport Infrastructure Efficiency Strategy
TIP	Transforming Infrastructure Performance
TLFF	Tropical Landscapes Finance Facility
TLS	Transport Layer Security
TRA	Traffic and Roads Agency
TRS	Total Return to Shareholders
TSO	The Stationary Office
UAVs	Unmanned Aerial Vehicles
UN	United Nations
UNCSD	United Nations Conference on Sustainable Development
UNDRR	United Nations Office for Disaster Risk Reduction

Abbreviations

UNECE	United Nations Economic Commission for Europe
UNESCAP	United Nations Economic and Social Commission for Asia and the Pacific
UNFCC	United Nations Framework Convention on Climate Change
UNISDR	United Nations Office for Disaster Reduction
UNWCDR	United Nations World Conference on Disaster Reduction
URT	United Republic of Tanzania
VA	Value Analysis
VE	Value Engineering
VM	Value Management
VOC	Volatile Organic Compound
VP	Value Planning
VPN	Virtual Private Network
WBS	Work breakdown structure
WCDR	World Conference on Disaster Reduction
WEF	World Economic Forum
WRAP	Waste and Resources Action Programme
WTP	Willingness-to-Pay
WSSD	World Summit on Sustainable Development
WWF	Worldwide Fund for Nature
2D	Two-Dimensional
3D	Three-Dimensional
4D	Four-Dimensional
5D	Five-Dimensional

1
INTRODUCTION

Edward Ochieng, Tarila Zuofa and Sulafa Badi

1.1 Learning outcomes of this chapter

In this chapter you will learn about:

- The aim and structure of this book, along with the background to key terminologies related to infrastructure systems, including what is meant by strategic infrastructure systems;
- Definition and scope of strategic infrastructure systems;
- Global infrastructure outlook;
- Trends in global infrastructure spending;
- The impact of the Covid-19 pandemic on construction and infrastructure project delivery;
- The impact of fiscal and monetary policies on infrastructure spending; and
- Need for sustainable infrastructure.

1.2 Primary aims and objectives of the book

The proposed book is aimed at helping policymakers and investors in selecting and prioritising their infrastructure needs beyond the constraining logic of political cycles, offering a practical set of recommendations for governments on attracting private capital for infrastructure projects while creating clear social and economic value for their citizens, and providing support to students, infrastructure researchers, practitioners and policymakers as they continue the challenging quest of improving and understanding the complex strategic planning and management of strategic infrastructure projects.

This book has been designed to accomplish three objectives:

- Demonstrate the importance of long-term strategic infrastructure planning;
- Ascertain the effectiveness of current infrastructure delivery approaches;
- Provide guidance for achieving sustainable infrastructure development that enhances the quality of citizens, maximises benefit realisation, enhances economic development and promotes a more effective and efficient use of the environment and financial resources.

1.3 The structure of the book

The book comprises the following seventeen chapters:

- **Chapter 1** introduces the background to strategic infrastructure systems, global infrastructure outlook, global infrastructure spending, impact of the Covid-19 pandemic on construction and infrastructure, value of fiscal and monetary policies on infrastructure spending and need for sustainable infrastructure.
- **Chapter 2** established the historiography, theory, and advances in our understanding of project benefits realisation in order to better facilitate the development of major infrastructure and construction projects.
- **Chapter 3** focuses on examining how resilience can be enhanced in major infrastructure projects.
- **Chapter 4** focuses on examining how the development and management of major sustainable infrastructure projects can be achieved.
- **Chapter 5** offers an opportunity to advance understanding of infrastructure development and the sustainability practices across the project life cycle of hosting mega-events, and within different sectors for a non-sporting event.
- **Chapter 6** has two broad aims. First, understand the importance and significance of wider stakeholders' involvement in infrastructure projects and second, discuss the processes of engaging stakeholders with appropriate reflection of their needs, requirements, impacts and satisfaction over the entire project life cycle.
- **Chapter 7** evaluates some of the management practices in collaborative infrastructure projects and assesses the modalities through which these management practices influence the outcome of these projects.
- **Chapter 8** explores some of the challenges, issues and processes used by decision makers and governments in the quest to deliver major infrastructure projects that are effective and efficient.
- **Chapter 9** appraises how the whole-life-cycle performance of infrastructure projects can be achieved and sustained.
- **Chapter 10** examines how the operations and maintenance strategies can be used to maintain infrastructure systems efficiently.
- **Chapter 11** examines how the financing and development of infrastructure projects can be achieved and sustained.
- **Chapter 12** identifies mechanisms for enhancing the effectiveness of the regulatory processes for infrastructure system development, through the auditing of the environment by evaluating political and regulatory risks, application of robust benchmarking evaluation systems, standards and contracts.
- **Chapter 13** outlines an operational model for infrastructure project delivery that emphasises behavioural aspects.
- **Chapter 14** examines how governments can enhance existing public-private partnerships models and practices in major infrastructure systems.
- **Chapter 15** primarily examines how infrastructure operators can apply design thinking principles on major infrastructure projects.
- **Chapter 16** explores the key cybersecurity issues and challenges facing oil and gas infrastructure systems as they become more digitised.
- **Chapter 17** delineates the concepts of digital innovation and appraises ways that the public sector can influence digital infrastructure innovation.

Introduction

1.4 Definition and scope of strategic infrastructure systems

Strategic infrastructure systems are the backbone that interconnect global economies. Functional strategic infrastructure systems create the greatest impact in terms of growth, social uplift and sustainability. It is generally presumed that for every dollar in capital spent on public infrastructure systems, the gross domestic product of a nation will increase by approximately US$0.5 to US$0.25 (WEF, 2012). Consequently, competitive economic merit clearly depends on a nation's infrastructure strategic vision and long-term planning; however, from the literature reviewed it was found that there is no succinct model available to assist governments in selecting and prioritising their infrastructure requirements beyond constraining logical political cycles. In fact, as demonstrated by Ochieng *et al.* (2017), infrastructure investments will only drive sustained economic growth when it is well aligned with a nation's priorities. As specified by Ochieng *et al.* (2017), critical infrastructure investments, whether they sustain existing infrastructures or construct new assets, are vital to economic growth. As shown in this book, several countries are not investing enough, which has hampered their progress and deferred an increased burden to the years ahead. In order to realise benefits from infrastructure systems, it is vital for policymakers to address the following three questions:

- What is the best way to select infrastructure ventures that will generate the utmost impact in terms of economic growth, sustainability and social improvement?
- Once selected, what is the best way for designated government departments to ensure that the ventures are delivered sustainably, effectively and efficiently?
- Do we have good governance structures that can demonstrate professionalism in the management of infrastructure projects and be used to monitor existing and new ventures?

As stated by OECD (2017), major infrastructure investments are often controversial. On the one hand, risks and uncertainties, particularly in relation to the appraisal of future requirements, make the decision-making process challenging. It is worth stressing that the provision of sufficient infrastructure is a very essential component of a nation's progression. This recognition is reflected in this book, and further reinforced by Chen and Bartle (2017), who suggested that a well-sustained, robust and efficient infrastructure system can be used to support and sustain a country's economy, improve the quality of life and strengthen global competitiveness. At this juncture, it is vital to address the following two questions:

- So, what does infrastructure mean?
- What is the standard classification of infrastructures?

From the reviewed literature, it was established that there is no standard or agreed-upon definition of infrastructure according to the present usage of the phrase, thus, within the context of this book the following working definitions have been adopted:

- A physical strategic infrastructure system or network that provides significant services to several sectors or societies and facilitates the overarching development of a country.
- A strategic infrastructure system is one that acts as an enabler for a nation's economic progression, social development and competitiveness.

Chen and Bartle (2017) showed that two approaches can be used to define infrastructure systems. One approach entails a narrow definition and refers to infrastructure as economic physical assets that can be used to support private business development. The second approach is a broader definition that views infrastructure as an inclusive physical asset needed to support both private economic activities and social services. The second approach comprises both social and economic infrastructure systems that are essential for societal development. According to Chen and Bartle (2017), social infrastructure entails schools, universities, hospitals, courts, prisons, parks, housing, recreational facilities, libraries, city halls and facilities. Economic infrastructure comprises roads, bridges, tunnels, airports, transit, ports, railways, energy production facilities, distribution networks, telecommunication systems, water and sewer systems and solid systems. Whereas, WEF (2012) went further and identified three types of infrastructure systems that overlap:

- **Economic infrastructure system:** ventures that will generate economic growth and enable the community to operate. Within this classification, WEF (2012) itemised the following infrastructures: transport facilities (air, sea and land), utilities (water, gas and electricity), flood defences, waste management and telecommunications network.
- **Social infrastructure system:** ventures that will assist in the provision of public services. Within this theme, the following were listed: social housing, health facilities, educational establishments and green infrastructure (multifunctional green space within and between urban areas, such as parks, gardens and green corridors), which improves social livelihoods and encourages biodiversity (Kumari and Sharma, 2016).
- **Soft infrastructure system:** public institutions that are needed to maintain the community. Within this theme, the following were identified: central government building laws, rules and systems that are in place to assist in the upkeep of the law and order, enhance educational attainment and address public health issues.

To provide a further granularity, economic, social and soft infrastructure systems have been further sub-divided into:

a. **Functional infrastructure:** an infrastructure system that will work and meet the requirements of the society. For instance, electricity grids that can cope with peak demand and motorways are rarely congested. Multi-purpose water dams can provide energy as well as water for irrigation and flood management.
b. **Strategic infrastructure:** ventures that are functional and create the greatest impact in terms of economic growth, social improvement and sustainability; that is, they provide the highest social benefit. Ochieng *et al.* (2017) exhibited that they can be split into:
    i. **Critical infrastructure:** these infrastructure systems support the socioeconomic development of a country. Examples include flood barriers, power generation and mass transit.
    ii. **Non-critical ventures:** these ventures are not critical; however, they accelerate economic growth or boost key environmental and social agendas. For instance, several countries have minimised their investments on fossil fuels and encouraged investment in renewable energy. In addition, some countries are using their broadband network in rural areas to stimulate rural economic progression.

Based on the above appraisal, the infrastructure services appraised in this book have been sub-divided into economic and social infrastructure systems (see Figure 1.1):

It is worth highlighting that, for social impact, economic infrastructure systems can improve society in a number of ways (WEF, 2012):

Introduction

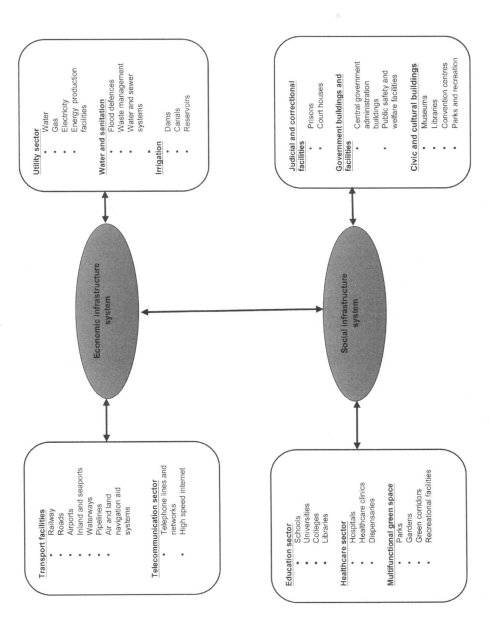

Figure 1.1 Standard classification of economic and soft infrastructure systems

- Transport systems allow the society to access a wider pool of job opportunities and can minimise the prices of commodities bought in shops, helping cut down the level of poverty and inequality.
- Connect homes to electricity supplies which can lead to better educational attainment and improved health and well-being at home and workplace.
- Provision of clean water supplies can lessen the spread of waterborne diseases and improved health equality.
- Accessibility of telephone, broadband and mobile phones makes it easier for society to communicate and undertake business.

From a sustainability and environmental perspective, it has been claimed that new economic infrastructure systems are seen as having a damaging effect on the environment, with increased pollution, the overuse of water resources, negative visual impacts and loss of land. However, as shown in this book, strategic economic infrastructures need not to damage the environment and can actually address certain environmental issues such as:

- Boosting electricity grids in areas with renewable energy potential, helping to widen the energy base and deliver a more sustained energy supply mix.
- Supporting road upgrades that will lead to less congestion, pollution and fuel consumption.
- Introduction of environmental laws and regulations that can be used to improve the quality of infrastructure systems.

1.5 Global infrastructure outlook

Infrastructure lies at the global nexus between economic growth, productive investments, employment creation and poverty alleviation. As demonstrated in this book, the functioning of our societies is based on an extensive interconnection of major infrastructures that ensure the effective functioning of physical assets providing goods and services and institutions from the perspective of governance. Thus, this book considers global strategic infrastructure as large capital assets that fulfil the economic, environmental and social aspects of societal and enterprise requirements. This category of assets may either be public or privately developed, owned and managed. The significance of infrastructure is broadly recognised and well researched; however, the lack of comprehensive historical information presents challenges in predicting how, where and when infrastructure investment in developing and developed economies will be required. As shown in Chapter 11, keeping pace with infrastructure needs remains a challenge and access to funding has been constrained. As populations and incomes grow in developed and developing economies, businesses are demanding more power and water to sustain their production processes. Moreover, they have a growing need for transport infrastructure to move society and goods (GIH, 2017). As observed from the reviewed literature, the past decade has seen growth rates in Asia surpass other continents of the world. Data examined from Global Infrastructure Hub (2017) suggested that the rate of growth in Asia is likely to slow down a little in the period to 2050, from an average of 5.3 percent over the next decade to an average of 3.7 percent. Nevertheless, the Asian continent is expected to account for almost half of global GDP by 2040, with noticeable inferences for accompanying infrastructure investment (GIH, 2017). Interestingly, at 4.2 percent, it has been suggested that Africa is expected to have the quickest growth over the next 20 years, though this is still expected to only cover 4.6 percent of the global economy in 2040.

Introduction

Infrastructure demand in the Middle East and North Africa (MENA) region are substantial, amounting to an annual requirement of US$100 billion, mostly in electricity generation, transport, water utility, sanitation systems, communication, and information technology. However, the region has been troubled with economic and sociopolitical turbulence in recent years. The GDP of oil-exporting countries in the region has been severely impacted by the sharp drop in oil prices by almost 50 percent of their 2014 levels (Arezki, 2020). To stabilise their economies in response to such an unprecedented decline in oil prices, several countries, notably those in the Gulf Cooperation Council (GCC), have introduced strict fiscal reform programmes such as the reduction in capital expenditures, the removal of fuel subsidies, and the introduction of value-added tax to generate revenue. Neighbouring non-oil exporting countries, despite benefitting from lower oil prices, also suffer from negative spillover effects due to reductions in investments, grants, and concessions (Arezki, 2020). Traditionally, most infrastructure financing in the MENA region is from governmental budgets; however, the tightening of public sector spending in recent years may encourage wider use of new forms of infrastructure financing such as public-private partnerships (PPPs) (Arezki, 2020).

At this juncture, it is worth emphasising that demographic changes are key in determining infrastructure demand, and are of course complicatedly aligned with economic growth forecasts. For instance, Africa's relatively strong rate of forecasted economic growth will be boosted by a strong population growth. The continent's population is expected to surpass two billion by 2040, an increase of almost 75 percent over the 2015 figure of 1.2 billion inhabitants (GIH, 2017). As revealed by Global Infrastructure Hub (2017), Africa's share of global inhabitant will increase from 16 percent to 22 percent by the end of the simulated period. Africa will be followed by Oceania; its population has been forecasted to reach 54 million by 2040, an increase of 40 percent. In Europe, it has been suggested that the population is expected to stall in that period at around 700 million inhabitants. Aside from population growth, the distribution of a nation's inhabitants plays a vital role in determining the amount and type of infrastructure that will be required. As nations become more prosperous, populaces have tended to gravitate towards urban areas to take advantage of the economic and social prospects they offer. Due to digital infrastructure transformation, urbanisation is forecasted to continue across all the regions but be sturdiest in countries where income levels will be lower as the proportion of the inhabitants living in urban areas will increase towards the high urbanisation rates observed in the Americas and Oceania (GIH, 2017). It is worth noting that urbanisation is often accompanied by an increase in the proportion of the populaces able to access utilities. Additionally, rising urban populaces typically influence city planning activities, leading to increased investment in road and public transport infrastructures.

1.6 Trends in global infrastructure spending

The construction sector is hugely significant in infrastructure investment, accounting for a disproportionate share of such spending relative to its economy-wide importance (Bivens, 2014). Moreover, because the construction sector is relatively labour intensive compared with many other forms of infrastructure spending (if not compared with economy-wide averages), it has a huge impact on jobs estimates spurred by such expenditure. As implied by McKinsey (2016), the world invests $2.5 trillion a year in the transportation, power, water and telecom systems on which businesses and inhabitants rely. However, this amount continues to fall short of the world's ever-expanding needs, which has resulted in lower economic growth and depreciation of the society essential services. The modelling results reviewed suggested that (GIS, 2017; McKinsey, 2016):

- From 2016 to 2030, the world needed to invest about 3.8 percent GDP, or an average of $3.3 trillion a year, in economic infrastructure just to support the expected rates of growth.
- Since the global financial crisis, infrastructure investment declined as a share of GDP in 11 of the G20 countries. Reductions were mainly in the European Union, the United States, Russia and Mexico. Canada, Turkey and South Africa maximised their infrastructure investments.
- Egypt, Ethiopia, Morocco, Senegal, Benin, Cote D'Ivoire, Ghana, Guinea, Rwanda and Tunisia fall under Compact With Africa ecosystem. These countries have been forecasted to invest a total of US $1.4 trillion in infrastructure between 2016 and 2040, or the US $57 billion per year.
- Institutional investors and banks had $120 trillion assets that could partially be used to boost infrastructure investments. Eighty-seven percent of this capital mainly originated from developed economies, while the largest needs were mainly in middle-income countries.
- With government funds increasingly squeezed to pay for social and economic infrastructure systems, government policymakers will turn to the private sector to fund some of the infrastructure projects.

By continent there is considerable disparity in the proportion of total fixed investment aligned to infrastructure investment; thus, the analysis presented in this book is based on disparity analysis which has been sourced from the literature. A number of studies including Fay *et al.* (2019) have keenly highlighted the infrastructure funding gap up to 2040 and even beyond. Just over 20 percent of total fixed investment in Africa is devoted to infrastructure, compared to 9 percent in the Americas (GIH, 2017). Since 2007, globally infrastructure investment has been dominated by two economic infrastructure systems: electricity and roads, which account for almost two-thirds of total spending. Telecoms and rail economic infrastructure systems have each contributed one-eighth of total spending, and a similar amount comes from investment in water, ports and airports merged. From the simulated data examined, it was found that the structure of infrastructure investment has remained largely consistent throughout 2007; however, 33 percent of investment in the electricity enabled it to slightly increase its share of the total from 34 percent in 2007 to 36 percent in 2015 (GIH, 2017). To meet their infrastructure requirements, countries will have to increase their spending as a proportion of GDP relative to what has been currently spent. It is worth mentioning that Global Infrastructure Hub (2017) appraisal was on regions, and that over half of the predicted infrastructure spending needs to 2040 is subsided by four countries: China, United States, India and Japan. China alone is estimated to account for one-third of global infrastructure investments.

In contrast to previous analysis, presented by McKinsey (2016) and Global Infrastructure Hub (2017), it is suggested that if current trends continue, the global infrastructure investment is likely to reach $3.8 trillion in 2040, an increase of 67 percent over 2015 value in real terms (GIH, 2017). As highlighted in the previous section, this reflects the economic growth and demographic shifts that are forecast over the timeframe to 2040. However, if nations were to meet their best-performing equals in terms of resources they will have to devote to infrastructure, the predicted value of infrastructure investment will need to rise to $4.6 trillion in 2040. It could, therefore, be insinuated that by 2040 there could be a gap of $820 billion between what governments will need to spend if current trends were to continue and what could be required if all nations were to match their best performing equals (GIH, 2017; McKinsey, 2016). To cover this gap, it has been indicated that globally countries will have to spend more than $97 trillion by 2040 to provide sufficient infrastructure to expanding populations (GIH, 2017).

Introduction

1.7 Covid-19: the impact on construction and infrastructure

According to Biorck *et al.* (2020), the infrastructure construction sector, and its wider ecology, erect buildings, infrastructure and industrial structures that are the underpinnings of our economies and vital to our daily lives. The infrastructure construction sector has successfully delivered challenging projects, from undersea tunnels to skyscrapers. However, the sector has performed insufficiently in many regards for an extended period of time. From examining the existing literature, one could allude that the Covid-19 pandemic may be another crisis that is likely to slow the growth of the sector. The impact of the pandemic on the global construction infrastructure sector has been varied depending on how severe country responses have been. A number of nations in the seven continents have used stimulus in the short-term to protect jobs and businesses from lockdown conditions. It could be argued that this will make some nations unable or unwilling to rapidly and substantially increase spending on infrastructure in the coming years to not exacerbate elevated government debt. It is worth highlighting that the global construction and infrastructure sector is already enduring the impact of delays in obtaining fabricated goods, plant and equipment from Covid-19 (Warwick *et al.*, 2020). PwC (2020), showed that the interconnectedness, complexity and goal nature of the global construction sectors supply chains and workforce affected the cost and schedule of infrastructure projects. It could also be hypothetically argued that the slowdown has had an impact on project start dates or stalled construction progress. Moving forward, the following actions could be adopted to mitigate risks aligned to the delivery of infrastructure projects (Biorck *et al.*, 2020; PwC, 2020; Ribeirinho *et al.*, 2020; Warwick *et al.*, 2020):

1. It will be important to have visibility over the project's third-party supply chain and access to that party's data to properly assess the likelihood of supply delays. The focus will have to be on most critical materials, equipment and products.
2. Most construction contracts on major infrastructure projects entail a force majeure clause although the application of a force majeure provision is on specific project tasks. Some of the clauses are general in nature while others are set out as exclusive or non-exclusive lists of project activities that amount to force majeure. It is worth noting that most standard contracts provide multiple courses of action for both contractors and principles to take to mitigate delays and loses. To address this problem, it will be important for parties to ensure that they have a good understanding of how specific provisions of their contractors are likely to operate, timeframes around notice requirements, terms describing triggering events and how practical steps can be adopted to navigate future challenges.
3. Travel restrictions on travel and social distancing will add to the existing supply chain challenges and impact the delivery of infrastructure projects. Infrastructure projects will inevitably experience delays and cost overruns that will impact infrastructure project delivery as a result. These and future challenges will force stakeholders to introduce new levels of resourcefulness to proactively manage the increased risk and adapt to the rapidly changing circumstances.
4. Clients will have to work closely with contractors, designers and suppliers to assess the actual and potential impacts on contract performance and look into how they will potentially mitigate project disruptions, in an effort to achieve the overriding aim of the project. For instance, clients, contractors, designers and engineers will have to rely more on digital collaboration tools as building-information modelling (BIM). Moreover, engineers and contractors will end up using 4D and 5D simulation to re-plan infrastructure projects

and re-optimise schedules. The case for tools that are proven to enhance productivity, such as 4D simulation, digital workflow management, real-time progress tracking and advanced schedule optimization, will become even stronger. Smart buildings and infrastructure that encompasses the Internet of Things (IoT) will enhance data availability and allow more efficient operations as well as new business models, such as performance-based and collaborative contracting. Moreover, digital technologies will lead to better collaboration, greater control of the value chain and shift toward more data-driven decision-making.

5. On the long term, an increase in off-site construction is likely to increase. Constructing in controlled environments will make even more sense in an integrated global world that requires close management of the movement and interaction of construction workers. Such rationale further strengthens the need for off-site construction, beyond the existing quality and speed benefits. It could be therefore implied that contractors will gradually push fabrication off-site and manufacturers will enhance their variety of prefabricated subassemblies. The shift toward a more controlled environment will be even more significant as the COVID-19 pandemic further unfolds. The next phase in the transition to efficient off-site manufacturing will entail the application of automated production systems; this will make construction more like automotive manufacturing.

6. Globally, government actions will be critical in ensuring a quick return to a healthy economy, but finding a compromise between spending and what is acceptable and affordable will be crucial. While investment in other areas of the economy will be required to generate a stimulus effect, infrastructure investment will be needed to address existing challenges and economic growth.

7. Safety and sustainability guidelines and possible standardisation of construction codes will be generated. Requirements for sustainability and work-site safety is likely to increase. Considering COVID-19 studies done so far, it could be suggested that new health and safety regulations will be needed.

As highlighted by IMF (2020), globally countries are projected to experience negative growth in 2020. It is worth noting that there are substantial differences across individual countries, reflecting the evolution of the pandemic and the effectiveness of containment strategies, as well as variation in economic structure. As noted by IMF (2020), in China, growth is projected at 1.0 percent in 2020 supported by policy stimulus. India's economy is forecasted to contract by 4.5 percent mainly due to a longer period of lockdown. Latin America, mainly Brazil and Mexico, have been forecasted to contract by 9.1 and 10.5 percent respectively, in 2020. The disruptions due to the pandemic have also affected Russia (6.6 percent), Saudi Arabia (-6.8 percent), Nigeria (-5.4 percent) and South Africa performance (-8.0 percent). Interestingly, in 2021 the growth rate for emerging economies and developing countries has been forecasted to strengthen to 5.9 percent, largely reflecting the rebound projection for China (8.2 percent). These projections suggest that governments will have to come up with long-term policy adjustments that can be used to ensure that the global construction and infrastructure sector is not severely impacted by the COVID-19 pandemic.

1.8 The impact of fiscal and monetary policies on infrastructure spending

Government policy can ultimately be expressed through its borrowing and spending activities. In this section, two types of government policy can affect infrastructure spending: monetary and fiscal policy. Monetary policy refers to central bank activities that are directed toward influencing the quantity of money and credit in an economy (IMF, 2020). By contrast, fiscal policy can be

Introduction

viewed as the government's decision about taxation and spending. Both policies can be used to regulate economic activity time, accelerate growth when an economy starts to slow and moderate an economy when an economy starts to overheat. It is worth noting that the overarching aim of both policies is normally the creation of an economic environment where growth is stable and inflation is stable and low. Globally, there have been recurrent calls to increase infrastructure investments (McKinsey, 2016). This is hardly a surprise, as increased global infrastructure investments could go a long way to resolving several challenges that a number of nations face. According to Bivens (2014), the impact of infrastructure investments on the overall level of activity depends on the level of productive slack in the economy, the viewpoint of monetary policy, and how the investments will be financed.

It is difficult to precisely predict the long-term effects of investments on the overall level of economic growth activities. Bivens (2014) asserts that one could reliably forecast the impact of infrastructure investments on the composition of labour demand. Even if these investments crowd out other forms of spending and do not affect the overall level of activity and employment, it remains the case that composition of employment complemented by additional spending on infrastructure could differ from that of the economic activity it potentially displaces. As disclosed by Bourne (2017), historical evidence shows significant positive effects of government infrastructure spending on productivity and provides minimal guidance on the worthiness of new projects today. Congestion and changing demand patterns do require new infrastructure investments, and government spending in certain areas can maximise growth. If the macroeconomic foundations of a country are in a poor state, it can adversely affect growth. Infrastructure spending can be used to provide essential public goods and grease the wheels of economic activity through improving mobility and enhancing the economy's capacity to grow (Bourne, 2017). As observed from the reviewed literature, economists and policymakers are divided on the true value of spending multipliers associated with both fiscal and monetary policies, not least because they tend to be time- and project-specific, while being difficult to approximate because of limited agreement on the appropriate methodological approach (Bivens, 2014). As specified below, a number of broad conclusions can be drawn regarding monetary and fiscal policies:

- **Country characteristics:** Nations with certain attributes, such as floating exchange rates and high levels of public debt, tend to have low fiscal multipliers in general (Bivens, 2014). Conversely, in nations with flexible exchange rates and an inflation-targeting regime, central banks will tend to raise interest rates that can be used to offset fiscal expansion.
- **State of the economy and monetary policy:** This theme shows that fiscal multipliers vary according to the state of the economy. Bivens (2012a) found that when the economy is in a downturn or recession, multipliers will be higher than when the economy is growing quickly or at full employment. Within this theme, the interaction between monetary and fiscal policies is even significant.
- **Timing and dose:** The third theme signifies that the efficacy of fiscal stimulus relies on having accurate data on the state of a country.
- **Longer-term effects:** Blinder (2006) stated that the overarching aim of fiscal expansions is to put resources to work for the short term. Interestingly, the stimulus theory suggests nothing about what the money should be spent on and how that might impact the economic activity in the medium and long term.

The reviewed literature shows that the majority of Keynesian economists have suggested that government investment spending has significantly bigger effects than either tax or government spending (Cogan *et al.,* 2010). Thus, within the context of this book, it could be implied that government investment spending can be an effective form of fiscal stimulus because infrastructure projects have a long lead time. The justifications for government provision or oversight of infrastructure projects can be split into three broad themes: (a) markets fail and require government correction; (b) the cost of government borrowing is cheap, and it is economical for nations to invest, and (c) nations can put social ambitious above narrow commercial interests. Most of the deliberations on infrastructure policy start with the evidence that infrastructure investment is necessary and then ask where the capital will come from to finance it; thus, it is vital for infrastructure policymakers to understand the influence of monetary and fiscal policies. Both monetary and fiscal policies are macroeconomic tools that can be used to influence major infrastructure investment decisions. Monetary and fiscal policies together have great influence over a country's national and local levels of infrastructure investment decisions. As shown in this book, both policies have a great influence on social and economic infrastructure systems.

1.9 Need for sustainable infrastructure

Infrastructure development, economic growth and climate change are intimately related (Qureshi, 2011). In the current context of increasing concerns about forecasts for global growth, infrastructure investment can play a crucial role by boosting global aggregate demand today and laying sturdier foundations for future growth. Weissman (2017) ascertained that sustainable infrastructure can refer to "green" or "smart buildings"; and it can comprise of a wide range of initiatives with a specific focus on energy, water, land management, green areas, smart technology and the use of sustainable, durable building materials. Arup (2020) verified that it is not simply the short-term provision of infrastructure that is of significance; planning and designing will take full account of its operational requirements and use. As established in this book, economic and social infrastructure systems must be sustainable if they are to benefit future generations and make a positive contribution to the upcoming society. If governments are to provide such infrastructure systems now, they will be investments that will pay off many times in years to come.

It is worth highlighting that sustainable infrastructure design is not just about new infrastructure. Sustainable infrastructure design entails rehabilitation, reuse or optimisation of existing infrastructure, which will be consistent with the standards of urban sustainability and global sustainable development (Arup, 2020; Weissman, 2017). This should incorporate infrastructure renewal, long-term economic analysis of infrastructure, energy use and reduced infrastructure costs, the protection of existing infrastructure from environmental degradation, material selection for sustainability quality, durability and energy conservation, minimising waste and materials, the redesign of infrastructure in light of global climate change and the remediation of environmentally damaged soils and water (Arup, 2020; Weissman, 2017). As shown throughout this book, sustainable infrastructure has coinciding benefits from physical, environmental, economic and social viewpoints. From an environmental viewpoint, sustainable infrastructure can abet climate resilience, which ultimately aids economic resilience. When equitable access is guaranteed, society will benefit from the infrastructure system because it will deliver services (such as power supplies, healthcare services and sewerage networks) that are vital for sustainable development (The Economist, 2019). Sustainable infrastructure can also assist nations in meeting their national objectives that will contribute to the overall 2 °C goal set in Paris. From a wider viewpoint, sustainable infrastructure can be used as a source of economic growth, social well-being and financial gains (Weissman, 2017).

From the reviewed literature, it has been observed that there is a focus of climate-resilient infrastructure in countries most at risk from the physical effects of climate change (Weissman, 2017). Low-lying nations such as Maldives, Netherlands and Singapore are utilising sustainable

Introduction

development principles to mitigate the worst possible impact of climate change, including flooding and rising sea levels. Infrastructure systems that can withstand the shocks and stresses experienced over its lifetime will provide resilience and protect development by providing a positive impact across the three pillars of sustainability: economic, environment and society. For instance, when resilient social and economic infrastructure systems ensure continuity of critical services such as power and water during a crisis, it offers greater stability to communities and reduced disruption to the society. Similarly, if both economic and social infrastructure systems are to be less frequently rebuilt and repaired, countries not only save money; they also use minimal resources. Japan is well known for its ability to construct highly resilient infrastructure systems that can sustain frequent severe earthquakes (The Economist, 2019). So what policies, changes, and laws will be instrumental in shaping the global sustainable infrastructure goals? As specified below, the following can be adopted (Qureshi, 2011; The Economist, 2019; Weissman, 2017):

1. **Articulating national strategies for sustainable infrastructure:** Nations need to articulate clear and wide-ranging strategies for sustainable infrastructure systems and integrate them in the overall policies for sustainable growth and development. There is a need for a wider articulation of policies on the direction of change and plans to address national infrastructure policies and market failures and other constraints to sustainable infrastructure development. From a policy viewpoint, integrated strategic models can be used to ensure coherence across individual public policy measures and provide clarity and confidence to both the public and private sectors.
2. **Adopting fundamental price distortions**: Remedial pervasive distortions in the pricing of natural resources and infrastructure services is key to enhancing the public policy environment for sustainable infrastructure. The main distortions are fossil fuel subsidies and the lack of carbon pricing, which both strongly bias infrastructure investment toward high-carbon sources of energy and undermine efficiency in energy use. The most significant action public policy can take to shift the inducement structure toward lower-carbon investment and development paths is to put a price on carbon emissions. As stated by Qureshi (2011), removing fossil fuel subsidies and taxing carbon emissions will rectify incentive distortions and mobilise additional fiscal resources that can be used to support sustainable infrastructure development.
3. **Enhancing the enabling environment**: Advancing sustainable infrastructure at scale and with the quality needed will require enhancements in the policy and institutional model governing investment in two vital respects. Firstly, there will be a need to boost investment planning, project preparation and management capabilities at local and national levels. Such capability enhancements will be mainly required in developing nations. Secondly, countries will be required to strengthen their regulatory and institutional models, particularly in infrastructure provision. It is worth noting that risks and transaction costs related to public policy are the main obstacles to private investment in infrastructure.
4. **Mobilising financing:** Increasing annual investment in infrastructure is currently a major challenge in most countries. As shown in the finance chapter (see Chapter 11), concerted mobilisation of public and private finance mechanisms will be required, especially through the use of innovative finance methods.

In the long term, the achievement of sustainable development infrastructure systems will require an amalgamation of sustained support and participation at the national level, local level, commitment from both the public and private sectors. In addition, there is a need for a legal and regulatory model that will allow growth while governing the economic, social and environmental standards that are key requirements of sustainability.

1.10 Chapter summary

This chapter has set the scene by providing the scope of strategic infrastructure systems, global infrastructure outlook, global infrastructure spending, impact of the Covid-19 pandemic on construction and infrastructures, value of fiscal and monetary policies on infrastructure spending and need for sustainable infrastructure. As shown in this chapter, the construction and infrastructure sector plays a significant role in driving economic growth in developed and developing economies and has a direct impact on GDP. In the current period of uncertainty, likely the construction and infrastructure sectors will face uneasiness from financiers. Lenders may not be willing to finance large scale infrastructure projects, and projects with a long delivery timeline will be under particularly intense scrutiny. The implications of the COVID-19 pandemic in the sector are wide ranging. But, there are numerous measures that the sector can adopt to mitigate and manage their construction operations. The following sections of the book appraise four primary themes of infrastructure systems: valued-based approach to infrastructure systems appraisal; enabling planning and execution; financing and contracting strategies for infrastructure systems; and digitising major infrastructure delivery. The primary themes have been divided into a series of secondary themes and provide the comprehensive knowledge necessary for assessing the value and benefits, sustainable major infrastructure project planning, stakeholder engagement, multi-partner collaborations, whole-life performance, regulatory process, public-private partnership, infrastructure financing, operation and maintenance, relationship risks, design thinking principles, innovation and technology.

References

Arezki, R. (2020). Developing public-private partnership initiatives in the Middle East and North Africa: From public debt to maximizing finance for development. *Journal of Infrastructure, Policy, and Development*, 4(1), pp. 73–86.

Arup (2020). Sustainable infrastructure design. Available from: https://www.arup.com/expertise/services/infrastructure/sustainable-infrastructure-design [cited 8 June 2020].

Biorck, J., Sjodin, E., Blanco, J.L., Mischke, J., Strube, G., and Ribeirinho, M.J. (2020). How construction can emerge stronger after coronavirus. Available from: https://www.mckinsey.com/industries/capital-projects-and-infrastructure/our-insights/how-construction-can-emerge-stronger-after-coronavirus [cited 2 June 2020].

Bivens, J. (2012a). Public investment: The next "new thing" for powering the economic growth. Economic Policy Institute, Briefing Paper No.338. Available from: http://www.epi.org/publication/bp338-public-investments/

Bivens, J. (2014). The short and long-term impact of infrastructure investments on employment and economic activity in the U.S. economy. Available from: https://www.epi.org/publication/impact-of-infrastructure-investments/ [cited 8 June 2020].

Blinder, A. (2006). The case against the case against discretionary fiscal policy. In: R. Kopcke, G. Tootell, and R. Triest (Eds.), *The macroeconomics of fiscal policy*, pp. 26–61. Cambridge, Mass: MIT Press.

Bourne, R. (2017). Would more government infrastructure spending boost the U.S. economy? Available from: https://www.cato.org/publications/policy-analysis/would-more-government-infrastructure-spending-boost-us-economy [cited 2 June 2020].

Chen, C. and Bartle, J.R. (2017). Infrastructure financing: A guide for local government managers. Available from: https://icma.org/documents/infrastructure-financing-guide-local-government-managers [cited 30 May 2020].

Cogan, J.F., Tobias, C., John, B.T., and Volker, W. (2010). New Keynesian versus old Keynesian government spending multipliers. *Journal of Economic Dynamics and Control*, 34, pp. 281–295.

Introduction

Fay, M., Hyoung, I.L., Sungmin, H., Mastruzzi, M., and Cho, M. (2019). Hitting the trillion mark: A look at how much countries are spending on infrastructure. Policy Research Working Paper 8730, World Bank, Washington, DC. Available from: http://documents.worldbank.org/curated/en/970571549037261080/pdf/WPS8730.pdf [cited 18 June 2020].

Global Infrastructure Hub (2017). Global Infrastructure outlook: Infrastructure investment needs 5o countries, 7 sectors to 2040. Available from: https://www.oxfordeconomics.com/recent-releases/Global-Infrastructure-Outlook [cited 30 May 2020].

IMF (2020). Fiscal policy: Taking and giving away: Available from: https://www.imf.org/external/pubs/ft/fandd/basics/fiscpol.htm [cited 3 June 2020].

IMF (2020). A crisis like no other, an uncertain recovery. Available from: https://www.imf.org/en/Publications/WEO/Issues/2020/06/24/WEOUpdateJune2020 [cited 2 July 2020].

Kumari, A. and Sharma, A.K. (2016). Infrastructure financing and development: A bibliometric review. *International Journal of Critical Infrastructure Protection*, 16, pp. 49–65.

McKinsey (2016). Bridging global infrastructure gaps. Available from: https://www.un.org/pga/71/wp-content/uploads/sites/40/2017/06/Bridging-Global-Infrastructure-Gaps-Full-report-June-2016.pdf [cited 30 May 2020].

Ochieng, E.G., Price, A.D.F., and Moore, D. (2017). Major Infrastructure Projects: Planning for Delivery. Basingstoke, UK: Palgrave Macmillan's Global Academic. Available from: https://he.palgrave.com/page/detail/major-infrastructure-projects-edward-ochieng/?sf1=barcode&st1=9781137515858 [cited 30 May 2020]

OECD (2017). Strategic infrastructure planning: International best practice. National Infrastructure Commission. Available from: https://www.itf-oecd.org/strategic-infrastructure-planning [cited 30 May 2020].

PwC (2020). Five actions can help mitigate risks to infrastructure projects amid COVID-19. Available from: https://www.pwc.com/gx/en/industries/capital-projects-infrastructure/publications/infrastructure-covid-19.html [cited 1 June 2020].

Qureshi, Z. (2011). The role of public policy in sustainable infrastructure. Available from: https://www.brookings.edu/wp-content/uploads/2016/07/public-policy-sustainable-infrastructure-qureshi-1.pdf [cited 6 June 2020].

Ribeirinho, M.J., Mischke, J., Strube, G., Sjodin, E., Biorck, J., Anderson, T., Blanco, J.L., Palter, R., and Rockhill, D. (2020). The next normal in construction: How disruption is reshaping the world's largest ecosystem. Available from: https://www.mckinsey.com/industries/capital-projects-and-infrastructure/our-insights/the-next-normal-in-construction-how-disruption-is-reshaping-the-worlds-largest-ecosystem [cited 9 June 2020].

The Economist (2019). The critical role of infrastructure for the sustainable development goals. Available from: https://unops.economist.com/wp-content/uploads/2019/01/ThecriticalroleofinfrastructureftheSustainableDevelopmentGoals.pdf [cited 6 June 2020].

Warwick, G., Webb, T., and Jackson, S. (2020). Covid-19 Australia: Managing the impact of a global pandemic on projects and construction. Available from: https://www.clydeco.com/insight/article/covid-19-australia-managing-the-impact-of-a-global-pandemic-on-projects-and [cited 1 June 2020].

WEF (2012). Strategic infrastructure: Steps to prioritise and deliver infrastructure effectively and efficiently. Available from: https://www.weforum.org/reports/strategic-infrastructure-steps-prioritize-and-deliver-infrastructure-effectively-and-efficiently [cited 30 May 2020].

Weissman, M. (2017). Sustainable infrastructure: A path for the future. Available from: https://www.nortonrosefulbright.com/en/knowledge/publications/0c89c7b4/sustainable-infrastructure-a-path-for-the-future [cited 6 June 2020].

PART I

Value-based approach to infrastructure systems appraisal

2
SIGNIFICANCE
The need for better benefits realisation in megaprojects

Kate Davis, Jeffrey Pinto and Francesco Di Maddaloni

2.1 Introduction

This chapter elaborates on the challenges of benefits realisation in major projects. The assumption that large-scale projects bring value and benefits to a wide range of project stakeholders is usually an implicit assumption underlying the willingness of governments to make significant investments in these ventures. This chapter will examine the historical roots of value management, the manner in which we can measure and understand project "value," and its implications for more effective project stakeholder management. We conclude the chapter by illustrating the significance of benefits management through an in-depth examination of an ongoing large-scale UK project: High Speed Two (HS2).

2.1.1 Chapter aim and objectives

The goal of this chapter is to establish the historiography, theory, and advances in our understanding of project benefits realisation in order better facilitate the development of major infrastructure and construction projects. It is believed that bringing megaprojects benefits either at the local, regional or national level represents a key, but challenging, task for project managers. Project managers are in need of a clearer understanding of realisable value that will enable them to cope with the uncertainty surrounding megaproject developments. By minimising the negative impact of such projects on both people and places and selecting the most beneficial and viable project for the wider communities, project managers and policymakers can catalyse their efforts and use of public resources.

2.1.2 Learning outcomes

The following learning outcomes have been identified for this chapter. After studying this chapter readers should know the following:

- The characteristics of public infrastructure and construction projects, also termed mega projects;
- How benefits realisation relates to megaprojects;
- Theoretical origins of value and benefits;
- The elements of project value management and how to interpret them;
- The main resources, competences, and capabilities when dealing with benefits and their objectives and constraints;
- How stakeholders are involved, managed and perceptions taken into account;
- How benefits realisation translates in practice; see Case Study: High Speed 2.

2.2 Overview of public infrastructure projects

Public infrastructure and construction projects can be major tools to enhance economic and social development (Jia et al., 2011; Kara et al., 2016). Therefore, it is not surprising that more and larger infrastructure projects are continuously proposed and introduced, with the global expenditure on infrastructure estimated to be US \$3.3 trillion a year for the period from 2016 to 2030 (McKinsey Global Institute, 2016). Infrastructure spending is mainly driven by large-scale projects, which have unique features in terms of their level of aspiration, lead times, complexity and stakeholder involvement (Barlow, 2000; Flyvbjerg, 2014). Therefore, it is typical that construction megaprojects are attracting more attention, as their growth results in an increased impact on people, budgets and urban spaces (Xue et al., 2015). According to Flyvbjerg (2014) and Hu et al. (2014), the terms "major project" or "major programme" are frequently used interchangeably to define large public projects when referring to megaprojects. When defining a "megaproject," the common characteristics in the literature include a strategically aligned set of multiple projects, costs in excess of \$500 million and completion times of more than five years (Major Project Association, 2014; Miller and Lessard, 2000). Notably, project managers are faced with increasing budget constraints, and, thus, the design, evaluation and selection of such highly costly projects has become particularly critical in turbulent economic conditions (Greenspan, 2004; Matti et al., 2017; NETLIPSE, 2016).

Although the likely benefits of megaprojects are largely recognised, the uncertainty surrounding their impact represents a key challenge for project managers and their parent organisations, especially because of the length of the life cycle of such projects (Marshall and Cowell, 2016; Zanni et al., 2017). The uncertainty of major infrastructure and construction projects is due to their complexity, i.e., "the property of a project which makes it difficult to understand, foresee and keep under control its overall behaviour, even when given reasonably complete information about the project system" (Vidal et al., 2011). Therefore, managing time and cost constraints is regarded as "firefighting" to keep afloat, which leads to unrealistic estimates in order to meet goals, while ignoring setting the real benefits in the feasibility stage (Flyvbjerg et al., 2003). It is recognised that benefits realisation is an important element for improving project performance (Laursen and Svejvig, 2016; Turner, 2014). Likewise, we believe that benefits realisation has a greater impact on project success, in which it is essential to minimise the waste of public resources by creating a better decision-making process that includes the needs and expectations of a broader range of project stakeholders and that leads towards more impactful megaprojects.

2.3 Cui bono? Defining the nature of project value

The terms *value* and *benefits* are sometimes used interchangeably, with several overlapping and, at times, ambiguous concepts such as "value" (Morris, 2013), "benefits" (Chih and Zwikael, 2015; Peppard *et al.,* 2007), "worth" (Zwikael and Smyrk, 2012), and "success" (Yu *et al.,* 2005). Additional concepts that are used quite often to voice these ideas include value creation (Andersen, 2014; Winter *et al.,* 2006a), benefits management (Ward and Daniel, 2012), and benefits realisation management (Bradley, 2010; Laursen and Svejvig, 2016). In their paper, "Taking Stock of Project Value Creation: A Structured Literature Review with Future Directions for Research and Practice," Laursen and Svejvig (2016) highlight inconsistent and sometimes murky terminology, including "Research-based view," "Contingency theory," "Principal-agent theory," "Transactional-cost theory" and Porter's "Value chain." They conclude that the project management literature rarely supports value creation for the funding organisation, highlighting an important distinction between *project management success* and *project success*. While the former relates to efficient output delivery, the latter is concerned with benefits realisation for the funding organisation. Thus, there is some semantic ambiguity in the distinctions between value and benefits. Part of the reason for this is the different foci of key project stakeholders and how they view the terms themselves.

"Value creation depends on the relative amount of value that is subjectively realized by a target user (or buyer) who is the focus of value creation – whether individual, organization or society – and that this subjective value realization must at least translate into the user's willingness to exchange a monetary amount for the value received." It follows from this definition that there is perceived use value, subjectively assessed by the user (or buyer), and then monetary exchange value, the price paid for the use value created (Bowman and Ambrosini, 2000). Value management traces its roots to the use of structured cost reduction techniques in manufacturing operations. During World War II, US manufacturing and strategic materials were prioritised for armaments, leaving other organisations in search of alternative materials and methods for producing goods. Finding processes and cheaper materials that allowed for the manufacturing of goods with no loss in quality became a goal of US companies and gave rise to a structured process that eventually coined the use of the term "value analysis" (VA). Value analysis was a means for industrial engineers to critically evaluate plant flow operations, employ cheaper materials, identify redundant or "non-value-adding" processes, and improve the overall efficient use of resources to maximise output. This leads to the general working definition (Kelly and Male, 1993p. 8): Value analysis is "an organised approach to the identification and elimination of unnecessary cost." Unnecessary cost is defined as cost that provides neither use nor life nor quality nor appearance nor customer features (Kelly and Male, 1993).

The second stage rose to prominence in the 1960s and shifted the focus from process improvement for existing products to the analysis of evolving designs in manufacturing and construction, a concept known as "value engineering" (VE), which was based on applying manufacturing principles as widely as possible, including infrastructure and construction. The formation of the Society of American Value Engineers (SAVE) in 1959 established the term *value engineering*, which came into common use as the preferred term, and is the term most used in the United States today. The first recorded use of a value incentive clause in a construction contract was in 1963 by the United States Navy Bureau of Yards and Docks. A characteristic of North American value engineering from its inception is the team approach to function definition and creativity through application of a logical, sequential approach to the study of value.

There are seven suggested phases in the VE process, following these steps (Miles, 1961):

1. Orientation – determining what is to be accomplished, what the customer really wants, and what the desirable characteristics of the finished product are.
2. Information – gathering as much information about the project as possible at the outset. Critical elements in information collection include:

    a. Clients' needs and wants – the fundamental requirements and the "wish list"
    b. Project constraints – factors that impose discipline on the project team; e.g., site conditions, timing, regulations, etc.
    c. Budgetary limitations – the amount that can be committed to the project
    d. Time constraints – life-cycle stages and impact on the project's completion

3. Speculation – ideas are generated through brainstorming to problem solution
4. Analysis – the whole life cost of each idea is estimated and they are jointly ranked for acceptability
5. Development and Planning – the project development schedule is established through work breakdown structures and network creation
6. Schedule Execution – the project is executed according to the original plans
7. Status Summary and Analysis – critical evaluation of the project is undertaken, with suggestions for improvement of the immediate project or for future development

The third stage of value management, which began in the 1990s, widened the scope of the service to include the analysis of the organisational and business strategies, which gave rise to the requirement for products and services. This emphasis on value planning (VP) highlighted the employment of strategic planning principles to address concepts of value as they related to new product development, new service introductions, and other strategic initiatives that organisations undertake. Strategic choices now required that firms address the manner in which strategic, real options maximised firm value, through cost reductions and/or benefits maximisation. In this sense, "value" moved from its earliest orientation as a production execution concept to a strategic task, migrating from the shop floor to the executive suites.

Value management (VM) derives its power from being a team-based, process-driven methodology that uses function analysis to analyse and deliver a product, service or project at optimum whole life performance and cost without detriment to quality. Value management developments were initially dominated by North American thinking (Dell'Isola, 1988; Fallon, 1980; Kaufmann, 1990; Miles 1972, 1989; Mudge, 1990; O'Brien, 1976; Parker, 1985; Zimmerman and Hart, 1982). From a European context VM is seen as a style of management. Bringing together the information from the three European value standards, it is a methodology whose goal is to reconcile differences in view between stakeholders and internal and external customers as to what constitutes value. It does this through a structured, systematic, analytical functioned-oriented and managed process involving a representative, multidisciplinary team brought together in a participatory workshop situation (Male *et al.*, 2007). Figure 2.1 offers a simplified timeline of the development of the various elements in VM.

2.4 Value as satisfaction versus consumption

For this chapter, we will use the terms "value" and "benefits" somewhat interchangeably. In the broadest sense, a benefit is the improvement resulting from a change (outcome) that is perceived as positive by one or more stakeholders (adapted from Bradley, 2010: xiii; Office of Government

Significance

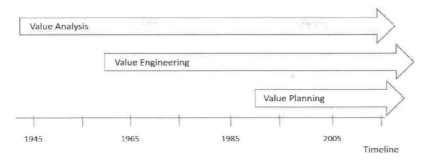

Figure 2.1 Timeline of the development of value management

Commerce, 2009: pp. 21–22). Value is often represented as a simple formula or ratio of needs satisfaction over resource usage; that is, attempting to satisfy user requirements while minimising the application of resources required to satisfy those needs. The fewer the resources used or the greater the satisfaction of needs, the greater the value (Venkataraman and Pinto, 2008). Thus, the concept of value relies on the relationship between the satisfaction of many differing needs and the resources used in doing so. Stakeholders and internal and external customers may all hold differing views of what represents value. The aim of VM is to reconcile these differences and enable an organisation to achieve the greatest progress towards its stated goals with the use of minimum resources (see Figure 2.2).

It is important to realise that value may be improved by increasing the satisfaction of need even if the resources used in doing so increase, provided that the satisfaction of need increases more than the increase in use of resources. Value management is distinct from other management approaches in that it simultaneously includes attributes which are not normally found together. It brings together, within a single management system, management style, positive human dynamics, and consideration of external and internal environment ("What is Value Management," n.d.).

2.5 Elements of project value management

Project benefit/value management is an emerging research area that emphasises the strategic roles of projects organisations, and describes the benefit management process within projects (e.g., Breese *et al.*, 2015). Although we actively seek to better understand benefits management and indeed, the nature of benefits that derive from projects in general, it is also the case that there is both an ad hoc understanding of benefits themselves as well as conflicting and multiple conceptualisations of project benefits among scholars. In a recent paper, Serra and Kunc (2015) argue that assessments of project benefits/value concern two interrelated but distinct

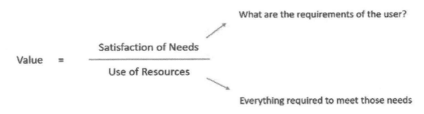

Figure 2.2 Project value

elements: project performance itself, often identified as efficiency measures of delivery according to predetermined metrics of budget, schedule, and requirements; and project success, which evaluates how well projects deliver benefits that meet wider business goals, thus creating value (Cooke-Davies, 2002). The argument is commonly made that project assessments are still too wedded to traditional metrics of project performance defined in the narrower framework, and not sufficiently broadened to account for a more inclusive and expansive idea of project success (Zwikael and Smyrk, 2012). The problem, of course, is that value from projects is often derived from these latter concepts, the ideas that most traditional project management fails to teach. Thus, when we focus too heavily on simple "project management success" metrics of cost, schedule, etc., we neglect the other, intangible elements that most directly address value realisation.

Ika (2009) dichotomises the value-related aspect of project assessment into 1) project/product success – satisfaction of end user and benefits to stakeholders and project staff, and 2) "strategic project management," which he identifies as business success, or the achievement of the client's strategic objectives. Similar models show how project success relates to value realisation: project success (outcomes and benefits) and project corporate success (achieving strategic objectives) (Camilleri, 2011); ownership success (benefits minus costs) and investment success (financial return) (Zwikael and Smyrk, 2011). These and other authors have adopted a model of value realisation that takes into consideration both tactical and strategic elements; that is, the short-term realisation of direct project outcomes (marketplace or technical success of the venture) and subsequent strategic advantages from the project.

Serra and Kunc (2015) offer a "chain of benefits" model that describes the development of a causal set of benefits from the results of projects (see Figure 2.3). This conceptual model suggests that benefits realisation starts with successful project completion, prompting business changes that not only yield immediate desired outcomes (tactical project success, in our parlance) as well as intermediate benefits. Business changes can also create side effects, which are the negative outcomes from change. A negative outcome might be the need for recruiting additional personnel with advanced skill sets (for IT projects) or cost increases from new regulations or safety requirements. Serra and Kunc (2015) argue that these side effects and consequences can also realise further intermediate benefits, which, in turn, contribute to the achievement of end benefits (Bradley, 2010) and end benefits directly contribute to the achievement of one or more strategic objectives of the organisation. Usually, end benefits result from changing processes composed of sets of projects that are managed together as a programme, which, because of the role programme management plays, allows the organisation to coordinate work in a synergic way to generate greater benefits than individual projects could (Thiry, 2001).

In 2016, the *International Journal of Project Management* dedicated a special issue on project benefit management, highlighting the need for future studies in this research area. Building upon the rise in interest in benefits realisation, there has been a steadily increasing interest in programme and portfolio management as vehicles for translating individual project benefits into a broader idea of generating corporate value (Pellegrinelli *et al.,* 2011). As Figure 2.3 implies, the advantages of using portfolio management lie with its emphasis on prioritising the most desirable (optimum) mix of projects and larger programmes to maximise value impact, within the realms of risk and cost (as shown in Figure 2.2). Thus, the use of portfolio management for benefits realisation is that it enables organisations to not only emphasise "doing projects right" but also "doing the right projects."

Significance

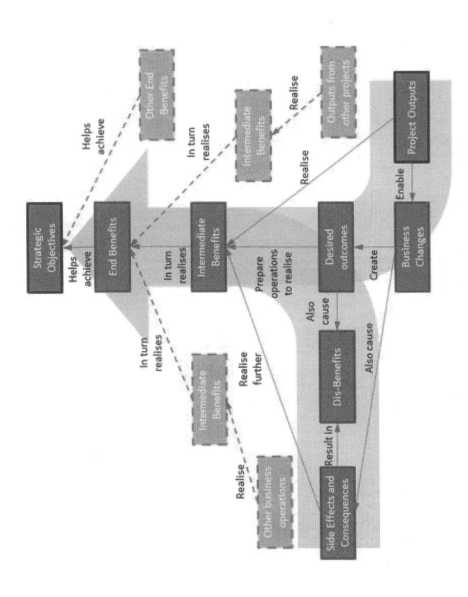

Figure 2.3 Chain of benefits
Source: Serra and Kunc, 2015; p. 56

2.6 Propositional elements in understanding project value management

Following Goodpasture's (2002) perspective, we can identify five fundamental concepts that must be embraced in order to manage projects for value. The first concept suggests that *projects derive their value from the benefits that the organisation accrues by achieving its stated goals*. Remember that projects are typically initiated as a perceived solution to a goal, need, or opportunity. Thus, when we want to determine the degree to which a project is being managed for "value," it is first critical to ensure that the project falls in line with organisational goals. Projects that are being run counter to a firm's stated goals (e.g., customer satisfaction, commercial success, or improving health and safety) already fail the first test of value. Inherent reasons for this have been attributed to a rogue sponsor with a fear of failure (https://onlinepmcourses.com/rogue-project-sponsor/). This results in project objectives aligning with a hidden agenda and the position of power being abused to meet their own goals (Helm and Remington, 2005). We cannot maintain the façade that a project is "valuable" when it clashes with the company's stated or supported goals.

Second, projects can be viewed as investments made by management in that they consume resources and time, and therefore, *projects are expected to provide returns with associated benefits*. Any investment comes with an expected return for the risk undertaken. When an organisation takes the step of investing a significant amount of money in a project, they do so with the understandable expectation that the project will yield an acceptable return, based on their internal rate of return requirements, or measured against some societal standard for desired outcome. The third concept in Figure 2.4 emphasises that *there are inherent risks in projects as there is considerable uncertainty surrounding their outcomes*. These risks may be technical (Does the technology driving the project work?) or commercial (Will the project succeed in the marketplace?), they may involve health and safety issues (Can we manage the project within appropriate parameters of safety?), or some combination of all of the above. An acknowledgement of project risk is recognition that all projects convey "unknowns" due to the unique nature of each endeavour. While investors may not have the wherewithal to manage these project risks, they do tolerate them as the potential rewards associated with project outcomes may outweigh the negative impact the risks.

The fourth concept defines project *value as a function of the resources committed (investment made) and the extent of risks taken*. As Goodpasture (2002: p. 4) notes, "The traditional investment equation of "total return equals principal plus gain" is transformed into the project equation of "project value is delivered from resources committed and risks taken." Using these terms, we can see that value will always walk a narrow line between expected return on investment and risk. When the equation gets out of balance, when the perceptions of the organisation are that the expected return cannot make up for excessive levels of risk, the project ceases to produce value. The implication of this concept is that different projects require different levels of investment with varying levels of risk. Consequently, the value delivered by each of these projects will also vary.

The fifth and final concept in Figure 2.4 suggests that *project value is the outcome of striking a balance among the three key project elements: performance, resource usage, and risk*. So, were we to think like an accountant, we would add up the credit column to include drawbacks such as expenditure (resource usage) and risk recognised and accepted. Balanced against these "credits" is the company's expectation of project performance and positive outcomes. Naturally, the higher the expected performance of the project, the greater the resource usage and risk a company is willing to commit to the project. Goodpasture's (2002) perspective on value implies that an organisation is constantly reassessing value in two ways: first, they take an individualist approach that looks at value in terms of one project at a time. Each brings its own potential value, requiring top management to sift through the pros and cons for each opportunity when deciding on a project investment strategy or when forced to choose among competing project options. Second,

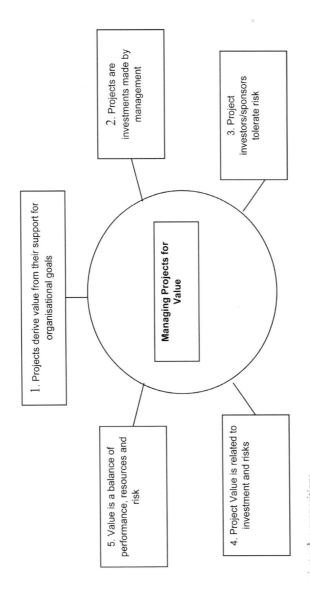

Figure 2.4 Five project value propositions
Source: Venkataraman and Pinto (2008: p. 164)

value is reaffirmed during the project's development cycle. A project may have shown promise of delivering value early in its initiation only to have that value brought into question later on. In this way, many projects are terminated short of delivery if the perception of value becomes negative. The metaphor of a set of balancing scales comes to mind: in one bowl we place our best guess as to a project's real benefits to the organisation and then weigh it against the risks and costs that we expect to accrue in consequence. Does the scale still tip in the direction of positive outcome? Then the project provides positive value for the organisation.

2.7 Benefits realisation: resources, competences, and capabilities

Benefits realisation for projects affects multiple stakeholders in multiple ways. That is, the "benefits" that an organisation and its stakeholders derive from their project activities must be weighed in the balance of the goals they seek and the likely outcomes, beyond profitability, that successful projects may offer them. Ashurst and Doherty (2003) formulated a view of benefits realisation in which they argued that firms gain benefits in three ways: resources, competences, and capabilities. For example, Barney's (1991) work on the "resource-based theory of the firm" argues that it is in an organisation's interest to invest in assets and other resources that offer a long-term competitive advantage. Following this argument, one way that an organisation realises benefits from projects is through the concomitant increase in resources (either material or human) that accrue from successful projects. Trained and increasingly competent project personnel, greater capital expenditures on future projects, and greater market share are ways in which firms can realise benefits in the form of resources. Prahalad and Hamel (1990) argue that these benefits are broader than simple resource advantages and play into enhanced competencies and firm capabilities.

Overall, it is possible to discern three broad categories of firm-level benefits that can be realised from successful projects (Ashurst and Doherty, 2003):

1. Resource-based benefits: Resources in the form of more "traditional" elements, including capital, people, and skills enhancement, as well as other, harder to measure resources, including credibility (reputational), intellectual property, and product/brand enhancement.
2. Competencies: When a firm manages and deploys its resources efficiently, it demonstrates greater competence to do the contracted or agreed-upon work. Successful projects allow firms to develop processes and procedures that enable them to engage in future project-based work with a level of skill that out-distances their competition. In effect, doing projects well is a forerunner of doing future projects well.
3. Capabilities: Sometimes viewed as a combination of resources and competences, capabilities enable an organisation to demonstrate competitive advantage. That is, organisations develop the benefit whereby other clients view them as having the capability to deliver superior solutions, products, or services through project activity.

Using this model, "benefits" to the organisation impact them on multiple levels and the overall combined effect can permeate the organisation through interaction. For example, as the above categorisation suggests, one added component of this viewpoint is the interrelationship of these various elements of benefits. For example, in the IT project setting, Santhanaman and Hartono (2003) demonstrated a clear link between an organisation's IS/IT capabilities, its overall performance, and its ability to secure a sustained advantage. The underlying point of this model is to recognise that benefits can be defined (and affect organisations) in multiple ways; most importantly, these ideas offer a complementary, rather than competing, model of benefits realisation. Firms

2.8 Management of stakeholders vs. management for stakeholders

In order to position the theoretical stance of our debate towards better benefits realisation and value co-creation, it is important to elucidate the two main and contrasting approaches of managing stakeholders. Scholars have highlighted two different and opposing stakeholder management approaches in the current literature: management-of-stakeholders and management-for-stakeholders (Freeman *et al.*, 2007). The first aligns with the instrumental formulation of stakeholder theory, which sees stakeholders as resource providers for the organisation and categorises them based on their potential ability to help or harm the organisation (Eskerod and Huemann, 2013). This approach is based on Salanick and Pfeffer's (1978) work which explains that stakeholders could be resource providers to the organisation, based on their interests.

The often-limited resources available within organisations have led to the predominance of the instrumental approach to stakeholder management in order to ensure that stakeholders comply with the organisation's needs (e.g., Johnson *et al.*, 2005; Mitchell *at al.*, 1997). From this perspective, the focus is narrowly on those vital or "primary" stakeholders, such as owners, suppliers, employees, and customers, who have historically obtained greater salience and attention from scholars and practitioners alike. In fact, it is well documented as to how managerial priority has been given to those salient or "primary" groups or individuals who have a formal contractual relationship with, or direct legal authority over, the organisation (Eesley and Lenox, 2006). However, the instrumental perspective has been long criticised by advocates of the normative core of stakeholder theory (Derry, 2012; Jones and Wicks, 1999).

Differing from the economic-based vision, a critical voice within stakeholder theory has acknowledged that business is always "moral in nature," where the focal organisation should involve gathering input from all the affected parties (Freeman, 1994; Jensen and Sandström, 2013). These principles, therefore, perceive the organisation as a connected set of relationships between stakeholders that is not built on principles of competition, but on cooperation and caring. In the pioneering work of Freeman's (1984) *Strategic Management: A Stakeholder Approach*, the central argument was that the organisation should not consider only those groups who can affect it but also those who are affected by its operations. Freeman (1984) was the first scholar who clearly identified the strategic importance of other groups and individuals to the organisation but, ironically, "the resulting work on stakeholder management has focused almost exclusively on the former: primary groups that are critical to the firm's survival in its current business" (Hart and Sharma, 2004, p. 9).

In this regards, management-for-stakeholders (Freeman *et al.*, 2007; 2010) links back to the normative formulation of stakeholder theory, which considers stakeholders as legitimate groups whose interests are respected and valued for consideration in their own right. Regardless of their ability to help or harm the organisation, and regardless of their level of power in the network of stakeholders, this holistic approach takes into account the marginalised or disempowered stakeholders, such as community groups, unions, consumer advocates, competitors, special interest groups, the media, and non-governmental organisations (Aaltonen *et al.*, 2008). In contrast to the instrumental approach, stakeholders are identified according to their interest in the focal organisation, and not vice versa. The management-for-stakeholders approach also explains that "firms have a normative [moral] commitment to advance stakeholder interests and that this commitment shapes firm strategy and influences financial performance" (Harrison and Freeman,

1999, p. 480). The aim of the corporate is thus focused on meeting and exceeding stakeholders' needs and expectations.

The frustration with developing a clearer understanding of project benefits management lies with the "accidental" nature of how many benefits are currently realised from projects. In 2009, the Association of Project Management (APM) Benefits Management SIG undertook a survey across APM members in the UK as part of the launch of the SIG. The results were fascinating and disturbing; the survey found that 60 percent of respondents described their organisation's approach to benefits management as informal or inadvertent (APM, 2009). Thus, decades after the establishment of professional project management organisations worldwide and on the heels of thousands of papers, books, and other published work on projects, the majority of project management professionals still operate in the dark with regard to understanding how to manage their projects for value.

2.9 Benefits realisation and stakeholder perceptions

Although the literature on megaprojects is moving forward, the classic project evaluation methods have been inefficient in capturing and including the views of a broader range of stakeholders and in balancing their economic and social needs and expectations (Eskerod and Huemann, 2013). The management and organisation literature illustrates various techniques that have helped public decision-makers cope with the growing uncertainty of their business environment, especially the complexity of the political, economic, social and technological changes (Porter *et al.*, 2004). Decisions made by project managers have a significant impact on the strategic value delivered by major programmes in the construction industry (Eweje *et al.*, 2012; Vuorinen and Martinsuo, 2018). However, although many models have been created to facilitate the process of managing major infrastructure and construction projects, the economic-based evaluation approaches such as net present value (NPV) are still by far the dominant methods used to evaluate this kind.

Due to the well-documented, complex, and uncertain nature of large infrastructure and construction projects, it is important to consider a stakeholder-oriented approach in the evaluation and approval of these highly risky projects in order to deliver the promised benefits to the broadest possible range of stakeholders. The main importance is not whether the project is finished in accordance with time and cost targets but that it produces an outcome at a time and cost that made it valuable to stakeholders (Turner, 2014). To further explain this point, it is important to note that the perceived final project outcomes are influenced by stakeholder perception (Davis, 2014; Di Maddaloni and Davis, 2018; Turner and Zolin, 2012). Moreover, the way stakeholders perceive project outcomes also change with time (Dalcher and Drevin, 2003; Turner *et al.*, 2009), and what really "fits" the unique characteristics of complex, long and expensive developments (scope, time, budget) of a megaproject are the benefits that it will produce to the wider community.

To illustrate, the Thames Barrier was "priced at £110.7 million in October 1973 (compared with initial estimates of £13–18 million) [and] was ultimately delivered at a cost of £440 million" (Dalcher, 2012, p. 648). Further, it took just under twice the estimated four years because of delays during the preconstruction phase. However, regardless of the delays, it is considered a great engineering achievement with the value of preventing floods and saving lives (Morris and Hough, 1987). On the other hand, Heathrow Terminal Five was completed successfully within time and cost constraints; however, British Airways had minor commissioning issues relating to check-in procedures for oversized baggage, leading to the later public and customer perception that the project was not able to deliver the promised benefits with consequent damage to the reputation of British Airways (Brady and Davies, 2009, 2010a, 2010b; Brady and Maylor, 2010).

Significance

This raises the question of whether a better focus on benefits realisation is required, especially for complex projects whose value is not immediately obvious at completion.

Involving a wider range of stakeholders is key to minimising benefit shortfalls and enhancing positive input through better stakeholder management procedures (Bourne and Walker, 2005; Cleland, 1986; Cleland and Ireland, 2007; Donaldson and Preston, 1995; Olander, 2007). However, an example where this is often missed is when megaprojects fail to align project objectives with those of the marginalised or disempowered stakeholders (Choudhury, 2014). Little has been done by managers and academics alike to achieve a people-centred vision for cities which enhances quality of life and produces prosperous neighbourhoods. Megaprojects should not be viewed as simply more expensive versions of normal projects; "mega" also relates to the skill level and attention required to manage and understand conflicting stakeholder interests and needs through the extensive project life cycle of major programs (Capka, 2004). In fact, findings from the literature show that a major challenge affecting large infrastructure developments is a lack of understanding of the various interest groups, the motivation behind their actions and their potential influence during the project life cycle (IFC, 2007; Miller and Olleros, 2001; Winch and Bonke, 2002).

During major projects, stakeholder needs are often different and a variety of disputes occur. Stakeholders' objectives, composition, relationship patterns and claims are unique and dynamic along different stages of the project (Windsor, 2010). In order to satisfy individual vested interests, stakeholders apply strategies to affect project decision-making. Understanding these strategies is helpful for project managers in forecasting stakeholders' likely behaviours (Frooman, 1999). Therefore, listening and responding to stakeholder interests and concerns is a process that helps project managers maximise stakeholder positive input and minimise any detrimental or negative impact (Bourne and Walker, 2005; Cleland and Ireland, 2007). Since Cleland (1986) brought the stakeholder concept into the project management field, the management of project stakeholders can be considered an established area in contemporary standards of project management (APM, 2013; PMI, 2013). However, often the project owner fails to take the opinions of other stakeholders into consideration and this will attract hostility towards the project. Therefore, a vast number of interests will be affected, both positively and negatively, throughout a construction project life cycle (Olander, 2007).

Yang (2013) focuses on stakeholder analysis and considers it either a process or an approach to support decision-making and strategy formulation. Conversely, Olander and Landin (2008, p. 561) state that the "stakeholder analysis process should be to identify the extent to which the needs and concerns of external stakeholders can be fulfilled, and analyse the possible consequences if they are not." Aaltonen (2011) states that stakeholder analysis in megaprojects is an interpretation process by project managers analysing the project stakeholder environment. Therefore, the importance of identifying exactly who the participants are also includes an accurate identification of the stakeholders' interests and their impact on the project (Achterkamp and Vos, 2008). Returning to our idea of stakeholders' influence on the delivery of project value, the more we can identify and categorise the various stakeholder interests, the better we are able to create value-laden projects for the widest possible audience.

In order to identify and prioritise stakeholders among different and competing claims, Mitchell *et al.,* (1997) developed the stakeholder "'salience model" based on three attributes of power, legitimacy and urgency. According to their typology, stakeholders belong to one of seven categories: "dormant," "discretionary," "demanding," "dominant," "dangerous," "dependent" and "definitive." This classification system indicates the amount of attention that project managers should give to stakeholders' needs and perceptions of value from project outcomes (Mitchell *et al.,* 1997). However, although many scholars cite this model in their work, important methods such as the "power/interest matrix" (Johnson *et al.*, 2005) and "stakeholder circle methodology" (Bourne and Walker, 2005) were developed from Mitchell *et al.'s* 1997 work reflecting the instrumental perspective of stakeholder theory, where prioritisation is necessary. Nonetheless, the model does not reflect the changing attitudes of the stakeholder dynamic through the different phases of the project life cycle (Olander, 2007) and neither that the resources, nor the network positions of stakeholders, can be considered static (Pajunen, 2006). The obvious implication is that project organisations face the very real conundrum of managing for value even in the face of transitory or shifting perceptions of what stakeholders seek from the project.

The challenge of delivering value is mirrored by the concomitant challenge of identifying, understanding, and developing strategies for managing project stakeholders based on their interests and perceptions of benefits to be realised. Literature shows growing attention to stakeholder attitudes towards a project. This attitude is captured by the model proposed by McElroy and Mills (2000), which distinguishes whether a stakeholder is an advocate or adversary of the project in five levels of "active opposition," "passive opposition," "not committed," "passive support" and "active support." Olander (2007) and Nguyen *et al.* (2009) propose a quantitative approach ("stakeholder impact index") to assess stakeholder impact by integrating more variables from Mitchell *et al.* (1997), Bourne and Walker (2005) and McElroy and Mills (2000). Moreover, a social network approach (Rowley, 1997) has been applied in stakeholder analysis for a small infrastructure project by Yang *et al.* (2011a), which considers the interaction among multiple stakeholders by examining their simultaneous influence to forecast the corresponding responses and organisational strategies (Rowley, 1997).

Based on an infrastructure project in Hong Kong, Li *et al.* (2012) consolidated a list of 17 stakeholder interests and different priorities in megaprojects of major stakeholder groups. What emerged is that in many cases stakeholders seek to prevent their vested interest from being jeopardised and an issue that is very important to one stakeholder group may be the lowest priority of another group (Li *et al.*, 2012). Some scholars focus on the link between spatial dynamics and stakeholder impact. This concept has been applied in the context of infrastructure planning by Dooms *et al.* (2013), which examines that stakeholder structure and interests vary with their spatial distance from the project, with stakeholders gaining higher salience as they become geographically closer to the project (Dooms *et al.*, 2013). However, although conceptual frameworks and analytical models have been suggested by stakeholder theory scholars, managerial priorities and concerns have been focused almost exclusively on those primary stakeholders important to the project's economic interests (Aaltonen and Kujala, 2010; Hart and Sharma, 2004).

Scholars have mainly distinguished primary stakeholders from secondary stakeholders, and classified them using the literature's prevailing stakeholder salience model proposed by Mitchell *et al.* (1997). Primary stakeholders are characterised by contractual relationships with the project, such as customers or suppliers, or have a direct legal authority over the project, such as governmental organisations. Secondary stakeholders do not have a formal contractual bond with the project or direct legal authority over the project (Eesley and Lenox, 2006), but they can influence the project (Clarkson, 1995). According to Aaltonen *et al.* (2008), while secondary stakeholders include community groups, lobbyists, environmentalists and other nongovernmental organisations, if they are excluded by project managers, they may engage in a set of actions to advance their claims, with negative consequences to direct operational costs and to the reputation of the focal organisation (Eesley and Lenox, 2006).

Much of the knowledge about stakeholder analysis practices in the megaproject context has been from the stakeholder impact perspective, especially on the impact that primary stakeholders can exert on project outcomes. In fact, the majority of prior project research has focused on

Significance

the management of those primary stakeholders important to the project's resources. Secondary stakeholders seek a claim for a legitimate role in project decision-making (Derakhshan *et al.,* 2019; Olander and Landin, 2008) and therefore, more time should be spent at the front end of a project (Pinto and Winch, 2016) and developing a stakeholder engagement plan which includes a broader range of stakeholders (Eskerod *et al.,* 2015a, 2015b: van den Ende and van Marrewijk, 2018).

In the last decade, major steps have been made by practitioners and academics towards a broader inclusiveness of stakeholders. In fact, the NETLIPSE research (Hertogh and Westerveld, 2009; Hertogh *et al.,* 2008), based on best practices and lessons learnt in large infrastructure projects in Europe, demonstrates the beneficial outcomes of involving stakeholders on an extended level in many megaprojects, such as the Øresund Crossing in Denmark, the West Coast Main Line in the UK, and the Bratislava Ring Road, the Lisboa-Porto High Speed Line and the North/ South Metro line in the Netherlands. These projects are clear examples of how organisations have seen local stakeholders' involvement as valuable and considered them as an important issue in any project (Buuren *et al.,* 2012; Hertogh and Westerveld, 2009; Hertogh *et al.,* 2008). The management of megaprojects needs to increase and enhance transparency, fairness and participation by considering and balancing the project's stakeholders' economic, ecologic, and social interests. Project managers need to consider a long-term perspective for ethical and sustainable development which will take into account the global, regional and local stakeholders (Eskerod and Huemann, 2013). It is noted that scarce managerial attention has been given to the process of managing the social and political impact of megaprojects affecting a broader range of project stakeholders.

Project management scholars have also linked benefits realisation to sustainable development (e.g., Sabini *et al.,* 2019; Silvius, 2017). Projects as a vehicle for change play a crucial role in the sustainable development of organisations and society, and recent debates have encouraged research in integrating broader societal objectives (sustainable developments) within projects (process and final goals) (Huemann and Silvius, 2017). The main argument is that benefit realisation helps to understand how sustainable development can be integrated in the management of projects, linking it to strategy. Keeys and Huemann (2017) show that the benefit co-creation process is an iterative process, shaping benefits throughout the project life cycle involving stakeholder engagement, adaptive process and emergence of benefits in context with a broad group of stakeholders. In turn, sustainable development allows businesses and their projects to deliver benefits to a broad group of stakeholders and, on the other hand, shapes the perceptions of how stakeholders make sense of organisations' activities (Di Maddaloni and Derakhshan, 2019).

Regarded as a high-level objective in constitutional documents and official policies of states, regional, and local governments (Ji and Darnall, 2018; Mossner, 2016), sustainable development has been generically defined as "development that meets the needs of the present without compromising the ability of future generations to meet their own needs" (WCED, 1987). In this definition, the values of solidarity and fairness between generations is thus evident. Along with this definition, recent literature emphasises the need for a holistic approach that integrates ecological, economic, and social dimensions when making decisions in organisations and society (e.g., Aarseth *et al.,* 2017). In 1997, Elkington introduced the triple bottom lines of sustainability as economic, social, and environmental. From Elkington's work, it is noticeable how the ecological, economic, and social dimensions (planet, profit, and people) are interrelated and influence each other. In this respect, sustainable development aims at reconciling economic, social, and environmental efforts through the elaboration of more comprehensive long-term strategies and societies' wider involvement in decision-making (Meadowcroft, 2013; Rickards *et al.,* 2014; Zeemering, 2018).

Through discussing and conceptualising 15 of the most representative megaprojects in the UK, Di Maddaloni and Davis (2018) have investigated the benefits and challenges of a more holistic approach of stakeholder management in large scale projects. The findings from their work emphasised the need for a "proactive" stakeholder management approach that takes into account both the views of primary and secondary stakeholders. Through building internal capabilities for secondary stakeholder management, organisations have to recognise the importance of creating the right vision for megaprojects and delivering not just assets but bringing extra values at national, regional or local levels. Therefore, by listening and taking on board the views of the affected people through informal and honest engagement, project managers can re-think their strategies for more sustainable megaprojects through time.

2.10 Chapter summary

We believe that enhancing a shared view of project objectives with a wider stakeholder group aids in achieving better project performance and is a key success factor for both project managers and policy makers in order to achieve benefits development. The focus on megaprojects' benefits has been from the national government's or the large public or private organisations' perspective (Mok *et al.*, 2015), in which the local context of these projects and related stakeholder management practices are often overlooked and therefore warrant investigation (Di Maddaloni and Davis, 2017). Due to the perceived benefit shortfalls of major infrastructure and construction projects, well-organised actions from "secondary stakeholder" groups have led to delays, cost overruns, and significant damage to the organisation's reputation (e.g., Hooper, 2012; Letsch, 2013; Teo and Loosemore, 2017; Watts, 2014). For instance, understanding and minimising the effect of megaprojects on people and places can help manage the project benefits by rethinking a more holistic approach that will take into account those stakeholders regularly affected by these projects, namely, the local community. By identifying connections and major assumptions on the influence of marginalised or disempowered stakeholders in megaprojects, this chapter remarks that stakeholder management as an essential process is designed to maximise positive inputs and minimise detrimental attitudes of all project stakeholders (Bourne and Walker, 2005; Cleland and Ireland, 2007).

2.10.1 Chapter discussion questions

1. **Consider a project you are familiar with; list and rank the benefits in order of importance to the strategy of the organisation.**
   - For example, the organisation may rank its profits, customer satisfaction, sustainability, and innovation as high.
2. **What is benefits management and why should it be considered important?**
   - "Benefits management is the identification, definition, planning, tracking and realisation of benefits. Benefits realisation is the practice of ensuring that benefits are derived from outputs and outcomes." (APM, 2019, online). Benefits management is important as it provides a structured approach for attaining organisational outcomes and successful delivery of projects and programmes.
3. **Who should be responsible for the benefits management in an organisation?**
   "The main roles and responsibilities relevant to benefits management are:
   - Senior Responsible Owner - responsible and accountable for programme or project success underpinned by delivery of expected benefits.

Significance

- Programme manager or project manager - responsible for ensuring proper day-to-day management with a strong focus on benefits realisation.
- Business change agent or benefits manager - oversight and direction of transitional arrangements into business as usual and the embedding of new capability to deliver expected benefits.
- Programme or project management office - responsible for maintaining a benefit documentation library for the programme or project including version control; the PMO may also be responsible for support and advice on benefits management and for reporting on progress towards benefits realisation.
- Organisational board - responsible for maintaining strategic oversight of the full range (portfolio) of benefits being projected across the organisation (Department of Finance, 2020, Online).

4. **When should benefits management start?**
   - Benefits management should start at the beginning of the project and be considered throughout the entire project.
5. **Which elements of project value management can be identified and how these can be interpreted?**

   - Project value concerns two interrelated but distinct elements: project performance itself, often identified as efficiency measures of delivery according to predetermined metrics of budget, schedule, and requirements; and project success, which evaluates how well projects deliver benefits that meet wider business goals, thus creating value (Cooke-Davies, 2002). Moreover, value realisation has to take into consideration both tactical and strategic elements; that is, the short-term realisation of direct project outcomes (marketplace or technical success of the venture) and subsequent strategic advantages from the project.

Five fundamental concepts must be embraced in order to manage projects for value:

1. Projects derive their value from the benefits that the organisation accrues by achieving its stated goals.
2. Projects are expected to provide returns with associated benefits.
3. There are inherent risks in projects as there is considerable uncertainty surrounding their outcomes.
4. Value is measured as a function of the resources committed (investment made) and the extent of risks taken.
5. Project value is the outcome of striking a balance among the three key project elements: performance, resource usage and risk. (Goodpasture, 2002).

2.11 Case study: High Speed 2 (United Kingdom)

The High Speed 2 (HS2) project, costing a projected £50 billion (with a new projected cost estimate of £65bn to £88bn), was initiated with the purpose of increasing the West Coast Main Line capacity and connecting the north of the UK to London and Europe. This was to be delivered in three phases, covering London to Leeds and Manchester via the West Midlands (Birmingham), and joining up with existing rail infrastructure to Liverpool, Newcastle, Edinburgh and Glasgow. This would be the UK's largest infrastructure project and encompass a number of major projects in their own right, such as land purchase and the redevelopment of London's Euston station.

With 18 trains an hour planned to run to and from London on the new railway, the Department of Transport has claimed HS2 will cut the Birmingham to London journey times from one hour and 21 minutes to 52 minutes. Once the next stage is complete, the journey time between Manchester and London will drop from two hours and seven minutes to one hour and seven minutes, and a trip between Birmingham and Leeds would fall from two hours to 49 minutes. Figure 2.5 illustrates the current proposed HS2 route.

The overall impact of the programme is to balance more the opportunities for the UK economy by linking the north and the south of the UK. The major intended social benefits to the project are to include increased seating capacity and supporting the longer-term need, a better and faster travel experience, improved safety for passengers and fewer car journeys, as well as offering a cost-effective alternative to air travel, thereby reducing environmental pollution.

2.11.1 Actual and forecasted costs

According to the Department for Transport progress report for 2020, the Department and HS2 Ltd have spent £7.4bn across the whole programme up through 31 March 2019, of which £6.3bn has been on Phase One. Around 44 percent (£3.287bn) has been spent on the acquisition of land and property as shown in Figure 2.6. The Department's emerging estimate, as of the first quarter of 2020, gives a potential cost of between £65bn and £88bn (2015 prices), 17–58 percent more than the available funding of £55.7bn agreed on with HM Treasury (National Audit Office, 2020).

The aim of the programme was for construction to be initiated in March 2020 and for it to be completed in full by 2033–2036. These targets have been readjusted since the review in August 2019 along with cost estimates. According to the expectations of both the Department for Transport and HS2 Ltd, partial Phase One services from Old Oak Common to Birmingham Curzon Street are to start between 2029 and 2033, with full service from Euston starting between 2031 and 2036. To date, it is not clear when full service to Leeds and Manchester will commence; however, HS2 Ltd estimates between 2036 and 2040.

2.11.2 Reasons for timescale and cost overruns

The current forecasted cost to complete the programme is significantly above the available funding and the programme will not be completed on time as shown in Figure 2.7.

There are lessons to be learned from the experience of HS2 for other major infrastructure programmes. Important reasons have been found to have an impact on time and cost deviations, questioning the real value of the project. These are in Table 2.1.

The above key learning points underlying the cost and schedule increases summarise the reasons why Phase One is now expected to cost more than the previous cost estimate in April 2017.

2.11.3 Delivery of expected benefits

Given the changes to timescales it is currently impossible to evaluate whether the chosen benefits will be realised. The focus of any analysis to date has centred on Phase One, which connects London to the West Midlands. Sixty-six percent of the land required has been purchased to date. Preparatory work has commenced to set in place the right infrastructure to reconfigure utilities and to carry out important archaeological work on 250 sites. Delays to the

Significance

Figure 2.5 Current proposed HS2 route
Source: National Audit Office (2020)

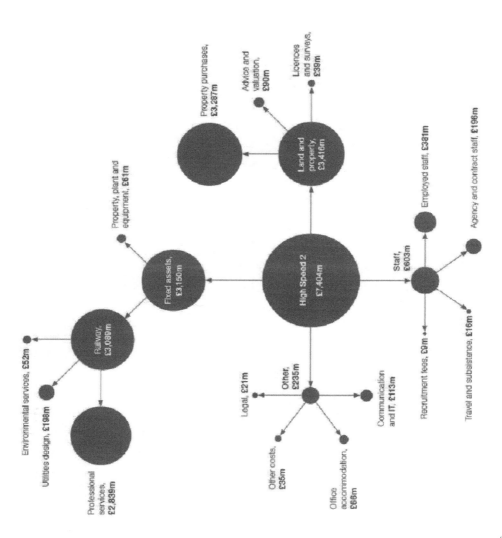

Figure 2.6 Actual costs
Source: National Audit Office (2020)

	Approved by the Department for Transport in November 2019	The Department's emerging estimates as at December 2019		
Cost (£ billion in 2015 prices)	**Phase One**	**Phase 2a**	**Phase 2b**	**Full programme**
Available funding†	27.1	3.5	25.1	55.7
Estimate range	31.0 to 40.0 [2,3,4]	4.5 to 6.5 [5]	29.0 to 41.0 [5]	65.0 to 88.0 [7]
Variance between cost estimate and available funding	+3.9 to 12.9 (14% to 47%)	+1.0 to 3.0 (+29% to 87%)	+3.9 to 15.9 (+15% to 63%)	+9.3 to 32.3 (+17% to 58%)
Opening date				
Original target	2026	2027	2033	2033
Current estimate	2029 to 2033 (partial) 2031 to 2036 (full)	2030 to 2031	2036 to 2040	2036 to 2040
Delay	+3 years to +10 years	+3 years to +4 years	+3 years to +7 years	+3 years to +7 years

Figure 2.7 Available funding and current forecast cost and schedule estimates

Source: National Audit Office (2020)

Table 2.1 Reasons for cost and time deviations

Element of costs and schedule estimate	Reasons for change in forecast
Main civil construction	Additional cost and time of constructing bridges, tunnels and earthworks.
Station design and building	Previous estimates of contractors' overheads and design costs were based on other programmes that underestimated the cost of HS2 stations.
Railway systems	Further development of the design has led to a better understanding of the work needed for systems.
Preparatory work	Site complexity and the volume of work needed has been greater than anticipated.
HS2 Ltd costs	HS2 Ltd incorrectly assumed land and property professional fees were included in the land and property budget. The lengthened schedule also means administrative costs will be spent over a longer period.
Utility diversions	A greater volume of work needed than first anticipated, particularly for site preparation.
Land and property acquisition	Updated surveyor estimates of actual properties to be acquired.

programme overall were caused largely by the underestimation of the complexity of its multiple sub-projects; in particular, that it is delivering infrastructure city to city and through urban areas with higher populations and higher disruption to services than previously thought. There were also some significant issues with the ground conditions encountered, which required the need for more detailed design.

Going forward with Phase One, there will be the need to consider the test, release, and management of benefits, while maintaining the momentum of future releases—all while integrating systems and teams that manage the day-to-day delivery with the project delivery and managing contractors who are delivering different elements of the project. HS2 is largely funded by taxpayers, who have been questioning the benefits and value of the project through the years. Those who oppose the scheme said the money would be better spent on improving Britain's current rail network, improving services outside of London first and foremost. There are also serious concerns over the impact HS2 will have on the environment, as the route cuts through some Britain's prized countryside.

2.11.4 Key stakeholders

The programme involves multiple stakeholder groups as shown in Table 2.2.

The recommendations made in the recent review include ensuring that the programme is reviewed on a regular basis, ensuring shared management information between different parties involved and ensuring that they have the correct capabilities and contractors managing different elements of the delivery. Crucially, there is a need to ensure that the costs do not spiral out of control and that the categories of benefit promised are being monitored on a regular basis. Some consideration will be made going forwards to cost savings that can be made through utilising some of the benefits of the current infrastructure, while balancing this with the benefits promised at the outset.

To date, the real value of the HS2 programme remains questionable. Undoubtedly, the HS2 project is an ambitious national programme, the construction of which will take decades. The Department for Transport, HS2 Ltd and government widely underestimated the task, leading to optimistic estimates being used to set budgets and delivery dates. In not fully and openly

Significance

Table 2.2 Key stakeholders

Stakeholder	Responsibilities
Government – Secretary of State, Department of Transport	Managing public interest and benefits
Investors	Investment and return management
Clients	Input of requirements and driving benefits
Programme board	Directing the programme
Executives of the HS2 company and board	Decision-making at company level
Network Rail	Delivery of the line
Suppliers and contractors	Delivering the existing network and new developments of the programme
Utility companies	Delivery of the line
Staff	Delivery of the line
Consumers and public	Users
Communities and groups (e.g., environmental groups)	Destroying countryside

recognising the programme's risks from the outset, the Department and HS2 Ltd have not adequately managed the risks to value for money. If risks had been recognised and managed earlier, the significant activity in a pressured environment over the past few years trying to understand and contain cost increases may not have been necessary.

2.11.5 Case study discussion questions

1. **Identify the benefits, disadvantages and outcomes for HS2.**

   ♦ **Benefits:** Increased seating capacity and support of the longer-term need, a better and faster travel experience, improved safety for passengers and fewer car journeys, as well as a cost-effective alternative to air travel, thereby reducing environmental pollution.
   ♦ **Disadvantages:** Destroying local countryside and businesses.
   ♦ **Outcomes:** Improving Britain's current rail network, improving services outside of London, more opportunities for the UK economy.

2. **Is it vital to differentiate each stakeholder's communication strategy, or can one size fit all? Discuss.**

   ♦ The reader here should discuss how the stakeholders identified in the case study will have differing levels of vested interest in the project. An appropriate stakeholder communication plan can be devised to assess the stakeholder's power, proximity and urgency to the project.

3. **Does an increase in budget result in an increase in benefits?**

   ♦ With HS2, the budget was massively underestimated with no increase in the foreseeable benefits. The end users will still get an increased seating capacity, a better and faster travel experience, and improved passenger safety. However, in order to complete the project, the budget had to increase.

4. **Is it worth investing possibly millions of pounds to build a strong brand image for HS2?**

   ♦ Strong brands attract more users and increase profitability. In turn this will enhance the value and benefits for stakeholders and generate a real rate of return.

5. **How can benefits and project management approaches help HS2 to work towards a successful project?**

♦ Readers should identify the benefits and project management approaches in the chapter and highlight that the application of a benefits management process on success criteria agreed by key stakeholders will promote better project management practices and subsequently have an effective impact on success.

References

Aaltonen. K. (2011). Project stakeholder analysis as an environmental interpretation process. *International Journal of Project Management* 29 (2), pp.165–183.

Aaltonen, K. and Kujala, J. (2010). A project lifecycle perspective on stakeholder influence strategies in global projects. *Scandinavian Journal of Management*, 26, pp.381–397.

Aaltonen, K., Kujala, J., and Oijala, T. (2008). Stakeholder salience in global projects. *International Journal of Project Management* 26, pp. 509–516.

Aarseth, W., Ahola, T., Aaltonen, K., Okland, A., Andersen, B. (2017). Project sustainability strategies: A systematic literature review. *International Journal of Project Management*, 35 (6), 1071–1083

Achterkamp, M.C. and Vos, J.F.J. (2008). Investigating the use of the stakeholder notion in project management literature, a meta-analysis. *International Journal of Project Management*, 26, pp. 749–757.

Andersen, E.S. (2014). Value creation using the mission breakdown structure. *International Journal of Project Management*, 32, pp. 885–892.

APM (2019). APM Body of Knowledge. 7th edition. Association for Project Management.

Ashurst, C. and Doherty, N.F. (2003). Towards the formulation of a "best practice" framework for benefits realisation in IT projects. *Electronic Journal of Information Systems Evaluation*, 6(2), pp. 1–10.

APM (Association for Project Management). (2009). Benefits Management – A strategic business skill for all seasons, (prepared by the APM Benefits Management Specific Interest Group).

Barlow, J. (2000). Innovation and learning in complex offshore construction projects. *Research Policy*, 29, pp. 973–989.

Barney, J.B. (1991). Firm resources and sustained competitive advantage. *Journal of Management*, 17(1), pp. 99–120.

Bradley, G. (2010). *Benefit realisation management: A practical guide to achieving benefits through change*, 2nd ed. Farnham: Gower.

Brady, T. and Davies, A. (2009). They think it's all over, it is now: Heathrow terminal 5. The Proceedings of EURAM 2009, *The 9th Conference of The European Management Review*, UK, May. University of Liverpool, Liverpool, UK.

Brady, T. and Davies, A. (2010a). From hero to hubris: reconsidering the project management of Heathrow's Terminal 5. *International Journal of Project Management* 28 (2), 151–157.

Brady, T. and Davies, A. (2010b). Learning to deliver a mega-project: the case of Heathrow Terminal 5. In: N. Caldwell and M Howar. (Eds.), *Procuring Complex Performance: Studies of Innovation in Product-Service Management*. New York: NY: Routledge.

Brady, T. and Maylor, H. (2010). The improvement paradox in project contexts: a clue to the way forward? *International Journal of Project Management* 28 (8), 787–795.

Breese, R., Jenner, S., Serra, C.E.M., and Thorp, J. (2015). Benefit management: Lost or found in translation. *International Journal of Project Management*, 33, pp. 1438–1451.

Bourne, L. and Walker, D. (2005). Visualising and mapping stakeholder influence. *Management Decision* 43 (5/6), 649–60.

Bowman, C. and Ambrosini, V. (2000). Value creation versus value capture: Towards a coherent definition of value in strategy. *British Journal of Management*, 11(1), pp. 1–15.

Buuren, A., Boons, F., and Teisman, G. (2012). Collaborative problem solving in a complex governance system: Amsterdam airport Schiphol and the challenge to break path dependency. *System Research and Behavioral Science*, 29, pp. 116–130.

Camilleri, E. (2011). *Project Success: Critical Factors and Behaviours*. Farnham: Gower.

Capka, R. (2004). Megaprojects they are a different breed. *Public Roads – U.S. Federal Highway Administration*, 68 (1), pp.1–10.

Chih, Y.Y. and Zwikael, O. (2015). Project benefit management: A conceptual framework of target benefit formulation. *International Journal of Project Management*, 33, pp. 352–362.

Choudhury, B. (2014). Aligning Corporate and Community Interests: From Abominable to Symbiotic. *Brigham Young University Law Review*, 2 (3), pp. 257–308.

Significance

Cleland, D.I. (1986). Project stakeholder management. *Project Management Journal*, 17 (4), pp.36–44.

Cleland, D.I. and Ireland, L.R. (2007). *Project Management: Strategic Design and Implementation*, 5th edition. New York, NY: McGraw Hill.

Cooke-Davies, T. (2002). The "real" success factors on projects. *International Journal of Project Management*, 20, pp. 185–190.

Dalcher, D. (2012). The nature of project management. *International Journal of Managing Projects in Business*, 5 (4), pp. 643–660.

Dalcher, D. and Drevin, L. (2003). Learning from information systems failures by using narrative and antenarrative methods. The Proceedings of the 2003 Annual Research Conference of the South African Institute of Computer Scientists and Information Technologists on Enablement Through Technology, Gauteng, South Arica.

Davis, K. (2014). Different stakeholder groups and their perceptions of project success. *International Journal of Project Management*, 32(2), 189–201.

Dell'Isola, A. (1988). *Value engineering in the construction industry*, 3rd ed. Washington, DC: Smith, Hinchman and Grylls.

Department of Finance. Available from: https://www.finance-ni.gov.uk/articles/programme-and-project-benefits-management [cited 23 March 2020].

Derekhshan, R., Mancini, M., and Turner, J.R. (2019). Community's evaluation of organizational legitimacy: Formation and reconsideration. *International Journal of Project Management*, 1, pp. 73–86.

Derakhshan, R., Turner, J. R., and Mancini, M. (2019). Project governance and stakeholders: A literature review. *International Journal of Project Management*, 37: 98–116.

Di Maddaloni, F. and Davis, K. (2017). The influence of local community stakeholders in megaprojects: Rethinking their inclusiveness to improve project performance. *International Journal of Project Management*, 35(8), pp. 1537–1556.

Di Maddaloni, F. and Davis, K. (2018). Project manager's perception of the local communities' stakeholder in megaprojects. An empirical investigation in the UK. *International Journal of Project Management*, 36, pp. 542–565.

Di Maddaloni, F. and Derakhshan, R. (2019). A leap from negative to positive bond. A step towards project sustainability. *Administrative Sciences*, 9(41), pp. 1–19.

Donaldson, J. and Preston, L. (1995). The stakeholder theory of the corporation: Concepts, evidence, and implications. *Academy of Management Review*, 20, pp.65–91.

Dooms, M., Verbeke, A., and Haezendonck, E. (2013). Stakeholder management and path dependence in large-scale transport infrastructure development: The port of Antwep case (1960–2010). *Journal of Transport Geography*, 27, pp.14–25.

Eesley, C. and Lenox, M.J. (2006). Firm responses to secondary stakeholder action. *Strategic Management Journal, Strategic Management*, 27, pp. 765–781.

Elkington, J. (1997). *Cannibals with forks—Triple bottom line of 21st century business*. Stoney Creek: New Society Publishers.

Eskerod, E. and Huemann, M. (2013). Sustainable development and project stakeholder management: What standards say. *International Journal of Managing Projects in Business*, 6 (1), pp. 36–50

Eskerod, P., Huemann, M., and Ringhofer C. (2015a). Stakeholder inclusiveness: Enriching project management with general stakeholder theory. *Project Management Journal*, 46 (6), pp. 42–53.

Eskerod, P., Huemann, M., and Savage, G. (2015b). Project stakeholder management — Past and present. *Project Management Journal*, 46 (6), pp. 6–14.

Eweje, J., Truner, R., and Muller, R. (2012). Maximizing strategic value from megaprojects: The influence of information-feed on decision-making by the project manager. *International Journal of Project Management*, 30, pp. 639–651.

Fallon, C. (1980). *Value analysis*, 2nd ed. Miles Value Foundation.

Fong, P.S-W. (2004). A critical appraisal of recent advances and future directions in value management. *European Journal of Engineering Education*, 29(3), pp. 377–388.

Flyvbjerg, B. (2014). What you should know about megaprojects and why: An overview. *Project Management Journal*, 45(2), pp. 6–19.

Flyvbjerg, B., Bruzelius, N., and Rothengatter, W. (2003). *Megaprojects and risk: An anatomy of ambition*. Cambridge: Cambridge University Press.

Freeman, R.E., Harrison, J.S., and Wicks, A.C. (2007). *Managing for Stakeholders: Survival, Reputation and Success*. New Haven, CT: Yale University Press.

Freeman, R.E., Harrison, J.S., Wicks, A.C., Parmar, B.L. and De Colle, S. (2010). *Stakeholder theory: The state of the art*. Cambridge, UK: Cambridge University Press.

Frooman, J. (1999). Stakeholder influence strategies. *Academy of Management Review*, 24, pp. 191–205.

Goodpasture, J.C. (2002). *Management projects for value*. Vienna, VA: Value Concepts Inc.

Greenspan, A. (2004). Risk and uncertainty in monetary policy. *The American Economic Review*, 94(2), pp. 33–40.

Harrison, J. F. and Freeman, R. E. (1999). Stakeholders, social responsibility, and performance: Empirical evidence and theoretical perspectives. *Academy of Management Journal* 42 (5): 479–85.

Hart, S.L. and Sharma, S. (2004). Engaging fringe stakeholders for competitive imagination. *Academy of Management Executive*, 18(1), pp. 7–18.

Helm, J. and Remington, K. (2005). Effective project sponsorship: An evaluation of the role of the executive sponsor in complex infrastructure projects by senior project managers. *Project Management Journal*, 36(3), pp. 51–61.

Hertogh, M. and Westerveld, E. (2009). Playing with complexity – Management and organization of large infrastructure projects. *NETLIPSE*. Amsterdam, The Netherlands.

Hertogh, M., Baker, S., Staal-Ong, P.L., and Westerveld, E. (2008). Managing large infrastructure projects –Research on best practices and lessons learnt in large infrastructure projects in Europe. *NETLIPSE*. Amsterdam, The Netherlands.

Hu, Y., Chan, A.P.C., Le, Y., and Jin, R. (2014). From construction management to complex project management: Bibliographic analysis. *Journal of Management in Engineering*, 11, pp. 1–11.

Huemann, M. and Silvius, G. (2017). Projects to create the future: Managing projects meets sustainable development. *International Journal of Project Management*, 35, pp. 1066–1070.

Ika, L.A. (2009). Project success as a topic in project management journals. *Project Management Journal*, 40(4), pp. 6–19.

Ji, H. and Darnall, N. (2018). All are not created equal: Assessing local governments' strategic approaches towards sustainability. *Public Management Review*, 20, pp. 154–75.

Jia, G., Yang, F., Wang, G., Hong, B., and You, R. (2011). A study of mega project from a perspective of social conflict theory. *International Journal of Project Management*, 29, pp. 817–827.

Jensen, T. and Sandström, J. (2013). In Defence of Stakeholder Pragmatism. *J Bus Ethics* 114, 225–237.

Johnson, G., Scholes, K., and Whittington, R. (2005). *Exploring corporate strategy: Text and cases*, 6th edition, Harlow, UK: Prentice Hall.

Jones, T. and Wicks, A. (1999). Convergent stakeholder theory. *Acad. Management Rev*. 24(2) 206–221.

Kara, M.A., Tas, S., and Ada, S. (2016). The impact of infrastructure expenditure types on regional income in Turkey. *Regional Studies*, 50(9), pp. 1509–1519.

Kaufman, J.J. (1990). *Value engineering for the practitioner*, 3rd ed. Raleigh, NC: North Carolina State Press.

Keeys, L.A. and Huemann, M. (2017). Project benefits co-creation: Shaping sustainable development benefits. *International Journal of Project Management*, 35, pp. 1196–1212.

Kelly, J., Male, S., and Graham, D. (2004). *Value management in construction projects*. Oxford, UK: Blackwell Science.

Laursen, M. and Svejvig, P. (2016). Taking stock of project value creation: A structured literature review with future directions for research and practice. *International Journal of Project Management*, 34, pp. 736–747.

Letsch, C. (2013). Turkey protest spread after violence in Istanbul over park demolition. *The Guardian*, 31st May 2013.

Li, T.H.Y., Ng, S.T., and Skitmore, M. (2012). Public participation in infrastructure and construction projects in China: From an EIA-based to a whole-cycle process. *Habitat International*. 36, pp. 47–56.

Major Projects Association (2014). A Fool with a Tool is still a Fool: Risk Management for Megaprojects and Major Programmes. *Said Business School*, Webinar, Feb 20.

Mason, M. (2010). Sample size and saturation in PhD studies using qualitative interviews. *Forum: Qualitative Social Research*, 11(3), pp. 1–19.

Male, S., Kelly, J., Gronqvist, M., and Graham, D. (2007). Managing value as a management style for projects. *International Journal of Project Management*, 25(2), pp. 107–114.

Marshall, T. and Cowell, R. (2016). Infrastructure, planning and the command of time. *Environmental and Planning C: Government and Policy*, 34(8), pp. 1843–1866.

Matti, C., Consoli, D., and Uyarra, E. (2017). Multi-level policy mixes and industry emergence: The case of wind energy in Spain. *Environment and Planning C: Politics and Space*, 35(4), pp. 661–683.

McElroy, B., and Mills, C. (2000). Managing Stakeholders. In: R. Turner. (Eds.). *People in Project Management*, pp. 99–119. Aldershot, UK: Gower.

Significance

McKinsey Global Institute. (2016). Bridging global infrastructure gaps. McKinsey and Company.

Meadowcroft, J. (2013). Reaching the limits? Developed country engagement with sustainable development in a challenging conjuncture. *Environment and Planning C: Government and Policy*, 31, pp. 988–1002.

Miles, L.D. (1961). *Techniques of value analysis and engineering*. New York: McGraw Hill.

Miles, L.D. (1972). *Techniques of value analysis and engineering*, 2nd ed. New York: McGraw Hill.

Miles, L.D. (1989). *Techniques of value analysis and engineering*, 3rd ed. Lawrence D. Miles Value Foundation.

Miller, R. and Lessard, D.R. (2000). *The Strategic Management of Large Engineering Projects*. Cambridge, MA: The MIT Press.

Miller, R. and Olleros, X. (2001). Project shaping as a competitive advantage. In R. Miller and D. R. Lessard (Eds.) *The Strategic Management of Large Engineering Projects – Shaping Institutions, Risks and Governance*. Cambridge, UK: MIT Press.

Mitchell, R.K., Agle, B.R., and Wood, D.J. (1997). Toward a theory of stakeholder identification and salience: Defining the principle of who and what really counts. *Academy of Management Review*, 22 (4), pp.853–886

Mok, K.Y., Shen, G.Q., and Yang, J. (2015). Stakeholder management studies in mega construction projects: A review and future directions. *International Journal of Project Management*, 33, 446–457.

Morris, P.W.G. and Hough, G.H. (1987). *The Anatomy of Major Projects: A Study of the Reality of Project Management*. Chichester, UK: John Wiley & Sons Ltd..

Mossner, S. (2016). Sustainable urban development as consensual practice: Post-politics in Freiburg, Germany. *Regional Studies*, 50, pp. 971–982.

Morris, P.W.G. (2013). *Reconstructing project management*. Chichester, UK: Wiley-Blackwell.

Mudge, A.E. (1990). *Value engineering: A systematic approach*. Pittsburgh, PA: J. Pohl Associates.

National Audit Office. (2020). *High speed two: A progress update*. London: Department for Transport and High Speed Two Ltd.

NETLIPSE. (2016). *10 years of managing large infrastructure projects in Europe: Lessons learnt and challenges ahead*. Amsterdam: Ovimex B.V. Deventer.

Nguyen, N.H., Skitmore, M., and Worg, J.K.W. (2009). Stakeholder impact analysis of infrastructure project management in developing countries: A study of perception of project managers in state-owned engineering firms in Vietnam. *Construction Management and Economics*, 27, pp. 1129–1140

O'Brien, J. (1976). *Value management in design and construction*. New York: McGraw Hill.

Office of Government Commerce. (2009). *Managing successful projects with PRINCE2. London:* Office of Government Commerce.

Olander, S. (2007). Stakeholder impact analysis in construction project management. *Construction Management and Economics*, 25, pp. 277–287.

Olander, S., and Landin, A. (2008). A comparative study of factors affecting the external stakeholder management process. *Construction Management and Economics*, 26, pp. 553–561.

Parker, D.E. (1985). *Value engineering theory*. New York: McGraw Hill.

Pellegrinelli, S., Partington, D., and Geraldi, J. (2011). Programme Management: An Emerging Opportunity for Research and Scholarship. In: P.W. G. Morris, J. Pinto, & J. Söderlund (Eds.), *The Oxford Handbook of Project Management* (pp. 252–272). Oxford, UK: Oxford University Press.

Peppard, J., Ward, J., and Daniel, E. (2007). Managing the realization of business benefits from IT investments. *MIS Quarterly Executive*, 6, pp. 1–11.

Pinto, J.K., and Winch, G. (2016). The unsettling of "settled science". The past and future of the management of projects. *International Journal of Project Management*, 34, pp.237–245.

Porter, A.L., Ashton, B., Clar, G., Coates, J.F., Cuhls, K., Cunningham, S.W., Ducatel, K., Van der Duin, P., Georghiou, L., Gordon, T., Linstone, H., Marchau, V., Massari, G., Miles, I., Mogee, M., Salo, A., Scapolo, F., Smits, R., and Thissen, W. (2004). Technology futures analysis: toward integration of the field and new methods. *Technological Forecasting and Social Change* 71, 287–303.

Prahalad, C.K. and Hamel, G. (1990). The core competencies of the corporation. *Harvard Business Review*, 68(3), pp. 79–91.

Rickards, L., Ison, R., and Funfgeld, H. (2014). Opening and closing the future: Climate change, adaption, and scenario planning. *Environment and Planning C: Government and Policy*, 32, pp. 587–602.

Sabini, L., Muzioo, D., and Alderman, N. (2019). 25 years of "sustainable projects": What we know and what the literature says. *International Journal of Project Management*, 37(6), pp. 820–838.

Santhanaman, R. and Hartono, E. (2003). Issues linking information technology performance to firm capability. *MIS Quarterly*, 27(1), pp. 125–153.

Serra, C.E.M. and Kunc, M. (2015). Benefits realisation management and its influence on project success and on the execution of business strategies. *International Journal of Project Management*, 33(1), pp. 53–66.

Silvius, A.J.G. (2017). Sustainability as a new school of thought in project management. *Journal of Cleaner Production*, 166, pp. 1479–93.

Teo, M. and Loosemore, M. (2017). Understanding community protest from a project management perspective: A relationship-based approach. *International Journal of Project Management*, 35 (8), 1444–1458.

Thiry, M. (2001). Sense making in value management practice. *International Journal of Project Management*, 19(1), pp. 71–77.

Turner, J.R. (2014). *Gower handbook of project management*, 5th ed. Farnham: Gower Publishing Ltd.

Turner, J.R., Zolin, R. and Remington, K. (2009) Monitoring the performance of complex projects from multiple perspectives over multiple time frames. In Proceedings of the 9th International Research Network of Project Management Conference (IRNOP), 2009-10-11-2009-10-13, Berlin

Turner, R., and Zolin, R. (2012). Forecasting Success on Large Projects: Developing Reliable Scales to Predict Multiple Perspectives by Multiple Stakeholders Over Multiple Time Frames. *Project Management Journal*, 45 (5), 87–99.

U.S. Department of Transportation, Federal Highway Administration (FHA). (2007). Highway Statistics 2007.

van den Ende, L. and van Marrewijk, A. (2018). Teargas, taboo and transformation: A neo-institutional study of community resistance and the struggle to legitimize subway projects in Amsterdam 1960–2018. *International Journal of Project Management*, 37(2), pp. 331–346.

Vidal, L.A., Marle, F., and Bocquet, J.C. (2011). Measuring project complexity using the analytic hierarchy process. *International Journal of Project Management*, 26(6), pp. 591–600.

Venkataraman, R. and Pinto, J.K. (2008). *Cost and value management in projects*. Hoboken, NJ: John Wiley and Sons.

Vuorinen, L., Martinsuo, M, 2018. Program integration in multi-project change programs: agency in integration practice. *International Journal of Project Management*, 36(4), pp. 583–599.

Vuorinen, L. and Martinsuo, M. (2019). Value-oriented stakeholder influence on infrastructure projects. *International Journal of Project Management*, 37(5), pp. 750–766.

Ward, J. and Daniel, E. (2012). *Benefits management: How to increase the business value of your IT projects*. West Sussex, UK: Wiley.

"What is value management," The Institute of Value Management. Retrieved from: http://www.ivm.org.uk/vm_whatis.htm.

Winch, G.M and Bonke S. Project stakeholder mapping: analyzing the interests of project stakeholders. In: Slevin, D.P., Cleland, D.I. and Pinto J.K., editors. The frontiers of project management research. Pennsylvania, USA: Project Management Institute. p. 385–403.

Winter, M., Andersen, E.S., Elvin, R., and Levene, R. (2006). Focusing on business projects as an area for future research: An exploratory discussion of four different perspectives. *International Journal of Project Management*, 24, pp. 699–709.

Xue, X., Zhang, R., Zhang, X., Yang, J., and Li, H. (2015). Environmental and social challenges for urban subway construction: An empirical study in China. *International Journal of Project Management*, 33, pp. 576–588.

Yang, J., Shen, G.Q., Bourne, L., Ho, C.M.F., and Xue, X. (2011a). A typology of operational approaches for stakeholder analysis and engagement. *Construction Management and Economics*, 29, pp.145–162.

Yu, A.G., Flett, P.D., and Bowers, J.A. (2005). Developing a value-centred proposal for assessing project success. *International Journal of Project Management*, 23, pp. 428–436.

Zanni, A.M., Goulden, M., Ryley, T., and Dingwall, R. (2017). Improving scenario methods in infrastructure planning: A case study of long-distance travel and mobility in the UK under extreme weather uncertainty and a changing climate. *Technological Forecasting & Social Change*, 115, pp. 180–197.

Zeemering, E.S. (2018). Sustainability management, strategy and reform in local government. *Public Management Review*, 20, pp. 136–153.

Zimmerman, L.W. and Hart, G.D. (1982). *Value engineering: A practical approach for owners, designers and contractors*. New York: Van Nostrand Reinhold.

Zwikael, O. and Smyrk, J. (2012). A general framework for gauging the performance of initiatives to enhance organizational value. *British Journal of Management*, 23, S6–S22.

Zwikael, O. and Smyrk, J. (2011). *Project management for the creation of organisational value*. London: Springer-Verlag.

3

MASTER PLANNING FOR RESILIENT MAJOR INFRASTRUCTURE PROJECTS

Tarila Zuofa

3.1 Introduction

As evidenced in this book, with ongoing huge investments to facilitate better infrastructure, major infrastructure planning remains a central point of discussion globally. Nonetheless, disruptions to existing infrastructure have increasingly become commonplace, leading to plunging opportunities for employment and hindering health and education in addition to restraining economic growth in both developing and developed nations. In March 2019, for instance, Cyclone Idai devastated communities across three African nations (Malawi, Mozambique and Zimbabwe). The International Committee of the Red Cross (ICRC) noted that the cyclone claimed over one thousand lives while several others were displaced because of severe damages to infrastructure and their means of livelihood (ICRC, 2019). A foremost observation from this cyclone incidence was that disaster preparedness is still grossly inadequate and governments in the affected countries simply lacked disaster preparedness. On the other side of the hemisphere, the mention of Hurricanes Harvey, Sandy, Snowvember, Irma and Maria bring very unpleasant memories associated with infrastructural breakdown, social structural failures and the impact that extreme weather events had on the affected locations. Despite the fact that Idai, Irma, Maria, and the others left very catastrophic damages, without being cynical, are they likely to be the last? This chapter is unable to provide a definite answer; nevertheless, it is able to make some noteworthy deductions from the above. Firstly, natural hazards can affect the seamless usage of infrastructure in developed and developing nations alike. More importantly, the lessons from these hostile events underscore a call to action for highly susceptible nations to develop resistance against further high-impact storms, coastal flooding and intense rainfall linked directly or indirectly to climate change. Beyond the impact of extreme weather events, uncertainties from the ongoing COVID-19 crisis also intensify the emergent demands for resilient and adaptable infrastructure that can effectively operate during periods of catastrophe and great uncertainty. Therefore, lasting and forward-looking strategies that support nations and key stakeholders in ensuring safer climates, while at the same time unlocking new economic opportunities through infrastructure development, are required.

The increasing use of the resilience concept in policy documents shows that the concept appeals to policy makers, political leaders and various others. In 2015, the UN Sustainable Development Goals (SDG) set a target to provide resilient infrastructure by specifically strengthening resilience and adaptive capacity to climate-related hazards and natural disasters in all countries and integrating climate change measures into national policies, strategies and planning. In Africa, the African Climate Resilient Infrastructure Summit aims to prepare the African continent to brace up for challenges of climate change impacts on infrastructure and attract international private investors and development agencies to invest in infrastructure resilient projects in Africa. Moving to Asia, several Asian super cities like Tokyo, Dhaka and Manila experienced serious flooding, cyclones and earthquakes in the past decade, leading to considerable investments by Asian Development Bank and others and making available considerable resources to assist in the introduction of several disaster risk strategies. Admittedly, from other continents not mentioned, several policy makers and business organisations even adopt the resilience concept as a catchphrase. However, much work remains to better understand and predict events that might affect major infrastructure projects. It is worth mentioning that investing in resilience and its scope is not only about developing climate resistant strategies for dams, roads or bridges, and power plants as well as other categories of infrastructure alone. Rather, it encompasses investing in the society, businesses, and communities who experience the impacts of changing climates and still require smarter developments that can optimise the delivery of better healthcare, education and livelihoods. Thus, the overarching aim of this chapter will be to examine how the resilience of infrastructure projects can be created and improved.

3.1.1 Chapter aims and objectives

This chapter focuses on examining how resilience can be enhanced in major infrastructure projects. Its main objectives are to:

- Develop an understanding of resilience and commonly associated concepts;
- Appraise current global initiatives towards facilitating resilient infrastructure;
- Depict the planning and design process for climate-resilient infrastructure; and
- Evaluate how resilience can be strengthened in various infrastructure sectors.

3.1.2 Learning outcomes

At the end of this chapter, readers will be able to:

- Critically discuss resilience and its commonly associated concepts;
- Articulate how current global initiatives can be used to facilitate resilient infrastructure;
- Describe the main planning steps for climate-resilient infrastructure; and
- Apply strategies to strengthen resilience in various infrastructure sectors.

3.2 Characterising key resilience concepts

The section provides brief descriptions of the key concepts associated with resilience. It provides insights into the main requirements for resilience from various domains to enrich our wider discussion of resilience in this chapter. The section also provides a summary of what the UK's National Infrastructure Commission (NIC, 2020) defined as the six aspects of resilience. As you read, you might notice significant connections in a few of the concepts.

3.2.1 Robustness

Robustness is defined as the ability of a system to be unresponsive towards changing environments (Fricke and Schulz, 2005). In essence, a system does not respond to any variations in the environment, nor changes any processes or properties when it encounters disturbances, but still achieves its desired outputs. For instance, a beam bridge might be designed with certain levels of tolerance to be robust enough to accept additional loading from increased wind or vehicular traffic. Even though such a design may be more cost effective when the disturbances are anticipated, the system may likely fail if pushed outside the system's tolerance by unexpected events. Robust infrastructure designs are usually suitable in situations where the uncertainties are reasonably more understood, typically in the near future. Thus, an infrastructure system might also be designed to be robust into the far future if uncertainties about the system are not likely to change during the system's life cycle.

3.2.2 Adaptability

Adaptability indicates where a system can be modified because of the actions triggered by a change agent (usually internal). Due to the extended life cycles of major infrastructure, it is foreseeable that changes may be required at some point in their period of function; hence, adaptable or flexible designs are normally used. It is worth noting that there are still disagreements on the definitions of "adaptability" and "flexibility" in engineering but Gilrein *et al.* (2019) demonstrated that they do not really mean the same. The installation of sensors and intelligent controllers in storm water management systems that enable stream and pipeline discharge rates to be automatically adjusted based on live data reflecting changes in either weather forecast or storms is one good illustration of adaptability in major infrastructure (Kerkez *et al.*, 2016).

3.2.3 Flexibility

From an engineering design perspective, Fricke and Schulz (2005) explained that flexibility represents a system designed in a manner that its requirements and performance can be altered in the future. This is typically achieved using platform or modular designs that make it easier to change when the need arises. While robustness and adaptability may support system recovery to desirable or normal levels, flexibility enables for changes in the "normal" state, usually achieved by upgrading the system. An illustration of flexibility can be drawn from the 25 de Abril Bridge that crosses over the Tagus River in Lisbon, Portugal. According to Gesner and Jardim (1998), this suspension bridge, originally built as a single deck for road traffic, was flexibly designed with strength to still accommodate a secondary railroad deck that was fitted several years later. Flexibility can also be demonstrated from those airports and other major infrastructure projects around the world that integrate contemporary designs and cutting-edge technology and allow future expansion with minimal disruption on existing operations.

Table 3.1 Aspects of resilience

Aspects of resilience	Brief explanation
Anticipate	Actions to prepare in advance to respond to shocks and stresses, such as collecting data on the condition of assets
Resist	Actions taken in advance to help withstand or endure shocks and stresses to prevent an impact on infrastructure services, such as building flood defences
Absorb	Actions that, accepting there will be or has been an impact on infrastructure services, aim to lessen that impact, such as building redundancy through a water transfer network to prepare for future droughts
Recover	Actions that help quickly restore expected levels of service following an event, such as procedures to restart services following an event such as a nationwide loss of power
Adapt	Actions that modify the system to enable it to continue to deliver services in the face of changes
Transform	Actions that regenerate and improve infrastructure

Source: NIC (2020)

In another vein, Table 3.1 briefly explains the six aspects of resilience identified by UK's NIC. The commission indicated that government, regulators and infrastructure operators need to consider these aspects for resilient infrastructure.

3.3 Resilience

In the engineering domain, Bruneau *et al.* (2003) defined resilience as the ability of a system to reduce the chances of shock, to absorb the shock if it occurs and to recover quickly after the shock and reinstate normal performance. Prior to this time, Holling (1973) introduced resilience in ecology as the capacity of a system to persist within a domain of attraction in the face of disturbances and changes in state variables, driving variables and parameters. From these earlier accounts until now, resilience has equally been metaphorically applied to illustrate the quality of being able to recover quickly or easily form, or resist being affected by shocks. Perhaps these perspectives have given rise to the widespread use of the term in technical and scientific literature with a diversity of interpretations. As demonstrated in Table 3.2 the term has remained in use by disciplines while different bodies have even developed standards and definitions for the concept.

Although several clear differences exist between definitions highlighted in Table 3.2, based on the perspectives followed in defining resilience, a few commonalities can be deduced. Some common aspects found in most definitions agree on:

- The presence or occurrence of an event that disrupts normal or expected function.
- A system coping with an event but still attempting to maintain or achieve desired function.
- Incorporating strategies for managing because of anticipation, absorbing or withstanding the effects of an incident.
- Adapting to maintain some level of functionality during an incident and still recovering to achieve an ultimate, desired level of function.

From the above, in simpler terms, resilience in the context of infrastructure can be considered as the immediate, mid- and long-term capacity of an infrastructure to resist, respond to, recover from and retain its essential functionality. This straightforwardly indicates that developing infrastructure systems that still deliver services even when some (or all) of its components have been

Master planning for resilient major infrastructure projects

Table 3.2 Definitions of resilience

Definition of resilience	Source
The ability of a system, community or society exposed to hazards to resist, absorb, accommodate and recover from the effects of a hazard in a timely and efficient manner, including through the preservation and restoration of its essential basic structures and functions.	United Nations International Strategy for Disaster Reduction–UNISDR (2009)
The ability to reduce the magnitude and/or duration of disruptive events. The effectiveness of a resilient infrastructure or enterprise depends upon its ability to anticipate, absorb, adapt to, and/or rapidly recover from a potentially disruptive event,	National Infrastructure Advisory Council–NAIC (2009)
The capacity of individuals, communities, institutions, businesses, and systems within a city to survive, adapt, and grow no matter what kinds of chronic stresses and acute shocks they experience.	100 Resilient Cities
Helping cities, organisations, and communities better prepare for, respond to, and transform from disruption.	Rockefeller Foundation (2015)

compromised, become vulnerable to natural hazards or are even destroyed will remain an essential consideration with the global infrastructure industry.

3.3.1 A global perspective on resilience

There has been a rising trend in disaster losses since the 1980s (World Bank, 2013), which has resulted in various causes of action at both regional and global levels. Nonetheless, the aftermath of the 2004 Indian Ocean earthquake and the subsequent tsunami appeared to herald a rethink in the history of global disaster risk management. The 2005 United Nations World Conference on Disaster Reduction (UNWCDR) in Japan provided a common platform for key stakeholders to develop a comprehensive agenda poised to mitigate disaster vulnerabilities. According to the UN International Strategy for Disaster Reduction (UNISDR) from this conference, the Hyogo Framework for Action: Building the Resilience of Nations and Communities to Disaster emerged (UNISDR, 2007). The Hyogo framework highlighted the need for, and identified ways of building, the resilience of nations and communities to disasters and established the following drivers:

- Challenges posed by disasters;
- The Yokohama Strategy: lessons learned and gaps identified;
- WCDR: objectives, expected outcome and strategic goals;
- Priorities for action 2005–2015;
- Implementation and follow-up.

Generally, the Hyogo Framework for Action (HFA) provided a critical direction in efforts to reduce disaster risk and contributed to the accomplishment of the UN's Millennium Development Goals. However, ten years later at the 2015 United Nations World Conference on

Disaster Risk Reduction, hosted in Sendai, Japan, it became evident that despite the adoption of the HFA, disasters continued to challenge efforts to achieve sustainable development especially in the developing world. Among many notable observations, the UN Office for Disaster Risk Reduction specifically disclosed that disasters, many of which are aggravated by climate change, are increasing in frequency and intensity (UNDRR, n.d.). Consequently, an "extended" HFA agreement was introduced. The aim of the new agreement (i.e. the Sendai Framework for Disaster Risk Reduction 2015–2030) is to achieve a significant reduction of disaster risk and damage to lives, livelihoods, health, physical, economic, cultural, social and environmental assets of people, communities and nations globally by 2030 (UNISDR, 2015a). Even though the 2030 target seems very ambitious, it is achievable. A very strong commitment and collaboration from various nations at different levels will be crucial for this achievement of the Sendai Framework for Disaster Risk Reduction 2015–2030.

In exploring the resilience of essential infrastructure systems, the seminal work of Hallegatte *et al.* (2019) noted that the lack of resilient infrastructure is harming people and businesses in untoward ways. Their study asserted the urgency of investing in resilient infrastructure irrespective of initial costs because of long-term benefits and profitability. Hallegatte *et al.* (2019) also advocated that good infrastructure management through coordinated actions is a necessity for achieving resilient infrastructure. As the global demand for infrastructure rises, UNISDR (2015b) disclosed that without significant improvements in infrastructure resilience, annual economic losses from natural disasters' damage to urban infrastructure can potentially rise to about US $415 billion by 2030. While there is agreement on the need to boost infrastructure, there are several debates on how to achieve this, given the challenges from various factors including climate change. Hence, scrutiny of climate resilient infrastructure has become increasingly important for societies and major infrastructure delivery stakeholders globally.

3.4 Climate resilient infrastructure

From rising numbers of storms, to more intense flood disasters, heat waves and accelerated melting of glaciers, our world has witnessed greater intensifying appearances of climate-related physical impacts in recent times. The Intergovernmental Panel on Climate Change (IPCC) and several other organisations have extensively documented the global effects of climate change in several reports (IPCC, 2019, 2018). When put together, these reports and the United Nations Framework Convention on Climate Change not only outline the devastating externalities that unmitigated global warming has on people and the environment (UNFCCC, 2015). To a degree, they also highlight the key role the infrastructure and construction sectors play in activating the necessary resources to reduce warming below 2 °C and preserve the earth's natural environment.

This book clearly lays out the need for more global investments in infrastructure as well as what is still required to ensure that any future investments remain fit for purpose. For instance, a dam or a seaport built in 2020 will be expected to be used for about 50 to 100 years depending on the design criteria. This also applies to a considerable proportion of other types of major infrastructure. With the impact of climate change continuously being confirmed by conspicuous changes in air and ocean temperatures, widespread snow and ice melting and the gradual rising of sea levels, time is needed to implement adjustments. There is also an increased likelihood and frequency of extreme climate driven events like floods and heat waves requiring more instantaneous responses. Therefore, it will be illogical to invest in any major infrastructure that does not take into consideration the susceptibility to climate change (see Vignette 3.1).

Vignette 3.1 The deep-water container terminal (DCT) in Gdansk, Poland

Located in Gdansk, this deep-water container terminal is the Baltic Sea's hub for larger cargo vessels from East Asia and one of Europe's largest maritime infrastructure projects (Nordic Investment Bank, 2016). According to the Nordic Investment Bank (NIB), before and during construction and also when in operation, the second berth at DGT Gdansk is required to implement actions to avoid or reduce significant environmental and social impacts identified in the environmental and social impact assessment. The facility is also required to comply with the Polish laws, the funding requirements by European Bank for Reconstruction and Development (EBRD) and the Nordic Investment Bank. Some of the specific actions include ensuring that the height of the quay wall takes into account sea level rise projects. Another consideration covers ensuring that the terminal creates communication channels with the Port Authority to receive information on extreme sea level and potential for waves to overtop port structures (European Bank for Reconstruction and Development, 2014).

Within this context, climate-resilient infrastructure is defined as deliberately planned, designed, constructed and operated infrastructure that anticipates, prepares and adapts to fluctuating climate conditions. From this definition, the typical significant characteristics of climate-resilient infrastructure is that it should be capable of withstanding, responding to and recovering rapidly from disruptions caused by climate conditions. It is worth mentioning that in defining climate resilient infrastructure, risks may not be fully eliminated, but the focus is on clear mitigation of climate-related disruptions. In some instances, this might be achieved by incorporating additional flexibility like shorter design life, retrofitting or replacement as the climatic phenomenon changes. Beyond having the capacity for climate adaptation, Meyer and Schwarze (2019) disclosed that climate resilient infrastructure should also be resilient to local and supra-local regulatory and other measures addressed to mitigating future climate change because of its impact on the living conditions of people and their economic and social advancement capabilities. In summary, the UK's government explained that existing infrastructure could be climate resilient by ensuring that maintenance arrangements incorporate resilience into the impacts of climate change over the infrastructure's lifetime. On the other hand, new infrastructure can be climate resilient by ensuring that the infrastructure is located, designed, built and operated with consideration for current and future climate changes (TSO, 2011). To achieve this, the following adaptation measures were proposed:

- Ensuring infrastructure is resilient to potential increases in extreme weather events such as storms, floods and heatwaves as well as extreme cold weather.
- Ensuring investment decisions take into account changing patterns of consumer demand because of climate change.
- Building in flexibility so infrastructure assets can be modified in the future without incurring excessive cost.
- Ensuring that infrastructure organisations and professionals have the right skills and capacity to implement adaptation measures (TSO, 2011).

3.4.1 *Planning and designing climate-resilient infrastructure*

Climate change has become a global phenomenon with potential dramatic effects; its integration during the planning and designing of infrastructure projects requires no further amplification. This is particularly accurate in developing and emerging countries where infrastructure needs are huge with scarce resources and where the augmentation of existing infrastructure investments must occur rapidly. It is also applicable in the developed countries because not taking into account climate change uncertainties might have disastrous consequences for any investments in long-lived infrastructure. As an illustration, the new railway for Crossrail has a minimum 120-year design life. To achieve this, Crossrail collaborated with the Environment Agency to develop infrastructure that will be protected from potential flooding until 2100 or beyond, even in the event of a possible failure of the Thames flood defences (Crossrail, 2018).

Globally, even as climate change may adversely affect many aspects of our lives, environment, business and public services, it is also an essential driver for modern dynamic infrastructure planning and design approaches that potentially address long-term climate change related susceptibilities. As exemplified by Crossrail, infrastructure usually have long-life expectancies (at least 50 years and more in the case of Crossrail) that require consideration of present and future climate conditions to support dedicated services over varying long periods. Should these considerations trigger changes about the way investors have been up until now making infrastructure investment decisions? For instance, in the United Kingdom, over 45 percent of the national infrastructure is financed through the private sector (Institution of Civil Engineers, 2018) while Asia requires about US $26 trillion until 2030 to cater to climate adjusted infrastructure investments (ADB, 2017). From the foregoing, there should certainly be concerns over the consequences of uncertain future climate change and its impact on the planning and design of major infrastructure among investors and other stakeholders. Collectively, several options might be potentially pursued by stakeholders, one being to delay the decision to invest in major infrastructure until a point where better clarity emerges. Another choice could be ignoring the possible uncertainties because of limited design and planning knowledge or accurate tools to incorporate future uncertainties. A final option may well be overspending scarce resources for a variety of potential outcomes at greater expense. However, none of these options is actually plausible; rather, innovative solutions for better planning and designing are required.

The Organisation for Economic Co-operation and Development (OECD) acknowledged that governments in several countries like Australia, Canada and the United Kingdom have all released technical guidance to ensure infrastructure design is resilient to climate change (OECD, 2018). However, studies by (Bloemen *et al.,* 2018 and Ranger *et al.,* 2013) articulated that the Thames Estuary 2100 Project (TE2100) remains one of the first major infrastructure projects to explicitly recognise and address the issue of the deep uncertainty in climate projections throughout the planning process. Ranger *et al.* (2013) summarised:

- The "decision–centric" process;
- The combination of numerical models and expert judgement to develop narrative sea level rise scenarios;
- The adoption of an "Adaptation Pathways" approach to identify the timing and sequencing of possible "pathways" of adaptation measures over time under different scenarios; and
- The development of a monitoring framework that triggers defined decision points.

Concerning the approaches adopted to enable TE2100 to cope with uncertainty, Ranger *et al.* (2013) concluded that the overall methodology exhibited dynamic robustness, which enhanced

Master planning for resilient major infrastructure projects

the development of flexible strategies that could evolve over time with fresh knowledge or as changing conditions. While an increasing trend in urbanisation and population growth will require new infrastructure planning and designing, ageing infrastructure will need replacement and retrofitting. Consequently, the techniques from TE2100 can be particularly applied during the planning and design of similar major infrastructure. In doing so, there is need to gather as much rich information to determine what is recognised or not about climate change by capturing various uncertainty sources. This facilitates firm decision-making, which compliments scenario planning. Of course, there are several key players required for the effective planning and designing of climate-resilient infrastructure like the government. Governments at various levels need to play an active role by providing relevant information that stakeholders had better understand by formulating advisory guidelines, regulations updating building codes and engineering design standards to factor changing climatic conditions. During planning and designing climate-resilient infrastructure, the affordability for infrastructure systems is set to become a progressively more important issue. As such, robust risk financing strategies will be required, both to fund investments from the onset or during adaptation and to pay for recovery when failures occur. In summary, for major infrastructure projects to be planned and designed for climate resilience, the systems have to be capable of absorbing turbulences, adapting for change and thriving for the future. All these can be related to robustness, adaptability and others mentioned when this chapter previously described resilience (see Section 3.2).

3.5 Appraising strategies for strengthening infrastructure resilience: a sectorial approach

From serving the most rudimentary needs to enabling very ambitious projects, infrastructure undoubtedly underpins all global efforts to enhance development and societal well-being. Huge investments are ongoing to guarantee better infrastructure; nevertheless, the quality and adequacy of infrastructure services will still differ widely across countries, as the population of people living in urban areas is set to rise in the future. For instance, projections by the United Nations suggest that 68 percent of the global population will reside in urban areas (UN, 2018). With already concentrated levels of economic activity in most urban areas, this will have enormous implications for how infrastructure in various sectors is planned in future. In certain flood prone cities, frequent urban flooding affects families and businesses by destroying their possessions, increasing the danger of water-borne infections associated with dysfunctional sanitation systems and interrupting economies.

The clustering of cities on coastlines and waterside settlements also raises concerns about vulnerability to storm surges. In addition, as the COVID-19 pandemic has revealed, there are likely implications for densely populated areas (Hamidi *et al.*, 2020). With all these susceptibilities and the magnitude of investments channelled towards future infrastructure in various sectors, exploring the place of resilience becomes imperative. The remaining part of this section briefly appraises some strategies that can be adopted to strengthen infrastructure resilience globally. Even though various infrastructure sectors are examined individually, it is very important to acknowledge their interdependency in reality, their multi-directional relationships and their connectivity when practically attempting to understand how they can be made more resilient.

3.5.1 *Cities and urban developments*

It is unquestionable that there are different explanations for the concept of resilient urban infrastructure and urban resilience. On a general note, urban resilience is depicted as the degree to which cities can resist, absorb and accommodate shocks through generation of new structures and processes. This has also led to globally diverse applications (see Vignette 3.2).

Vignette 3.2   Snapshot of resilient urban and city infrastructure developments in selected areas

ASIA

The Energy and Resources Institute (2011) noted that many Indian cities have taken bold steps in building resilient infrastructure in their bid to achieve sustainable development. Cities like Gorakhpur, Surat and Indore have developed their own specific resilience strategies. The Gorakhpur resilience strategy adopts an integrated approach that addresses institutional, social, behavioural and technical modalities of intervention (Energy and Resources Institute, 2011). To achieve this, the strategy underscores effective implementation of master plans while building in climate concerns. According to Rajasekar *et al.* (2012), the Surat and Indore resilience strategies adopt four major principles: building on current and planned initiatives; demonstrating resilience-building projects to leverage further action; multi-sectoral information generation, and shelf of projects; and building synergies between national and local institutions. Collectively, these Indian cities demonstrate a commitment to building resilient communities in a more inclusive method. Despite Japanese cities being more prone to external environmental shocks in the form of tsunamis and other natural disasters, they still appear to be the custodians of the most resilient infrastructure assets in the Asian Pacific region. Investing in resilient infrastructure has been vital in protecting Japanese cities. According to *PwC* (2013), a number of ground-breaking initiatives in Japanese cities focus on leveraging new technologies to develop safe, sustainable, energy-efficient communities.

THE MIDDLE EAST

According to Griffiths and Sovacool (2020), if Masdar City, United Arab Emirates, does work as currently planned, it would motivate the importance of integrating city planning, passive design, energy supply, transport, water and recycling efforts. Even though the above-mentioned measures enhance urban resilience and demonstrate responsibility towards environmental sustainability, Masdar City still faces challenges in its path ahead (Griffiths and Sovacool, 2020).

LATIN AMERICAN

UNISDR (2008) once observed that most Latin American cities experienced infrastructure delivery problems whenever they experience heavy flooding, earthquakes or cyclones. Despite these problems, KPMG (2013) disclosed that significant progress has been made in the construction of resilient infrastructure. Cities such as Mexico, São Paulo and Paraguay have all recorded significant progress in building resilient infrastructure through integrated urban development plans. In Arequipa, Peru, an innovative collaboration with Cerro Verde (a copper mining company) has resulted in a more resilient water supply for the city (Rodríguez Tejerina, 2015). Once known as the most violent city in the world during the 1980s and 1990s Medellín, Colombia, has experienced phenomenal transformation and become a more inclusive, vibrant, and resilient city. The Metrocable is a major deliverable of a multi-sector approach, resulting in urban transformation and resilience. Its transformation story now attracts strong global interest (Fundacion Idea, 2017).

EUROPE

European cities appear to play a crucial role in the strategies for achieving urban resilience. Galderisi (2012) explained that several European Commission recommendations on resilient infrastructure building continue to serve as guidelines that European cities have to observe to boost their resilience. Milan has joined cities around the world using nature-based solutions. According to ICLEI (2019), the air-filtering and cooling power of plants and trees is used for the vertical forest residential towers in Milan.

AFRICAN CITIES

As noted in the UN's The World's Cities in 2018, the city of Nairobi, Kenya's largest city and capital, has witnessed a rapid population growth (UN, 2018). According to Asoka *et al.* (2013), if this trend continues, it will exert excessive pressure on existing infrastructure and necessitate an expanded infrastructure base. In recent times, its city council has initiated several road re-surfacing and maintenance projects to address the impact of rains. Nonetheless, these projects still encounter implementation challenges as evidenced by the road infrastructure that continues to worsen (Asoka *et al.*, 2013).

Abuja, Nigeria's federal capital, has experienced rapid population growth that exerts pressure on urban infrastructure and services (Jinadu, 2004). The city presents a case good case for the development of resilient infrastructure; however, Onyenechere (2010) noted that governments at various levels have paid limited attention to opportunities for infrastructure development and provision. Some of the factors that might be working against the building of resilient urban infrastructure in Abuja were articulated by Usman and Tunde (2010).

If Rwanda achieves its urbanisation target of 35 percent by 2024, it will potentially become one of East Africa's most urbanised countries (Gubic and Baloi, 2019). The details contained in policy documents including "The Republic of Rwanda" (n.d.) highlight how key players and government are prioritising developing basic infrastructure in Kigali and other urban centres to achieve resilience.

Vignette 3.2 demonstrates that cities have a role to play in achieving resilience by putting in place mitigation, adaptation or other measures specifically targeted for urban areas. The array of examples demonstrates how some cities have been purposeful about achieving resilience while others have been less so. As articulated by the Rockefeller Foundation, delivering the "right" kind of urban infrastructure requires an understanding of what cities need today and anticipating what they might need in the future (Rockefeller Foundation, 2019). Nevertheless, while focusing on delivering the "right" kind of infrastructure, there is still an urgent need for cities to prioritise the design and implement resilient infrastructure towards ensuring cities are more resilient.

3.5.2 Power sector

Rentschler *et al.* (2019) provided evidence to suggest that natural hazards are among the leading causes of power outages around the world. However, using Bangladesh as an example, their study also identified that non-natural factors and system failures were also responsible for power outages. Irrespective of the causes, the effect of power outages is usually massive, ranging from domestic to social and economic activity disruption. Several strategies to strengthen the resilience of power infrastructure can be considered. One understandably and broad approach is to make the power infrastructure assets more robust (see section 3.2 for a brief description of robustness).

Another obvious but all-important strategy is maintenance. With power infrastructure like hydropower plants built to function for as long as 100 years, the relevance of regular maintenance cannot be overemphasised. Disappointingly, due to a number of factors including paucity of funds and political instability, the maintenance of power infrastructure is either performed below required standards or simply ignored in several African nations (McKinsey, 2020; PwC, 2015; Sy and Copley, 2017).

One noteworthy observation is that the issue of lack of maintenance is not only common to infrastructure in Africa or developing countries. To illustrate this, following heavy storms in 2017, the main and emergency spillway of the Oroville Dam in California threatened to fail, thereby leading to the evacuation of 180,000 people. *Los Angeles Times* reported that a lack of maintenance was the cause of the problem (*Los Angeles Times*, 2017). The independent forensic team report undertaken by the Association of State Dam Safety Officials (ASDSO) stated, "The Oroville Dam spillway incident was caused by a long-term systemic failure to recognise and address inherent spillway design and construction weaknesses, poor foundation bedrock quality, and deteriorated service spillway chute conditions" (ASDSO, 2018).

Advanced technology and innovations also provide strategies for strengthening the resilience of power infrastructure. Through current advancement in technology and innovations, automation, smart grids, advanced metering infrastructure, innovative communication and remote sensing all boost reliability, improve responsiveness and mitigate natural hazards. For instance, the Potomac Electric Power Company credited its ability to rapidly restore power after Hurricane Sandy to improved responsiveness and situational awareness by the use of advanced technology that allowed them to quickly locate outages (US Department of Energy, 2013; The White House, 2013). Finally, strategies for strengthening the resilience of power infrastructure can go beyond the technicalities of reinforcement and shock absorption. As articulated by Wang et. al. (2016), there are several non-technical strategies like improved organisational procedures such as effective leadership and financial preparedness that can still be used to strengthen the resilience of power infrastructure.

3.5.3 Transportation sector

Transportation infrastructure systems can include roads, airways, railways, water, pipeline transportation and all other infrastructure essential for the operation of various modes of transportation. Disruptions include daily operational variations, interruptions due to failures of infrastructure, vehicular obstructions, engineering works and severe weather conditions like heavy winds, rain or snowstorms to natural hazards such as hurricanes or earthquakes. Disruptions and disasters commonly lead to delays, temporary or permanent closures and interruptions of services associated with transport infrastructure.

Across various contexts, there is a growing recognition that lack of resilience can have serious socioeconomic consequences, especially in the context of interconnected infrastructure like transportation (Hallegatte *et al.,* 2019). Consequently, in recent times, significant attention has been given to improve the protection of critical infrastructure including transport systems in Europe; see *Rail Adapt* (International Union of Railways, 2017). In the United States as well, the *Presidential Policy Directive 21 − Critical Infrastructure Security and Resilience* formalised the call to enhance the nation's critical infrastructure functioning and resilience by acknowledging the importance of critical infrastructure, including transportation systems (The White House, 2013). From all these initiatives and general literature, it can be readily accepted that building resilient transport systems needs to go beyond applying hard measures such as stone barriers but also integrating better working practices and adequate operational buffers. The current state of practice in the transportation sector following several natural disasters and other disruptions also gives itself to the conclusion that developments in resilience require continuous advancement to help account for future uncertain events and emerging hazards. Miyamoto, an international global company that provides critical services that sustain industries and safeguards communities around the world, performed an assessment of critical infrastructure that are particularly vulnerable to natural hazards (Miyamoto, 2019). Table 3.3 excerpts key information on the transport sector.

As summarised in Table 3.3, a common resilience improvement strategy identified for railways is conducting daily testing and continual field operations for quality assurance. Furthermore, to strengthen infrastructure resilience within the transport sector, the improvement strategies highlighted have adopted engineering-based solutions and innovative application of technology to reduce the impacts of hazards on current, mid- and long-term capacity of the transportation infrastructure systems. Thus, it can be posited that measures for strengthening resilience in transportation infrastructure systems require a proactive phase that deliberately plans for resilience and a reactive phase that protects against possible disasters or disruptions. At the various phases of the life cycle of transportation infrastructure, due to the interdependency of transportation systems, multidisciplinary approaches for assessing inter-reliant resilience are required. For this reason, a combination of various data sources like geological, traffic, travellers, weather and mathematical models should be considered, especially during the planning phase. This provides an opportunity for improved multidisciplinary assessment methods and early warning systems for predicting disruptions and potential disasters.

The investigation of the reduction of disruption costs through engineering and planning solutions that can help strengthen transport infrastructure has also been studied in recent times. According to Rozenberg *et al.* (2019), in practice, for these solutions to be effective, resilience must be integral to decision-making at different levels: resilience of infrastructure assets, resilience of infrastructure services and resilience of infrastructure users. Another thing to add is that generally, with transportation infrastructure, the level of resilience required should always be correlated to the intensity of use, the availability of substitutes and the economic importance of the system or service it provides. For example, since country roads and busy highways have different intensities of use, various levels of planning for resilience integration will be required. Finally, as the technology for connected and autonomous vehicles (CAV) speedily gains momentum and starts to transform the face of the existing transportation infrastructure, the impact of CAV technology on transportation system resilience not covered in this chapter becomes a promising research area.

3.5.4 *Water infrastructure*

Part of the resolutions adopted by the General Assembly on 27 July 2012 articulated that water is at the core of sustainable development as it is closely linked to a number of key global challenges (UN, 2012). As the world experiences a growing population, there is an increasing need to balance the competing industrial demands on water resources in a manner that means the public still have an adequate amount for their needs. Furthermore, in uncertain times where the effects of climate change on water resources remain distressing, strategies to strengthen resilience in the water infrastructure must be prioritised. Within water infrastructure research, Butler *et al.* (2014) made these proposals: design resilience (a set of design principles for the infrastructure); technology-based resilience (using technology devices to limit flood damage and speed recovery); and specified resilience (agreed performance in terms of pressure or flow) as the dimensions of resilience in water infrastructure. Based on certain principles (see Figure 3.1), water infrastructure managers were sensitised on the importance of and measures to build the resilience of water

Table 3.3 Resilience improvement strategies for selected transportation infrastructure

Infrastructure category	Hazard	Improvement strategy
Railway	Earthquake hazard	Seismic switch installation to safely stop and restart railway operations
		Perform routine and regular maintenance for all components and fix any observed problems.
		Conduct the higher-level QA protocol (e.g., daily testing and continual and detailed field inspection).
		Assess geotechnical components and reinforce where seismic deficiency exists
	Liquefaction	Drainage and drainpipe installation to reduce the amount of water in the soil
		Soil improvements and densification application by several measures like cement mixing and soil compaction
		Use deep foundations, such as pile, wall pile, or piled raft
		Conducting higher-level QA protocols
	Wind	Replace building envelopes and retrofitting roof-wall connections
		Strengthen the building-type components for a higher design wind speed and load
		Ensure that all equipment are properly anchored using positive mechanical attachments that are wind-rated
		Check wind resistance of railway bridges and piers and retrofit them if necessary
	Flood	Elevate the components (e.g., equipment) and install watertight barriers
		Install flood-monitoring sensors to notify operators when flooding occur or reaches certain water levels for each component
		Relocate critical components and equipment to a flood-safe location
		Duplicate the path of critical components and equipment to provide disaster redundancy
Highways on grade	Earthquake	Adopt a two-tier design approach
		Properly compact the underlying embankment
		Use earthquake-resistant embankments or foundations
		Perform a check and ensure slope stability of the embankment
	Landslide	Landslide mitigation
		Hydrological solutions: Add drainage to reduce water pressure; prevent water from entering the hillside by diversion
		Mechanical/structural solutions: use tiebacks or soil nails, shotcrete the surface and construct retaining walls and steel-wood walls
		For erosion control, add steel netting, geomats and coconut fibre mesh
	Liquefaction	Dynamic compaction
		Vibro-compaction to stabilise granular soil
		Grouting
Secondary urban roads	Earthquake	Use a two-tier design approach fully operational for a 100-year earthquake with no major damage for a 1,000-year event
		Properly compact the underlying material
		Use earthquake-resistant foundations
	Liquefaction	Retrofitting of existing urban roadways on liquefiable soils are not cost-effective.
		Liquefaction performance can be enhanced by several methods to improve the quality of the underlying soil
	Flooding	Use barriers where possible
		Improve drainage
		Maintain roads

Source: Miyamoto (2019)

service provision to natural hazards and climate risks while ensuring systems still safeguard service provision (Stip *et al.*, 2019).

Figure 3.1, depicting the partial outcomes from Stip *et al.* (2019), demonstrates that the best way to build resilience is by combining infrastructure and institutional measures, and supply augmentation and demand management with measures to increase system efficiency. According to Stip *et al.* (2019), when choosing resilience measures, water systems managers should consider these six principles.

3.6 Towards resilient infrastructure

3.6.1 Any alternative tools for measuring resilience success?

As major infrastructure choices potentially define the lifespans of infrastructure investments and have wider implications for stakeholders, measuring or tracking their performance in terms of resilience will always be of significant interest. However, factors like physical, social, organisational or institutional considerations as well as concerns over practicality have been cited as difficulties associated with measuring resilience in current studies (Prior, 2014; Sun *et al.*, 2020). Recently, Bennon and Sharma (2018) performed a review of twelve commonly used standards and assessment tools available to investors and developers to measure or report on the sustainability and resilience of their infrastructure investments. Among the tools they evaluated were Standard for Sustainable and Resilient Infrastructure (SuRe), Civil Engineering Environmental Quality Assessment and Awards (CEEQUAL), International Finance Corporation (IFC) Performance Standards, Equator Principles and World Bank EHS Guidelines, Greenhouse Gas (GHG) Protocol Accounting and Reporting Standard and United Nations Sustainable Development Goals. Even though Bennon and Sharma (2018) provided a detailed account on how the twelve tools can be applied practically, Hallegatte *et al.* (2019) struggled to establish how investors and decision-makers would identify:

- The performance dimension (i.e., how natural risks and climate change will affect the return on financial products); and
- The environment, social and governance dimensions (i.e., how will financial products contribute to economic, social, and environmental sustainability) based on several current tools.

Accordingly, Hallegatte *et al.* (2019) disclosed that the World Bank Group is committed to proposing a resilience rating system, which can report the resilience characteristics of projects to investors and decision-makers better. When in use, the World Bank Group's rating system will factor in natural disaster and climate change risks, resilience, estimated benefits and profits as well as the broader implications on communities and economies for public and private infrastructure investments. Unlike other tools that produce cumbersome data, the World Bank Group's rating system will not create new data but will translate technical information in project documentation into a simple rating. It will be based on two dimensions of resilience:

- Resilience of investments and projects to measure the extent to which projects consider disaster and climate risks;
- Resilience built through the investments and projects, directed at specific components purposefully designed with the objective of building the resilience of beneficiaries into consideration (Hallegatte *et al.*, 2019).

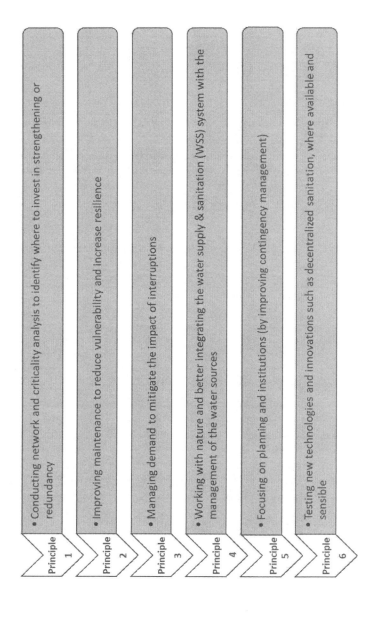

Figure 3.1 Principles for achieving resilience in water infrastructure
Source: Stip *et al.* (2019)

3.6.2 Advancing the aspiration for resilient major infrastructure

The benefits of resilient infrastructure are well established in this chapter and beyond. Since resilience needs to be integrated into all infrastructure assets and systems throughout their life cycles, it requires the engagement of several stakeholders including government, public institutions and agencies, communities, data providers, developers, vendors, investors and operators from the onset. Therefore, as highlighted in Figure 3.2, inclusive planning among stakeholders is required to share knowledge and capacity required for projects. The inclusive planning approach also includes integrating social and ecological resilience into planning decisions where appropriate. As part of the planning process, there is a need to consider a major infrastructure project beyond individual elements but rather as complex and interconnected systems. Note that the failure of one infrastructure component can elicit simultaneous failures or a cascading collapse of other parts of the system. One crucial factor for planning is the availability of data. In many developing countries, the data needed to understand hazards are lacking; therefore, inclusive planning needs to accommodate investments in data and tools to provide the necessary underpinning for planning decisions. Through all the above progress, major infrastructure planners should embrace alternative approaches that accommodate for wider uncertainties but still facilitate adaptive decision-making.

This chapter's central message is that resilience needs to be integrated into the life cycles for major infrastructure. For this to be realised, resilient design directives become compulsory. Central governments have a crucial role to develop and enforce national technical codes and standards that propagate the need for resilience. Admittedly, some nations will contend with situations where standards have been replicated from other countries without explicitly being adjusted to reflect their own peculiar circumstances. In such cases, this means adjusting international best practices to reflect local circumstances and explicitly articulating the precise roles and responsibilities of households, communities, the private sector, other tiers of government and the international community in strengthening resilience. Overall, central government needs to set out clear, proportional and realistic standards for the levels of infrastructure resilience it anticipates under different circumstances. It is crucial to have clear standards for the resilience

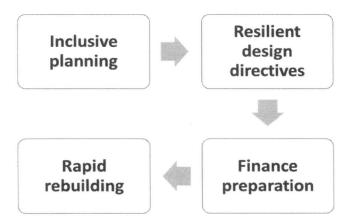

Figure 3.2 Advancing the aspiration for resilient major infrastructure

of major infrastructure, as this helps antedate potential shocks and stresses and prepare for any potential disruptions or failures.

Despite the role of public funding in the provision of infrastructure, there is ample evidence to suggest the willingness of private investors to upturn their infrastructure investment in future. Nonetheless, for this to be feasible, central government still has a role to play here by introducing policies or incentives needed to mobilize investments in resilient infrastructure. Decision-makers should entrench provisions to frequently monitor infrastructure resilience in response to anticipated climate change and population demographics. This responsibility for monitoring resilience should be designated during the planning process. Finally, based on the initial studies on resilience it was established that not only the strength characteristics of the systems with respect to disturbances are important but that capacity building and facilitating recovery from disturbances are of key importance for resilience. From a rudimentary risk management perspective, it is impractical to eliminate all risks; therefore, rapid rebuilding should be the option where extreme events affect infrastructure. Even though disaster and other extremities are not desirable, and possibly inevitable, they are yet an opportunity to build better and rapidly when the need for replacement arises.

3.7 Chapter summary

Even though the world is witnessing significant investments to expedite better infrastructure, there are still concerns about disruptions to existing infrastructure and the potential implications for future major infrastructure. This chapter examined the concept of resilience and its application in major infrastructure. It established that despite resilience being open to diverse interpretations, primarily with the infrastructure sector, resilience encompasses the immediate, mid- and long-term capacity of an asset to resist, respond to, recover from and retain its essential functionality after an unfavourable event. The chapter reviewed current global initiatives to drive resilience and established that stronger commitment and collaboration from various nations remain key to achieving a more resilient global infrastructure sector. The relevance of climate resilience was discussed and the key considerations during the design and planning phases were explored. Subsequently, various infrastructure sectors were explored with potential strategies for driving resilience were articulated. The chapter established the interdependency of most infrastructure sectors and indicated that this is vitally necessary to achieve resilience in practice.

3.7.1 Chapter discussion questions

1. Within the context of developing, developed and emerging economies, appraise the fundamentals for achieving more resilient transportation and power sectors.
2. You are the project director for an organisation that supports the adoption and incorporation of resilience in terms of floods, fire and measures to limit the deteriorating nature of cities. Using key information from Vignette 3.2 and other sources, discuss the key activities that drive the future planning of major infrastructure in your current location.
3. According to India's Ministry of Shipping, there is a constant drive to develop the country's ports and trade related infrastructure to accelerate growth in the manufacturing industry and to support the "Make in India" initiative (Ministry of Shipping, n.d.). Reflecting on the key information provided in Vignette 3.1 and wider sources of information, evaluate key planning and design considerations for Greenfield port developments in India.

4. The Brookings Institution once reported that Romania's infrastructure is in appalling state (The Brookings Institution, 2018). To the chagrin of the country's government, BBC News even reported that an entrepreneur built a one-metre-long stretch of motorway in north-eastern Romania in 2019 (BBC News, 2019). Critically evaluate options for developing resilient infrastructure in Romania and countries with similar infrastructure shortages.

3.8 Case study: Remembering the Haitian 2010 earthquake

When Haitians started their day on the 12th January 2010, they could not have imagined the devastation that was about to take place in their country by the end of that day. Recorded as Haiti's most devastating natural disaster in recent history, Amnesty International disclosed that the earthquake resulted in 200,000 deaths and left approximately 2.3 million people homeless because of collapsed buildings and other infrastructure (Amnesty International, 2014). The World Bank also reported that damages and losses stemming from the 7.0 magnitude earthquake were estimated at US $7.8 billion, representing 120 percent of Haiti's GDP and reconstruction needs were estimated at US $11.3 billion (World Bank, 2019). The unprecedented impact of the earthquake left the governance, the social and economic state of the nation, comatose. Nonetheless, following a request made by the government of Haiti (GoH) in March 2010, the Inter-American Development Bank (IDB), the United Nations (UN) and the World Bank, along other contributing donors, established a multi-donor fund called the Haiti Reconstruction Fund (HRF) (UN, 2010; Haiti Reconstruction Fund, n.d.). The role of the HRF is to support the GoH's post-earthquake action plan for the recovery and development of Haiti and related initiatives. Some advantages of the multi-donor approach are that it increases co-ordination by pooling resources from many donors in support of the government's recovery plan and draws on the comparative advantages of proven international and locally active partners. Based on GoH's request, the World Bank and the International Development Association (IDA) serve as trustees for the fund (Haiti Reconstruction Fund, n.d.). Ten years later, the *Miami Herald* stated that the Haitian Parliament has not voted on a new quake-resistant building code. It added that some of the expensive and ambitious projects promised, such as the general hospital and Caracol Industrial Park, are yet to realise their potential. Regrettably, the economy, which saw some growth after an estimated US $7.61 billion in humanitarian and reconstruction aid was supplied following the earthquake, seems to now be in ruins (Miami Herald, 2020).

3.8.1 Case discussion questions

1. From a major infrastructure planning perspective, identify the main lessons from the Haitian earthquake.
2. Using the results from question 1, reflect on which lessons are most crucial for developing, developed and emerging economies that have experienced similar disasters in more recent times. Propose steps to support the realisation of resilient infrastructure by 2030.
3. You are appointed to lead a consortium that will oversee the remaining rebuilding of Haitian major infrastructure and institutions. Discuss the most important steps and considerations you can take to strengthen the resilience of transportation and water sector infrastructure in Port-au-Prince and its environs.
4. Propose action plans for GoH and similar nations to plan for climate-resilient infrastructure projects following natural disasters.

References

ADB (2017). Meeting Asia's infrastructure needs. Available from: https://www.adb.org/publications/asia-infrastructure-needs [cited 13 May 2020].

Amnesty International (2014). Facts and figures document. Available from: https://www.amnesty.org/download/Documents/4000/amr360032014en.pdf [cited 20 July 2020].

ASDSO (2018). Independent forensic team report Oroville Dam spillway incident. Available from: https://damsafety.org/sites/default/files/files/Independent%20Forensic%20Team%20Report%20Final%2001-05-18.pdf [cited 11 April 2020].

Asoka, G.W., Thuo, A.D., and Thuo, M.M. (2013). Effects of population growth on urban infrastructure and services: A case of Eastleigh neighbourhood Nairobi, Kenya. *Journal of Anthropology and Archaeology*, 1(1), pp. 41–56.

BBC News (2019). Romania's mini-motorway built to shame a nation. Available from: https://www.bbc.co.uk/news/world-europe-47582694 [cited 22 July 2020].

Bennon, M. and Sharma, R. (2018). State of the practice: Sustainability standards for infrastructure investors. Available from: https://www.guggenheiminvestments.com/GuggenheimInvestments/media/PDF/WWF-Infrastructure-Full-Report-2018.pdf [cited 19 July 2020].

Bloemen, P., Reeder, T., Zevenbergen, C., Rijke, J., and Kingsborough, A. (2018). Lessons learned from applying adaptation pathways in flood risk management and challenges for the further development of this approach. *Mitigation Adaptation Strategies for Global Change*, 23, pp. 1083–1108.

Wang. X., Brown, R., Prudent-Richard, G., and O'Mara, K. (2016). Enhancing power sector resilience: Emerging practices to manage weather and geological risks. Available from: https://openknowledge.worldbank.org/bitstream/handle/10986/26382/113894-ESMAP-PUBLIC-FINALEnhancingPowerSectorResilienceMar.pdf?sequence=1&isAllowed=y [cited 11 June 2020].

Bruneau, M., Chang, S.E., Eguchi, R.T., Lee, G.C., O'Rourke, T.D., Reinhorn, A.M., Shinozuka, M., Tierney, K., Wallace, W.A., and Von Winterfeldt, D. (2003). A framework to quantitatively assess and enhance the seismic resilience of communities. *Earthquake Spectra*, 19(4), pp. 737–738.

Butler, D., Farmani, R., Fu, G., Ward, S., Diao, K., and M. Astaraie-Imani, M. (2014). A new approach to urban water management: safe and sure. *Procedia Engineering*, 89, pp. 347–354.

Crossrail (2018). Sustainability summary 2018. Available from: https://learninglegacy.crossrail.co.uk/wp-content/uploads/2018/07/Sustainability-Summary-2018.pdf [cited 14 July 2020].

Energy and Resources Institute (2011). Mainstreaming urban resilience planning in Indian cities a policy perspective. Available from: http://acccrn.net/sites/default/files/publication/attach/38.%20Mainstreaming%20Urban%20Resilience%20Planning%20copy.pdf [cited 23 July 2020].

European Bank for Reconstruction and Development (2014). Deepwater Container Terminal, Gdansk, Poland, Environmental and Social Action Plan. Available from: https://www.ebrd.com/english/pages/project/eia/45805esap.pdf [cited 14 April 2020].

Fricke, E. and Schulz, A.P. (2005). Design for changeability (DfC): Principles to enable changes in systems throughout their entire lifecycle. *Systems Engineering*, 8(4), pp. 342–359.

Fundacion Idea (2017). Urban resilience in Latin America: A brief guide for city policymakers. Available from: https://website-c230-consultores.nyc3.digitaloceanspaces.com/IDEA/files/UrbanResilience_PolicyBrief_170407_English_1536704685.pdf [cited 18 July 2020].

Galderisi, A. (2012). The resilient city. Enhancing urban resilience in the face of climate change. *Journal of Land Use, Mobility and Environment*, 5(2), pp. 69–87.

Gesner, G.A. and Jardim, J. (1998). Bridge within a bridge. *Civil Engineering*, 68(10), pp. 44–47.

Gilrein, E.J., Carvalhaes, T.M., Markolf, S.A., Chester, M.V., Allenby, B.R., and Garcia, M. (2019). Concepts and practices for transforming infrastructure from rigid to adaptable. *Sustainable and Resilient Infrastructure*, pp. 1–22.

Griffiths, S. and Sovacool, B.K. (2020). Rethinking the future low-carbon city: Carbon neutrality, green design, and sustainability tensions in the making of Masdar City. *Energy Research and Social Science*, (62), pp. 1–9.

Gubic, I. and Baloi, O. (2019). Implementing the new urban agenda in Rwanda: Nation-wide public space initiatives. *Urban Planning*, 4(2), pp. 223–236.

Haiti Reconstruction Fund (n.d.). Haiti reconstruction fund. Available from: https://www.haitireconstructionfund.org/background [cited 21 July 2020].

Hallegatte, S., Rentschler, J., and Rozenberg, J. (2019). Lifelines: The resilient infrastructure opportunity. Sustainable Infrastructure. Available from: https://openknowledge.worldbank.org/handle/10986/31805 [cited 12 December 2019].

Hamidi, S., Sabouri, S., and Ewing, R. (2020). Does density aggravate the COVID-19 pandemic? *Journal of the American Planning Association*.

Holling, C.S. (1973). Resilience and stability of ecological systems. *Annual Review of Ecology and Systematics*, 4(1), pp. 1–23.

ICLEI (2019). Resilient cities, thriving cities: The evolution of urban resilience. Available from: http://e-lib.iclei.org/publications/Resilient-Cities-Thriving-Cities_The-Evolution-of-Urban-Resilience.pdf [cited 19 July 2020].

ICRC (2019). *Cyclone Idai: Facts and figures.* Available from: https://www.icrc.org/en/document/cyclone-idai-facts-and-figures-0 [cited 20 December 2019].

Institution of Civil Engineers (2018). State of the nation 2018: Infrastructure investment. Available from: https://www.ice.org.uk/getattachment/news-and-insight/policy/state-of-the-nation-2018-infrastructure-investment/ICE-SoN-Investment-2018.pdf.aspx [cited 12 May 2020].

International Union of Railways (2017). Rail Adapt. Adapting the railway for the future. Available from: https://uic.org/IMG/pdf/railadapt_final_report.pdf [accessed 15 July 2020].

IPCC. (2018). IPCC special report: Global warming of 1.5°C. Geneva: Intergovernmental Panel on Climate Change. Available from: https://www.ipcc.ch/site/assets/uploads/sites/2/2019/06/SR15_Full_Report_High_Res.pdf [cited 25 March 2020].

IPCC. (2019). Climate change and land: An IPCC special report on climate change, desertification, land degradation, sustainable land management, food security, and greenhouse gas fluxes in terrestrial ecosystems. Geneva: Intergovernmental Panel on Climate Change. https://www.ipcc.ch/site/assets/uploads/2019/08/Fullreport-1.pdf. [cited 1 April 2020].

Jinadu, A.M. (2004). Urban expansion and physical development problem in Abuja: Implications for the national urban development policy. *Journal of the Nigerian Institute of Town Planners*, 17, pp. 15–29.

Kerkez, B., Gruden, C., Lewis, M., Montestruque, L., Quigley, M., Wong, B., and Pak, C. (2016). Smarter storm water systems. *Environmental Science and Technology*, 50(14), pp. 7267–7273.

KPMG (2013). Resilience: With a special feature on Latin America's infrastructure market, Insight: The global infrastructure magazine/Issue No. 5. Available from: https://assets.kpmg/content/dam/kpmg/pdf/2014/12/insight-resilience-issue-5-2013.pdf [cited 18 July 2020].

Los Angeles Times (2017). Oroville Dam spillway had two dozen problems that may have led to mass failure, report says. Available from: https://www.latimes.com/local/california/la-me-oroville-dam-report-20170510-story.html [cited 19 June 2020].

McKinsey (2020). Solving Africa's infrastructure paradox. Available from: https://www.mckinsey.com/industries/capital-projects-and-infrastructure/our-insights/solving-africas-infrastructure-paradox [cited 20 July 2020].

Meyer, P.B. and Schwarze, R. (2019). Financing climate-resilient infrastructure: Determining risk, reward, and return on investment. *Frontiers of Engineering. Management*, 6, pp. 117–127.

Miami Herald (2020). Ten years after Haiti's earthquake: a decade of aftershocks and unkept promises. Available from: https://www.miamiherald.com/news/nation-world/world/americas/haiti/article238836103.html [cited 21 July 2020].

Ministry of Shipping (n.d.). Port modernization and new port development. Available from: http://sagarmala.gov.in/project/port-modernization-new-port-development [cited [22 July 2020].

Miyamoto (2019). Increasing infrastructure resilience background report. Available from: http://documents1.worldbank.org/curated/en/620731560526509220/pdf/Technical-Annex.pdf [accessed 17 July 2020].

National Infrastructure Advisory Council (2009). Critical infrastructure resilience final report and recommendations. Available from: https://www.cisa.gov/sites/default/files/publications/niac-critical-infrastructure-resilience-final-report-09-08-09-508.pdf [cited 29 December 2019].

NIC (2020). Anticipate, react, recover. Resilient infrastructure systems. Available from: https://www.nic.org.uk/wp-content/uploads/Anticipate-React-Recover-28-May-2020.pdf [cited 4 June 2020].

Nordic Investment Bank (2016). NDPTL grant helps kick off port expansion in Gdansk. Available from: https://www.nib.int/who_we_are/news_and_media/articles/1883/ndptl_grant_helps_kick_off_port_expansion_in_gdansk [cited April 14 2020].

OECD (2018). Climate-resilient infrastructure. Available from: http://www.oecd.org/environment/cc/policy-perspectives-climate-resilient-infrastructure.pdf [cited 11 January 2020].

Onyenechere, E.C. (2010). Climate change and spatial planning concerns in Nigeria: Remedial measures for more effective response. *Journal of Human Ecology*, 32(3), pp. 137–148.

Prior, T. (2014). Measuring critical infrastructure resilience: Possible indicators, risk and resilience report 9. Available from: https://ethz.ch/content/dam/ethz/special-interest/gess/cis/center-for-securities-studies/pdfs/SKI-Focus-Report-10.pdf [cited 12 July 2020].

PwC (2013). Rebuilding for resilience fortifying infrastructure to withstand disaster. Available from: https://www.pwc.com/gx/en/psrc/publications/assets/pwc-rebuilding-for-resilience-fortifying-infrastructure-to-withstand-disaster.pdf [cited 15 July 2020].

PwC (2015). A new Africa energy world a more positive power utilities outlook. Available from: https://www.pwc.com/gx/en/utilities/publications/assets/pwc-africa-power-utilities-survey.pdf [cited 21 March 2020].

Rajasekar, U., Bhat, G.K., and Karanth, A. (2012). Tale of two cities: Developing city resilience strategies under climate change scenarios for Indore and Surat, India. Available from: http://acccrn.net/sites/default/files/publication/attach/TaleofTwoCities_TARU_0.pdf [cited 20 July 2020].

Ranger, N., Reeder, T., and Lowe, J. (2013). Addressing 'deep' uncertainty over long-term climate in major infrastructure projects: Four innovations of the Thames Estuary 2100 Project. *EURO Journal on Decision Process*, 1(3–4), pp. 233–262.

Rentschler, J., Obolensky, M., and Kornejew, M. (2019). Candle in the wind? Energy system resilience to natural shocks. Available from: http://documents1.worldbank.org/curated/en/951761560795137447/pdf/Candle-in-The-Wind-Energy-System-Resilience-to-Natural-Shocks.pdf [cited 13 May 2020].

Resilient Cities 100. Defining resilient cities. Available from: https://www.100resilientcities.org/ [cited 12 November 2019].

Rockefeller Foundation (2015). City resilience framework. Available from: https://www.rockefellerfoundation.org/wp-content/uploads/City-Resilience-Framework-2015.pdf [cited 12 January 2020].

Rockefeller Foundation (2019). Urban resilience infrastructure: An imperative in a climate uncertain world. Available from: https://www.rockefellerfoundation.org/blog/urban-resilience-infrastructure-imperative-climate-uncertain-world/ [cited 18 July 2020].

Rodríguez Tejerina, M. (2015). Sustainable cities in Latin America. Available from: https://www.iddri.org/sites/default/files/import/publications/wp1615_en.pdf [cited 18 July 2020].

Rozenberg, J., Alegre, X.E., Avner, P., Fox, C., Hallegatte, S., Koks, E., Rentschler, J., and Tariverdi. M. (2019). From A Rocky Road to Smooth Sailing: Building Transport Resilience to Natural Disasters. Available from: https://openknowledge.worldbank.org/bitstream/handle/10986/31913/From-A-Rocky-Road-to-Smooth-Sailing-Building-Transport-Resilience-to-Natural-Disasters.pdf?sequence=1&isAllowed=y [accessed 15 July 2020].

Stip, C., Mao, Z., Bonzanigo, L., Browder, G., and Tracy, J. (2019). Water infrastructure resilience – Examples of dams, wastewater treatment plants, and water supply and sanitation systems. Available from: http://documents1.worldbank.org/curated/en/960111560794042138/pdf/Water-Infrastructure-Resilience-Examples-of-Dams-Wastewater-Treatment-Plants-and-Water-Supply-and-Sanitation-Systems.pdf [cited 19 July 2020].

Sun, W., Bocchini, P., and Davison, B.D. (2020). Resilience metrics and measurement methods for transportation infrastructure: The state of the art. *Sustainable and Resilient Infrastructure*, 5(3), pp. 168–199.

Sy, A. and Copley, A. (2017). Closing the financing gap for African energy infrastructure: Trends, challenges, and opportunities. Available from: https://www.africa50.com/fileadmin/uploads/africa50/Documents/Knowledge_Center/Closing_the_African_Infra_Financing_Gap_-_Brookings_2017.pdf [cited 22 March 2020].

The Brookings Institution (2018). Romania: Thriving cities, rural poverty, and a trust deficit. Available from: https://www.brookings.edu/blog/future-development/2018/06/05/romania-thriving-cities-rural-poverty-and-a-trust-deficit/ [cited 22 July 2020].

The Republic of Rwanda (n.d.). 7 Years Government Programme: National Strategy for Transformation (NST1) 2017 – 2024. Available from: http://www.minecofin.gov.rw/fileadmin/user_upload/NST1_7YGP_Final.pdf [cited 25 April 2020].

The White House (2013). Economic benefits of increasing electric grid resilience to weather outages. Available from: https://www.energy.gov/sites/prod/files/2013/08/f2/Grid%20Resiliency%20Report_FINAL.pdf [cited 15 April 2020].

The White House (2013). Presidential policy directive – Critical infrastructure security and resilience. Available from: https://obamawhitehouse.archives.gov/the-press-office/2013/02/12/presidential-policy-directive-critical-infrastructure-security-and-resil [accessed 14 July 2020].

Master planning for resilient major infrastructure projects

TSO (2011). Climate resilient infrastructure: Preparing for a changing climate Available from: https://assets.publishing.service.gov.uk/government/uploads/system/uploads/attachment_data/file/69269/climate-resilient-infrastructure-full.pdf [cited 11 May 2020].

UN (2012). Resolution adopted by the general assembly on 27 July 2012. Available from: https://www.un.org/ga/search/view_doc.asp?symbol=A/RES/66/288&Lang=E [accessed 14 July 2020].

UN (2018). The world's cities in 2018. Available from: https://www.un.org/en/events/citiesday/assets/pdf/the_worlds_cities_in_2018_data_booklet.pdf [cited 19 July 2020].

UN International Strategy for Disaster Reduction (2007). Hyogo framework for action 2005-2015: Building the resilience of nations and communities to disasters. Available from: https://www.preventionweb.net/files/1037_hyogoframeworkforactionenglish.pdf [cited 20 April 2020].

UN Office for Disaster Risk Reduction (n.d.). What is the Sendai framework for disaster risk reduction? Available from: https://www.undrr.org/implementing-sendai-framework/what-sf [cited 20 April 2020].

UN. 2018. World Urbanization Prospects 2018. New York: United Nations department of economic and social affairs, population division. Available from: http://esa.un.org/unpd/wup/ [cited 23 July 2020].

UNFCCC. (2015). *The Paris Agreement*. Bonn: United Nations Framework Convention on Climate Change. Available from: https://unfccc.int/process-and-meetings/the-paris-agreement/the-paris-agreement [cited 4 April 2020].

UNISDR (2015a). Sendai framework for disaster risk reduction. Available from: https://www.preventionweb.net/files/43291_sendaiframeworkfordrren.pdf [cited 22 April 2020].

UNISDR (2015b). UNISDR ANNUAL REPORT 2015. Available from: https://www.unisdr.org/files/48588_unisdrannualreport2015evs.pdf [cited 20 January 2020].

United Nations (2010). International conference raises almost $10 Billion as more than 130 donors contribute Towards a New Future for Haiti. Available from: https://www.un.org/press/en/2010/ga10932.doc.htm [cited 20 July 2020].

United Nations International Strategy for Disaster Reduction (2009). 2009 UNISDR terminology on disaster risk reduction. Available from: https://www.unisdr.org/files/7817_UNISDRTerminologyEnglish.pdf [cited 27 December 2019].

US Department of Energy (2013). Comparing the impacts of Northeast Hurricanes on energy infrastructure. Available from: https://www.energy.gov/sites/prod/files/2013/04/f0/Northeast%20Storm%20Comparison_FINAL_041513b.pdf [cited 15 April 2020].

Usman, B.A. and Tunde, M.A. (2010). Climate change challenges in Nigeria: Planning for climate resilient cities. *Environmental Issues*, 3(1), 101–10.

World Bank (2013). Building resilience. Integrating climate and disaster risk into development. Available from: http://documents1.worldbank.org/curated/en/762871468148506173/pdf/826480WP0v10Bu0130Box37986200OUO090.pdf [cited 20 April 2020].

World Bank (2019). Rebuilding Haitian infrastructure and institutions. Available from: https://www.worldbank.org/en/results/2019/05/03/rebuilding-haitian-infrastructure-and-institutions [cited 20 July 2020].

4

ACHIEVING SUSTAINABLE MAJOR INFRASTRUCTURE PROJECTS

Development and management

Tarila Zuofa

4.1 Introduction

Infrastructure plays a crucial role in the well-being and affluence of societies. For instance, economic and soft infrastructure systems like hospitals, schools, transportation, electricity, telecommunication, drinking water supply, treatment and disposal of wastewater, and information and communications technology are the mainstay for economic development, competitiveness and inclusive growth globally. It is therefore important to understand all aspects between the relationship of major infrastructure to economic growth and sustainability. Greater efficiencies created by sustainable major infrastructure can result in waste reduction, reduced energy consumption, limited depleted resources and minimised air pollution. However, it is commonly accepted (Bhattacharya *et al.,* 2019a; Rozenberg and Fay, 2019, New Climate Economy Report 2018) that the majority of the existing infrastructure around the world are still not sustainable in the long term. Hence, there is an increasingly urgent need to implement changes either by modification of existing infrastructure or by the provision of additional infrastructure.

The guiding principles and approaches to sustainability and climate change agreed upon at the Rio Earth Summit in 1992, the various International Conferences on Financing for Development, United Nations Sustainable Development Summit, 2015, as well as the 2015 United Nations Climate Change Conference (COP21) in Paris, all provided speechmaking opportunities on the relevance of creating a more sustainable world. Nevertheless, beyond the rhetoric commonly associated with these landmark events, the major challenge is swinging into action and implementation to yield anticipated results. Currently, the global community faces several noteworthy challenges such as being able to rekindle global growth, the timely delivery of sustainable development goals (SDGs) and investing in the future of the planet through resilient climate action. In addition to the above are the recent threats and opportunities triggered by COVID-19 with its implication for our built, social and economic landscape. Hence, addressing all these significant challenges has gained even greater urgency. At the heart of the aforementioned is the ambition to invest in sustainable infrastructure since the major themes emerging from the summits previously mentioned indicate that sustainable infrastructure is strategic to delivering on both the development and climate agendas. Looking to the future, how can things be shaped to achieve and manage sustainable major infrastructure projects better?

All these present wider opportunities for deepening our understanding of sustainable development and infrastructure. Therefore, an understanding of how various aspects of infrastructure relates to sustainability becomes very necessary. Proper and sustainable infrastructure project development and management are vitally important and a well-thought-out plan is crucial to the delivery of major infrastructure projects globally. Forward thinking infrastructure stakeholders also acknowledge the relevance of shifting infrastructure investments toward more sustainable options that address the earlier mentioned challenges and are still consistent with attaining low carbon output and climate resilience. This chapter begins by explaining the concept of sustainable development and how the guiding principles from major summits are operationalised to the sustainability of major infrastructure. It also considers the prerequisites for sustainability and its relevance in major infrastructure development and management. The importance of project leadership in sustainable infrastructure development is discussed along with benchmarking and sustaining continuous improvement of major infrastructure. It concludes by discussing a framework for achieving sustainable major infrastructure projects.

4.1.1 Chapter aims and objectives

This chapter focuses on examining how the development and management of major sustainable infrastructure projects can be achieved. Its main objectives are to:

- Develop an understanding of the principles for sustainable development and the dimensions of infrastructure sustainability;
- Appraise current global initiatives that enhance sustainable major infrastructure;
- Evaluate the prerequisites for developing and managing sustainable major infrastructure;
- Depict how leadership facilitates the development of sustainable infrastructure;
- Establish how benchmarking can be integrated for sustainable major infrastructure development and management; and
- Present a framework for achieving sustainable major infrastructure projects development and management post COVID-19.

4.1.2 Learning outcomes

At the end of this chapter, readers will be able to:

- Critically discuss the principles for sustainable development and sustainability;
- Articulate how global initiatives facilitate the development of sustainable major infrastructure;

- Describe the key requirements for developing and managing sustainable major infrastructure projects;
- Gain insights into leadership's role in facilitating sustainable infrastructure projects;
- Apply a benchmarking strategy for major sustainable infrastructure projects; and
- Propose a framework for achieving sustainable major infrastructure projects in the future.

4.2 Towards sustainable development and sustainable infrastructure

Even though there were earlier discussions on sustainable development and sustainability in the 1970s, the trailblazing publication of the Brundtland Commission's report defined sustainable development as "development that meets the needs of the present without compromising the ability of future generations to meet their own needs." About 35 years later, a plethora of definitions now exist for sustainable development. From their evaluation, Zuofa and Ochieng (2016) observed that sustainable development is likened to a very broad concept that might still be widely questioned and open to numerous interpretations. Nevertheless, in their respective contexts, several definitions of sustainable development (Mensah, 2019) focused more on the natural environment, social and economic development. Majority of these definitions equally acknowledge that the environment's ability to meet present and future needs have need of sustenance. The definitions also underscore the relevance of providing the basic needs of the world's underprivileged people to allow them to experience a reasonably comfortable way of living.

Several contemporary global policies and agendas have emerged in the past three decades. In recent times, these include the landmark international agreement in 2015 adopted at the United Nations Sustainable Development Summit and the 2016 High-level Political Forum on Sustainable Development, both geared to adopting sustainable development measures based on the 2030 Agenda. The 2030 Agenda integrates in a balanced manner the economic, social and environmental dimensions of sustainable development with a commitment to eradicate poverty and achieve sustainable development by 2030 worldwide, ensuring that no one is left behind. Bhattacharya and Jeong (2018) explained that these policies heralded newer understandings that sustainable development, poverty reduction and climate change are complementary and entwined. These policies and agendas also draw attention to the benefits of transiting to low-carbon alternatives. When assessed critically, through the provision of a range of essential services, major infrastructure investments made over the next few years will be decisive and have great potentials to drive sustainable development and the attainment of the 2030 agenda. Sustainable infrastructure is equally an important component required for achieving the agenda and unambiguously articulated in at least three SDGs:

- **SDGs 6:** ensure availability and sustainable management of water and sanitation for all
- **SDG 7:** ensure access to affordable, reliable, sustainable and modern energy for all
- **SDG 9:** build resilient infrastructure, promote inclusive and sustainable industrialisation and foster innovation

What this suggests is that as global needs grow at swift rates, it is also imperative to build high-grade infrastructure that ensure universal participation in the successful implementation of the SDGs. Comparable to sustainable development, several approaches have been taken to develop a common meaning of sustainable infrastructure. Even though these approaches have provided additional clarity, Zuofa and Ochieng (2016) also posited that the concept of sustainable infrastructure is yet to be understood fully and subject to different interpretations. A common meaning of sustainable infrastructure will enable approaches that are more concerted

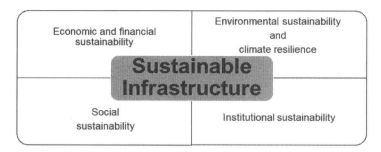

Figure 4.1 Four dimensions of infrastructure sustainability

by providing clearer objectives for major infrastructure projects and facilitate the identification of key actions at each stage of the project cycle to bring together various stakeholder groups in more coordinated ways.

Bhattacharya *et al.* (2016) defined sustainable infrastructure as infrastructure that is socially, economically and environmentally sustainable. As observed from other definitions, it might be worth concluding that the applications of the sustainable infrastructure concept depends on several factors, including the sector and the relevant geographical context. The Inter-American Development Bank (IDB) (2018a) identified that the concept of sustainable infrastructure needs to help drive transformational change rather than becoming a trivial buzzword to repackage long-standing ways of simply developing, operating, and investing in infrastructure. This chapter considers sustainable infrastructure to be infrastructure projects that are planned, designed, constructed, operated and decommissioned in a manner to ensure economic and financial, social, environmental (including climate resilience), and institutional sustainability over the entire life cycle of the project (IDB, 2018a). As shown in Figure 4.1, from this definition, sustainable infrastructure has four dimensions of sustainability: economic and financial sustainability, environmental sustainability and climate resilience, social sustainability and institutional sustainability.

The Inter-American Development Bank also formulated guiding principles that provide additional clarity for each of these dimensions of sustainability (see Table 4.1). The purpose of providing major infrastructure includes the expansion of electricity distribution networks, connecting people and commerce through better transport infrastructure, access to clean water to improve people's health. Therefore, the dimensions of sustainability covered in Table 4.1 are pertinent as they provide a coherent guide on coverage that can be applied for any major infrastructure project. For succinctness, the summaries used for the various sustainability dimensions are condensed in the table.

In defining major sustainable infrastructure, it is also imperative to consider high-quality infrastructure that are sustainable and resilient with potentials to intensify economic activities, generate new employment, empower gender participation and support an inclusive participation in the successful implementation of Agenda 2030. The UK's Crossrail is a good illustration of an infrastructure project with characteristics poised to deliver an excellent service fit for future generations while still making positive national and global economic, environmental and social contributions. To achieve this, Crossrail developed a holistic strategy that covers seven major sustainability themes: economic progress, sustainable consumption and production, addressing climate change and energy, the physical environment, improving health, protecting people's safety, security and health and promoting greater equality of opportunity for social inclusion (Crossrail, 2018).

Achieving sustainable major infrastructure projects

Table 4.1 Guiding principles for dimensions of sustainability

Sustainability dimension	Summary of guiding principle
Economic and financial	Infrastructure is economically sustainable if it generates a positive net economic return, considering all benefits and costs over the project life cycle, including positive and negative externalities and spill overs. In addition, the infrastructure must generate an adequate risk adjusted rate of return for project investors.
Environmental sustainability including climate resilience	Sustainable infrastructure preserves, restores, and integrates the natural environment, including biodiversity and ecosystems. It supports the sustainable and efficient use of natural resources, including energy, water, and materials. It also limits all types of pollution over the life cycle of the project and contributes to a low carbon, resilient, and resource efficient economy. Sustainable infrastructure projects are (or should be) sited and designed to ensure resilience to climate and natural disaster risks.
Social sustainability	Sustainable infrastructure is inclusive and should have the broad support of affected communities; it serves all stakeholders, including the poor; it contributes to enhanced livelihoods and social well-being over the life cycle of the project.
Institutional sustainability	Institutionally, sustainable infrastructure is aligned with national and international commitments, including the Paris Agreement, and is based on transparent and consistent governance systems over the project cycle.

Source: Inter-American Development Bank (2018a)

4.3 Global initiatives and the call for sustainable infrastructure

4.3.1 An overview of global sustainability initiatives

An OECD Report, *Infrastructure to 2030 Mapping Policy for Electricity, Water and Transport*, highlighted that parts of infrastructure systems in OECD countries are ageing rapidly. The report also noted that increasingly stringent and complex accessibility to public finances has become a major challenge (OECD, 2007). However, it still identified several strategies to augment the current capacity to meet future infrastructure needs. Some of the recommendations articulate measures that governments collectively and individually can adopt to develop more initiatives that are sustainable.

In addition to the above, there are a few comparable narratives from several other global initiatives. For example, since creation by the G20 in 2014, the Global Infrastructure Hub has collaborated with governments, the private sector, multilateral development banks (MDBs) and international organisations to promote the development of resilient and sustainable infrastructure. Recent G20 summits equally highlighted the urgent need for the development of global sustainable infrastructure. The Global Infrastructure Connectivity Alliance (GICA) was formed at the 2016 G20 Summit in Chengdu, China. Its main objective was to enhance cooperation and synergies of existing and future global infrastructure and trade facilitation programs to improve connectivity among countries (GICA Secretariat, 2017). To achieve its mandate, GICA identified resources and gaps, shared good practices, mapped connectivity initiatives and monitored connectivity. During the 2018 G20 summit in Argentina, the role of infrastructure as a key driver of economic prosperity, sustainable development and inclusive growth was still emphasised. Accordingly, the summit validated the "Roadmap to Infrastructure as an Asset Class" and the "G20 Principles for the Infrastructure Project Preparation Phase." More recently, in 2019, the Japanese G20 President stressed the significance of the Quality Infrastructure Investment Agenda and its potential to reap economic, social and environmental co-benefits that extend beyond the physical value of infrastructure assets. Looking ahead, the 2020 summit, scheduled to be held in Riyadh, Saudi Arabia, aspires to build on the strong legacy of the G20 under the theme "Realising Opportunities of the 21st Century for All." Even though the collective roadmap for the G20 Saudi Presidency aims at empowering people, safeguarding the planet and shaping new frontiers (G20, 2019). Since sustainable infrastructure usually embraces holistic processes to restore and maintain harmony between the natural and built environments, in manners that allow people live in balanced economic environments, an overview of Saudi Arabia's G20 Presidency's focus still underscores the relevance of sustainable infrastructure towards attaining its collective roadmap.

Finally, Mercer and Inter-American Development Bank (IDB, 2017) identified 30 initiatives comprising those with sustainable infrastructure as a core focus in their mission statements and those focused on infrastructure funding and investment. The 30 initiatives were further categorised:

- **Five influencers:** Provide thought leadership and research relating to sustainable infrastructure or work to influence public or industry policy and/or the financial system to align infrastructure investment plans with intended nationally determined contributions (INDCs) and other environmental/social outcomes. The list of the five influencers include New Climate Economy (NCE) and Energy Transitions Commission (ETC).
- **Thirteen mobilizers:** Including Africa50, Matchmaker and Sustainable Development Investment Partnership (SDIP), they all work with governments to develop "bankable" projects and/or convene investors to channel more funds into sustainable infrastructure projects.
- **Twelve tool providers:** Including Bloomberg New Energy Finance (BNEF) and Global Infrastructure Basel (GIB), they seek to enable integrated environmental or social analysis of infrastructure projects and/or the investment process.

While noting that many initiatives make every effort to promote sustainability, the report disclosed the relevance of improved collaboration and advocated the need to clarify, convene, commit and collaborate as steps to further align, support and leverage the identified initiatives (IDB, 2017).

4.3.2 *The global call for sustainable infrastructure*

Over time, the weight of eco-friendly evidence has steadily grown to support the case that exceptional levels of economic advancement and wealth are resulting in climate change variations and human inequalities. Global warming and climatic disturbances are identified as being triggered by pollution created by greenhouse gases (GHGs). Hence, the objective of the United Nations Framework Convention on Climate Change has been to "stabilise greenhouse gas concentrations in the atmosphere at a level that would prevent dangerous anthropogenic (human-induced) interference with the climate system" (UN, 1992). Contemporary advances in technology now also offer faster assessments for pollution, coastal erosion and deforestation while the impact of using non-renewable natural resources and the consequences of not managing the use of renewable resources is more accurately assessed. For example, Ali *et al.* (2018) disclosed that the construction industry in the United Kingdom has been the highest contributor to construction waste (62 percent) as compared to other sectors. However, from the *Resources and Waste Strategy: At a Glance* (DEFRA, 2018) and other initiatives it is evident that the UK construction industry aims to contribute to waste reduction or elimination by adopting new policies and practices.

As established in this prime, infrastructure is the mainstay for global economic development, competitiveness and inclusive growth. A number of studies have underscored the positive relationship between infrastructure investment and economy gains globally. In Africa, for instance, Jedwab et. al. (2017) demonstrated how railway infrastructure investment in Kenya produced lasting economic gains through trade cost reduction and market integration. Similarly, the Olkaria power plant, one of the world's largest geothermal infrastructure facilities, increased geothermal contribution to the Kenyan national energy mix by 51 percent (World Bank, 2015a). From the operations of this power plant, the supply of reliable and clean energy in Kenya has improved while the cost of electricity to consumers dropped (World Bank, 2015a). Yoshino and Pontines (2015) studied the Southern Tagalog arterial road (STAR) in Philippines. Their report suggested that infrastructure investments like highways indirectly boost tax and non-tax revenues through their abilities to reduce cost of transportation and enhance the activity of businesses along its passage. Donaldson and Hornbeck (2016) examined the historical impact of railroads on the American agricultural sector and noted its positive impact on market integration and economic development; roads improve access to jobs, reduce trade costs and barriers, and catalyse growth of agricultural and industrial clusters. The Worldwide Fund for Nature (WWF) noted that road projects proposed as part of the One Belt One Road Initiative (BRI) East–West and North–South corridors can provide improved transport infrastructure and make substantial contributions to Myanmar's social, economic and infrastructure development (WWF, 2017). Notwithstanding the illustrations cited to demonstrate the positive relationship between various infrastructure and economic enhancement, current infrastructure investment trends (see Figure 4.2) and the New Climate Economy Report (2018) still suggest the inadequacy of global investment in infrastructure,

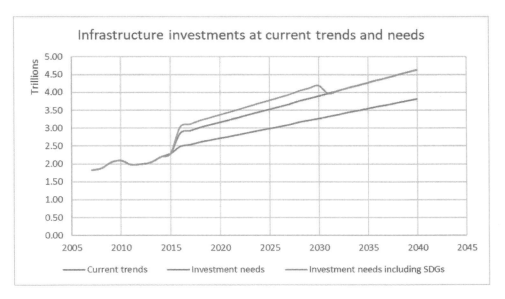

Figure 4.2 Global infrastructure trends and needs, including investments to meet SDGs

Source: Original adapted from Global Infrastructure Hub (2020)

Likewise, analysis from McKinsey (2016) demonstrated that inadequate infrastructure is historic in both emerging and advanced economies; hence, augmenting existing and investing in new infrastructure remains valid globally. With infrastructure assets like bridges and dams requiring long lifespans, and populations in developing countries and all countries' urban settlements rising, the challenges are to ensure robust technology and a reliable energy supply for customers, all while dealing with the challenges of climate change. Infrastructure provision will remain an essential priority for governments and investors alike. Despite the fact that the world is in dire need of more infrastructure, caution needs to be applied to ensure that inefficient and unsustainable options for delivering infrastructure are discarded (see Vignette 4.1).

Vignette 4.1   A tale of two major infrastructure projects

Myingyan is a city in Myanmar and the capital of the Myingyan township. It is also the location of the Sembcorp Myingyan Independent Power Plant that officially opened in March 2019. Infrastructure Asia (n.d.) explained that Sembcorp Myingyan IPP was funded via multilateral institutions including the Asian Infrastructure Investment Bank (AIIB) and international commercial lenders like DBS Bank and OCBC. With a contracted capacity of 225 megawatts, it will help to alleviate Myanmar's severe power deficit (Sembcorp, 2019). The project is Myanmar's first power plant to integrate both gas-fired and solar power generation. Using solar panels to generate renewable electricity for onsite use reduces the reliance on gas turbines for its operations and decreases GHG emissions. The plant has also created jobs for locals with Myanmar nationals comprising 95 percent of the workforce (Sembcorp, 2019). Job creation validates the positive value infrastructure projects bring to communities since this develops human capacity and provides a means of livelihood for several others.

Located on Congo's River in the Democratic Republic of Congo-DRC, the US $80 billion Grand Inga Dam might become the world's largest hydropower scheme if successfully completed. With an estimated capacity of 40 gigawatts, this major infrastructure could significantly boost energy supply on the African continent. As Warner *et al.* (2019) disclosed, the eight separate dams (Inga 1-8) are envisioned to light up and power Africa. The project will possibly aid various business sectors and directly or indirect stimulate economic growth. Despite the encouraging potentials from the project, Showers (2012) raised serious concerns over the adverse environmental impact in relation to the project since the Congo River feeds on the Atlantic, noting that any flow disruption could have dismal implications for the region and world. There are also uncertainties on whether local communities will eventually benefit from generated electricity and apprehensions on how agricultural activities can cope with possible upstream salt-water penetration.

As Bhattacharya and Jeong (2018) indicated, sustainable infrastructure is accepted as a critical foundation to deliver on sustainable growth, promote climate resilience and limit global warming. The ongoing COVID-19 pandemic further intensifies the mounting need for robust and flexible infrastructure that can effectively operate during seasons of crisis. Given this big opportunity, it is imperative that when nations initiate infrastructure projects, they strive to introduce projects that are sustainable, technologically advanced and resilient. Therefore, the design and construction of future infrastructure is central to reversing some of the previously mentioned trends. Key to achieving this is ensuring that the economic, social and environmental impact of infrastructure projects remains subject to critical scrutiny. This is addition to the integration of

Achieving sustainable major infrastructure projects

ethical and value-driven processes during critical decisions from the onset to enable potential issues associated with inclusivity to be properly evaluated and mitigated where necessary (a ready contrast can be drawn from the projects in Vignette 4.1). The implication is that holistic sustainability must be embedded as a primary element of the entire design process and needs to be reflected throughout the life cycle of future major infrastructure projects globally.

4.4 Prerequisites for implementing major sustainable infrastructure

It is not sufficient to just build new bridges, port facilities, dams, roads and other infrastructure that potentially drive economic growth and improve well-being. As mentioned in the previous section, caution needs to be applied to ensure that inefficient and unsustainable options for delivering infrastructure are discarded. From the definition and attributes previously discussed, the implementation of sustainable infrastructure depends on certain prerequisites (see Figure 4.3), actions and the commitment of various interested stakeholders.

Innovation and technology integration: As observed by the World Economic Forum, integrating new technologies during the design, construction and operational phase of infrastructure assets yields significant cost reduction and improved functionality (WEF, 2020). Aside from these benefits, artificial intelligence (AI), advanced data analytics, fintech, cloud computing, 5G, new materials, renewable energy technology and 3D printing are among the several innovations changing the global infrastructure landscape because they optimise project delivery, reduce community disruption, minimize environmental harm and increase safety (WEF, 2020). With other new solutions in the infrastructure sector, like decarbonisation, material processing at reduced temperature, the use of recycled materials and low clinker cement, there are ample illustrations to indicate that technology development and innovation are key to achieving sustainable infrastructure. More specifically, several studies including Oke *et al.* (2017) underscored the benefits of innovation and technology development on sustainable infrastructure. Their study examined the general benefits of nanotechnology on sustainable construction by focusing on applying nanotechnology to traditional construction materials. Similarly, when completed in 2040, Singapore's Tuas Mega Port will provide a positive illustration of how innovation drives sustainability. The

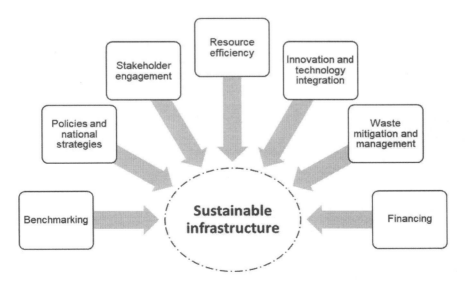

Figure 4.3 Prerequisites for implementing major sustainable infrastructure

Strait Times (2019) disclosed that beyond becoming the world's largest fully automated terminal equipped with automated wharf and yard functions as well as full-electric automated guided vehicles, the port will be operated by intelligent data driven operations management systems and smart engineering and power management platforms. In the words of Prime Minister Loong, "Tuas Port is also an opportunity to peer over the horizon and rethink the future of shipping. Because the port will be on a completely Greenfield site, we can design from a clean slate and make innovation and sustainability key features" (*Strait Times*, 2019). Collectively, the aforementioned illustrations and other studies (Henson, 2011, Shurrab *et al.*, 2019) exemplify how through innovation and advancements in technology, more sustainable infrastructure projects and systems are derived.

Waste mitigation and management: Waste produced during construction, refurbishment and demolition of various infrastructure assets is now a global issue that negatively affects the environment and raises consternations about activities within the construction industry. According to Lu *et al.* (2018), in most developed countries construction contributes 20–30 percent of solid waste ending up in landfills. In the United Kingdom, Department for Environment, Food and Rural Affairs DEFRA (2013) estimated that 44 percent of total waste to UK's landfill is generated by construction related activities. The case of the UK is not isolated as similar patterns are identifiable in other nations—for instance, 29 percent in the United States (Yu *et al.,* 2013); 25 percent in Hong Kong (Lu and Tam, 2013), 30–40 percent in China (Huang *et al.,* 2018) and 25 percent in Canada (Yeheyis *et al.,* 2013). While these figures raise concerns for sustainability and the construction industry as a whole, due to the significant benefits of waste minimisation, like the preservation of natural resources and reduced carbon footprint, the subject of waste mitigation and management has attracted significant attention. As a result, there has been a substantial body of knowledge on construction waste (including Ajayi and Oyedele, 2018; Osmani *et al.,* 2008; Shi *et al.,* 2016; Yates, 2013). Collectively, these studies identify the various sources of waste and advocate that effective mitigation and management of waste can be achieved through various strategies such as modification of design, material procurement and other construction activities during the whole project life cycle. In general, therefore, as a prerequisite for sustainable infrastructure, the construction sector now makes a major contribution by minimising the amount of waste generated during construction and maximising the amount of waste arising that can be recycled, all while concurrently using recycled materials from other sectors.

Resource efficiency: The construction industry has been closely associated with improper utilisation of resources and contributing to the greenhouse effect. For instance, construction activities commonly use concrete and cement because of their strength and durability. However, they exert a substantial negative impact on the environment in terms of carbon dioxide emissions and are responsible for the exhaustion of energy resources (Kartik *et al.* (2003). With greater awareness on the increased value of resources and the importance of using them efficiently, Chi *et al.* (2020) indicate that the crucial issue is the question of how to maximize the positive role of construction while minimising its negative impacts. Consequently, resource efficiency and the adoption of resource efficient methods are high up on the industry's agenda. For this reason, the report *Rethinking Timber Buildings* highlights the time and efficiency savings that can result from using mass timber as a sustainable and safe alternative to more commonly used materials (Arup, 2019). There are several other illustrations demonstrating efforts to achieve resource efficiency, but in summary, WRAP (2015) articulated diverting priority materials from landfills and minimising resource use in products and buildings as preferred alternatives.

Stakeholder engagement: Several stakes have to be considered in development and management of any major infrastructure project. Stakeholders represent the individuals or organisations with interests in terms of either legitimacy, power or urgency considered for the success of any project and in its functional environment. Due to the involvement of a multiple stakeholders, the delivery of infrastructure projects is also highly complex. Even the success or failure of infrastructure projects is largely sometimes dependent on the perception of stakeholders rather than attaining the technical characteristics. For instance, the public or other stakeholders can view dams, airports or other ongoing or completed infrastructure projects perceived as successful from the contractor's perspective as failures. Therefore, the incorporation of stakeholders' interests through engagement has the potential to help the achievement of a balance between the technical requirements and the meeting of the societal requirements for sustainable infrastructure.

It is worth mentioning that for some categories of project, such as railways, certain stakeholders like the public may not have formal powers to affect the project decision-making process but would still have informal powers that, when exercised, can press more powerful stakeholders like the government into changing their perceptions of the project. To this end, Rangarajan *et al.* (2013) integrated several tools and processes to describe a methodology that identifies and classifies stakeholders and how to analyse their interests, needs, issues, and instabilities for railway and other infrastructure projects in the United States. Their study highlighted the importance of identifying uncertainties, needs, issues, and risks associated with the transportation planning projects. The study underscored the significant role various stakeholders play in developing sustainable infrastructure. It generally suggested that engaging with stakeholders during the early phases increases mutual confidence between communities, business and local government and improves the infrastructure services delivery design towards sustainable solutions. Clear and straightforward communication about the impact of projects, interdependencies and payoffs, whether positive or negative, and its outcomes all create community trust and buy-in, which facilitate the necessary strength for projects. Therefore, the key message for major infrastructure projects is simple; embracing stakeholder engagement strategies like constant communication, transparent disclosure of information and strengthening community development where applicable are practical ways for establishing connections between stakeholders and tackling issues that need to be considered to attain holistic sustainability. Through this, a full knowledge of potential stakeholders at risk of exclusion from receiving the benefits of infrastructure is generated and where required mitigation measures can be more straightforwardly introduced to enhance social inclusivity.

Policies and national strategies: With the current global infrastructure deficit, the implementation of sustainable infrastructure requires strong and concerted efforts at national levels before replication to other levels. Across the world, several nations rely on arrangements and guidance by national governments in pursuit of various objectives. Recognising that the aspiration for sustainable infrastructure requires broader partnership among several stakeholders, different nations have adopted various policies and strategies. As explained by Qureshi (2016), countries need to articulate clear and comprehensive strategies for sustainable infrastructure and embed them in overall strategies for sustainable and inclusive growth and development. These national strategies need to be driven and supported by plans in key infrastructure sectors and subnational jurisdictions that are important providers of infrastructure (Qureshi, 2016).

In Canada, for instance, through its long-term infrastructure plan, the government will be investing over C$180 billion in infrastructure until 2028. According to Infrastructure Canada (2018), the Canadian 12-year plan that commenced in 2016 focuses on making significant long-term investments in public transit (C$28.7 billion), green infrastructure (C$26.9 billion), social infrastructure (C$25.3 billion), infrastructure for rural and northern communities (C$2 billion) and trade and transportation infrastructure (C$10.1 billion). As observed in the investment figures highlighted above, social and green infrastructure are the mainstay of the plan for the five priority investment streams; nonetheless, there is still an elaborate emphasis on the role of sustainability. Rwanda presents another illustration of the relevance of articulating clear and comprehensive national strategies for sustainable infrastructure in a developing nation. Recognising the central role infrastructure in economic transformation, Gubic and Baloi (2019) identified that Rwanda has made continuous investments to unlock constraints for economic transformation consequently; Rwanda's *Vision 2050* aspires to take Rwanda to high living standards by the middle of the 21st century and high quality livelihoods (The Republic of Rwanda, n.d.; World Bank, 2020). To achieve this vision, the broad priority is developing modern infrastructure and livelihoods and this impetus is significantly addressed through infrastructure sustainability. In Australia, strategic planning for the assessment of infrastructure needs at a national level falls under the purview of Infrastructure Australia (IA), an independent body formed by the federal government under the *Infrastructure Australia Act 2008*. As evidenced in *Prioritising Reform: Progress on the 2016 Australian Infrastructure Plan* (Infrastructure Australia, 2018) and various Australian Infrastructure Audits (Infrastructure Australia 2016, Infrastructure Australia 2019), the multi-stage process overseen by Infrastructure Australia is a good example of how strategic infrastructure planning should be informed by a strong evidence base that promotes physical assets and policy reforms. For instance, sustainability in terms of social inclusion and the environment is more prominent as the latest Australian Infrastructure Audit (in 2019) covers social infrastructure, stressing its significance to the country's growth and development.

Even though the relevance of national policies and strategies as prerequisites for sustainability is non-contentious, it must be acknowledged that different nations have different infrastructure requirements which might reflect in their policies and national strategies. For instance, an advanced nation like the United Kingdom with a massive infrastructure stock has different infrastructure priorities when compared to a developing country like Moldova, which is merely attempting to provide basic infrastructure. Nevertheless, the World Bank indicates that with the right policies in place, even low-income nations can meet their infrastructure objectives (World Bank, 2019). The caution, in the above, being "with the right policies," which explicates why governments have to play an active role even in countries where ownership of much of the available infrastructure might be in private hands. At the most fundamental level, this is because the long-lasting nature of infrastructure projects makes them more susceptible to any modifications in national policies and strategies.

Financing: As established in various chapters, there is a growing consensus among the various global economy and fiscal development policy stakeholders that new infrastructure projects must be developed with stronger commitments to resilience, sustainability and accountability. Despite governments holding definitive responsibilities for ensuring that infrastructure projects align with their national policies and strategies for sustainability, not every government readily has the financial resources to accomplish this. For example, Shand (2019) noted that meeting the demand for sustainable infrastructure encounters a myriad of difficulties directly associated with financing for Ghana and many other countries across the global south. With the global south and north collectively requiring significant infrastructure investment due to either underinvesting, ageing infrastructure or rapid urbanisation, the responsibility for bankrolling sustainable infrastructure goes beyond the direction of government. Indeed, this creates a greater role for potential investors, especially private sector participation in infrastructure development and financing. As disclosed in the UN 2020 *Financing for Sustainable Development Report*, since the adoption of the Addis Ababa Action Agenda (AAAA), the financing landscape changed dramatically (UN, 2020). Among the changes was the creation of a more formalised role for the private sector in infrastructure financing. It is worth stating that for both governments and private investors to accelerate their investment in sustainable infrastructure, an appropriate mix of financing platforms and vehicles are needed. Currently, there are multilateral synergies in progress; multilateral development banks act as mobilisers of private finance for sustainable infrastructure since they possess organising power and abilities to support client governments during the infrastructure project preparation. The Asian Infrastructure Investment Bank (AIIB), for instance, began in 2016 as a China-led multilateral institution with an ambition of investing in sustainable infrastructure and being "lean, clean and green." Similarly, China's Belt and Road Initiative (BRI) was introduced in 2013 by President Xi Jinping as an ambitious development campaign. According to the State Council of the People's Republic of China (2015), the BRI will:

> "Help align and coordinate the development strategies of the countries along the belt and road, tap market potential in this region, promote investment and consumption, create demands and job opportunities, enhance people-to-people and cultural exchanges, and mutual learning among the peoples of the relevant countries, and enable them to understand, trust and respect each other and live in harmony, peace and prosperity."

As projected by the World Economic Forum, BRI will lend as much as $8 trillion for infrastructure in 68 countries (WEF, 2017). With potential environmental and social impacts from large-scale infrastructure projects as well as the economic viability of infrastructure investments and the need to maintain cordial international relations, Ahmad *et al.* (2018) are hopeful that the BRI should provide tremendous opportunities to drive strong, sustainable infrastructure and inclusive growth in partner countries if continuously managed properly. The ambition of filling the sustainable infrastructure investment gap will no doubt require appropriate financial instruments and vehicles adapted to the sustainable infrastructure paradigm. As nations embrace various ambitious goals to transform their economies in sustainable and inclusive manners, there is a need to look away from simply spending more to spending better on the right objectives (Rozenberg and Fay, 2019). Laying a robust foundation of sustainable infrastructure through financing remains pivotal to achieving these ambitious transformational goals among nations globally.

Benchmarking: Decisions are needed whether to provide strategic metrics for analysing sustainable construction practices (Presley and Meade, 2010), improving the quality of rail freight rolling stock (Cullinane *et al.*, 2016), or developing indices to assess potential improvements for the sustainable development of buildings in the long and short term (Hassan, 2016). As fully explained later in this chapter (see Section 4.6.1), the quest to deliver sustainable infrastructure has led to the development of a plethora of benchmarking guidelines and protocols. Some like G7 Ise-Shima Principles for Promoting Quality Infrastructure Investment provide high-level definitions and principles for the quality of infrastructure investment and serve as a basis for other tools. Additional tools and rating systems, such as SuRe and Envision, offer comprehensive approaches for sustainable infrastructure through well-defined structures and a clear set of indicators that are quantifiable and monitored. Even though some of these tools were originally intended to assess specific project phases, their use has been extended to cater to different project phases. Given the urgent need to scale up sustainable infrastructure, benchmarking is required to provide a deep understanding of the processes, parameters and skills that deliver superior infrastructure. Hence, either commonalities by way of shared definitions or understanding as well as common frameworks are constantly needed to ensure that infrastructure delivery efforts remain well aligned to promote sustainability.

Due to insufficient space to present the full details of all aspects highlighted in Figure 4.3, only brief exposés were provided in this section. Nevertheless, in addition to the discussion above, there are other prerequisites for sustainable infrastructure identified in literature: green supply chains (Balasubramanian and Shukla, 2017), effective project management approach (Agyekum-Mensah *et al.,* 2012) and intra-organisational leadership (Opoku *et al.,* 2015). KPMG added environmental, social and governance principles (ESG), sustainable finance, true value and climate change and decarbonisation as four key enablers for achieving sustainable infrastructure (KPMG, n.d.). Even though the details of a comprehensive list of prerequisites for implementing sustainable infrastructure might be inconclusive and too numerous to mention here, all the areas covered still provide information on crucial considerations for infrastructure decision-makers.

4.5 The importance of project leadership in sustainable infrastructure development

In describing leadership, it is important to acknowledge that several definitions of the term are discussed in literature (Daft, 2015). For example, Doh (2002) defined leadership as an executive position in an organisation and that it is a process that has influence on others. Yukl (2006) outlined leadership as the process of influencing others to understand and agree about what needs to be done and how to do it, and the process of facilitating individual and collective efforts to accomplish shared objectives. From these definitions, it can be noted that leadership is considered as a process of influencing others. The definitions also suggest that leadership occurs within groups and involves the realisation of goals. Tabassi and Abu Bakar (2010) add that leadership is not just a process but also a process that involves influences, that occurs within a group context, and that involves personal discovery and development as well as involvement in goal attainment. The definitions above imply that any person in an organisation could possibly be a leader if they are involved in a process of influence.

Extant literature (Bass and Avolio, 1994; Turner and Baker, 2018; Yang *et al.,* 2011) provides a platform for an understanding of various leadership theories and styles of leaders. Even though it is not intended to provide a comprehensive review of these theories and leaderships styles, Table 4.2 summarises common leadership theories and provides some structure for the remaining discussion on leadership.

It is worth mentioning that Table 4.2 and other theories of leadership are accurate in one way or other. Some theories focus on the nature of the leader, their personality and traits, while others

Table 4.2 Summary of common leadership theories and their main features

Leadership theory	*Summary of main features*
The traits theory	Concentrates on the characteristics or approaches of individual leaders (Tannenbaum and Schmidt, 1973; Yukl, 1998).
Leader–member exchange (LMX)	Considers that leaders form differentiated patterns of relationships with their subordinates (Erdogan and Bauer, 2014; Kang and Stewart, 2007).
The behavioural theory	Deals with the styles adopted by leaders for particular tasks and shows specific behaviours related to effective leadership (Bryman, 1992).
The situational (contingency) theory	Highlights how interactions with external environments shape leadership action (Hersey and Blanchard, 1999).

concentrate on identifying the different roles and responsibilities, rather than characteristics, of leaders. Other theories view leadership as specific to the situation and are based on the notion that different situations necessitate different leadership styles (Ayiro, 2009; Bonebright *et al.*, 2012). As noted by Fairholm (2002) and Turner and Baker (2018) leadership theories have progressed to now include leadership traits and style, leader–member exchange, leadership behaviour, contingency approaches, transformational leadership, charismatic leadership theory, great man leadership theory and shared leadership.

Unlike in permanent organisations, when undertaking infrastructure projects, the teams usually face uncertainties and challenges that require proper leader and team collaboration. Since the relationships between the projects' team leaders and other members of the project teams last for relatively short periods, the team leaders require strong ties with their teams for effective interactions. Toor and Ofori (2006) described different possible types of leadership styles, such as transactional, transformational, charismatic, democratic, autocratic, consultative, laissez faire, authoritative, participative, tyrant, task oriented, relationship oriented, production oriented, employee oriented, authority compliance and team management. From the above discussion, leadership is a complex process that embraces a wide collection of behaviours, styles and traits, but its purpose is successfully influencing others to accomplish goals.

Earlier in this chapter, sustainable infrastructure was identified as a significant factor in advancing societies and achieving their sustainable development goals. However, as observed from literature (Abrahams, 2017 and Aghimien *et al.*, 2019; Djokoto *et al.*, 2014), the transition towards the adoption of sustainability practices presents several challenges including that of leadership (Opoku *et al.*, 2015). Yet leadership plays a very significant role in the construction industry as a result; Ofori and Toor (2008) identified leadership as a key success factor required for achieving sustainability in the construction sector. Tabassi and Abu Bakar (2010) indicated that effective leadership style is critical to all successful projects and organisations while influential reports by Egan (1998) and Latham (1994) asserted the relevance of leadership and its significant role in driving change in the construction industry. Despite all these assertions on the relevance of leadership, a Chartered Institute of Building (CIOB report, *Leadership in the Construction Industry*, disclosed a stark lack of leadership within the construction industry (CIOB, 2008). The report also highlighted the need to reassess the leadership qualities and skills required by the construction industry. To develop great leaders, the report advocated a greater focus on the softer skills of relationship management: creativity and emotional intelligence (CIOB, 2008). It is worth mentioning that the results from yet another report by CIOB raised concerns about leadership in the construction industry. *The Green Perspective: A UK Construction Industry Report on Sustainability* concluded that the industry does not have good leadership on issues of sustainability (CIOB, 2007). Since leadership is about envisioning and influencing the future, the findings from these reports indicate an urgent need for the construction sector to embrace leadership that gives primacy to sustainability.

4.5.1 Enhancing sustainability practices through sustainability leadership

Within the context of this chapter, sustainability practices refer to any practices aimed at achieving sustainability and sustainable development during the delivery of projects. They comprise an assemblage of practice features implemented by one or more stakeholders who, in unambiguous contexts, are motivated by the desire to achieve sustainable value for their projects. From the literature, several sustainability practices are identifiable, with some being applicable to infrastructure projects. Examples include environmental practices such as waste management and energy efficiency (Yusof *et al.*, 2016), environmental training programs and pollution reduction

(Chen *et al.,* 2016). Others consist of social practices like the social commitment and social participation of construction firms (Huang and Lien, 2012). In an era of increasing global awareness about the importance of addressing climate change, smarter solutions, reducing greenhouse gas emissions and other sustainability objectives, Khalfan *et al.* (2002) indicated that sustainable objectives are only accomplished if practices are informed by knowledge — for example, having an understanding of which sustainability practices are to be prioritised and executed, when and how they need to be implemented, and by who. Therefore, within the construction and infrastructure sector, in practical terms sustainability practices could still encompass adopting deliberate steps towards reusing and repurposing waste by diverting waste from landfill to reusing wastewater for irrigation. In summary, Opoku *et al.* (2015) stated that sustainability practices involve the whole project cycle (i.e., pre-construction, construction, and post-construction) and may embrace waste management, procurement, sustainable design, whole life costing, and utilisation of materials and resources.

With infrastructure projects taking on new dimensions and the ever-dynamic global construction industry shifting to meet new and varied needs, Spencer Stuart (2015) sought to gain leadership insights for the global infrastructure industry through surveys and interviews with key stakeholders. Their report revealed a sector on the cusp of significant change and predicted an emerging leadership profile that underscores the following:

- **Strategic thinking:** The ability to consider all available options with a long-term lens and draw on the best ideas, wherever they come from, is more critical than ever as projects become more complex and companies take a larger share of the risk.
- **Commercial and financial mind-set:** The growth of public-private partnerships and the growing financial stake of projects require a strong business mindset and commercial savvy.
- **Collaboration and influencing skills:** Successfully leading an infrastructure business today requires executives to be able to unlock knowledge and capabilities across the entire organisation and to lead cultural change that places consumer awareness and social impact high on the agenda.
- **Stakeholder management:** Infrastructure leaders must be able to nimbly manage the needs of an increasingly broad set of stakeholders — both private and public — and speak credibly to all.

Giving the nature of the construction and infrastructure sector, these predictions are tenable. Therefore, the industry needs to identify and implement prime responses with respect to both the opportunities the changes offer and the challenges they portend. Being mindful that sustainability is marked by a high degree of subjectivity from stakeholders and aptly designated a wicked problem (Pryshlakivsky and Searcy, 2013; Lans *et al.,* 2014), the implementation of sustainability practices require leadership action toward shared envisioning of economic and financial, social, institutional, environmental and climate resilience concerns among numerous stakeholders. For example, Whitehead (2014) outlined the sustainability journey of Balfour Beatty PLC from a position of naivety to where it scored highly on sustainability. Of note was the relevance of top management advocacy as well as an appropriate conceptual understanding of the initiative at every level of leadership. This led to the appointment of a sustainability director who championed sustainability discussions at board level at Balfour Beatty. For this reason, what is required is leadership that facilitates the development of collaborative sustainability practices while fostering cross-sector collaboration required to address the global concerns on sustainability. This chapter describes the above type of leadership as sustainability leadership. Citing the Cambridge Institute of Sustainability Leadership's definition, Visser and Courtice (2011) described sustainability leaders as individuals who are compelled to make a difference by deepening their awareness of themselves in relation to the world around them. In doing so, they adopt new ways of seeing, thinking and interacting that result in innovative, sustainable solutions. Building on the above and Yukl (2006), enhancing sustainability practices for infrastructure project delivery requires leadership that influences others to understand and agree about what needs to be done and how to do it, and the process of facilitating individual and collective efforts to accomplish shared sustainability objectives. A noteworthy point articulated by Visser and Courtice (2011) is that sustainability leadership is not a separate school of leadership, but a particular combination of leadership characteristics applied within a definitive context. Accordingly, they observed that if sustainability leadership were to be aligned with any mainstream school of leadership theory; perhaps the contingency theory would be most relevant because of the context of sustainability and the aspiration for leadership that can facilitate a more sustainable future (Visser and Courtice, 2011).

4.6 Systems and tools for stimulating sustainable infrastructure

As illustrated in Table 4.3, the growing global recognition of sustainable infrastructure's primacy and the establishment of knowledge-sharing platforms provided the incentive for the emergence of a number of tools to guarantee the sustainability of infrastructure projects. In addition, environmental impact assessments (EIAs), cost-benefit analyses (CBAs) and environmental and social impact assessments (ESIAs) have been widely deployed as project level tools. OECD (2006) explained that the original theoretical framework of CBAs did not take account or include environmental sustainability but subsequent iterations have integrated it mostly through the valuation of environmental assets. However, Sheate (2010) still disclosed that sustainability is a foundational principle in the theory of EIAs and ESIAs and is intended to influence formal decision-making processes with clear objectives and a transparent process but the principal purpose of CBAs, EIAs, and ESIAs is at the single project level.

Besides these generic tools, there are other project-level sustainability assessment and rating tools (see Table 4.3) with sustainability rating systems applicable at specific project phases. According to the Institute for Sustainable Infrastructure (ISI), there is a lack of guidance on incorporating sustainability concerns at the policy and upstream planning phases of infrastructure development (ISI, 2015). This lack of guidance places restrictions on the effectiveness with which sustainability becomes incorporated at later phases. In recognition of the need for a more holistic approach, several alternatives like the Envision Rating System for Sustainable

Table 4.3 Sustainability rating systems and phase applicability

Sustainability rating systems and tools	Applicability
Envision rating system for sustainable infrastructure (Envision)	Design
Infrastructure sustainability rating scheme (IS)	Design
Infrastructure voluntary evaluation sustainability tool (INVEST)	Upstream planning and design
Standard for sustainable and resilient infrastructure (SuRe)	Design
Sustainable transport appraisal rating (STAR)	Upstream planning and design
Hydropower sustainability assessment protocol (HSAP)	Upstream planning and design
SE4All regulatory indicators for sustainable energy tool (RISE)	Upstream planning
Inter-American Development Bank (IDB) Safeguards and Policies	Design and financing
International Finance Corporation (IFC) Performance Standards	Design and financing
World Bank (WB) Environmental and Social Framework and Policies	Design and financing

Source: Inter-American Development Bank (2018b)

Infrastructure have been developed. This rating system, developed between the Zofnass Program for Sustainable Infrastructure at the Harvard University Graduate School of Design and the Institute for Sustainable Infrastructure (ISI), supports the implementation of sustainable infrastructure and incorporates best practices from industry (ISI, 2015). Envision comprises five categories of impact (i.e., quality of life, leadership, resource allocation, natural world and climate and risk) and 60 different sustainability criteria across these categories. Collectively, these criteria, referred to as "credits," address various environmental, social, and economic impacts from project design, construction and operation. It also scores various project characteristics against the credits to ascertain the number of points projects can achieve against specified measures of performance (ISI, 2015).

Even though the majority of the previously mentioned rating systems offer highly structured methods to assess how projects meet sustainability goals, they still possess several inherent deficiencies that have a bearing on their legitimacy and application. With Envision, for instance, since points per criteria tend to be subjective, they do not always align with evidence on how the public actually values the impact. There are also issues with data concentrated credit and point systems. As a final point because rating systems usually require designs to be completed and may require projects to be in operation for evaluation consequently, rating systems potentially provide value in decisions at planning stages, but they are mainly geared towards evaluating projects within completed designs and might not account for long-standing impacts across sectors.

4.6.1 *Sustainability benchmarking and continuous improvement*

Over the last two decades, benchmarking has gained popularity as a business management tool that brings change and improvement to an organisation's business processes through a process of learning from the successes of other organisations and applying similar practices in their own organisations. As Mittelstaedt (1992) and Spendolini (1992) illustrated, several definitions of benchmarking have been widely used. For instance, Wireman (2004) defined benchmarking as the search for an industry's best practices that lead to superior performance. Ahren and Parida (2009) indicated that benchmarking supports management in their pursuit of continuous improvement of their operations and supports the development of realistic goals and strategic targets.

In another vein, Anderson and McAdam (2004) explained that benchmarking is a structured process established by the development of the systematic process model providing a common language within the organisation. Dey (2002) outlined the objectives for benchmarking as follows:

- To identify key performance measures for each function of a company's operations;
- To measure the internal performance levels of the company as well as the performance levels of the competitors;
- To compare performance levels in order to identify areas of competitive advantage and disadvantages; and
- To implement programs for closing the gap between the internal operations and other companies.

When put together, the definitions and objectives of benchmarking establish its relevance for business expansion, and how it provides a learning base and offers a strategy for performance improvement within organisations and government. With increasing infrastructure demands, a growing mandate for leaner alternatives, waste reduction and the aspiration to improve quality while maximising effectiveness of infrastructure, benchmarking is also considered a facilitator for innovation since it supports a better understanding of the external environment and promotes organisational learning (Spendolini, 1992). However, to achieve innovation via benchmarking, organisations need to overcome their internal boundaries in order to evaluate opportunities and threats in the external environment.

Pisu *et al.* (2012) explored three different approaches to benchmarking infrastructure. The first approach benchmarks based on growth contributions by using macroeconometric techniques to estimate the impact of the existing infrastructure capital stock on growth and to infer its growth-maximising level. However, this approach fails to consider the impact of infrastructure on dimensions of social welfare. One caveat with benchmarking based on growth contributions is that more infrastructure spending does not always translate to better outcomes. The other approach utilises cost-benefit analysis of new projects to benchmark the performance of infrastructure by measuring the pre- and post-cost-benefit analyses of infrastructure projects. This approach takes into account desirable and undesirable outcomes of the infrastructure, thereby providing a welfare perspective not offered by the previous approach. However, it does not compare the performance of the new infrastructure with existing ones. The final approach by Pisu *et al.* (2012) benchmarks the social efficiency of infrastructure service provisions based on the existing capital stock and taking into account positive and negative externalities using a set of selected metrics based on data envelopment analysis (DEA). According to Bhanot and Singh (2014), DEA is a non-parametric mathematical programming technique that allows for the concurrent evaluation of multiple inputs and outputs to calculate a single comprehensive measure of efficiency. It is widely used for performance measurement and benchmarking because it overcomes many drawbacks of traditional performance measurement systems (George and Rangaraj, 2008) and allows customisation based on local priorities for the infrastructure. An examination of the literature reveals the application of DEA in a variety of infrastructure settings such as irrigation (Phadnis and Kulshrestha, 2012), railways and rail (Bhanot and Singh, 2014) ports (de Koster *et al.,* 2009).

4.6.2 *Attaining best practice in benchmarking major infrastructure projects*

To better appreciate their usefulness and to appraise the range of valued or desired benefits, the performance of infrastructure projects regularly comes under scrutiny by various stakeholders globally (WEF, 2019; ADB, 2017). In most instances, the outcomes provide insights that challenge the status quo. An illustration can be drawn from the United Kingdom where the International Centre for Infrastructure Futures (ICIF) disclosed that infrastructure performance indicators were criticised for not sufficiently integrating the broader context of societal, environmental and economic needs (ICIF, 2015). For example, while a performance indicator like the cost-benefit analysis provides an outlook of the impacts of projects (positive and negative) and is a major decision-making tool for competing projects, it is open to misrepresentation or poor articulation. As a result, The UK's Institute for Government (IFG) advocated that multi-criteria analysis might prove more effective than traditional methods like cost-benefit analysis during the assessment and appraisal of infrastructure (IFG, 2017). With the objectives of benchmarking well established in the previous section, what lies ahead is embracing practises that encourage reliable benchmarking in the infrastructure sector.

Through utilising its Transforming Infrastructure Performance (TIP) programme, the Infrastructure and Projects Authority (IPA) developed a top-down benchmarking methodology. The primary aim of this methodology is to encourage better and more consistent benchmarking across infrastructure projects among both government departments and client organisations (IPA, 2019). Figure 4.4 provides a summary of the IPA benchmarking methodology guidance.

As illustrated (see Figure 4.4) the benchmarking methodology guidance comprises seven major stages. According to IPA (2019), **Stage 1** commences with a confirmation of the project objectives and subsequently links the objectives to a benchmark. IPA recommends benchmarks

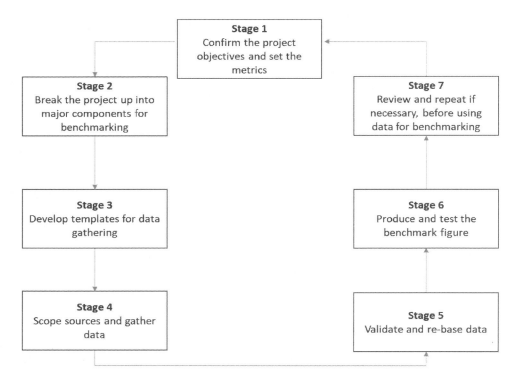

Figure 4.4 Benchmarking guidelines
Source: IPA (2019)

to measure whole life project performance as well as project costs. In **Stage 2**, the project is broken into major components for benchmarking. For example, in a port facility, these could include the terminal building, quay construction, harbour and fairway dredging, etc. As part of stage 2, the benchmark highlights that the identified components can be compared to those from a range of other projects to produce a benchmark indicative asset cost (BIAC) or benchmark indicative non-asset cost (BINAC). Using experiential knowledge, the components are robustly benchmarked against agreed metrics. To reduce reinventing the wheel, IPA suggests that BIAC and BINAC templates should be shared across organisations and their project teams.

Stage 3 entails the collaborative development of templates with delivery partners. This process facilitates improved stakeholder understanding of project metrics calculations. During **Stage 4**, various sources are scoped and data is gathered. As explained by the IPA, data can be sourced internally by the project team, externally using collaborations or via procured data as well as internationally. The use of international data for benchmarking is advantageous and increases the range of assessments. It is particularly useful where there is limited local data or if projects are undertaken in regulated sectors like the energy or nuclear industry. Irrespective of the source of data, the safe and secure handling of information must be clearly considered. As part of **Stage 5**, the gathered data is validated, cleansed and rebased. Data is usually rebased to make all records in the dataset analogous and consistent. Regularly undertaking these activities enhances the integrity and quality of any benchmarks being used for decision-making. In **Stage 6**, the benchmark figures are produced and tested. The figures produced are a reflection of all the components developed at Stage 3 and clearly describe the project performance. The final stage, **Stage 7,**

encompasses reviews and repetition where required before using data for benchmarking. In terms of practical significance, the IPA noted that its benchmarking methodology was tested and proven as a credible concept. For this reason, the Department for Transport (DfT) integrated the methodology as part of the Transport Infrastructure Efficiency Strategy (TIES) benchmarking initiative. In the light of increasing infrastructure demand, the TIES presents a timely opportunity for infrastructure stakeholders to collaborate and drive efficiency by improving understanding of costs and performance through benchmarking (TIES, 2019).

4.7 Achieving sustainable major infrastructure projects development and management post-COVID-19

To achieve the development and management of sustainable major infrastructure, many sectors of infrastructure must be considered as interconnected systems. This is because the degree of sustainability of certain infrastructure systems can have direct and indirect impacts on others. Thus, the successful implementation of sustainable construction principles, as advocated earlier in this chapter, requires effective integrated actions as well as the commitment of various stakeholders. While debates are ongoing over precise definitions and understanding of it, in this chapter sustainable infrastructure is regarded as consisting of the following dimensions:

- Economic and financial sustainability;
- Social sustainability;
- Environmental sustainability (including climate change and resilience); and
- Institutional sustainability.

The aftermath of COVID-19 has further uncovered structural, social, environmental, and economic problems that several nations may have unconsciously neglected for several years. All these provide serious evidence to question the feasibility of achieving sustainable infrastructure in line with various global targets like the SDGs and Paris Agreement of the United Nations Framework Convention on Climate Change. From a social perspectivem for instance, COVID-19 exposed the disparity in access to health services in certain nations while several others were stretched beyond their limits. Economically, many developing and developed nations alike observed an upsurge in unemployment rates. In addition, the sharp increase in mobile communication as an outcome of cancelled domestic and international business travel echoed the importance of cyber-safe remote technology infrastructure. Turning the focus away from the former, as nations get set to relaunch their economies, sustainability is deemed imperative especially when it comes to designing, creating or maintaining existing major infrastructure. To demonstrate how sustainable infrastructure can aid the post-COVID-19 recovery, the WEF (2020) provided several noteworthy articulations and disclosed that:

- Technologically advanced, sustainable and resilient infrastructure can pave the way for an inclusive post-COVID-19 economic recovery.
- Low and middle-income countries could see $4 return for every $1 spent on building infrastructure that focuses on long-term resilience.
- Governments should ensure investments go to infrastructure projects that are sustainable, technologically advanced and resilient (WEF, 2020).

From the earlier sections, it was well established that a variety of initiatives are ongoing to optimise the quality and sustainability of major infrastructure projects. According to Bhattacharya *et al.*

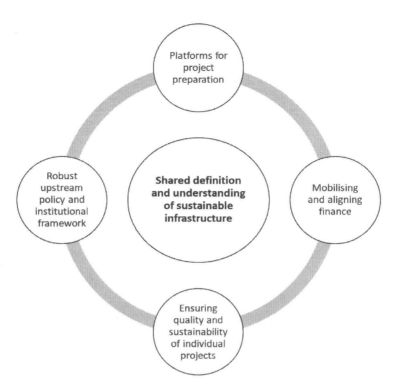

Figure 4.5 Integrated framework for delivery of sustainable infrastructure

Source: Bhattacharya *et al.* (2019b)

(2019b), these initiatives highlight prominence of several aspects including governance and public sector management, project prioritization and procurement, public private partnerships, project preparation platforms and facilities, climate sustainability and resilience, and mobilisation of private finance. Consequently, Figure 4.5 depicts an integrated framework that links and better aligns some efforts that can help unlock and scale up investments, ensure projects are sound, and mobilize and apply finance from all sources most effectively and sustainably.

Of central relevance in the framework is having a clear and shared definition and understanding of sustainable infrastructure and an open infrastructure database. From previous sections, it was incontestable the concept of sustainable infrastructure is still not commonly understood or agreed upon among stakeholders. Unfortunately, the multiplicity of approaches to sustainability usually creates confusion and create obstacles to attracting investments.

However, a shared definition and understanding encourage more simplified implementation, thereby making certain that stakeholders are heading in the direction of the same objectives that in turn make the measurement of progress and feedback less ambiguous and effective. The framework also acknowledges the need for robust policies and institutional underpinnings. Oftentimes, the policies and institutional underpinnings for delivering on infrastructure projects reflect various considerations like peculiar national challenges and aspirations; hence, they become multifaceted. As explained by Bhattacharya *et al.* (2019b), meeting the sustainable infrastructure challenge requires moving decision-making upstream to integrate policy objectives across sectors and to optimise for social, economic and environmental outcomes. For this to be attained, wide and inclusive engagement with stakeholders will be needed to establish buy-in to a long-term view.

Another crucial aspect of the framework is ensuring quality and sustainability of individual projects and advocating that life cycle planning from design to decommissioning should be based on sustainability criteria. Collectively, these criteria and associated assessment tools ensure that projects do no social or environmental harm and have good governance arrangements. Over time, most tools have undergone recent modifications to underscore the relevance of sustainability and resilience Bhattacharya *et al.* (2019b). It is also imperative to develop effective structures that facilitate financing of projects; the International Institute for Sustainable Development (IISD) disclosed that lack of awareness and capacity constraint leading to inadequate deal flow are two common issues that encumber greater private infrastructure investment at the national or subnational levels (IISD, 2017). However, Project Preparation Facilities (PPF) have been used to develop bankable and investment ready projects; nevertheless, not all PPFs have clear and long-term strategies (WEF, 2015). All the same, platforms for project preparations advocate the need to develop robust platforms that facilitate upstream financing and project readiness. As a result, MDBs established SOURCE, a multilateral project development platform that augments project preparation, engages stakeholders and supports the collection of data (Bhattacharya *et al.,* 2019b). In conclusion, with the urgency of the scale and shift in infrastructure investment necessary to deliver the sustainability agenda comes a commensurate challenge to mobilise finance not just at accurate scales but also to finances well aligned with sustainability criteria. Consequently, the final part of the framework acknowledges the importance of mobilising and aligning finance.

4.8 Chapter summary

It is well documented that the world needs more infrastructure but not just any infrastructure. As demonstrated in this chapter, to achieve the economic, social and environmental object-ives embodied by the Sustainable Development Goals (SGDs), infrastructure must be sus-tainable. Scholars have reviewed current actions proposed at the global and national levels to achieve this and noted that while positive results have emerged, more action is still needed to achieve and surpass the SDGs. Hence, decision-makers have identified and considered several prerequisites for infrastructure. In an era of increasing awareness of sustainability, the need for purposeful leadership that clearly embraces sustainability was emphasised as the way forward for major infrastructure development and management. The chapter also discussed the import-ance of benchmarking since major infrastructure projects are complex with their benefits open to being regularly scrutinised; hence, it evaluated a consistent approach to benchmarking that improves project delivery and performance. The chapter concluded by discussing a framework that integrates major areas with the view to unlocking and scaling up infrastructure investments through effective and sustainable financing options.

4.9 Chapter discussion questions

1. Within the context of developing, developed and emerging economies, appraise the prerequisites for sustainability in major infrastructure projects. Observe for commonalities and areas of variance and evaluate the most crucial aspects for policy makers and infrastruc-ture development organisations.
2. Various illustrations on the outcomes of major infrastructure projects were discussed in this chapter. Reflect on the key information from the Grand Inga Dam and Sembcorp Myingyan projects provided in Vignette 4.1 and other sources of information. Examine the value infrastructure projects bring to local communities in terms of social advancement.

Propose how inclusiveness can be maximised for major sustainable infrastructure projects in future.
3. As a project manager for a major infrastructure project of your choice, critically discuss how you can integrate sustainability awareness in your leadership and project delivery strategy.
4. Assume you are a consultant to the National Infrastructure Monitoring Agency of the country you reside in. Utilise the framework discussed in Section 4.6.2 to benchmark the current performance of an ongoing major public infrastructure project that you are familiar with.

4.10 Case study: The Eko Atlantic city project

4.10.1 Preamble

According to the World Bank, several towns and cities along the West African coastline have become prominent and important contributors to their national gross domestic products because of coastal activities like shipping, oil and gas, fishing, tourism and commerce (World Bank, 2015b). This has led to population concentrations along the coastline. As a result, many capitals and major towns have become coastal. At the same time, AFDB (2018) noted that this geographical space is undergoing annual coastal retreats of one to two metres. This level of coastal erosion has devastating effects for infrastructure and inhabitants. The projected rise in sea levels due to climate change further exacerbates these risks. Lagos, Nigeria, is one of such cities already experiencing all of the above (AFDB, 2018).

With a current population of 22 million inhabitants, the New Climate Economy Report (2018) considered Lagos to be the most populous city on the African continent and one of the fastest growing cities in the world. The location of several financial institutions and multi-national

Figure 4.6 Partial pictorial view of Eko Atlantic City

companies' headquarters make it one of the most globally connected cities in Africa. Like many rapidly growing cities, economic growth and development of Lagos has been hampered by a dearth of functional infrastructure. Over the past 100 years, the shoreline of Victoria Island, the main business hub of Lagos, retreated significantly with catastrophic consequences on the infrastructure of communities adjacent to the sea (AFDB, 2018). Several temporary coastal protection measures have been in place to mitigate the coastal erosion, but in 2006, the protective beach disappeared resulting in flood damage, loss of lives and destruction of infrastructure along Ahmadu Bello Way and Bar Beach (AFDB, 2018). As explained from the Eko Atlantic City Milestones, this necessitated the birth of the construction of the "Great Wall of Lagos," a sea wall that "wraps" around the city and protects it from the effects of the surging Atlantic (Eko Atlantic Milestones, n.d.).

4.10.2 The Eko Atlantic City development project

Officially referred to as the Nigeria International Commerce city, but with other assumed names like "the new gateway to Africa," Africa's first smart city or simply Eko Atlantic. This project has been anticipated to be the gateway to the African continent and will alleviate infrastructural and congestion issues Lagos city is currently facing (Eko Atlantic City, n.d.; Ventures Africa, 2012). The estimated financial cost of the project is $6 billion and the project developers, South Energyx, believe Eko Atlantic City will withstand the worst storms that the Atlantic can gather for the next two centuries. The entire project is expected to be completed in two distinct phases:

- **Phase 1:** Shoreline protection and reclamation activities of over 1000 hectares of land.
- **Phase 2:** Production of a development platform and master plan to enable the construction of a new mixed-use development and city (Eko Atlantic City). The master plan (2015–2040) will distinguish the six phases of development of individual plots.

Figure 4.7 Part view of Eko Atlantic City

Ajibade (2017) identified several major players including the following:

- Nigerian and international financial institutions (KBC; BNP Paribas Fortis, European; First Bank, Guaranty Trust Bank, FCMB, and Access Bank, Nigeria).
- Dredging International, Belgium; Royal Haskoning, Netherlands; Dal Al-Handasah and MZ Architects, Lebanon, as the consulting, architecture and engineering firms.
- Private property developers (ESLA International, Lebanon; Orlean Invest, Nigeria).
- Marketers (Diya, Fatimilehin and Co, estate firm, Nigeria).

Meanwhile, AFDB (2018) identified several local, regional, national and international stakeholders for the project such as the local communities, Lagos state ministries, the federal Ministry of the Environment, international NGOs. To demonstrate the commitment of Eko Atlantic City to environmental friendliness, roads are built using block stones rather than asphalt, energy efficient materials will be used for construction and developers will build to the International Finance Corporation's Green EDGE standard (Gbenga, 2020).

From a comprehensive impact assessment by teams of qualified international specialists using standard methods and techniques, it was suggested that the project will have minimal adverse environmental effects on the majority of receptors. Despite these promising accounts, Ajibade (2017) disclosed that there is no disagreement on the detrimental impacts of erosion and storm surges on the coast of Lagos but no agreement on whether the proposed Eko Atlantic City adaptation project would lead to urban resilience and sustainability.

4.11 Case discussion questions

1. Identify the major lessons from the Eko Atlantic City project.
2. From the results mentioned above, which lessons are most crucial for future similar projects in other developing, developed and emerging economies?
3. As a project manager, what will be your approach to achieve the required balanced perspective on the dimensions of infrastructure sustainability for the Eko Atlantic City project?
4. If your organisation was appointed to continue the development of the Eko Atlantic City project, what steps will you take to:

   - Facilitate individual and collective efforts to accomplish shared sustainability objectives among the main delivery partners.
   - Integrate the broader context of societal, environmental and economic requirements while gauging the performance of the project.

5. Concerns have been raised about the sustainability of Eko Atlantic City. In addition, unprecedented challenges like COVID-19 have revealed how well cities are developed and managed as well as the impact this brings for functionality during crisis periods.

   - Discuss strategies the main developers can still adopt to improve wider stakeholder confidence in the project outcomes and benefits.
   - Propose action plans that can be adopted by the policy makers in Nigeria and other nations to enhance the development and management of similar infrastructure projects.

References

Abrahams, G. (2017). Constructing definitions of sustainable development. *Smart and Sustainable Built Environment*, 6(1), pp. 34–47.

AFDB (2018). Available from: https://www.afdb.org/fileadmin/uploads/afdb/Documents/Environmental-and-Social-Assessments/Nigeria_Eko_Atlantic_ESIA_ESMP_Summary.pdf [cited 10 May 2020].

Aghimien, D.O., Aigbavboa, C.O., and Thwala, W.D. (2019). Microscoping the challenges of sustainable construction in developing countries. *Journal of Engineering, Design and Technology*, 17(6), pp. 1110–1128.

Agyekum-Mensah, G., Knight, A., and Coffey, C.H. (2012). 4Es and 4 Poles model of sustainability: Redefining sustainability in the built environment. *Structural Survey*, 30(5), pp. 426–442.

Ahmad, E., Neuweg, I., and Stern, N. (2018). China, the world and the next decade: Better growth, better climate. London: Grantham Research Institute on Climate Change and the Environment and Centre for Climate Change Economics and Policy, London School of Economics and Political Science. Available from: http://www.lse.ac.uk/granthaminstitute/wp-content/uploads/2018/04/Ahmad-et-al_China-the-world-and-the-next-decade_Better-growth-better-climate-1.pdf [cited 25 April 2020].

Ahren, T. and Parida, A. (2009). Maintenance performance indicators (MPIs) for benchmarking the railway infrastructure: A case study. *Benchmarking: An International Journal*, 16(2), pp. 247–258.

Ajayi, S.O. and Oyedele, L.O. (2018). Critical design factors for minimising waste in construction projects: A structural equation modelling approach. *Resource Conservation and Recycling*, 137, pp. 302–313.

Ajibade, I. (2017). Can a future city enhance urban resilience and sustainability? A political ecology analysis of Eko Atlantic City, Nigeria. *International Journal of Disaster Risk Reduction*, 26, pp. 85–92.

Ali, S.F., Ali, A., and Bayyati, A. (2018). Achieving sustainability in the UK construction by reducing waste generation. Psychology and sustainable construction: Searching the gap between psychology and construction for a sustainable built environment. Wolverhampton, UK, 19–20 Dec. Available from: https://openresearch.lsbu.ac.uk/item/86836 [cited 11 February 2020].

Anderson, K. and McAdam, R. (2004). A critique of benchmarking and performance measurement lead or lag. *Benchmarking: An International Journal*, 11(5), pp. 465–83.

Arup (2019). Rethinking timber buildings. Available from: https://www.arup.com/perspectives/publications/research/section/rethinking-timber-buildings [cited 27 January 2020].

Asia Development Bank (2017). Meeting Asia's infrastructure needs. Available from: https://www.adb.org/sites/default/files/publication/227496/special-report-infrastructure.pdf [cited 13 March 2020].

Ayiro, L.P. (2009). An analysis of emotional intelligence and the performance of principals in selected schools in Kenya. *Advances in Developing Human Resources*, 11(6), pp. 719–746.

Balasubramanian, S. and Shukla, V. (2017). Green supply chain management: An empirical investigation on the construction sector. *Supply Chain Management: An International Journal*, 22(1), pp. 58–81.

Bass, B.M. and Avolio, B.J. (1994). *Improving organizational effectiveness through transformational leadership*. Thousand Oaks, CA: Sage.

Bhanot, N. and Singh, H. (2014). Benchmarking the performance indicators of Indian Railway container business using data envelopment analysis. *Benchmarking: An International Journal*, 21(1), pp. 101–120.

Bhattacharya, A., Meltzer, J., Oppenheim, J., Qureshi, Z., and Stern, N. (2016). Delivering on sustainable infrastructure for better development and better climate. Available from: https://www.brookings.edu/wp-content/uploads/2016/12/global_122316_delivering-on-sustainable-infrastructure.pdf [cited 18 November 2019].

Bhattacharya, A. and Jeong, M. (2018). Driving the sustainable infrastructure agenda in emerging markets. Available from: https://economic-policy-forum.org/wp-content/uploads/2018/11/GIZ-paper-20180630_final.pdf [cited 23 December 2019].

Bhattacharya, A., Casado, C.C., Jeong, M., Amin, A., Watkins, G., and Zuniga, M.S. (2019b). Attributes and framework for sustainable infrastructure consultation report. Available from: https://publications.iadb.org/publications/english/document/Attributes_and_Framework_for_Sustainable_Infrastructure_en_en.pdf [cited 10 May 2020].

Bhattacharya, A., Gallagher, K.P., Muñoz Cabré, M., Jeong, M., and Ma, X. (2019a) Aligning G20 infrastructure investment with climate goals and the 2030 Agenda, Foundations 20 Platform, a report to the G20. Available from: https://www.foundations.org/wp-content/uploads/2019/06/F20-report-to-the-G20-2019_Infrastrucutre-Investment.pdf [cited 22 December 2019].

Bonebright, D.A., Cottledge, A.D., and Lonnquist, P. (2012). Developing women leaders on campus: A human resource–women's center partnership at the University of Minnesota. *Advances in Developing Human Resources*, 14(1), pp. 79–95.

Bryman, A. (1992). *Charisma and leadership in organization*. London: Sage.

Chen, P.H., Ong, C.F., Hsu, S.C. (2016). Understanding the relationships between environmental management practices and financial performances of multinational construction firms. *Journal of Cleaner Production*, 139, pp. 750–760.

Chi, B., Lu, W., Ye, M., Bao, Z., and Zhang, X. (2020). Construction waste minimization in green building: A comparative analysis of LEED-NC 2009 certified projects in the US and China. *Journal of Cleaner Production*, 256, pp. 1–10.

CIOB (2007). The green perspective a UK construction industry report on sustainability. Available from: https://policy.ciob.org/wp-content/uploads/2007/08/The-Green-Perspective-A-UK-construction-industry-report-on-sustainability-2007.pdf [cited 11 December 2019].

CIOB (2008). Leadership in the construction industry. Available from: https://www.ciob.org/sites/default/files/CIOB%20research%20-%20Leadership%20in%20the%20Construction%20Industry%202008.pdf [cited 11 December 2019].

Crossrail (2018). Sustainability summary. Available from: https://learninglegacy.crossrail.co.uk/wp-content/uploads/2018/07/Sustainability-Summary-2018.pdf [cited 13 November 2019].

Cullinane, K., Bergqvist, R., Cullinane, S., Zhu, S., and Linkai, W. (2016). Improving the quality of Sweden's rail freight rolling stock: The use of DEA in benchmarking and pricing. *Benchmarking: An International Journal*, 24(6), pp. 1552–1570.

Daft, R.L. (2015). *The leadership experience*, 6th ed. Stamford, CT: Cengage Learning.

de Koster, M.B.M., Balk, B.M., and van Nus, W.T.I. (2009). On using DEA for benchmarking container terminals. *International Journal of Operations and Production Management*, 29(11), pp. 1140–1155.

DEFRA (2018). Resources and waste strategy: At a glance. Available from: https://www.gov.uk/government/publications/resources-and-waste-strategy-for-england/resources-and-waste-strategy-at-a-glance [cited 14 December 2019].

Department for Environment, Food and Rural Affairs (Defra) (2013). Waste prevention programme for England: Overview of evidence–a rationale for waste prevention in England. Available from: https://assets.publishing.service.gov.uk/government/uploads/system/uploads/attachment_data/file/264909/wpp-evidence-overview.pdf [cited 20 December 2019].

Dey, P.K. (2002). Benchmarking project management practices of Caribbean organizations using analytic hierarchy process. *Benchmarking: An International Journal*, 9(4), pp. 326–356.

Djokoto, S.D., Dadzie, J., and Ohemeng, E.A. (2014). Barriers to sustainable construction in the Ghanaian construction industry: Consultants perspectives. *Journal of Sustainable Development*, 7(1), pp. 134–143.

Donaldson, D. and Hornbeck, R. (2016). Railroads and American economic growth: A "Market Access" approach. *The Quarterly Journal of Economics*, 131(2), pp. 799–858.

Egan, J. (1998). Re-thinking construction: Report of the construction industry task force, department for environment transport and the regions, London.

Eko Atlantic Milestones (n.d.). Eko Atlantic Milestones. Available from: https://www.ekoatlantic.com/milestones/Eko-Atlantic-Milestones-Issue-1.pdf [cited 10 May 2020].

Erdogan, B. and Bauer, T. (2014). Leader-member exchange (LMX) theory: The relational approach to leadership. In: D. Day (Ed.), *The Oxford handbook of leadership and organizations*, pp. 407–433. New York, NY: Oxford University Press.

Fairholm, M. R. (2002). Defining leadership: A review of past, present, and future ideas. Available from: http://www.strategies-for-managing-change.com/support-files/definingleadershipreview.pdf [cited 11 May 2020].

G20 (2019). Overview of Saudi Arabia's G20 2020 Presidency. Available from: https://g20.org/en/g20/Documents/Presidency%20Agenda.pdf [cited 15 May 2020].

Gbenga, A.M. (2020). Eko Atlantic sets the stage for potential green city status with first IFC edge certification. Available from: http://venturesafrica.com/eko-atlantic-sets-the-stage-for-potential-green-city-status-with-first-ifc-edge-certification/ [cited 25 March 2020].

George, S.A. and Rangaraj, N. (2008). A performance benchmarking study of Indian Railway zones. *Benchmarking: An International Journal*, 15(5) pp. 599–617.

GICA Secretariat. (2017). G20 global infrastructure connectivity alliance 2017 work plan. Available from: https://www.bundesfinanzministerium.de/Content/DE/Downloads/G20-Dokumente/GICA-2017-work-plan.pdf?__blob=publicationFile&v=2 [cited 17 January 2020].

Global Infrastructure Hub (2020). Forecasting infrastructure investment needs and gaps. Available from: https://outlook.gihub.org/ [cited 30 January 2020].

Gubic, I. and Baloi, O. (2019). Implementing the new urban agenda in Rwanda: Nation-wide public space initiatives. *Urban Planning*, 4(2), pp. 223–236.

Hassan, O.A.B. (2016). An integrated approach to assessing the sustainability of buildings. *Journal of Engineering, Design and Technology*, 14(4), pp. 835–850.

Henson, K. (2011). The procurement and use of sustainable concrete on the Olympic Park. Learning legacy: Lessons learned from the London 2012 Games construction project. Available from: http://learninglegacy.independent.gov.uk/documents/pdfs/procurement-and-supply-chain-management/01-concrete-pscm.pdf [cited 27 April 2020].

Hersey, P. and Blanchard, K.H. (1999). *Leadership and the one-minute manager.* New York: William Morrow.

Huang, C.F. and Lien, H.C. (2012). An empirical analysis of the influences of corporate social responsibility on organizational performance of Taiwan's construction industry: Using corporate image as a mediator. *Construction Management Economics*, 30(4), pp. 263–275.

Huang, B., Wang, X., Kua, H., Geng, Y., Bleischwitz, R., and Ren, J. (2018). Construction and demolition waste management in China through the 3R principle. *Resources Conservation and Recycling*, 129, pp. 36–44.

IDB (2017). Building a bridge to sustainable infrastructure. Mapping the global initiatives that are paving the way. Available from: https://www.mercer.com/content/dam/mercer/attachments/private/nurture-cycle/gl-2016-responsible-investments-building-a-bridge-to-sustainable-infrastructure-mercer.pdf [cited 30 January 2020].

Infrastructure and Projects Authority (2019). Best practice in benchmarking. Available from: https://assets.publishing.service.gov.uk/government/uploads/system/uploads/attachment_data/file/783525/6.5341_IPA_Benchmarking_doc_FINAL_Web_050319.pdf [cited 14 March 2020].

Infrastructure Australia Act (2008). Infrastructure Australia Act. Available from: https://www.legislation.gov.au/Details/C2014C00639 [cited 14 April 2020].

Infrastructure Asia (n.d.). Sembcorp Myingyan independent power plant. Available from: https://www.infrastructureasia.org/en/projects/sembcorp-myingyan-power-plant [cited 12 January 2020].

Infrastructure Australia (2016). Australian infrastructure plan priorities and reforms for our nation's future. Available from: https://www.infrastructureaustralia.gov.au/sites/default/files/2019-06/Australian_Infrastructure_Plan.pdf [cited 16 April 2020].

Infrastructure Australia (2018). Prioritising reform progress on the 2016 Australian infrastructure plan. Available from: https://www.infrastructureaustralia.gov.au/sites/default/files/2019-06/prioritising-reform-report.pdf [cited 15 April 2020].

Infrastructure Australia (2019). An assessment of Australia's future infrastructure needs the Australian infra-structure audit 2019. Available from: https://www.infrastructureaustralia.gov.au/sites/default/files/2019-08/Australian%20Infrastructure%20Audit%202019.pdf [cited 16 April 2020].

Infrastructure Canada (2018). Investing in Canada. Canada's long-term infrastructure plan. Available from: https://www.infrastructure.gc.ca/alt-format/pdf/plan/icp-pic/IC-InvestingInCanadaPlan-ENG.pdf [cited 17 March 2020].

Institute for Government (2017). How to value infrastructure: Improving cost benefit analysis. Available from: https://www.instituteforgovernment.org.uk/sites/default/files/publications/IfG%20Report%20CBA%20infrastructure%20web%20final1.pdf [cited 14 March 2020].

Institute for Sustainable Infrastructure (2015). Envision rating system for sustainable infrastructure. Available from: https://research.gsd.harvard.edu/zofnass/files/2015/06/Envision-Manual_2015_red.pdf [cited 28 March 2020].

Inter-American Development Bank (2018b). IDBG framework for planning, preparing, and financing sustain-able infrastructure projects: IDB sustainable infrastructure platform. Available from: https://publications.iadb.org/en/idbg-frameworkplanning-preparing-and-financing-sustainableinfrastructure-projects-idb-sustainable [cited 25 March 2020].

Inter-American Development Bank (2018a). What is infrastructure sustainability? A framework to guide sustainability across the project cycle. Available from: https://publications.iadb.org/publications/english/document/What_is_Sustainable_Infrastructure__A_Framework_to_Guide_Sustainability_Across_the_Project_Cycle.pdf [cited 10 November 2019].

International Centre for Infrastructure Futures (2015). A critique of current infrastructure performance indicators: Towards best practice. Available from: https://research.ncl.ac.uk/ibuild/outputs/reports/ICIF&iBUILD%20Indicators%20Interim%20Summary%20Report%20Final.pdf [cited 14 March 2020].

Jedwab, R., Kerby, E., and Moradi, A. (2017). History, path dependence and development: Evidence from colonial railways, settlers and cities in Kenya. *The Economic Journal*, 127, pp. 1467–1494.

Kang, D. and Stewart, J. (2007). Leader-member exchange (LMX) theory of leadership and HRD: Development of units of theory and laws of interaction. *Leadership and Organization Development Journal*, 28(6), pp. 531–551.

Kartik, H.O., Russell, L.H., and Ross, S.M. (2003). HVFA concrete – an industry perspective. *Concrete International*, 25(8), pp. 29–34.

Khalfan, M.M.A., Bouchlaghem, N.M., Anumba, C.J., Carrillo, P.M., and Glass, J. (2002). Managing sustainability knowledge for a sustainable build environment. Available from:https://www.irbnet.de/daten/iconda/CIB10194.pdf [cited 1 June 2020].

KPMG (n.d.). Achieving sustainable infrastructure. Available from: https://home.kpmg/xx/en/home/insights/2020/01/enabling-sustainable-infrastructure.html [cited 25 April 2020].

Lans, T., Blok, V., and Wesselink, R. (2014). Learning apart and together: Towards an integrated competence framework for sustainable entrepreneurship in higher education. *Journal of Cleaner Production*, 62, pp. 37–47.

Latham, S.M. (1994). Constructing the team: Report of the Government/Industry review of procurement and contractual arrangements in the UK construction industry, HMSO, London.

Lu, W. and Tam, V. (2013). Construction waste management policies and their effectiveness in Hong Kong: A longitudinal review. *Renewable and Sustainable Energy Reviews*, 23, pp. 214–223.

Lu, W., Chen, X., Peng, Y., and Liu, X. (2018). The effects of green building on construction waste minimization: Triangulating "big data" with "thick data." *Waste Management*, 79, pp. 142–152.

McKinsey (2016). Bridging global infrastructure gaps. Available from: https://www.mckinsey.com/industries/capital-projects-and-infrastructure/our-insights/bridging-global-infrastructure-gaps [cited 13 February 2020].

Mensah, J. (2019). Sustainable development: Meaning, history, principles, pillars, and implications for human action: Literature review. *Cogent Social Sciences*, 5(1), pp. 1–21.

Mittelstaedt, R.E. (1992). Benchmarking: How to learn from best-in-class practices. *National Productivity Review*, 11(3), pp. 301–315.

New Climate Economy Report (2018). Unlocking the inclusive growth story of the 21st century: accelerating climate action in urgent times. Available from: https://newclimateeconomy.report/2018/wp-content/uploads/sites/6/2018/09/NCE_2018_FULL-REPORT.pdf [cited 12 June 2020].

OECD (2006). Cost-benefit analysis and the environment: Recent developments. Available from:https://www.oecd-ilibrary.org/environment/cost-benefit-analysis-and-the-environment_9789264010055-en [cited 12 February 2020].

OECD (2007). Infrastructure to 2030. Mapping policy for electricity, water and transport. Available from: https://www.oecd.org/futures/infrastructureto2030/40953164.pdf [cited 15 January 2020].

Ofori, G. and Toor, S.R. (2008). Leadership: A pivotal factor for sustainable development. *Construction Information Quarterly*, 10 (2), pp. 67–72.

Oke, A.E., Aigbavboa, C.O., and Semenya, K. (2017). Energy savings and sustainable construction: Examining the advantages of nanotechnology. *Energy Procedia*, 142, pp. 3839–3843.

Opoku, A., Cruickshank, H., and Ahmed, V. (2015). Organizational leadership role in the delivery of sustainable construction projects in UK. *Built Environment Project and Asset Management*, 5(2), pp. 154–169.

Opoku, A., Cruickshank, H., and Ahmed, V. (2015). Organizational leadership role in the delivery of sustainable construction projects in UK. *Built Environment Project and Asset Management*, 5(2), pp. 154–169.

Osmani, M., Glass, J., and Price, A. (2008). Architects' perspectives on construction waste reduction by design. *Waste management*, 28, pp. 1147–58.

Phadnis, S.S. and Kulshrestha, M. (2012). Evaluation of irrigation efficiencies for water users associations in a major irrigation projects in India by DEA. *Benchmarking: An International Journal*, 19(2), pp. 193–218.

Pisu, M., Hoeller, P., and Joumard, I. (2012). Options for benchmarking infrastructure performance. OECD economics department working papers, no. 956. Available from: https://www.oecd-ilibrary.org/docserver/5k9b7bnbxjwl-en.pdf?expires=1594808430&id=id&accname=guest&checksum=638AFAF262133EFA0AFD7C9CE99BDC2A [cited 12 March 2020].

Presley, A. and Meade, L. (2010). Benchmarking for sustainability: An application to the sustainable construction industry. *Benchmarking: An International Journal*, 17(3), pp. 435–451.

Pryshlakivsky J. and Searcy C. (2013). Sustainable development as a wicked problem. In: S. Kovacic, A. Sousa-Poza (Eds.), *Managing and engineering in complex situations. Topics in safety, risk, reliability and quality*, 21, pp. 109–128. Dordrecht: Springer.

Qureshi, Z. (2016). Meeting the challenge of sustainable infrastructure: The role of public policy. Available from: https://www.brookings.edu/wp-content/uploads/2016/07/Sustainable-Infrastructure-Policy-Paperwebfinal.pdf [cited 25 March 2020].

Rangarajan, K., Long, S., Tobias, A., and Keister, M. (2013). The role of stakeholder engagement in the development of sustainable rail infrastructure systems. *Research in Transportation Business and Management*, 7, pp. 106–113.

Rozenberg, J. and Fay, M. (2019). Beyond the gap: How countries can afford the infrastructure they need while protecting the planet. Available from: https://openknowledge.worldbank.org/handle/10986/31291 [cited 15 November 2019].

Sembcorp (2019). Sembcorp celebrates the official opening of its Sembcorp Myingyan power plant in Myanmar. Available from: https://www.sembcorp.com/en/media/media-releases/energy/2019/march/sembcorp-celebrates-the-official-opening-of-its-sembcorp-myingyan-power-plant-in-myanmar/ [cited 15 January 2020].

Shand, W. (2019). New Climate Economy (NCE), 2020. The finance landscape in Ghana: Mobilising investment in sustainable urban infrastructure. Available from: https://urbantransitions.global/wp-content/uploads/2020/01/The_Finance_Landscape_in_Ghana_final.pdf [cited 4 June 2020].

Sheate, W.R. (2010). The evolving nature of environmental assessment and management: Linking tools to help deliver sustainability. In Tools, Techniques and Approaches to Sustainability. In: W.R. Sheate (Ed.), *Tools, techniques and approaches for sustainability: Collected writings in environmental assessment policy and management*, pp. 1–29. Singapore: World Scientific.

Shi, C., Li, Y., Zhang, J., Li, W., Chong, L., and Xie, Z. (2016). Performance enhancement of recycled concrete aggregate-A review. *Journal of Cleaner Production*, 112, pp. 466–472.

Showers, K. (2012). Grand Inga: Will Africa's mega dam have mega impacts? Available from: https://www.internationalrivers.org/campaigns/grand-inga-dam-dr-congo [cited 15 March 2020].

Shurrab, J., Hussain, M., and Khan, M. (2019). Green and sustainable practices in the construction industry: A confirmatory factor analysis approach. *Engineering, Construction and Architectural Management*, 26(6), pp. 1063–1086.

Spencer Stuart (2015). Building the future leadership insights for the global infrastructure industry. Available from: https://www.spencerstuart.com/-/media/pdf-files/research-and-insight-pdfs/arcl_infratalentsolution_101415_final.pdf [cited 23 December 2019].

Spendolini, M.J. (1992). *The benchmarking book*. New York: Amacom.

Strait Times (2019). Tuas Port to be world's largest fully automated terminal when completed in 2040. Available from https://www.straitstimes.com/singapore/tuas-port-to-be-worlds-largest-fully-automated-terminal-when-completed-in-2040-pm-lee [cited 4 June 2020].

Tabassi, A.A. and Abu Bakar, A.H. (2010). Towards assessing the leadership style and quality of transformational leadership: The case of construction firms of Iran. *Journal of Technology Management in China*, 5(3), pp. 245–258.

Tannenbaum, R. and Schmidt, W.H., (1973). How to choose a leadership pattern. *Harvard Business Review*. May-June, pp. 162–180.

The International Institute for Sustainable Development (2017). Project preparation facility: Enabling local governments access to private finance. Available from: https://www.iisd.org/sites/default/files/publications/project-preparation-facility-government-access-private-finance.pdf [cited 14 May 2020].

The Republic of Rwanda (n.d.). 7 Years Government Programme: National Strategy for Transformation (NST1) 2017–2024. Available from: http://www.minecofin.gov.rw/fileadmin/user_upload/NST1_7YGP_Final.pdf [cited 25 April 2020].

The State Council of the People's Republic of China (2015). Full text: Action Plan on the Belt and Road Initiative. Available from: http://english.www.gov.cn/archive/publications/2015/03/30/content_281475080249035.htm [cited 4 April 2020).

The Transport Infrastructure Efficiency Strategy (2019). One year on report. Available from: https://assets.publishing.service.gov.uk/government/uploads/system/uploads/attachment_data/file/782158/ties-one-year-on-report.pdf [cited 14 March 2020].

Toor, S.R. and Ofori, G. (2006). An antecedental model of leadership development, in: Proceedings of joint international symposium of CIB working commissions W55/W65/W86, Rome, Italy.

Turner, J.R. and Baker, R. (2018). A review of leadership theories: identifying a lack of growth in the HRD leadership domain, *European Journal of Training and Development*, 42 (7/8), pp. 470–498.

UN (1992). United Nations framework convention on climate change. Available from: https://unfccc.int/files/essential_background/background_publications_htmlpdf/application/pdf/conveng.pdf [cited 12 December 2019].

UN (2020). Financing for sustainable development report 2020. Available from: https://www.iau-hesd.net/sites/default/files/documents/fsdr_2020.pdf [cited 27 April 2020].

Ventures Africa (2012). Ventures Africa Eko Atlantic City: A mammoth new development on the coastline of Lagos. Available from: http://venturesafrica.com/eko-atlantic-city-a-mammoth-new-development-on-the-coastline-of-lagos/ [cited 10 May 2020].

Visser, W. and Courtice, P. (2011). Sustainability leadership: Linking theory and practice, *SSRN Working Paper Series. Available from*: https://www.cisl.cam.ac.uk/resources/sustainability-leadership/sustainability-linking-theory-and-practice [cited 25 November 2019].

Warner, J., Jomantas, S., Jones, E., Ansari, M.S., de Vries, L. (2019). The fantasy of the grand Inga Hydroelectric Project on the River Congo. *Water*, 11(3), pp. 1–14.

WEF (2017). China's $900 billion New Silk Road. What you need to know. Available from: https://www.weforum.org/agenda/2017/06/china-new-silk-road-explainer/ [cited 25 April 2020].

WEF (2020). How sustainable infrastructure can aid the post-COVID recovery. Available from: https://www.weforum.org/agenda/2020/04/coronavirus-covid-19-sustainable-infrastructure-investments-aid-recovery/ [cited 4 June 2020].

WEF (2020). How sustainable infrastructure can aid the post-COVID recovery. Available from: https://www.weforum.org/agenda/2020/04/coronavirus-covid-19-sustainable-infrastructure-investments-aid-recovery/ [4 May 2020].

Whitehead, C. (2014). Towards a sustainable infrastructure company. Available from: https://www.icevirtuallibrary.com/doi/pdf/10.1680/ensu.14.00043 [cited 13 November 2019].

Wireman, T. (2004). *Benchmarking best practice in maintenance management*. New York: Industrial Press Inc.

World Bank. (2015a). Kenya's geothermal investments contribute to green energy growth, competitiveness and shared prosperity. Available from: https://www.worldbank.org/en/news/feature/2015/02/23/kenyas-geothermal-investments-contribute-to-green-energy-growth-competitiveness-and-shared-prosperity [cited 15 December 2019].

World Bank (2015b). Living on the edge: Saving West Africa's coastal assets. Available from: https://blogs.worldbank.org/africacan/living-on-the-edge-saving-west-africas-coastal-assets [cited 2 May 2020].

World Bank (2019). Beyond the gap: How countries can afford the infrastructure they need while protecting the planet. Available from: https://openknowledge.worldbank.org/handle/10986/31291 [cited 16 April 2020]

World Bank (2020). Future drivers of growth in Rwanda. Innovation, integration, agglomeration, and competition. Available from: http://documents1.worldbank.org/curated/en/522801541618364833/pdf/Future-Drivers-of-Growth-in-Rwanda-Innovation-Integration-Agglomeration-and-Competition.pdf [cited 1 June 2020].

World Economic Forum. (2015). Africa strategic infrastructure initiative: A principled approach to infrastructure project preparation facilities. Available from: http://www3.weforum.org/docs/WEF_African_Strategic_Infrastructure_Initiative_2015_IPPF_report.pdf [cited 5 May 2020].

World Economic Forum. (2019). The Global Competitiveness Report 2019. Available from: http://www3.weforum.org/docs/WEF_TheGlobalCompetitivenessReport2019.pdf [cited 13 March 2020].

WRAP (2015). At the forefront of the circular economy wrap. Available from: http://www.wrap.org.uk/sites/files/wrap/Bus_Plan_2011_Final_WEB_2.pdf [cited 4 February 2020].

WWF. (2017). Greening China's Belt and Road Initiative in Myanmar: Rapid assessment of opportunities and risks for Myanmar's natural capital from China's Belt and Road Initiative. Available from: https://wwfasia.awsassets.panda.org/downloads/BRI_Final_Digital_090118.pdf [cited 13 January 2020].

Yang, R.L., Huang, C.F., and Kun-Shan Wu, K.S. (2011). The association among project manager's leadership style, teamwork and project success. *International Journal of Project Management*, 29(3), pp. 258–267.

Yates, J.K. (2013). Sustainable methods for waste minimisation in construction. *Construction Innovation*, 13(3), pp. 281–301.

Yeheyis, M., Hewage, K., Alam, M.S., Eskicioglu, C., and Sadiq, R. (2013). An overview of construction and demolition waste management in Canada: A lifecycle analysis approach to sustainability. *Clean Technologies and Environmental Policy*, 15(1), pp. 81–91.

Yoshino, N. and Pontines, V. (2015). The "Highway Effect" on Public Finance: Case of the STAR Highway in the Philippines. ADBI Working Paper 549. Tokyo: Asian Development Bank Institute. Available from: http://www.adb.org/publications/highway-effect-public-financecase-star-highway-philippines/ [cited 13 January 2020].

Yu, A.T.W., Poon, C.S., Wong, A., Yip, R., and Jaillon, L. (2013). Impact of construction waste disposal charging scheme on work practices at construction sites in Hong Kong. *Waste Management*, 33(1), pp. 138–146.

Yukl, G. (1998). *Leadership in organizations*, 4th ed. New Jersey, NY: Prentice Hall.

Yukl, G. (2006). *Leadership in organizations*. New York, NY: Elsevier.

Yusof, N.A., Abidin, N.Z., Zailani, S.H.M., Govindan, K., and Iranmanesh, M. (2016). Linking the environmental practice of construction firms and the environmental behaviour of practitioners in construction projects. *Journal of Cleaner Production*, 121, pp. 64–71.

Zuofa, T. and Ochieng, E.G. (2016). Sustainability in construction project delivery: A study of experienced project managers in Nigeria. *Project Management Journal*, 47(6), 44–55.

5

PLANNING FOR SUSTAINABLE INFRASTRUCTURE DEVELOPMENT DURING MEGA-EVENTS

An Expo 2020 case study

Samih Yehia and Ashly Pinnington

5.1 Introduction

5.1.1 Mega-event and sustainable infrastructure development

Expo is a mega-event defined as "a global event dedicated to finding solutions to fundamental challenges facing humanity by offering a journey inside a chosen theme through engaging and immersive activities" (BIE, 2015). Mega-events have considerable significance in terms of the exchange, transfer and diffusion of information, values and technologies. The infrastructure development and elevation represent the core for hosting such events. The expectation is that hosting an event such as Expo 2020 will promote the economic competitiveness of Dubai and will be a showcase for place marketing. A group of researchers have argued that mega-events share the same features of the reparative nature with an international exposure that can offer fascinating spectacles for a specific period, enhance the globalised built environment, and give the hosting city privilege and legacy (Hiller, 2000; Ritchie, 1984; Roche, 2000). Getz (1997) defined the mega-event as a planned occurrence of limited duration that has a long-term impact on the host area by increasing the tourist volume, the publicity and international exposure, infrastructure development and the organisational development, which in turn increase the destination's capacity and attractiveness. O'Reilly *et al.* (2008) have characterised mega-events as "global properties."

This chapter seeks to understand the practices that help maintain a balance between the three pillars of sustainable development (economic, social, environment) and to discover the success factors that can influence sustainability. Furthermore, as legacy and sustainability have been competing in importance during the last few mega-events, the authors intend to demonstrate how these two concepts could complement each other rather than compete for priority. The rationale for this approach lies in the nature of mega-events, which require a long period of preparation but are implemented within a limited project period. While some people believe that the journey in its entirety has to be sustainable, others hold that the key objective has to be planning for the legacy of hosting these events. The final challenge addressed in this research lies in the overall goal of hosting a mega-event so as to create a positive legacy, and to advance sustainability by using such events not only to elevate the infrastructure but also to do so through a sustainable development process.

This chapter collects and analyses different governmental and private-sector settings and documents. To investigate the issues of transport, and the current capacity and efficiency of the transport system, the study covers various modes of public transport, sea transport, infrastructure plans and airports, roads and metro railways. The construction methods and standards employed are based on comparisons to several international standards, including ISO, LEED, ASTM, EN, and BS. The rationale for analysing current construction practices is that this provides a clear indication of the extent of sustainability and unsustainability in relation to materials used, recycled content and the construction waste management system. The energy and water supply are other major subjects that should be considered specifically in the United Arab Emirates (UAE), which has made significant progress in the production of clean energy and a reduction in dependency on non-renewable energy sources.

The chapter offers an opportunity to advance understanding of the infrastructure development and the sustainability practices across the project life cycle of hosting mega-events, and within different sectors for a non-sporting event. It studies sustainability practices by examining all aspects of hosting the mega-event as an overall programme. The specific focal case studies are generated based on the highest contributing sectors – construction, transport, and utilities.

Davenport and Davenport (2006) emphasised that the infrastructure and transport arrangements usually cause the main ecological threat of having mass tourism during a limited period. The rationale behind this is because such an event will require infrastructure which is not a necessity for the local community after event completion (hotels, buildings, roads, upgrading of the airport terminal for increased air travel, more taxis and cars on the roads, overuse of water resources, sewage facilities, and litter). This pressure on resources can lead to an increase in the carbon footprint, quick deprivation of natural resources, and disruption to the lifestyle of the local community and wider society. Some of those environmental degradation impacts are substantial and often irreversible and may also have social ramifications (Heikkurinen, Clegg, Pinnington, Nicolopoulou and Alcaraz, 2019).

5.1.2 Chapter aim and objectives

This chapter aims to explore the challenges of hosting a sustainable mega-event through infrastructure development and discover how hosting a mega-event can enable construction, utilities and transport to achieve greater sustainability outcome and create long lasting legacy. The main objectives are to:

- Provide a list of recommendations for hosting sustainable mega-events through sustainable infrastructure development that overcome the common shortfalls experienced in past events;
- Examine how to achieve sustainability commitments during the critical stages of event planning;
- Identify the optimal sustainability practices for Dubai during its preparations for hosting its first mega-event;

- Introduce the role of planning and leadership in creating long-lasting legacy;
- Demonstrate the role of subjective and proactive legacy plans in achieving overall sustainability for the infrastructure;
- Appraise the three most critical infrastructure sectors that impact mega-event sustainability.

5.1.3 Learning outcomes

The following learning outcomes have been identified for this chapter. Readers will be able to:

- Identify the key themes of mega-event sustainability;
- Describe the significance of building green major infrastructure development;
- Explain the role of leadership in reducing carbon emissions;
- Exemplify the importance of planning for legacy to achieve sustainable infrastructure; and
- Describe how to integrate green principles within the design phase of a mega project.

5.2 The core area of mega-event sustainability

The mega-event became an international attraction with high participation and willingness shown by the majority of countries involved. Yan (2013) indicated that such an event serves as an instrument in facilitating community-building and fostering urban renewal in order to provide a better quality of life and environment. Government bodies are more frequently using this event as a national or regional development tool as it serves to attract more budgets for infrastructure development, branding and promoting the hosting city or country as a tourist destination, while emphasising environmental considerations and sustainability concerns as an essential cornerstone in tourism (Yan, 2013). However, Pelhan (2011) indicated that the only barrier to the more extensive dissemination of sustainability through the mega-event industry is still the economic barrier.

Deng and Boom (2011) describe the role of hosting Shanghai's Expo 2010 in the infrastructure development of the Huangpu River area, a 5.28-kilometre development with 2.5 million square metres of construction that came as part of the plan known as the Third Riverfront Development triggered by hosting Expo 2010. The development took place along a riverfront of 8.3 kilometres; development included a double-deck pedestrian walkway and convenient access for public transit, which was completed ahead of the Expo opening. Another study by Deng and Poon (2014) indicated that the legacy strategy used in this development was a utilitarian one where, after the event, the exhibition structures were repurposed into public, institutional or commercial use. Hosting Expo 2010 was part of a massive riverfront renewal spanning 75 kilometres of the Huangpu riverfronts initiated in 2002, and it catalysed change for the full area.

5.2.1 Transportation

As the mega-event life cycle includes intensive preparation and infrastructure development ahead of the event as well as major spectators' activities during the event, the transportation of people and goods plays a significant part in the overall programme's sustainability outcomes. The mobility of people and goods is the core of any mega-event. Countries' candidacy for hosting a mega-event will not be accepted without a plan for infrastructure upgrading along with transport strategy on how people and goods are going to be moved. The infrastructure development is frequently cited as one of the most critical motives for the countries or cities to bid to host a mega-event (Guala and Turco, 2009; Hall, 2012; Horne and Manzenreiter, 2006. However, these mobility projects tend to receive less media attention than the construction of stadiums or venues do. Mobility projects have to be tailored based on the need for the event and the adaptability of those projects to serve the local residences after the event. Hiller (2006) indicated that the post-event stage would require a pre-event plan for how to adapt the mega-event infrastructure to the needs of the residents. Malhado *et al.* (2013) argued that urban mobility has the potential to significantly contribute to the economic regeneration of the host cities by attracting national and international investors and encouraging them by providing business environment improvements.

May *et al.* (2017) view the sustainability framework for urban transport against six principles:

- The economic efficiency, as the system has to have a budget and an investor. If so, the project should undertake a positive feasibility study and have a viable business model that will serve as an economic booster and element for investment attraction.
- The liveability of the streets and neighbourhoods.
- The transport system has to respect the environment.
- Equity and social inclusion as the transport system should be able to serve the upper community layers by providing a satisfactory cost-effective transport solution, as well as availability, accessibility, and quality of service.
- Transport system has to be safe and trusted by society.
- The sustainable urban transport system should contribute to economic growth and be self-sustaining.

The European Commission (2013) stakeholders from the European Union (EU) member states worked together to set out the sustainable objectives for a viable mega-event transport included the following objectives:

- **Accessibility:** A sustainable transport system should aim to provide accessibility to all categories of inhabitants, commuters, visitors, and businesses.
- **Health and safety:** A sustainable transport system has to have a limited hazardous impact on health, safety, and security of the citizens. This confidence in the system will encourage the citizens to rely on it.
- **Pollution:** Noise emissions, greenhouse gas emissions, and energy consumption have to be kept at the minimum level with continuous research and studies on how to continue to reduce pollution.
- **Efficiency:** A sustainable transport system should be efficient and cost-effective for the transporting of both goods and passengers, and be reliable in timing and quality of the provided service.
- **Urban development:** A sustainable transport system has to enhance the attractiveness and quality of the urban environment.

5.2.2 Utilities

ISO 2012121 (2012) highlighted the importance of having an energy-saving and sustainable energy supply plan in order to reduce the use of fossil fuels and the resultant environmental impacts. Energy is part of the resource utilisation plan; if it fails, the full sustainable system will be at risk. Under these conditions, the energy supply plan for a mega-event has to be detailed and clear, and it should continue after the closing of the event. The experience generated by hosting a sustainable mega-event through energy production and emission reduction should shape the future practices of the hosting city.

5.2.3 Construction

In order to promote sustainable construction at the programme level, Shi *et al.* (2012) created a checklist that should be taken into consideration. This checklist includes many essential questions: whether the programme has an agreed definition of the "sustainable construction goals," whether the "sustainable construction department" is set at the programme level with a clear function and role, and whether the programme has an innovative and sustainable "land use plan." Furthermore, many other elements in the sustainable construction plan also have to be engaged: the "water and energy saving design guidelines," the "reduction on environmental loadings plan," and the "legacy plan." In addition, the "performance-tuning design" will be considered as extra support for the construction sustainability plan along with the construction usage of water and electricity plans. In the sustainable construction plan, the programme should have a material saving and reuse plan along with a plan for sustainable transportation. At the later stages, the programme should have an environment, water-saving and energy-saving plan for the operation phase. Lastly, with the continuous improvement process in sustainability, the sustainable construction programme should have a flexible performance-tuning plan along with an energy-saving plan for the demolition process and waste management.

Shi *et al.* (2012) claimed that the sustainability theme in Expo 2010 featured significant considerations of sustainable construction. All of the buildings in the pavilion were prefabricated; the work was completed with a zero waste approach; the walls and roofs were insulated with 170 mm polymeric foam materials; all the windows were triple glazed; the construction cost per square metre was very competitive; high-level air filtering equipment was installed; the buildings were constructed with 30,000 m^2 of solar photo-voltaic panels able to generate 2.8 MW of electricity per year; and vertical greening systems were incorporated to save natural resources and minimise the volume of environmental loadings, while still providing a comfortable place for living and working. Similarly, Hult (2013) highlighted that Expo 2010 represented a significant node in the network of transnational imaginaries of a sustainable urban future. Construction for a mega-event has multiple benefits for the local construction communities as it will attract multi-national expertise, improve the local workforce by working to comply with international standards, and solve the problems of some of the social challenges.

Comparably, Deng and Boom (2011) used the preparation period for Expo 2010 to explain the pressure in construction during the peak period preparation in August 2006, which was 1,350 days away from the opening day in May 2010. They described the situation by stating that the number of construction crews at that time was around 50,000. Many projects were undertaken synchronously like laying the new subway lines, the expansion of the international airport terminals, the development of train stations, and the renewing of the 5.28 square kilometres of the Huangpu River on both banks. Considering the scale, the nature and the diversity in projects and the schedule, the organiser was in a real race against the "ticking clock" while many critical decisions were made on what to preserve, dismantle, reuse, and construct (Mir and Pinnington, 2014; Rees-Caldwell and Pinnington, 2013). Managing such a programme is always a primary challenge to hosting a mega-event. Furthermore, Shi *et al.* (2012) proposed that sustainable construction should also address issues like land-use planning, energy conversation, reduction of environmental loading, high quality of service design, and performance-tuning design. Some previous mega-events followed these considerations to some extent; the 1992 Olympic Games in Barcelona were a chance for environmental recovery of a declining industrial area when 5.2 kilometres of coastal area were regenerated. Sydney's Olympic Park was built in a formerly derelict industrial area full of toxic waste and transformed to a major sporting and recreational centre of the city (Iraldo *et al.,* 2014).

5.3 Building green major infrastructure development

The objective of the construction sector is to provide a reliable, efficient, and sustainable sector that can supply the UAE with the required facilities like infrastructure, hotels, buildings, villas, offices, and public transport at internationally accepted standards. Expo 2020 construction site and services building is part of the construction process to host this event. The application of the latest green design requirements was followed. Staples (2018) reported that the Sustainability Pavilion, one of the main buildings in the Expo 2020 site, is designed to generate 22,000 litres of water per day and 4 GWh of electricity per year, which exceeds the building requirement to operate. In addition to that, Expo management aimed to ensure that all the permanent buildings at the site would receive Leadership in Energy and Environmental Design (LEED) Gold Standards, which were developed by the United States Green Building Council. In the legacy phase, this specific building is going to be recycled as a centre for science and for children. The design of this pavilion aimed to achieve LEED platinum; therefore, the bidding contractors for this pavilion were given a set of sustainable strategies that their bids must adhere to.

Dubai leadership extended the development of the 4.38 square kilometers site of District 2020 into a more comprehensive national development of Dubai South and linked this new mega-development via the extension of Dubai Metro and four major UAE highways. By doing so, Dubai is planning to transform this area into a new national business facility and a new Dubai Exhibition Centre. In 2018, the workers on site exceeded 15,000 and logged in over 16 million hours of work; however, at the peak the construction workers are expected to be around 35,000 as reported by Badam (2018). Dubai is executing this mega-project based on the same model as other mega-projects completed earlier.

5.4 Role of leadership in reducing carbon emissions

Mega-events have the power to spread messages that inspire people to respond to challenges that are common to humanity: the environment, racism, hunger, discrimination, global warming, poverty, pandemics, violence and many more. As mega-events are being hosted in countries with many social and economic issues, having a cause that will serve the hosting destination in particular and the world in general becomes a necessity. In some cases, those messages didn't achieve set goals. However, it still represents a starting point to discuss contemporary challenges

This chapter endeavours to present the situation of Dubai and the UAE through the development plan and plans for sustainability and assess how Dubai was progressing with or without the mega-event. A significant point in this development plan was the Dubai Strategic Vision 2015, announced on 3 February 2007, almost five years before Dubai secured Expo 2020 and three years before bidding for the event. The decision-makers in the UAE understood that this wouldl require the implementation of international standards and best practices in the details of the residents' lives, work, institutions, and society. The ruler of Dubai, H.H. Sheikh Mohammed, announced this vision under the theme "Dubai ... where the future begins." Vision 2015 from Dubai Strategic plan (2015) was a re-envisioning of the Vision 2010 plan as Sheikh Mohammed stated that in 2015 "we achieved the plan set for 2010" and Vision 2015 was a revision of Vision 2010 by following the same logic and taking into account the global financial crisis. H.H. Sheikh Mohammed said:

> In 2000, the plan was to increase GDP to $30 billion by 2010. This figure was exceeded in 2005, with GDP reaching $37 billion. The plan also included an increase of income per capita to $23,000 by the year 2010. In 2005 the average income per capita reached

$31,000. In other words, in five years we exceeded the economic targets that were originally planned for a 10-year period.

Vision 2015 included many initiatives in order to ensure sustainable development of the infrastructure and environment sectors. The plan sets out four strategic objectives:

1. **Sustainable urban development:** The construction sector in the UAE is the spine of the country's development. The plan calls for strategic urban planning that optimises the use of the land in order to preserve the natural resources while developing. This involves comprehensive and integrated planning of the elements of urban development. In addition, this planning promotes the policies for providing affordable houses for the local Emiratis, ensures public services and facilities for growth, and plans to upgrade and ensure enforcement of the existing labour housing policies.
2. **Sustainable energy, electricity, and water supply:** The strategic plan aims to develop an innovative framework to integrate policy, secure a sustainable supply, and implement initiatives in order to manage the demand.
3. **Sustainable transport:** The Vision aims to provide an integrated road and transportation system that helps people and goods movements while improving the safety levels for all the system users. In order to achieve that, the Vision aims to address the congestion problems, accommodate future needs, increase the share of public transport, implement initiatives to reduce private vehicles, increase road capacity and road network systems, manage the demand, consolidate accident and emergency management, and improve drivers' behaviour.
4. **Sustainable environment:** The Vision sets environmental targets that aim to ensure a safe and clean environment which involves aligning environmental regulations with international standards, developing and applying the enforcement mechanisms, integrating environment-related issues with the development policies and programmes, and raising the environmental awareness level.

5.5 Importance of planning for legacy to achieve sustainable infrastructure

Since hosting a mega-event requires upgrading the infrastructure of the hosting city or cities to meet the requirement of the mega-event awarding committee, the development often will include "industrial relocation," "urban development," "infrastructure renovation" and "inward investment" as expounded by Roche (1994). The process of achieving these goals involves significant development activity, which can lead to decrease unemployment. By reducing unemployment, the purchasing power of residents increases and will affect many other sectors, even those that have no social or economic relation with the mega-event. Through application of a random growth model, Hotchkiss *et al.* (2003) demonstrated the positive impact derived from hosting a mega-event on the employment level, while Mason and Paggiaro (2012) highlighted the role of positive emotion during the mega-event in delivering a high-quality mega-event with long-lasting impact on behavioural intentions and visitors' satisfaction.

Coates and Humphreys (2003) consider that being concerned with just the direct financial benefits of hosting a mega-event is a case where the economic benefits should be considered weak at best. The investment in hosting the mega-event should target the triggering of a development plan, attract international investment, and generate an appropriate investment environment. Those elements will speed up the development cycles and inject cash flow in the market mechanism. Cornelissen *et al.* (2011) stated, "[The] mega-event served to speed up developments. Direct economic effects of the mega-event are represented by the domestic and foreign investment triggered by the event, the infrastructure work related to the event, and the new income generated from spectators and participants." Cornelissen *et al.* (2011) reflected how the mega-event can generate income through the development of event venues, from increases in government tax income, and by the growth stimulated in ancillary sectors like leisure consumption, new tourist attractions, and unique construction to serve this new sector. Cornelissen and colleagues further contend that a newly constructed "iconic" building can become a landmark and a part of a city's character, which will enhance the overall image of the city. In addition, such an iconic building can become an "aesthetic focal point" functioning as an incentive for continuous urban development and entertainment facilities.

The idea of the mega-event legacy emerged after critics debated the appropriateness of governments approving large amounts of spending and consumption for hosting mega-events. The detractors in these debates often argue that investment is diverted from the many cities and towns which will not be part of the mega-event, and that this affects the overall concept of sustainable development. Use of this terminology of sustainability and legacy creates a larger framework for the overall outcomes of the mega-event and can reduce criticism of the event.

5.6 Integrating green principles within the awarding system for hosting sustainable mega-event

The mega-event opening day is the deadline of the mega-infrastructure and the development plan which is mainly initiated seven years ahead of the event. This process is highly resource-consuming, requires substantial investment in several fields, and brings significant risks that hosting cities may not be able to accommodate. As this chapter focuses on the sustainability and legacy of hosting such events, it examines the practices that took place in Dubai after winning the bid for Expo 2020.

The World Summit on Sustainable Development (WSSD) raised the issue of sustainability in the transport sector since the United Nations (UN) considers transport as a critical sector for sustainable development (UNCSD 2012). Pitts and Liao (2009) stated that hosting the Olympic Games leads to massive consumption of resources and energy which mainly comes as a result of travel. Even so, transport is not the only concern in sustainable environment practices; Ma *et al.* (2011) consider the sustainable environment in mega-events as a holistic system that has to consider the protection of natural resources and cultural heritage, building facilities and infrastructure development, energy production, saving and consumption sustainable techniques, water, wastewater, and waste management methods. All these points and more should be taken into consideration by every successful environmental sustainability plan. However, it is widely debated that achieving sustainability is actually not commercially viable as it costs the hosting city much more to achieve it.

5.7 Applying green considerations within the design stage

The 2012 Olympic Games in London was the first mega-event endowed with certification event sustainability (CES) while the first Expo with CES was Milan's Expo 2015 (Guizzardi & Prayag 2017). This Expo obtained CES because of its environmental management system (EMS), which complied with ISO 20121 (Guizzardi *et al.,* 2017). Getz and Page (2016) stated that despite the fact that literature on sustainability in the mega-event has grown exponentially over the past 25 years, a review of 85 event-impact studies shows the increasing importance of mega-events in our society with only a minimal focus on the environmental impacts of those events. A review by Mair and Whitford (2013) proposed that the literature of mega-events should focus on three main topics: event impacts, the link between events and tourism, and the event types and definitions. They concluded that the socio-cultural and environmental impacts of hosting mega-events are under-researched in terms of economic impact. Henderson (2011) indicates that the term "sustainable" is currently considered alongside many other similar terms like "greening," "environmentally friendly," "corporate social responsibility," "ecology" and "eco-friendly." These terms are used interchangeably with "sustainable" when taking a stand against the abuse of our surroundings in the pursuit of commercial activity (Al-Reyaysa, Pinnington, Karatas-Ozkan, and Nicolopoulou, 2019).

Achieving a sustainable mega-event is a challenging goal that requires interdisciplinary action that has to be constructed on multiple perspectives and employ interrelated science. Pelhan (2011) clarified that hosting a sustainable mega-event will require responsible sourcing, solid waste management strategies, event legacy projects and a vision to use the mega-event as a catalyst for behaviour change. Dodouras and James (2004) suggested that the path to achieving the set sustainability agenda requires strong integration on both horizontal (cooperation between sectors) and vertical (cooperation between levels) dimensions. Sustainability as a mutual goal requires the collaborative engagement of project stakeholders. Dodouras and James (2004) argued that the challenges of sustainability should be considered in both the long term and the short term. The significant challenges to achieving such a goal commence with changes in attitudes and practices in the hosting society, and shift how economies are operated.

Iraldo *et al.* (2014) remarked that Expo 2015 in Milan was the world's first mega-event with a design deemed to meet ISO 210121 compliance with the event sustainability management system. This Expo also conforms to the European Regulation n.1221/2009/EC for objective input. Its theme, Feeding the Planet, Energy for Life, considered the methods that humanity will use to feed itself and the planet, which includes research and joint efforts to share sustainable models of production and consumption, following a multidisciplinary approach.

Dubai's plan for its Expo legacy is backed by strong connections with leading local companies like Emaar Properties, the Dubai World Trade Centre (DWTC), the Emirates airline, Dubai Ports (DP), and many more successful local companies. In 2015, Emaar Properties launched a global competition among 13 leading international architects to design the theme park. Fahy (2016) reported that Bjarke Ingels Group (BIG) who created the design for the Opportunity Pavilion had a design philosophy that reflects a "belief that contemporary urban life is a result of the confluence of cultural exchange, global economic trends and communication technologies." Deulgaonkar (2016) stated that BIG is known for its innovative approach to architecture and they are working on many master plans internationally including the new headquarters for Google. Foster and Partners, an international design firm based in London, won the Mobility Pavilion contract, and Grimshaw Architects won the Sustainability Pavilion contract. Those three pavilions will be the centrepieces of the two-square-kilometre Expo site; this will surround the central Al Wasl Plaza, which represents the heart of Expo 2020. Hellmuth, Obata + Kassabaum (HOK) architects designed the master plan of the Expo 2020 site (Deulgaonkar, 2016). However, by recruiting all those international companies, Al Sammarae (2019) indicated that all the major pavilions are designed by leading international companies that have swept the boards and won the design competitions among all the local and regional architects. Local architects, sadly, will not have the legacy of designing any major part of the Expo 2020, and this is a missed opportunity.

5.8 Sustainability of the mega-event life cycle

In order to consider "the sustainable" viewpoint, any business model, including the hosting of mega-events, should consider the economic perspective that will achieve the required sustainability level without imposing a high cost or affecting the quality of services provided (Marchet *et al.*, 2013). For this reason, debates on sustainable development often return to the Sustainable Development Triple Bottom Line (SDTBL), coined first by John Elkington in 1999 and adopted by a large number of authors (e.g., Gibson, 2001; Pope *et al.*, 2004; Sherwood *et al.*, 2005). The rationale for the SDTBL comes from the arguments that each business model should prepare a tripartite bottom-line strategy. The profit bottom line represents the *profit and loss* account of the business model; the *people* component, the second, measures how socially responsible the business model is, and the third is the *planet* account, which measures how environmentally sustainable the business model is. Carter and Rogers (2008) indicate that the sustainability "three-pillar" model is what will secure long-term economic viability of the sustainability system and will in turn become part of ordinary daily practices and routines of stakeholders instead of having to be an imposed system. Through the SDTBL, mega-event organisers are more likely to secure the achievement of sustainability targets in different areas, including transport, and these targets will have a long-lasting impact.

Dealing with the challenges of delivering sustainable mega-events should always take into consideration the stakeholders' complexity along with the readiness of the hosting city to pursue sustainability. The plan for achieving sustainability targets should be generated based on the existing sustainability practices within the relevant countries before adopting any system from other countries. Furthermore, additional vital practices that should be considered to prevent making the same mistake involve learning from previous case studies explaining the challenges they faced in achieving the sustainability targets and establishing an avoidance plan. It is widely known that achieving sustainability targets in programme management may lead to further spending; that is why wise programme management should identify the critical behaviours that have a significant impact on the overall sustainability of the mega-events and should invest in reducing identified polluters for positive long-lasting impact. The reason for these deliberations is due to the risk of high spending on the event simply for the sake of showing the world that the hosting city complies with general expectations; yet those acts could conversely reduce the competitiveness of the city, resulting in a backlash after the closing of the mega-event.

Dodouras and James (2004) found that the positive environmental impact of the mega-event from the construction perspective is by making the sector greener, initiating infrastructure projects in the public transport sector, upgrading the water and sewage services, and building new airport terminals. Preuss (2013) asserted that hosting a mega-event is one significant opportunity for the hosting city to develop a green economy through signalling sustainability considerations. Henderson (2011) proposed that the concept of offsetting environmental impact emerged as a result of notional ideas suggesting that the choice of one progressive action may compensate for the action of another destructive one. Henderson further stated that a good example of this will be the reduction of the CO_2 from the atmosphere to offset that emitted by travel. Some of the most used methods to offset CO_2 are through the construction of large solar panel farms, replacing fossil fuel power plants that burn carbon fuels such as coal or oil with natural gas, initiating the internal process to become part of the international programme of certified emission reductions (CERs) credits, and constructing more green buildings.

5.9 Chapter discussion questions for further reflection

Hosting a mega-event is always accompanied by significant infrastructure projects that require resource management and intense planning, particularly if the construction is going to be sustainable. The construction will take place concurrently in multiple locations – including roads,

Planning for sustainable infrastructure development during mega-events

airports, metro lines, utilities supply, hotels, and venues. Such construction conditions require compromises in decisions as sustainable construction is always challenging. This chapter identified many of the practices that are considered vital for sustainable mega-event infrastructure development:

1. Sustainable green design
2. Economic efficiency
3. Respect for environment
4. Accepted by society
5. Reliability
6. Design sustainability of the infrastructure lasts after the event completion

Bourdeau (1999) argued that sustainable construction is the approach that the construction industry uses to respond to sustainable development requirements. Kibert (2016) indicated that "sustainable construction" has been applied interchangeably with other related terms like "green" or "high-performance" construction. Hill and Bowen (1997) proposed that sustainable construction contains four pillars: social, economic, biophysical, and technical. In addition to the benefits listed earlier, building these projects through green processes significantly affects the future of the country, common norms in construction industry practice, and sets a better example for the future generations.

Preuss (2013) explained that any mega-event motivates people to consider new approaches to how it is run. This chapter identified a number of the success factors for sustainable mega-event infrastructure development. It has highlighted the importance of planning for sustainability and legacy during the design stage. Balancing the importance of the sustainable development pillars and developing the infrastructure with leadership vision are key elements to achieve in sustainable infrastructure development.

Infrastructure development in Dubai is closely linked to the economic growth of the city along with the capacity to attract significant numbers of international investors. Hosting a mega-event can have a triggering effect on attracting financial investment. The chapter has provided a blueprint for creating a sustainable legacy and sets an example for planning for legacy in urban developments. District 2020 demonstrates how Dubai is attempting to balance sustainability with legacy; in as much as Dubai's leadership wants to build this project in a sustainable way, the legacy considerations are what really drive the current motive of building venues that will remain long after the event has occurred. The sales of the units inside the project are ongoing and people who have purchased them expect to take ownership of their units immediately after the event. Dubai public and private sectors are continually accumulating experience in planning, marketing, constructing, and delivering mega-projects. The rapid increase in the resident population in Dubai during the past 30 years was mentioned earlier.

This chapter has attempted to advance knowledge and understanding of sustainable infrastructure development and set a course for achieving sustainability while hosting a mega-event. This mega-event project is part of Dubai's portfolio of mega-events and, despite a recent slowdown in the financing of many construction projects, Dubai is dealing with the impacts of the pandemic on this rescheduled mega-event. This chapter has elaborated on the idea of awarding mega-events only to cities and countries that have an overall development plan such that the mega-event can be integrated within the portfolios of many other projects. Previous experience in hosting destinations implies that any country or city bidding for a mega-event should have an ongoing development plan with a secured budget of at least four times the cost of hosting the event. Possession of substantial financial resources for an overall development plan reduces the risk of hosting a mega-event.

Mega-events can become catalysts for change, including in the voluntary sector of communities. The culture of volunteering for jobs such as crowd management and event guides was not a common norm within Emirati society. However, the mega-event is bringing this to life and encouraging it in a conservative Muslim society. Such change in social practices is happening in many areas of construction, utilities and transport, including electrical cars, solar power generation, construction regulation, public transport, and innovation in air and sea transport.

1. What 10 practices do you consider most important for sustainable mega-event infrastructure development, in any two of the following:
    a. Sustainable green design
    b. Economic efficiency
    c. Respect for environment
    d. Accepted by society
    e. Reliability
    f. Design sustainability of the infrastructure lasts after the event completion
2. List three to five success factors for sustainable mega-event infrastructure development in each of the following:
    a. Construction
    b. Utilities
    c. Transport

1. Write a two-page brief for government officials working in a city planning authority on "Legacy Planning in Urban Development - City Mega-Events."
2. Select three to five areas of technological change and innovation and explain how they will influence sustainable infrastructure development by 2030.

5.10 Expo 2020 case study

Expo 2020 is following the strategy of the utilitarian legacy plan by repurposing the exhibition's structures for public, institutional, or commercial use afterwards (Deng and Poon, 2014). Expo 2020 announced its clear legacy plan in 2017. However, Deng and Poon (2014) contest that, based on the experiences from previous Expos (Seville, Lisbon, Hannover, and Zaragoza), the immediate transformation of the whole site into the legacy plan was virtually impossible. The main obstacle behind achieving this was due to the inevitable retrofitting process and the lag period between the merging of the built legacies into the established urban fabric and social system function. Emaar Properties work on avoiding this by planning many huge property developments around the Expo site called Dubai South, an infrastructure development, and District 2020.

District 2020 is the official name of the Expo site after the completion of Expo. It will retain Expo 2020's conference and exhibition centre through the development of the facilities into a new exhibition centre being developed by Dubai World Trade Center (DWTC) Limited Liabilities Company, the significant regional exhibition company. Design Mena (2016) indicates that the Sustainability Pavilion will be transformed into a children's science centre. District 2020 will feature a diverse selection of academic institutions, museums, and galleries. In addition, 135,000 square metres will be transformed into commercial space with flexible spaces ranging

Planning for sustainable infrastructure development during mega-events

from hot desks and co-working spaces for small and medium enterprises into multi-building companies for large corporations. The 65,000 square metres will be for residential use, another 45,900 square metres will serve as parkland and 10 kilometres will be set aside for cycling tracks (Morgan, 2017). The location of the site will encourage people to live in this area; it is located in between four major highways, neighbours Al Maktoum International Airport and is close to the upcoming Star Mall project.

The District 2020 legacy plan of Expo 2020 will include walkways and biking paths to invite people to be mobile and explore the city on foot or by bicycle. The telecommunication infrastructure will include smart services that embrace the latest available technologies, which will enable both a seamless virtual and physical experience. The dedicated metro station will be part of project named "Route 2020" for Dubai Metro line with a promise to have the transition as soon as Expo draws to a close (Shahbandari, 2017). Al Wasl Plaza will hold shows and concerts while also providing a relaxing space for people. District 2020 is scheduled to commence on 11 April 2020 right after the completion of Expo 2020 as indicated by District 2020 (2017). Morgan (2017) indicates that District 2020 will reuse 80 percent of the Expo 2020's site infrastructure through Dubai's legacy strategy; also worthy of note is that all these buildings will meet or exceed LEED Gold Standards, an international prestigious certificate for green buildings.

The legacy considerations in Expo 2020 reflect how far Dubai is planned to extend sustainable benefits well beyond the value of hosting the mega-event as an event. The Expo 2020 is embedded within the total development plan of the city and many mega-projects are under construction. Those mega-projects are scattered over many sectors and include road infrastructure, residential buildings, airports, new communities, metro, and much more. Crossing next to the project site of Expo 2020 was a repetitive experience to see the impact that this project will have. District 2020 is a reflection of how Dubai is attempting to balance sustainability with legacy; in as much as Dubai's leadership wants to build this project in a sustainable way, the legacy considerations are what really drive the current motive of building venues that will remain long after the event. Due to the large number of ambitious construction and infrastructure initiatives, the UAE public and private sectors are continually accumulating experience in planning, marketing, constructing, and delivering projects (Abdalla and Pinnington, 2012; Yehia, 2020).

5.10.1 *Sustainable construction*

The practices and regulations that were repeated by the participants and collected from the secondary data in how Dubai regularised the construction sector through the Dubai Strategic Plan 2015 were classified. This plan aimed to make the green building practices mandatory in order to adapt the best environment-friendly international standards, to keep Dubai as a lively city, to maintain a sustainable development process, and to achieve a clean pollution-free environment. Those practices and regulation are categorised as follows:

- Resource effectiveness: energy
- Resource effectiveness: water
- Accessibility and indoor environmental quality
- Ecology and planning
- Management
- Transport
- Pollution
- Waste management
- Construction materials

Dubai's leadership aims to create and maintain buildings that perform highly in the challenging environment of the Gulf Cooperation Council (GCC) countries. The discussion on water and energy resource effectiveness and transportation are going to be kept for the second and third case studies. By having a vision to enhance public health and enhance the urban planning to be more sustainable, Dubai adopted those rules in order to create an excellent city that provides high standards of success and sustainable supply for the needs of the existing population and future generations.

1. **Accessibility and indoor environmental quality**: The fresh air supply in the ventilated or air-conditioned spaces was observed in all of the construction sites visited before and during the study. The social pillar of sustainability is respected in most of the designs. The optimal rate is to have 12 litres per person per second if smoking is banned. In case smoking is permitted, the fresh air supply rate should be 32 litres per person per second. The research also showed that many new requirements applied in the last few years include building compliance with volatile organic compound (VOC), which was defined by the green building regulation to be below 300 micrograms/mt^2, the formaldehyde to be below 0.08 parts per million, and the suspended particulates to be below 150 micrograms/mt^2. Those requirements have to be achieved and tested through a certified air testing company accredited by the Dubai Municipality before occupation of the building. In addition, it is common in all the buildings in Dubai to pay service charges for maintenance of common areas and security. Those companies are setting inspection programmes for all types of machinery, which reduces the cost of energy operation and increases the lifetime of the machinery. The green building regulation targets a thermal comfort level of temperature to be between 22.5 and 25.5 degrees Celsius while humidity is to be 30–60 percent. All the contractors should ensure that the heating, ventilation, and air conditions system (HVAC) should be able to achieve the above figures for at least 95 percent of the year. By this, the Dubai construction practices are respecting the social pillar of sustainability, being economically viable and still respecting the environmental targets in reducing the carbon footprint. In addition to this, Expo 2020 is designed to be a pedestrian-friendly zone.
2. **Ecology and planning**: The study found that a regulation from the Dubai government is that 25 percent of the planted area of a building or villa should utilise plants and trees that can adapt to Dubai's climate.
3. **Management:** Dubai Expo 2020 site is at present one of the largest construction projects in the country. With 190 countries that have confirmed their participation in the event, thousands of workers (40,000 workers during peak construction) are building the structure for the 4.38 km^2 site that will be transformed into District 2020. Expo 2020 appointed consultants to monitor sustainability goals and achievements and report the results globally in order to ensure honesty in achieving the commitments. On the other side, technology in construction was well observed as well. The Building Information Modelling (BIM) process is an intelligent 3D model-based process that helps architecture, engineering, and construction (AEC) professionals to gain tools for a more efficient plan and design and to construct and manage buildings and infrastructure. BIM helps in producing a digital representation for the complete physical and functional characteristics of a built asset.
4. **Pollution:** In 2017, the municipality of Dubait launched Air Quality Strategy 2012–2017, with a budget of Dh500 million, aiming to make Dubai among the world's best air quality cities through different initiatives and air quality monitoring (Gulf News, 2017).

5. **Waste management:** The Expo 2020 construction site diverted more than 370,000 tonnes of constructed waste generated through the construction process from landfill to the municipality of Dubai and Bee'ah (Sharjah-based company) in order to recycle the construction materials. Dubai currently has a plan to divert 50 percent of the waste going to landfills into waste segregation plants that are being built in Al Bayada and Al Gusais. These plants started operations in 2018 according to Saseendran (2018), who reported that the plants have contracts for selling recyclables inside and outside the country. The two plants together will have the processing capacity of between 3,000 and 5,000 tonnes of domestic waste daily. By 2020, Dubai aims to divert more than 50 percent of the waste that would normally go to landfill sites to the waste segregation plants. The balance of the waste will be used in another plant that is being developed under the technology of waste-to-energy plants called "Wastenizer." The project will have the capacity to treat 5,000 tonnes of solid waste, which will generate 185 MW of electricity through innovative, effective self-ignition catalyst e-stones driven by renewable energy. By having three plants in operation, Dubai will be able to divert 75 percent of total waste from landfills within the next 3 years into those two types of waste treatments. However, it is important to state that Dubai plans to steer 100 percent of the city's waste away from landfills by 2030.

The construction part of hosting a mega-event is one of the pressuring sectors in the sustainability model of any country. Going through this case study showed that Dubai is managing this part correctly. Dubai has learned significantly from the previous failures and successes of mega-events. Dubai learned how to face the criticism of high spending and justified spending billions in order to host a time-limited event by showcasing the legacy part of the Expo 2020 through a new city called District 2020. The selected location is in an empty area of desert located between Dubai and Abu Dhabi, which is on the opposite side of the typical development areas between Dubai and Sharjah. Dubai is also getting benefits from this hub in order to develop the area surrounding the city through new significant developments like Dubai South, Al Maktoum Airport, Dubai Industrial City, and Dubai Logistics City, among other initiatives. Dubai already has a strong strategic development plan; inserting a mega-event within this development plan makes it more viable, attractive, and profitable. Expo 2020 became a catalyst of change for Dubai South, and this part has become the most attractive location for investment in Dubai. Dubai is developing projects at values that exceed Expo 2020 costs at the same time. This is a significant reason why Dubai was able to accommodate the spending on this project within the UAE budget without disturbing the development of the country or putting pressure on its financial resources. The UAE government is managing the wealth of the country in a proper way through the federal and local governments.

5.10.2 Sustainable utilities

Expo 2020 is forming a platform to encourage creativity, innovation and collaboration in the three main areas of the sub-themes of this event – opportunity, mobility, and sustainability. However, hosting a mega-event for six months places tremendous pressure on the utilities for the hosting destination, which requires a proper legacy plan to benefit from this increase in development for the future generation. The challenge of meeting this extra capacity may require building what the country does not need afterwards, so the development plan should only be set out to serve the event. The sole supplier of water and electricity in the Emirate of Dubai is the Dubai Water and Electricity Authority (DEWA), an entity owned entirely by the government. The DEWA owns, operates, and maintains power stations and desalination plants, aquifers, power and water transmission lines, and power and water distribution networks in Dubai. This represents a massive job for one entity; failure to be sustainable will not only affect the utility supply or Dubai but will affect the sustainability model of the whole UAE.

DEWA aims to be among the best utilities provider through efficiency, reliability, green economy and sustainability. The clean energy will provide seven percent of Dubai's total power output from clean energy by 2020, and this is expected to reach 25 percent in 2030 and 75 percent in 2050. Currently, the DEWA power station and water desalination stations are mainly fuelled by natural gas. DEWA's plans to reduce the consumption on the demand side are being implemented before Expo 2020. DEWA aims to reduce the power and water consumption by 30 percent by 2030 through the following eight programmes, as set by Al Tayer:

1. Building regulations: It was discussed in the previous case study about how the municipality's regulations are being implemented to have more green buildings through sustainable designs, materials selection and sustainability considerations.
2. Building retrofits: DEWA established a demand side management company called Etihad ESCO. The role of this company is to support the improvement of energy efficiency in over 30,000 existing buildings in Dubai by retrofitting them. This plan's value and accumulative costs are estimated at US $8.1 billion up to 2030 while the present value of saving is estimated at US $22.3 billion, which should leave the DSM plan with a positive net economic impact of net present value of US $14.1 billion.
3. District central cooling: DEWA owns 70 percent of the Emirates Central Cooling System Corporation (EMPOWER), a major provider of district cooling services in the region.
4. Water reuse.
5. Efficient irrigation.
6. Specifications of energy efficiency: Selecting the construction materials that will reduce the water and electricity consumption.
7. Outdoor lighting.
8. Shams Dubai initiatives to install solar panels on houses and buildings.

The sustainable supply and production of utilities is a challenging task that requires a vision, leadership, and collaboration between different entities. DEWA and Expo 2020 discovered this at an early stage and decided to partner together in order to provide this global event with electricity and water during the six months of the event, excluding the electricity generated on site. The power supply of the event is going to be produced in Mohammed bin Rashid Al Maktoum Solar Park, the largest single-site solar park in the world, which intends to generate 5,000 megawatts (MW) of clean energy by 2030. This park is enabling DEWA and Dubai to meet the increase in Dubai's electricity requirements in a sustainable way and continue the improvement of the electricity infrastructure. Expo 2020 management is looking to have this event as a showcase for the possibilities in renewable energy by supplying 50 percent of the required power through different renewable energy sources. Those sources include the Sustainability Pavilion (see the construction case study). The investment in such infrastructure is putting pressure on the funding sources, with an investment of US $1.16 billion allocated by DEWA for infrastructure projects to support Expo 2020.

The efforts of DEWA and Dubai leadership in achieving sustainable supply was reflected by the Dubai Declaration of 2017, which included the launching of Mohammed bin Rashid Al Maktoum Solar Park. It represents a concrete example of the efforts that are being taken in order to achieve sustainable utilities. It is the largest single-site solar park in the world based on the

Planning for sustainable infrastructure development during mega-events

Independent Power Producer (IPP) model. The park plans to have a total capacity of 1,000 MW by 2020 and 5,000 MW by 2030. It was very impressive to learn how the UAE built such an environment for investing in solar energy much faster than most of the countries in the region.

The water supply represents a big challenge in the UAE with most of the water used coming from the desalination of the Arabian Gulf seawater through Jebel Ali Power and Desalination Complex. DEWA is committed to maintaining water quality for the stakeholders and the marine water resources. However, the quality of the seawater intake can be interrupted through the rise of seawater temperatures, oil spills, algal blooms, seasonal seaweeds, and high turbidity due to industrial development, as of the DEWA Sustainability Report (2015). If the water intake is of low quality, the cost of production and the amount of energy required increase. This requires proper management and continuous monitoring systems which DEWA processes. DEWA currently produces 470 million Imperial gallons per day (MIGD) for 750,596 customers.

The three main desalination processes are (i) multi-stage flashing (MSF), (ii) the multi-effect desalination process (MED), and reverse osmosis desalination (RO). DEWA initially depended on the MSF technology with a limited portion to come through the RO technology. In March 2018, DEWA awarded a contract of US $237 million for a seawater reverse osmosis-based desalination plant in Jebel Ali. The plants will be commissioned by May 2020 in order to meet the reserve margin criterion for the peak water demand in Expo 2020 and beyond. The new plant is based on RO, which requires 90 percent less energy than the MSF does. The power for this station is going to come from the MBR Solar Park, which will help reduce the footprint of the desalination process and contribute to Dubai's Cleaner Energy Strategy 2050.

By having this RO plant in progress by 2020, the carbon footprint of the water desalination process will be significantly reduced and DEWA will present a showcase about the ability to adapt to the weather of the UAE in supplying enough water of a high quality without affecting the environmental sustainability pillars. In order to maintain the sustainability system in Dubai for the utilities supplies, the study adopted the overall considerations of sustainability from DEWA in order to create the following sustainability checklist:

1. **Economic aspects**

    ◆ Availability and reliability of water and electricity
    ◆ Economic and market presence
    ◆ System efficiency
    ◆ Programmes for demand-side management
    ◆ Applying best procurement practices

2. **Social aspects**

    ◆ Health and safety observance for all stakeholders
    ◆ Stakeholder happiness
    ◆ Emergency plans and readiness for disaster
    ◆ Continuous training and education for different stakeholders
    ◆ Emiratisation
    ◆ Anti-corruption
    ◆ Labour/management good relationship
    ◆ Labour practices

3. **Environmental aspects**

- Research and development in green energy
- Compliance with local laws and international standards for the environment
- Plans to reduce emissions
- Supplier environment awareness and assessments
- Environmental impact of product and service
- Effluents and waste

DEWA scored high in the above checklist except on two points (effluents and waste and anti-corruption); availability of water and electricity was the highest. The leadership of the UAE and Dubai along with the Dubai Supreme Council of Energy and DEWA was able to present an example of a sustainable utilities provider that can meet the requirements of an event the size of Expo 2020 while still being sustainable. Serving Dubai stakeholders in a sustainable way will build a strong legacy for the city. The decision to not write a specific paragraph on the legacy in utilities is based on the reality that those investments in infrastructure will be there for generations to come and will be part of the city's development legacy even more than the event's legacy. After this case study research period, the vision of the Supreme Council of Energy is making Dubai a role model for the world in energy security and efficiency, which also is being implemented through various projects ongoing in 2019. Dubai plans ahead for excess capacity and reserves, thus showing how responsible the people making the decisions in the city are. The Dubai leadership has engaged in numerous actions demonstrating how their commitment to sustainability is enlightened and progressive as evidenced through the practices and achievements of the utilities industry.

5.10.3 Sustainable mobility

Sustainable transport is the third challenge that any mega-event may face, and remains one of the most pressured sectors in the overall sustainability considerations. Dubai's Road and Transport Authority (RTA) is the sole agency responsible for providing and meeting all transport, roads and traffic requirements in Dubai, between Dubai and other Emirates, and between Dubai and the neighbouring countries in order to ensure an advanced sustainable integrated ground transport system that serves the vision of Dubai and meets the goals of the government's agenda. Dubai Airports is the agency responsible for air transport while the Dubai Ports World is the agency responsible for the sea transport. The RTA was founded in 2005 and is responsible for the following: buses, taxis, inter-city transport, roads engineering, registration and licensing, marine rransport, commercial ads on the right of way, public buses, roads beautification, roads and parking, and rail projects. Thus, the RTA is responsible for providing and managing all the ground transportation means in Dubai, and making it more sustainable. This is advanced by setting Dubai's future through the right policies and legislations, adopting new technology and implementing best practices to serve the RTA stakeholders. Dubai Airports is responsible for providing a safe and sustainable environment for the air transport sector while Dubai Ports World operates 78 marine and inland terminals in 40 countries along with Jebel Ali Port. In this chapter, the study assesses how sustainable the transport system is in the three areas, to what extent Dubai's development plan has helped the city to plan for a legacy of Expo 2020, and how this event acts as a catalyst for change in this area. The transport sustainable stakeholder management framework was identified and checked against how far it has been considered.

The sustainability in transport is very challenging for every hosting destination. In this case, Dubai has to be ready to host 25 million visitors: that represents the entire Australian population passing through Expo 2020's gates during the six months of the event. This massive challenge is

Planning for sustainable infrastructure development during mega-events

addressed in Expo 2020 through the mobility theme of the event as Dubai is looking to build a strong legacy in transport during this event. Dubai's infrastructure is planned to amaze the world in the future. The sustainability considerations in transport are a vital sector for the UAE Agenda 2030 for sustainable development, which viewed transportation as one of the strategic sectors in the UAE along with education, health, water, renewable energy, space, and technology.

With a vision to have "safe smooth transport for all," the RTA is putting the sustainability strategy into action in order to achieve health and safety, green economy and environmental sustainability in transport. Dubai's government allocated 21 percent from the overall 2018 expenditures as a budget for the infrastructure projects, which reflects the importance given by the leadership to transport. As the UAE is hosting Expo 2020 for the first time in the region, H.H. Sheikh Mohammed bin Rashid announced this figure during a visit to the Dubai Metro Red Line to Expo 2020 site and expressed that investment in infrastructure is the main driver for the economy of Dubai. He added that those infrastructure projects in Dubai and the UAE are a key part of the country's comprehensive development plan, which will play a pivotal role in enhancing the economic environment. In addition to this, he added that happiness and welfare of the community are high-priority strategic objectives for the government. By having high-quality infrastructure, investors and tourists will be attracted further to Dubai, which will enhance the country's position as a selected destination for living and working in. H.E. Mattar Al Tayer, the Director General and Chairman of the Board of Executive Directors of RTA, added that this large investment reflects the leadership's determination to improve such projects and meet the ambitious objectives of Dubai Transport. The RTA represents the government arm of the ground transport in Dubai. This agency set many ambitious objectives that the research presents below.

The RTA fleet will reach 2,085 before Expo 2020. This investment is contributing in making high-quality public transport an ideal mobility choice, which is expected to raise the share of public transport to 30 percent by 2030. Intelligent Traffic Solutions (ITS) by the RTA target to manage the demand side and reduce congestion in Dubai's roads through electronic traffic systems, intelligent traffic studies and design, operation through observation of the traffic movement, active sensors, set traffic databases, and implemention of a toll collection system. The residents of Dubai have doubled in the last 12 years but traffic congestion has reduced significantly. In March 2019, the RTA revealed that the traffic control hub located in Al Barsha is nearly a quarter completed. This smart traffic system that cost US $160.6 million will help the RTA to raise the coverage of the smart system from 11 percent to 60 percent. This will reduce the time taken to detect accidents and congestion on the roads by ensuring that all the roads covered by the smart system will receive a quick response. This system will be vital during Expo 2020 by providing instant traffic information to all stakeholders via messaging signs and smart apps.

The RTA plan named "almasar" is a five-year strategy, from 2017 to 2022, that aims to pave the internal roads of 16 residential areas with plantation and landscaping, which will contribute to the environment and green economy strategy of the city. The RTA stakeholder engagement process involves identifying all the stakeholders being affected and potentially being affected and assessing how the RTA can influence them. The RTA's aim through stakeholder engagement is to understand their concerns, build strategic future decision on their feedback and ensure sustainability activities.

The pillars of sustainability are being addressed in the RTA's strategy by ensuring customer satisfaction that will raise people's happiness, give accessibility for people of determination, and enhance customer health and safety through advanced infrastructure and continuous improvement. RTA should keep looking at these factors while transporting around 1.49 million passengers daily, with a growth in roadways by 354 percent since 1991. The country will have 5.3 million vehicle registrations by 2020, 500 million rail passengers by 2020 and 292 million taxi passengers.

This rapidly growing transport network requires planning and management of resources along with flexible designs and leadership support.

The boundaries of environmental sustainability for the RTA are set through leadership requirements contained within Vision 2020, Dubai Plan 2020, Expo 2020 and the UAE's commitment in 2017 to the SDGs. From this, the RTA is continuously improving its environmental performance. In addition, the RTA is being sustainable through providing a safe and healthy workplace for employees and contractors, improving safety on the roads, and limiting the impact on the environment. In 2017, the RTA exceeded the set target of reducing the greenhouse gas emissions per passenger (kg CO_2/passenger) to 1.06. Further, the RTA has set targets to achieve 1051 kWh/lighted lane km for the rate of improvement in street lighting efficiency and was able to achieve 876 in 2018, which represents a challenging figure that is yet to be improved.

The RTA is fostering environmental sustainability in transport by different initiatives like converting 50 percent of its vehicles into electric or hybrid by 2020 to reach 2,280 cars (from 147 cars in 2015). The innovation in bus operations helped the RTA to see a remarkable reduction in the footprint per passenger from 0.54 litres/passenger in 2014 to 0.44 litres/passenger in 2017. The RTA is also complying with the Paperless Strategy 2020 by increasing the number of non-face-to-face channels for licensing agency services, which will help to reduce the number of visits to customer services; this resulted in a saving of 19,571 tCO_2 in 2017 from being emitted. Dubai Metro has been in operation since 2009 while Dubai Trams entered service in 2014. Both initiatives of the Rail Agency by the RTA were able to contribute to sustainable transport through average daily ridership of over 550,000 passengers for the Metro and 17,000 for the tram. This has helped offset 341,000 tCO_2 annually and helped Dubai to achieve better energy efficiency compared to 75,000 tCO_2 in 2010.

The RTA strives to reduce the emissions from the transport operations in a challenging industry that is based on mobile combustion and electricity consumption and represents almost 99 percent of the total carbon emissions. The Dubai Supreme Council provided emission calculators to give specific figures for the emission factors for RTA tCO_2 emissions.

With all of the RTA's efforts, the 2017 figures represent an increase compared to the 2014 figures where the total emissions were 780,106. However, the reason behind that is due to the increase of transport operations between those two years. The study looked to find the figures for total greenhouse gas (GHG) emissions from public transport per passenger and was able to find that these figures dropped from 1.26 $kgCO_2$ e/passenger in 2014 to 1.07 $kgCO_2$ e/passenger in 2017. Those figures represent a great achievement and reflect the seriousness of the sustainability steps being taken by the RTA.

The RTA has achieved all those above sustainability steps by the good backup in the economic sustainability. The efficient public transportation plan and investment strategy is reflected through the success of the RTA in financing this massive infrastructure development without any serious setbacks. Users of public transport in several European and Asian countries can confirm that the level of luxury in Dubai public transport is above that of transport in the other locations. Public transport in Dubai is a world-class transport infrastructure that meets the demands of different stakeholders at affordable costs. The RTA is providing this massive development in infrastructure and still maintains good revenue, which contributes to the UAE economy in different ways. In 2017, the RTA officially announced an increase of 10.5 percent in revenues while achieving 100 percent of projects that were planned. Dubai Water Canal is one of the most expensive projects completed in the area not related to Expo 2020 through a partnership between RTA and two main developers in Dubai (Meydan and Merass) at a total cost of US $735 million. This represents the productive partnership with the private sector. Completing this project while preparing for Expo 2020 reflects the vision and financial position of Dubai's government and the RTA. It has three vehicle bridges and a ramp along with five footbridges. This project extends the waterfront of Dubai by 6.4 kilometres and improves the quality of water in Dubai Creek by 33 percent. It has become a new attractive hub for tourists. Financing RTA targets in a sustainable manner reflects the leadership's strategy for mobility in Dubai. Without this, the RTA will incur big debts that they will not be able to pay in the years to come.

The population of Dubai is expected to double by 2030, which represents a challenge for the public transport services and gives a rationale for the legacy considerations of RTA projects related to Expo 2020. Dubai has to be able to serve 25 million visitors in a period of six months and to maintain the services on the demand for urban mobility. The RTA's organisational structure is based on the multiple agency model principle, which simplifies the decision process.

The RTA strives to implement world-class transport policies and legislations in order to provide public transport experience through a sustainable approach. On 25 January 2019, the RTA became the first UAE government entity to obtain ISO certification in facility management, which reflects the compliance of the agency with international standards of asset management and sustainability. This certification is in addition to having ISO 14001 certification for its environment management system. The RTA manages the demand side as well through different policies and legislations that favour mass transit rather than single-occupant vehicles through a proactive system. The RTA strategy goes beyond infrastructure projects to enhance the road network, develop the public transport systems, increase cyclist networks, implement different policies to overcome congestion, and promote sustainable transport and enhance traffic safety awareness. The RTA disseminates this heavy load across multiple agencies in order to enhance system, process and governance as following:

i. **Strategy and Corporate Governance** (SCG) sector aims to sustain the organisation's excellence initiatives.
ii. **Corporate Technology Support Services** (CTSS) sector is the proactive agency that works towards meeting the requirements of Dubai's government in initiatives like Government Plan, Dubai Smart City Plan, Expo 2020, and Dubai 2021 plan. The CTSS aim to adopt best-quality services to customers' practices, implement technological solutions, ensure best levels of integration, information security, and resources optimisation, and governance of technical system.
iii. **Corporate Administrative Support Services** (CASS) sector aims to raise the stakeholders' happiness and ensure pioneering services.
iv. **Public Transport Agency** (PTA) is responsible for building the public transport network that sets plans for ground and marine public transport through different means.
v. **Traffic and Roads Agency** (TRA) is responsible for providing a seamless travel experience and connecting Dubai through planning, designing, constructing and maintaining the road networks.
vi. **Rail Agency** (RA) isresponsible for establishing and delivering the best railway facilities across Dubai and building a modern network for years to come including developing, operating, maintaining and selecting locations.
vii. **Licensing Agency** (LA) is responsible for licensing drivers and vehicles and works to improve drivers' performance and vehicle safety in order to provide a better transport environment.
viii. **Dubai Taxi Corporation** (DTC) is responsible for providing different kinds of services and customer care to meet highest levels of transportation quality standards.

In 2017 the RTA announced officially that the assets' performance targets achieved 99.4 percent, the optimised assets' value reached 1.2 percent and the assets managed efficiently and effectively reached 101.9 percent. Those figures reflect enhanced efficient and effective assets management for the overall public and ground transport system in Dubai. Overall, the Dubai road network has been extended by 62 percent from 8,715 km in 2005 to 23,084 km in 2018, and the rate of fatalities reduced from 21.9 per 100,000 population in 2005 to only 2.3 per 100,000 population in 2018 as of RTA official figures. Expo 2020 remains a challenge with the number of visitors expected yet Dubai has the right transport agency to deal with the transport challenge in a sustainable method. Dubai's transport strategy reflects the leadership's vision in transforming Dubai into an international hub that applies the latest technology in mobility as well as providing an attractive business environment. The transport system observation in Dubai confirms that this city has great leadership behind the considerations of sustainability, the respect of the passengers, the management of transport and traffic, and the continuous development of those systems.

Dubai Ports (DP) World is the maritime logistics company that is headquartered in the UAE and provides remarkable connectivity and infrastructure across transport modes. DP World has global operations with 78 marine terminals over 50 related business located in 40 different countries across six continents including Jebel Ali Port. DP World's vision in contributing to the future of the world trade is targeted by connecting communities through trade and economic prosperity.

The year 2018 was viewed as the period of strategic growth for DP World in multiple fields including smarter trade using data-driven logistics, designing acquisitions strategies to increase the global business footprint, using innovation initiatives to find competitive trade solutions, and adopting sustainable value creation for all the stakeholders. DP World is engaging with all the stakeholders through the company's vision by changing the organisational structure in order to implement new technological changes. Furthermore, the gross capacity of DP World throughout the business portfolio grew by 2.9 percent to reach 90.8 20-foot equivalent units (TEU) compared to 88.2 TEU in 2017. The sustainable growth of the company is targeting 100 million TEU by 2020, DP World's plans to expand the company's portfolio further through different investments, acquisitions, and partnerships led the company to invest around US $3.5 billion in 2018. The return on capital employed (ROCE) is a key measure of how well the investment strategy is delivering value for shareholders. In 2018 the ROCE was 8.4 percent compared to 7.1 percent in 2014. The earnings per share (EPS) also increased from US $.081 in 2014 to US $1.53 in 2018. Those figures reflect the economic sustainability of the company. In the environmental considerations, DP World is committed to preventing and minimising any negative impact on the environment. The company believes that attitudes of doing responsible business are the only ways to ensure sustainable corporate success. DP World became the first company of its type to join the World Ocean Council, which combines efforts from different stakeholders in order to protect the oceans. In 2018 the carbon intensity decreased to 14.9 percent from 15.8 percent in 2014 despite the company's growth and corresponding increase in energy use. The figure that reflects this improvement is viewed by the reduction of kilograms of carbon dioxide equivalent per 20-foot equivalent unit; this figure decreased by five percent between 2014 and 2018. The energy consumption, which also remains a key performance indicator in reducing the carbon footprint of the company, was reduced by 13 percent in the last four years. In 2018, the company was able to offset more than 55,738 tonnes of CO_2 emissions by using renewable energy sources to invest in low-carbon fuels like liquefied natural gas. This reduction is aligned with the government strategy to achieve a green economy. DP World is executing a solar power programme in the UAE which will be able to generate clear energy to power 4,600 homes on

Planning for sustainable infrastructure development during mega-events

completion. In 2018, the company launched the first green storage and warehouse facilities with plans to install 88,000 solar panels in the first phase.

Dubai Airports is the agency responsible for operating and developing the two main airports of the city: Dubai International (DXB) and Dubai World Central (DWC), which was formally known as Al Maktoum International Airport. The vision of Dubai Airports is to be one of the world's most active airport companies by creating infrastructure that adopts a continuous development strategy, amazes the customer by providing unique experiences, sets new standards of travelling experience, and leads the innovation process in this industry. Air transport contributes to the annual GDP of Dubai by 28 percent; this is leading to positive impacts on the social and economic development. Dubai Airport strives to follow ethical conduct in its business through honest, integrity, and respect for all the stakeholders.

Dubai Airport gives high attention to the social and environmental pillars of sustainability by first providing a healthy work environment through adopting advanced safety and well-being practices with their employees along with an attractive and rewarding career system. This was observed by visiting both airports and chatting with some employees who confirmed this aspect of well-being and asserted that they are receiving continuous training for skills development, which will help them improve their career. Second, Dubai Airport's stakeholders are also encouraged to develop a sustainable approach in work. Third, in the environment pillar, Dubai Airport worksto reduce the impact of its operations on the carbon footprint of the city. Dubai Airport recycles thousands of items (paper, cartons, plastic and aluminium cans) across both airports, has implemented efficient lighting, and uses flow arrestors in order to control the water consumption.

Dubai's two main carriers work jointly: Emirates and Fly Dubai. Emirates, one of the world's largest international airlines, is an independent entity that serves 155 airports in 83 countries starting from Dubai. The operating revenue was delivering the group the 30th consecutive year of profits, which increased from US $22.4 billion in the financial year 2013–2014 to US $25.08 billion in 2017–2018. While reducing costs, Emirates was able to carry 58.5 million passengers and invest US $2.7 billion in new aircraft and equipment, acquisition of companies, modern facilities, and employees' initiatives in 2018. In addition, the company announced an agreement to purchase 40 new Boeings at a cost of US $15.1 billon with a delivery date of 2022. Along with that, another agreement of US $16 billion has been set with Airbus for 36 additional aircraft. Those new investments, along with the already massive existing investments, lay the infrastructure to help carry the 25 million visitors for Expo 2020.

5.11 Chapter summary

The three case studies in this chapter present the sustainability and legacy considerations adopted by Dubai in general and the management of Expo 2020 by giving equal priority to the different pillars of sustainability. They offer proof that producing infrastructure sustainability requires a joint effort from different entities with strong leadership and proper design at each step. Furthermore, the cases show that Dubai is doing what it used to do before Expo 2020, yet now it has a new purpose to combine the efforts around it. It is not the first time that Dubai is executing a mega-project, or developing a modern transport system, or investing in utilities to achieve a world–class level. Dubai was doing so before Expo 2020 and attended to several sustainability considerations even when this term was not as popular as it is today. Yet this event, with seven years of preparation, has created goals that each entity or department wants to perform with excellence to contribute to the success of Dubai in this event.

The chapter presents the argument that executing a sustainable project is not as expensive as it used to be. Having a market that is ready for and aware of the importance of doing so represents the core of achieving a sustainable mega-event. Dubai's planning infrastructure maintained a steady development in this direction before Expo and used the event to accelerate this infrastructure development once preparation for Expo 2020 began. The social participation in the preparation of Expo 2020 also represents a new development for the social pillar in the UAE along with all the other initiatives to improve the unbalanced figures of gender in UAE society presented earlier.

The tangible and intangible legacy for Expo 2020 will not be determined easily from now. We have to wait for many years after the event to present the actual outcomes. However, from the way that Dubai is planning, we can expect that the event is going to leave a legacy in research for a sustainable future and international joint efforts for solving humanity's common difficulties, and it will continue to position the UAE as the gateway of the region. The framework of hosting a sustainable mega-event will include many points: first, a supportive leadership that believes in sustainability considerations and wants to contribute to the world in this direction; second, the sustainability pillars have to be equally prioritised as ignoring one pillar will affect the overall system; third, as much as a legacy plan is important, the way that legacy is reached is important and has to be planned in a flexible way to adopt future development and changes; fourth, the mega-event should come as part of an ongoing development plan. The event owner should not accept any bid that will use the mega-event to initiate such plans as it will never be sustainable; and fifth, the mega-event management teams should be aware about the shortfalls of previous events and ensure that plans for the new events will not repeat similar previous mistakes.

5.12 Review and case discussion questions

The results of the three case studies on Expo 2020 demonstrate the role of legacy in the sustainability plan and reveal how the hosting destination can create a long-lasting impact. A mega-event boosts the hosting city's GDP, increases foreign investment, creates employment and enhances the economy. However, those benefits come with risks of increased inflation rates, mass relocations and changes in socio-economic groups, pollution, reduction in social aid, and increased congestion. Heavy investment in infrastructure, pressure on resources, intensive development of specific areas, and challenges to the hosting destination's sustainability model are the main reasons for initiating the discussion about the sustainable mega-event. Consequently, sustainability is the end goal of Expo 2020, set within the three pillars of social, economy and environment.

Previous researchers (Dodouras and James, 2004; Pelham, 2011) addressed the challenges of hosting mega-events through resource utilisation and responsible sourcing, urban legacy project and vision, deploying the mega-event as a catalyst of behaviour change, stakeholder engagement, and achieving a carbon-neutral strategy. We have argued that Dubai is emerging as an innovator in inter-urban planning; Lauermann (2019) asserts that communicates knowledge of sustainability that cities hosting mega-events can use as a model for visualising and designing sustainability. The three case studies show how Dubai is going beyond those considerations by providing a new level of planning to host a sustainable mega-event in the three case-study sectors along with implementation of a strong legacy plan.

5.12.1 Recommendation 1: hosting a sustainable mega-event requires developing a deep understanding of a mega-event's complex, multiple impacts

Expo 2020 is contributing to the overall sustainability plan of the city by having joint goals that different government entities are trying to serve following high standards of excellence. This approach is improving overall sustainability performance indicators of the mega-event design and planning led by the Dubai Strategy 2021. Expo 2020 is a site location with related buildings and infrastructure that will serve the UAE economy for years to come. District 2020 is planned to become a business hub for the region, growing the economy and stimulating developments in the transport sector, hosting more international companies, creating new communities, attracting more foreign investment, establishing a new research centre, supporting responsible business, and applying the latest technology and systems The infrastructure support in utilities and transport will offer a long-lasting legacy for the site and develop a business hub that will serve as a gateway for the region.

During the planning stage, it was envisaged that over the course of the six-month event, around 25 million visitors (70 percent international) would pass through Expo 2020 gates. EY Consultancy calculated the Expo impact in three periods: the seven-year build-up to the Expo event, Expo running for six months, and the subsequent ten-year legacy period. The gross value added (GVA) to the economy for the pre-Expo period was predicted to be US $10.2 billion, with a further US $6.1 billion expected for the Expo 2020 event through spending on hotels, hospitality and business services. The legacy period is anticipated to generate approximately US $16.9 billion as the legacy infrastructures begin to pay back the initial investment. Expo 2020 is expected to generate direct results of US $15.1 billion out of the overall impact of US $33.2 billion (Matthew Benson, EY Middle East and North Africa). These figures represent the actual contribution of Expo 2020 to the economy. Tourism is a strong contributor to the Dubai economy that represents US $8.3 trillion per year or 10.4 percent of global GDP. Hosting the mega-event will encourage more people to visit Dubai and to repeat this visit after the end of the event.

The expanded Dubai Exhibition Centre (DEC) is a key player in the legacy period and is expected to attract around 1.6 million visitors per year to the site along with the expected 1.1 million overseas visitors who will travel to see the Expo site post-event. These figures represent a respite from the risk that Dubai is overbuilding for Expo with government-related entities (GREs) like RTA, DP World, DEWA and Dubai Airports spending billions on investment. The second and third case studies found significance evidence supporting the economic pillar of sustainability and government organizations' ability to overcome the long-standing debt problems by showing their ability to invest effectively in the utilities and transport industries, pay back the investment costs, and continue to remain profitable.

5.12.2 Recommendation 2: mega-events require intensive planning

Expo 2020 succeeded in increasing the volunteering spirit in the UAE with more than 30,000 volunteers committed to contributing to the event's success. These volunteers come from diverse nationalities, ages and backgrounds. The Expo team communicated that they targeted 30,000 volunteers; however, the actual number of volunteers has exceeded 50,000 people from 200 nationalities. This social achievement represents a strong outcome for Expo 2020. Furthermore, a study on global migration trends for the high-net-worth individuals (HNWIs) estimates that Dubai is in fifth place internationally in attracting HNWIs, who are people with at least US $1,000,000 net worth assets. HNWIs are usually attracted to modern destinations with low crime rates, good schools, good business opportunities, and a modern society. Dubai is able to meet these criteria and Expo 2020 has further affirmed Dubai's status as an attractive destination for HNWIs. A World Bank report for 2017–2018 ranked the quality of roads in the UAE as number one in the world. Transport is a crucial driver of economic and social development, which helps connect people to jobs, education, and health services, and ensures an efficient supply of goods and services. Transport accounts for 65 percent of global oil consumption and 23 percent of the world's energy-related CO_2 emissions. More than 1.25 million people are killed globally in road accidents each year, and air pollution of motorised road transport attributes to 185,000 deaths per year (World Bank 2018). Dubai's transport system and Expo 2020 are both designed to increase UAE traffic mobility and safety performance.

DEWA and RTA stakeholder engagement programmes aim to raise stakeholder satisfaction through ensuring social sustainability. Hosting Expo 2020 requires further collaboration with such large contributors to the national economy. Expo 2020 has been planned to support 905,200 job-years between 2013 and 2031; a figure that represents more than 10 percent of the UAE population. The long-term investment is expected to be US $33.4 billion as the impact on the UAE during those years will result in economic dividends that will benefit businesses, both large and small, across a range of sectors for years to come. The small and medium enterprises secured contracts of around US $1.3 billion, which support 12,600 job-years. This reflects the vision of Expo 2020 in fostering innovation and supporting small business as per the EY consultancy report for the economic impact of Expo 2020.

5.12.3 Recommendation 3: developing sustainable infrastructure for hosting a mega-event requires multi-stakeholder engagement occurring on multiple levels

The numerous environmental considerations of Expo 2020 include mandatory green building compliance, design of the buildings, Expo 2020 venue innovations in generating energy, adaptation of a passive energy strategy, selection of construction material, managing waste on-site, using recycled materials, and many more. All of these practices on the Expo 2020 construction site reflect the intention of Dubai's government to host a sustainable event that leaves a positive legacy.

The second case study presented the initiatives that DEWA is making to achieve a sustainable clean supply of energy, commencing from Sheikh Mohammed bin Rashid Al Maktoum Solar Park, the largest single-site solar park in the world, adaptation of RO water desalinisation, use of energy mixes, increasing dependence on green coal and natural gas, and managing the demand side. The third case study also shows multiple green initiatives including public transport with a competitive, efficient infrastructure, Smart Dubai, reliance on innovation in transport, ensuring a pioneering service, providing multiple means of transport, reducing congestion, reducing stakeholders' travel distance, promoting electrical and hybrid vehicles, reducing transport emissions, investing in new vehicles and plans for better environmental performance, and adopting advanced technology.

5.12.4 Recommendation 4: leadership engagement on environmental impacts increases attention to its attainment

The role of the design and planning for achieving sustainability and building a long-lasting legacy was explored in the three cases. The first case study shows how the design of District 2020 will inherit 80 percent of Expo 2020's physical and digital infrastructure. This constitutes 200,000 square metres of LEED Gold Standard structures built for Expo 2020 that will be repurposed to create a new community. District 2020 will offer a dynamic environment for businesses of all sizes with sustainable flexible buildings built to reduce power and water consumption for an anticipated community of 90,000 people. In the same area, multiple social, cultural and educational facilities have been integrated within the design of the development, including Al Wasl Plaza. Another 2,300,000 Square metres of floor area will be made available for third parties to develop residential, commercial, hospitality, education and mixed-used spaces. The transition period was planned to start immediately after the closure of the event in April 2022 and future occupants were intended to receive possession of their units in October 2022. Since March 2020, the Expo 2020 event schedule has had to adapt flexibly to the exigencies of the global pandemic. Nevertheless, the three case studies present multiple examples and evidence for the power of design and project management in achieving a sustainable mega-event. The *construction case* presents the development of the Expo site, the Dubai Municipality's design guidelines named green building regulations, the multiple ideas for recycling materials on the construction site, and design considerations to save energy and water. In the *transport case*, the study presents how the public transport design will contribute to the sustainability of the transport system in general, how planning for the future has given Dubai multiple competitive edges over similar cities, how the city has become a centre of air and sea transport, and how plans set earlier were capable of expansion. Those considerations reflect the real value of design capability in the city. In the *utilities case*, DEWA is designing an energy mix that will provide a sustainable supply for years to come. Furthermore, the urban and infrastructure development, along with the economic and social legacy of Expo 2020, have been comprehensively planned, and creating sustainable outcomes after it has finished is key. Dubai's flexible design and legacy considerations demonstrate leadership in realising the vision of Expo 2020. The case findings support the assertion that this mega-event project can meet the three pillars of sustainability, executing the power of planning and design in the success of the mega-event and its legacy. District 2020 represents an ambitious plan for a unique experience of legacy considerations that stand out from all previously hosted mega-events.

5.12.5 Recommendation 5: sustainable design and planning are a key factor

Expo 2020 is raising the bar for sustainability and legacy plans. Milan's Expo 2015 was the first mega-event to obtain CES certification through the applied EMS and compliance with ISO 20121. Dubai's Expo 2020 is bigger in size but has maintained its design commitment, while Expo 2015 shrank by approximately 60 percent of the planned size, resulting in an event one-fifth the size of Shanghai's Expo 2010 (Shi et al. 2012). Expo 2020 is building the largest car park from recycled materials, setting new standards for green buildings, building the first self-sustaining building, obtaining a minimum LEED gold certificate in all buildings, and still developing a strong legacy. Power supplied to the site comes from sustainable energy resources, the UAE population is engaged through multiple initiatives, the site's health and safety requirements are high, workers' welfare is monitored, and contractors and SMEs are connected through a transparent portal to operate in a cost-effective way delivering effective economic performance. Expo 2015 faced substantial financial losses, resulted in incomplete structures, an absence of transparency, and overall destroyed the opportunity for Italy to re-launch its image from that of a struggling economy. The major themes and structures of Expo 2020 were completed almost a year ahead of the scheduled event, reinforcing the image of Dubai as a central business and transport hub for the region, as well as projecting confidence to existing and new financial partners.

Expo 2020's legacy plan responds to the question of how to host a sustainable mega-event while building a strong legacy. District 2020 will provide quality infrastructure for business, comprising approximately 86 buildings for office and residential use; and Al Wasl Plaza will remain open to host visitors post-Expo. Overall, 80 percent of structures built for Expo 2020 will be repurposed through the District 2020 plan, which employs the latest smart systems to reduce energy consumption, monitor and control building functions and collect data. Siemens, the strategic partner of Expo 2020, is the main solutions provider and has located its regional company headquarters on the site.

5.12.6 Recommendation 6: legacy consideration should be planned during the design stage

The leadership's determination backed by a sustainable design that considered key success factors for the three sustainability pillars and a strategy with embedded sustainability considerations helped to secure the Expo 2020 bid. Dubai's Expo 2020 site was designed to reduce the usual shortfalls previous mega-events have experienced. Dubai's Expo 2020 team endeavoured from the outset to learn from past best practices in developed countries. The aim was to build facilities that will reduce the impact on the environment for years to come. Dubai's leadership contributed to the sustainability over the Event Life Cycle (ELC) by setting a vision and monitoring its project achievement by each government organization involved. With leadership vision, a developing country can be sustainable over the ELC through project management that embeds sustainability in each step, creates a project governance process to follow the application of those plans, and designs any new buildings or facilities with equal project importance ascribed to the three pillars of sustainability. Project execution has to be followed up by intensive ongoing assessment during the construction process, also ensuring business excellence and satisfaction of all stakeholders.

5.12.7 Recommendation 7: leadership emphasis on sustainability is mandatory for sustainable infrastructure development of mega-event

Successful leadership in sustainable mega-projects assigns equal importance to goal achievements in all three pillars of sustainability. The development plans for RTA and DEWA are based on renewable and sustainable energy resources that can reduce emissions, provide sustainable supply, be economically feasible, and serve future generations. This chapter on Expo 2020 shows how delivering an exceptional sustainable mega-event requires ensuring the economic, social and environmental pillars are upheld and planned project goals are achieved to generate successful outcomes.

1. List in rank order the 10 top positive and negative impacts of mega-events. How can project leaders and project teams optimise positive outcomes and minimise negative ones?
2. List five major considerations when planning the design and delivery of mega-events. What aspects of construction, resources and transportation require especially intensive planning?
3. How can sustainable infrastructure for hosting a mega-event maximise multiple stakeholder engagement?
4. In what ways could local politicians and project managers encourage contractors' project teams to engage with environmental impacts?
5. What are the best methods available for planning sustainable mega-projects?
6. What contingencies and issues of legacy are hard or impossible to plan during the design stage? An example of an unanticipated event in 2020 is the COVID-19 pandemic.
7. Identify ways that leadership on the three pillars by local government officials is essential for sustainable infrastructure development of any mega-event.

References

Abdalla, H.G. and Pinnington, A.H. (2012). Transformational leadership in a public sector agency. In T.= Dundon and A. Wilkinson (Eds.). *Case studies in global management: Strategy innovation and people.* Sydney, Australia: Tilde University Press, Chapter 21, pp. 184–194.

Al Hashimy, R. (2016). Infrastructure works for Expo 2020 Dubai site to begin this summer. *Emirates 24/7* [online] 26 June. [Accessed 19 February 2018]. Available at: http://www.emirates247.com/news/emirates/infrastructure-works-for-expo-2020-dubai-site-to-begin-this-summer-2016-06-27-1.634242.

Al Sammarae, R. (2019). Expo 2020 Dubai's three thematic districts reach completion. *Architect.* [online] 21 aMay. [Accessed 1 March 2018]. Available at: https://www.middleeastarchitect.com/43146-expo-2020-dubais-three-thematic-districts-reach-completion

Al-Reyaysa, M., Pinnington, A.H., Karatas-Ozkan, M., and Nicolopoulou, K. (2019). The management of corporate social responsibility through projects: A more economically developed country perspective. *Business Strategy and Development*, 2(4), pp. 358–371.

Badam, R. (2018). UAE's national tree to take pride of place at Expo 2020 Dubai. *The National.* [online] 18 October. [Accessed 28 April 2019]. Available at: https://www.thenational.ae/uae/government/uae-s-national-tree-to-take-pride-of-place-at-expo-2020-dubai-1.781994.

Bourdeau, L. (1999). Sustainable development and the future of construction: A comparision of visions from various countries. *Building Research and Information*, 27(2), pp. 354–366.

Carter, C.R. and Rogers, D.S. (2008). A framework of sustainable supply chain management: Moving toward new theory. *International Journal of Physical Distribution and Logistics Management*, 38(5), pp. 360–387.

Coates, D. and Humphreys, B. (2003). Professional sports facilities, franchises and urban economic development. *Public Finance and Management*, 3(1), pp. 335–357.

Cornelissen, S., Bob, U., and Swart, K. (2011). Towards redefining the concept of legacy in relation to sport mega-events: Insights from the 2010 FIFA World Cup. *Development Southern Africa*, 28(3), pp. 307–318.

Davenport, J. and Davenport, J.L. (2006). The impact of tourism and personal leisure transport on coastal environments: A review. *Estuarine, Coastal and Shelf Science*, 67(1), pp. 280–292.

Deng, Y & Boom, W (2011). Mega-challenges for Mega-event flagships. *Architectural Engineering and Design Management*, vol. 7(6), pp. 23–37.

Deng, Y. & Poon, S.W. (2014). Mega-event flagships in transformation: learning from Expo 2010 Shanghai China. *Journal of Engineering, Design and Technology*, vol. 12 (4), pp 440–460.

Design Mena (2016). Dubai awards $2.9bn deal to extend metro to Expo 2020 site [online]. *Design Mena.* [Accessed 20 April 2018]. Available at: http://www.designmena.com/thoughts/dubai-awards-2-9bn-deal-to-extend-metro-to-expo-2020-site?hilite=%22expo%22%2C%222020%22

Deulgaonkar, P. (2016). Global architects to design Dubai Expo 2020 theme pavilions. *Emirates 24/7* [online]. [Accessed 15 April 2018]. http://www.emirates247.com/business/corporate/global-architects-to-design-dubai-expo-2020-theme-pavilions-2016-03-12-1.624019.

DEWA Sustainability Report (2015). Government of Dubai. [Accessed 19 April 2018]. Available at: https://ecgi.global/sites/default/files/dewa-sustainability-report-2015.pdf

District 2020 (2020). [online] 30 May. [Accessed 4 November 2018]. Available at: https https://www.district2020.ae

Dodouras, S. and James, P. (2004). Examining the sustainability impacts of mega-sports events: Fuzzy mapping as a new integrated appraisal system, The 4th International Postgraduate Research Conference in the Built and Human Environment, Salford University. Manchester. 29 March–2 April.

Dubai Strategic Plan (2015). Highlights Dubai Strategic plan (2015) [Accessed 18 May 2018]. Available at: http://www.dubaiblog.it/wp-content/uploads/2010/11/Dubai_Strategic_plan_2015.pdf

European Commission (2013). *Together towards competitive and resource-efficient urban mobility.* Brussels, EC.

Elkington, J. (1999). *Cannibals with forks: The triple bottom line of 21st century business.* Oxford: Capstone.

Expo 2020 (2019) [Accessed 25 April 2018]. Available at: https://www.expo2020dubai.com

Fahy, M. (2016). Dubai Expo2020: Spectacular designs chosen for centrepiece pavilions. *The National* [online]. [Accessed 20 April 2018]. Available at: https://www.thenational.ae/business/property/dubai-expo-2020-spectacular-designs-chosen-for-centrepiece-pavilions-1.167983?videoId=5587256436001

Getz, D. (1997). *Event management and event tourism.* New York: Cognizant Communication Corporation.

Getz, D. and Page, S. (2016). Progress and prospects for event tourism research. *Tourism Management*, 52(1), pp. 593–631.

Gibson, R. (2001). *Specification of sustainability-based environmental assessment decision criteria and implications for determining "significance" in environmental assessment* [online]. Canadian Environmental Assessment Agency Research. [Accessed 12 December 2015]. Available at: http:\\static.twoday.net/.

Guala, A., & Turco, D. (2009). Resident perceptions of the 2006 Torino Olympic Games, 2002-2007'. *Sports Management International Journal*, Vol. 52), pp. 21–42.

Gulf News (2017). Dh500m projects to make Dubai's air quality the best. *Gulf News.* [online] 26 May. [Accessed 12 November 2018]. Available at: https://gulfnews.com/uae/environment/dh500m-projects-to-make-dubais-air-quality-the-best-1.2033539

Guizzardi, A., Mariani, M., and Prayag, G. (2017). Environmental impacts and certification: Evidence from the Millan World Expo 2015. *International Journal of Contemporary Hospitality Management*, 29(3), pp. 1052–1071.

Hall, C. (2012). Sustainable mega-events: Beyond the myth of balanced approaches to mega-event sustainability. *Event Management*, 16(2), pp. 119–131.

Heikkurinen, P., Clegg, S.R., Pinnington, A.H., Nicolopoulou, K., and Alcaraz, J.M. (2019). Managing the Anthropocene: Relational agency and power to respect planetary boundaries. *Organization and Environment*, first published online on 17 October 2019.

Henderson, S. (2011). The development of competitive advantage through sustainable event management. *Worldwide Hospitality and Tourism Themes*, 3(3), pp. 245–257.

Hiller, H. (2000). Mega-events, urban boosterism and growth strategies: An analysis of the objectives and legitimations of the Cape Town 2004 Olympic Bid. *International Journal of Urban and Regional Research*, 24(2), pp. 449–458.

Hiller, H. (2006). Post-event outcomes and the post-modern turn: The Olympics and urban transformations. *European Sports Management*, 6(4), pp. 317–332.

Hill, C. & Bowen, P. (1997). Sustainable Construction: Principles and a Framework for Attainment, *Construction Management and Economics*, vol.15(3), pp.223–239.

Horne, J., & Manzenreiter, W. (2006). An Introduction to the Sociology of Sports Mega-events. *The Sociological Review*, vol. 54(2), pp.1–24.

Hotchkiss, J., Moore, R., and Zobay, S. (2003). Impact of the 1996 summer Olympic Games on employment and wages in Georgia. *Southern Economic Journal*, 69(3), pp. 691–704.

Hult, A (2013). Swedish Production of Sustainable Urban Imaginaries in China, *Journal of Urban Technology*, vol. 20(1), pp. 77–94.

ISO international Organisation for Standardization (2012). *The event sustainability management system* [online]. [Accessed 12 May 2017]. Available at: www.iso.org/iso/iso20121

Iraldo, F., Melis, M., and Pretner, G. (2014). Large-scale events and sustainability: The case of the universal exposition Expo Milan 2015. *Economics and Policy of Energy and the Environment*, 17(3), pp. 139–165.

Kibert, C. (2016). *Sustainable construction: Green building design and delivery*. 4th ed. New Jersey: John Wiley and Sons, Inc.

Lauermann, J. (2019). Visualising sustainability at the Olympics. *Urban Studies*, pp. 1–18.

Ma, S., Egan, D., Rotherham, I., and Ma, S. (2011). A framework for monitoring during the planning stage for a sports mega-event. *Journal of Sustainable Tourism*, 19(1), pp. 79–96.

Mair, J. and Whitford, M. (2013). An exploration of events research: Event topics, themes and emerging trends. *International Journal of Event and Festival Management*, 4(1), pp. 6–30.

Malhado, A., Araujo, L. & Ladle, R. (2013). Missed Opportunities: Sustainable Mobility and the 2014 FIFA World Cup in Brazil, *Journal of Transport Geography*, vol. 31(1), pp. 207–208.

Marchet, G., Melacini, M., and Perotti, S. (2013). Environmental sustainability in logistics and freight transportation literature review and research agenda. *Journal of Manufacturing Technology Management*, 25(6), pp. 775–811.

Mason, M. and Paggiaro, A. (2012). Investigating the role of festival scape in culinary tourism: The case of food and wine events. *Tourism Management*, 33(6), pp. 1329–1336.

May, A., Boehler-Baedeker, S., Delgado, L., Durlin, T., Enache, M., & Van Der Pas, J. (2017). Appropriate national policy frameworks for sustainable urban mobility plans. *European Transport Research Review*, 9(1), pp. 7

Mir, F.A. and Pinnington, A.H. (2014). Exploring the value of project management: Linking project management performance and project success. *International Journal of Project Management*, 32(2), pp. 202–217.

Morgan, J. (2017). Officials unveil Expo 2020 Dubai's District 2020 legacy project [online]. Construction week online. [Accessed 20 April 2018]. Available at: http://www.constructionweekonline.com/article-46220-officials-unveil-expo-2020-dubais-district-2020-legacy-project/.

O'Reilly, N., Lyberger, M., McCarthy, L., and Séguin, B. (2008). Mega-special-event promotions and intent to purchase: A longitudinal analysis of the Super Bowl. *Journal of Sport Management*, 22(4), pp. 392–409.

Pelhan, F. (2011). Will sustainability change the business model of the event industry? *Worldwide Hospitality and Tourism Themes*, 3(3), pp. 187–192.

Pitts, A. and Liao, H. (2009). *Sustainable olympic design and urban development*. London and New York: Routledge.

Pope, J., Annandale, D., and Morrison-Saunders, A. (2004). Conceptualizing sustainability assessment. *Environmental Impact Assessment Review*, 24(6), pp. 595–616.

Preuss, H. (2013). The contribution of the FIFA World Cup and the Olympic Games to green economy.' *Sustainability*, 5(8), pp. 3581–3600.

Rees-Caldwell, K. and Pinnington, A.H. (2013). National culture differences in project management: Comparing British and Arab project managers perceptions of different planning areas. *International Journal of Project Management*, 31(2), pp. 212–227.

Ritchie, J. (1984). Assessing the impact of hallmark events: Conceptual and research issue. *Journal of Travel Research*, 23(1), pp. 2–11.

Roche, M. (1994). Mega-events and Urban Policy, *Annals of Tourism Research*, vol. 21(1), pp. 1–19.

Roche, M. (2000). *Mega-events and modernity: Olympics and expos in the growth of global culture*. London: Routledge.

Saseendran, S. (2018). 2 new Dubai plants to help divert 50% of waste going to landfills. Gulf News. [online] 24 March. [Accessed 19 May 2018]. Available at: https://gulfnews.com/uae/environment/2-new-dubai-plants-to-help-divert-50-of-waste-going-to-landfills-1.2193605

Shahbandari, S. (2017). Dubai Metro Route 2020: Shaikh Mohammad launches drilling work. [online] *Gulf News*. [Accessed 18 December 2017]. Aavailable at: https://gulfnews.com/news/uae/transport/dubai-metro-route-2020-shaikh-mohammad-launches-drilling-work-1.2112061

Sherwood, P., Jago, L., and Deery, M. (2005). Triple bottom line evaluation of special events: Does the rhetoric reflect reporting? Third International Event Management Conference, Sydney: Charles Darwin University, pp. 632–645.

Shi, Q., Zuo, J. & Zillante, G. (2012). Exploring the management of sustainable construction at the programme level: a Chinese case study, *Construction Management and Economics*, vol. 30(6), pp. 425–440.

Strategic Planning Guide (2008). Strategic planning guide. The Executive Council. 30 March 2008.. [Online]. [Accessed 18 December 2015]. Available at: https://www.mbrsg.ae/getattachment/9c9117b8-a132-495f-b920-54ea5051f630/Dubai-Government-Guide-to-Strategy-Implementation.aspx

World Bank (2018). CO2 emissions (metric tons per capita). [Accessed 15 April 2015]. Available at: https://data.worldbank.org/indicator/EN.ATM.CO2E.PC

Yan, Y. (2013). Adding environmental sustainability to the management of event tourism. *International Journal of Culture, Tourism and Hospitality Research*, 7(2), pp. 175–183.

Yehia, S.N. (2020). *Managing Sustainable Global Events: Sustainability Practices of Expo 2020*. Unpublished PhD Thesis, Dubai: The British University in Dubai.

6

STAKEHOLDER ENGAGEMENT IN MAJOR INFRASTRUCTURE PROJECTS

Hemanta Doloi

6.1 Introduction

Infrastructure projects are the backbone of society and often result in not only transforming the physical space but characterising the community at large. Adequate involvement of the broader community in the development of infrastructure projects is highly crucial for doing it right and making it happen in the societal contexts. One of the overarching objectives in developing modern infrastructure projects is 'sustainability' or 'triple bottom line (TBL)'. Yet, anecdotally, there is no good example where such a concept has ever resulted in desirable outcomes in projects. While the concept of TBL predominately refers to a measure of effective use of resources and reduction of greenhouse gases in the mainstream literature (Lockie, 2001; Mandarano, 2009; Wittig, 2013), a clear consensus on the measure of social dimensions leading to social sustainability outcomes is not quite prominent within the construction industry practice. Given the built environment demands 40–50 percent of global resources and generates a proportional amount of waste, the construction industry, which contributes 5–10 percent of national GDP globally, has a prominent role to play in managing, controlling and meeting the sustainable outcomes across the projects. The term TBL, first coined by Elkington (1998), suggests that sustainability can be achieved in the intersection of social, economic and environmental performance, where a decision will not result in economic benefit, but also affects environment and society in a positive way. While the economic and environmental dimensions of sustainability are reasonably adopted in most modern projects, today's leaders are significantly lacking a holistic approach to quantification of socially sustainable outcomes in projects. In this chapter, the author argues that social sustainability and the underlying performance of infrastructure projects is a function of stakeholders being managed well for the whole life of the project.

Due to the virtue of infrastructure projects being the societal backbone, these mega projects not only support the growth and well-being of the community but also bring about unique changes in the infrastructure landscapes. It is of utmost importance that planning of the infrastructure projects is based on the needs of the society taking broad views of the factors such as location-based relevance, target community and changes in demography, and internal and external market forces.

The sustainability concept and the success and failure of infrastructure projects are interlinked. In sustainability measurement, achieving a balance through the TBL is crucial. Objective measures of TBL over the project development life cycle is the most essential requirement for understanding the true performance of capital intensive infrastructure projects. While economic and environmental parameters within the TBL framework are relatively well documented in the mainstream literature, comprehensive understanding of social performance in projects is not well. A viable project both economically and environmentally may still fail if the societal norms are not addressed from the outset and it is perceived by the community at large to be the wrong project in the wrong context.

Thus, inadequate attention to the multiple stakeholders including their varied interests, impacts and satisfactions could attribute to the failure of infrastructure projects from a community perspective. As the size of project increases, the composition and complexity of stakeholders also increases. The complexity is further increased due to these factors: the risks and uncertainties associated with megaproject infrastructure environments; the large number of stakeholders involved at the various stages of project development; the inflexible multi-role administrative governance structure that leads to high public attention and disputes, and the inability of project managers to accurately identify and prioritise stakeholders as the projects grow in scope and complexity. Thus, the overarching aim of this chapter will be to examine how the multiple stakeholders involved in large infrastructure projects can be effectively engaged, managed and enabled for delivering wider societal benefits.

6.1.1 Chapter aim and objectives

This chapter has been developed with two broad aims: first, to understand the importance and significance of wider stakeholders' involvement in infrastructure projects; and second, to discuss the processes of engaging stakeholders with appropriate reflection of their needs, requirements, impacts and satisfaction over the entire project life cycle.

Referring to the Project Management Body of Knowledge (PMBOK, 2017), projects are unique entities and are characterised by numerous factors ranging from type, location, cost, quality, to development environment. As infrastructure projects are usually developed in relation to a number of dependent sectors or other related projects, the characteristics of these projects are highly diverse. Infrastructure projects involve a multitude of stakeholders from multiple sectors and backgrounds. For instance, a project developed through public–private-partnership procurement routes will have a large representations of public, private, legal, and financial stakeholders with varied stakes and interests in projects. One of the key requirements in such projects is to know the stakeholders and identify the stakes of all involved over the entire project life cycle. Once the stakeholders are identified, the next big challenge is to map them out with reference to the project and their underlying interests and impacts. Traditionally, numerous practices are adopted by the authorities for engaging stakeholders in the infrastructure projects. Among the practices, town hall meetings or co-design workshops are quite common in most public projects, at least in Australian contexts. In the traditional practice of stakeholder engagement, project issues are presented to the communities in the town hall meetings where authorities present the available options. Then the attendees are given options for making written submissions within a specified timeframe. However, such responses, especially in the open-ended submissions, are often quite subjective and not helpful for ascertaining the interests and impacts at individual levels. Despite the existence of such prevailing practices, anecdotally the community at large finds themselves at lost, which results in resentments and dissatisfactions in public projects. This is because the process of engaging stakeholders in such projects is disengaging and informal. Most of the time, top-down imposition of design solutions coupled with prejudice in the decision-making prevail over open-ended idea solicitations at the early stage of the project. Stakeholders' engagement in such practices often results in sub-standard benefits or no benefits at all in the societal perspective.

Drawing from the inefficiencies in the current practices of stakeholders' engagement, the objectives of the chapter include:

- Establishment of a framework for robust engagement of the stakeholders in infrastructure projects;
- Mapping of stakeholders in relation to their stake, interests, and impacts in specific project contexts;
- Assessment methodology including performance measurement of stakeholders' engagement practice in realising stipulated benefits in societal perspective;
- A roadmap for improving stakeholder engagement practices in infrastructure projects.

6.1.2 Learning outcomes

The following learning outcomes have been identified for this chapter. Readers will be able to:

- Gain an insight into the identification of stakeholders and their interests;
- Describe the stakeholder engagement processes and their inter-relationships;
- Articulate the process of social network analysis in the context of stakeholder engagement;
- Understand the processes of integrating the stakeholders and their stakes for improving the social performance of infrastructure projects.

6.2 Defining stakeholders

As defined in PMBOK (2017), stakeholders include all the people who are impacted by or can impact the project in a positive or negative way. While such a definition entails a broad inclusivity of the community in general terms, estimation of accurate assessment of the impacts on each person is a difficult task. The typical four-phase project stakeholder management processes defined in PMBOK (2017) do not articulate any clear mechanism for quantifying impacts on stakeholders in individual contexts. The four phase includes *identification, planning, management* and *control*. Referring to the PMBOK guide, the first step of identification of stakeholders comprises strict scrutiny of the stakeholders in relation to the scope of project. The process of identification of stakeholders is interlinked with the project phases as well. For instance, the stakeholder base of a project at its inception stage would be vastly different from the tender phase. Similarly, varied contexts as the project proceeds over the life cycle also play a significant role in relation to stakeholder involvement in projects. Focusing on the cost estimation context, Doloi (2011) demonstrated the dynamics of stakeholders, their roles and interests across inception, tendering and initial construction stages. Due to the virtue of infrastructure projects being complex, numerous contexts prevail. With the changes of stakeholder base and underlying characteristics, accurate assessment of the attributes requires appropriate tools and processes. Some of the key attributes that require structural measurement and quantitative estimates include such stakes, interests, impacts, and acceptance threshold including degree of satisfaction (Doloi, 2018). These attributes, if managed well, have the potential to yield an acceptable level of performance, ensuring better inclusivity of projects from the societal perspective. Thus, engagement and coordination of stakeholders' interests and meeting the societal needs and requirements are pivotal considerations in the context of achieving social sustainability outcomes in infrastructure projects. However, process of engagement and coordination of the stakeholders vary vastly in the current best practice including mainstream literature (Bahadorestani *et al.*, 2019; Collinge, 2020; Jing *et al.*, 2011; Wu *et al.*, 2019). Drawing on this knowledge gap, this chapter highlights an alternative framework for enhancing the stakeholder engagement practice and achieving the best value outcomes in infrastructure projects.

6.3 The identification of stakeholders and their interest

Decision-making on infrastructure planning is a complex process and integration of social issues exerts further challenges due to conflicting requirements of the stakeholders in the community. New infrastructure and urban development must be rationalised in their planning, design and operations ensuring fitness-for-purpose and adequate societal value contributions (e.g., local resource utilisation, community wealth creation, protection of nature and heritage) within the community (Leshinsky, 2008). Social inclusivity requirements, especially in the procurements of public projects, are being conceptualised as Social Procurement by many organisations including the Victorian State Government in Australia (Victorian Government, 2020). In order to ensure value creation and complete social acceptance, infrastructure projects must be planned and developed by aligning the needs and requirements of the wider community proactively. One major weakness in the traditional approach is the poor inclusion of the stakeholders' concerns and lack of any model for allowing consolidation of the consultation feedback in an objective decision-making process (Lockie, 2001). Instead, the stakeholders are often considered a nuisance and their interests treated as constraints in achieving the project objectives. Consequently, projects suffer serious setbacks due to political, social, environmental and community challenges and fail through statutory processes (Wittig, 2013).

In viewing success and failure, traditional measures bounded by the iron triangle of cost, time and quality are not quite appropriate in modern infrastructure projects. Our increasingly sophisticated society comprises not only a highly well-connected community but also active and responsible citizens who love to be part of public policies and decision-making, especially when it comes to shaping their built environment. The "not in my backyard" concept prevails most in society when a community not only wishes to be informed but also to be involved in decision-making associated with public goods and services that impact public lives.

On the stakeholder identification front, while noting every single person directly or indirectly connected with the project is important, grouping of the stakeholders based on common characteristics is more useful. Focusing on social sustainability in infrastructure projects, Almahmoud and Doloi (2015) demonstrated three key stakeholders groups prevailed in most projects. These groups are *industry or business community*, *neighbourhood community* and *end-user community*. The industry community is the group of stakeholders comprising the business roles who is responsible for delivering the project as per the business objectives being stipulated in the project contract. Neighbourhood community is the group of stakeholders who will share or live by the project or surround the built environment. This community not only witnesses the change in the physical space resulting from the project but also adapts the surroundings as their living environment. User community is represented by the stakeholders who are the ultimate beneficiaries of the project. By way of adapting to the end facilities delivered by the project, the user community finally gives approval to the projects. End users' acceptance of the project depends on the benefits being realised through improved goods and services, which becomes the testament of the project being a failure or a success from a public or community perspective (Almahmoud and Doloi, 2015).

While these three categories of stakeholders encompass a wide stakeholder base, the classification did not address the stake and interest particularly well. Referring to the PMBOK (2017) guide, categorisation of stakeholders by their power and interest is also an important dimension in stakeholder engagement process. With this in mind, the three categories asserted in Almahmoud and Doloi (2015) require an additional dimension encompassing the *governance community* as one of the key stakeholders, especially in the infrastructure projects. The governance community refers to the stakeholders or stakeholder groups with political power within the jurisdiction of the project. Examples of governance community could range from local council or state government to a relevant section or department or the government at the federal level. Referring to the literature, the stake is defined as a function of interest, impact and influence in most projects (Doloi, 2011, 2013, 2018). However, there is not much evidence on how the stake is measured with an accurate reflection of the underlying issues associated with a particular stakeholder or a stakeholder group. Most of the time, the measurement scale is subjective and based on general perceptions. In such a practice, an appropriate strategy for managing stakeholder expectations is often difficult to implement in project specific conditions.

Figure 6.1 depicts four key groups of stakeholders with perceived level of interest, impact and influence respectively. As seen, the stake is a function of interest, impact and influence and all three attributes are meant to be independent of each other. Interest measures someone's direct or indirect benefits resulting from the project. A person with high interest is not necessarily highly impacted. Impact is a measure of how stakeholders could potentially be affected bu to the project being completed, both at first and well after the physical alteration has taken place. Typically, both positive and negative impacts could trigger high interest from a stakeholder in a project. The measure of influence is about the power that could result from changing a decision or successfully making an alteration in the project configuration.

6.4 Tools and techniques for effective stakeholder engagement

Stakeholder engagement is not a new concept, and there has been a lot of research published that articulates the processes, tools and techniques in the mainstream literature. A few selected pieces of literature are summarised below.

Delgado (2007) asserted nine key tools for successful engagement of stakeholders in infrastructure projects undertaken by the local governments in Australia. These tools are open dialogue (listening circles), visioning, asset based community development, world café, deliberative democracy or citizen deliberative councils, parish mapping, open space technology, photo voice and participatory learning and action approaches (PLA). During the planning phase of infrastructure projects, local governments use these tools in a variety of ways. Open forums, web-based communication or even town hall meetings are common events among local governments where

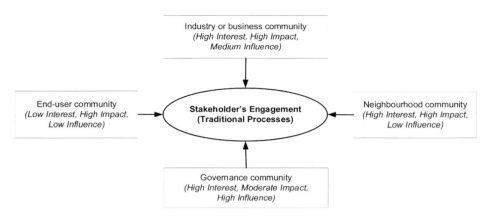

Figure 6.1 Four broad categories of stakeholders and underlying characteristics

the general public is invited to weigh in on a range of project solutions before the best option is chosen. For the sake of brevity, while these tools are not discussed in details here, targeted engagement and context-specific communication practices are at the core of all the tools for achieving successful outcomes in projects.

Baker (2012) described how managing stakeholders is a structured step by step approach leading to an effective communication management plan that includes clear expositions of status, progress, and forecasts over the project life cycle. The first step is stakeholder identification, and the author suggested five key tools: brainstorming, interviews, historical stakeholders base, contracts with vendors and suppliers and social network analysis. Use of a nominal group technique, Crawford slip method and affinity diagrams are also suggested as useful tools in the brainstorming process. For the second step, stakeholder classification, the research suggested to utilise two models: power-interest grid and salience model. In the power-interest grid, a 2x2 matrix comprising power and interest is analysed to derive appropriate communication strategies. In the salience model, three dimensions – power, urgency and legitimacy – are used to derive up to eight different stakeholder groups. These eight groups – dormant, discretionary, demanding, dominant, dangerous, dependent, definitive and non-stakeholder – are then analysed to devise unique strategies to deal with and tailor the management approach in individualised contexts. Resulting from the first two steps, the rest of the three steps – strategy, plan and execution – are subsequently followed to achieve the best value outcomes in most infrastructure projects.

Investigating the complicated relationships between stakeholders' power, interests, and public engagement outcomes, Yu and Leung (2018) developed an integrated structural model highlighting the significance for striking a balance in public engagement in infrastructure projects. The results showed that the outcome of public engagement in projects is the function of power and interest of the stakeholders. Wider stakeholders, including project teams, media and nongovernmental organizations commensurate with the power and interest, should be engaged in the decision-making process.

Investigating in a comprehensive project stakeholder typology model (PSTM) based on stakeholder salience attributes, Bahadorestani et al. (2019) revealed the importance of 15 different types of stakeholder engagement typology based on four key attributes, namely potency, legitimacy, urgency, and proximity. Utilising a Venn diagram for understanding the overlapping areas, the research asserted that stakeholder engagement typologies should be treated independently even though perceived overlapping may exist.

Based on an in-depth analysis of stakeholder engagement practice in a hospital project, Collinge (2020) revealed its theoretical and practical complexity in the management processes. The research asserted that effective stakeholder engagement lies within the responsibility of the project team and should be managed with appropriate organizational structure and underlying scope of work being undertaken in the project. Corporate social responsibility (CSR) and ethical standards of the organisations need to reflect stakeholder engagement strategies exclusively for the specific projects.

In short, effective engagement is not simply a process-driven exercise; its approach to projects is strategy based and context specific. Projects are unique entities and the extent of the stakeholder base depends on the development contexts and operating environments of the projects. Understanding the intersection points between the project, stakeholders' interests, impacts and stakes is an important first step for devising effective strategies for engagement and management of multifaceted stakeholders in infrastructure projects.

6.5 Formulation of appropriate stakeholder engagement strategies

Based on the research by the author and with subsequent publications on the topic (Almahmoud and Doloi, 2015; Doloi, 2011, 2012; Doloi, 2018), a clear pattern has emerged on alternative but objective stakeholder engagement processes in public projects. Figure 6.2 depicts a schematic diagram of a stakeholder engagement process applicable in most projects. The remainder of the section explains the processes highlighting the progressive steps in the context of the project's life cycle.

6.5.1 Identification of stakeholders, project contexts and issues

As seen, the process requires not only the identification of the stakeholder base at the outset but also clear exposition of the contexts and issues associated with the project. As the stakeholders and the project are intricately linked, the stake of a particular stakeholder is a function of numerous connected elements in relation to the contexts and issues in the project. Without sufficient breakdown of the project issues and specific contexts, the connection points of the stakeholders are not possible to ascertain. For instance, some of the broader issues under the financial context could be return on investment or cost benefit assessments. Similarly, a few prevailing issues under the environmental context could be embodied energy or carbon emission standards, for example. In this section, a sincere attempt must be made to identify all the stakeholders in the entire infrastructure project life cycle. With reference to the work breakdown structure of the project, all the major project issues under various contexts should be listed out before proceeding to the next section. This process requires careful scrutiny of both stakeholders and project managers so that a clear roadmap can be established for engaging stakeholders with relevant issues in specific contexts.

6.5.2 Mapping of stakeholders with project issues

In this section, the stakeholders are required to be placed in such a way that their connection points with the project issues are clearly established under each of the relevant contexts. This process is known as mapping whether directional or unidirectional maps are constructed that depict the flow of information or cause-effect phenomena. Usually, such mapping requires stakeholders' direct input. Upon presenting the issues and possible implications, stakeholders are able to determine the connection points and visualise the project. While there may not be any particular process for extracting the key project issues for stakeholder mapping, project business case forms a good basis for deciphering the project for visualising by the stakeholders from their interest and impact perspectives.

6.5.3 Stakeholder interest, impact and expectations

The focus of this process is to receive the stakeholders' input on all the project issues that they perceive from interests, impacts, expectations and demand perspectives. How a stakeholder becomes interested in a particular infrastructure project issue and how such issue could impact them need to be ascertained objectively. This is possible only when the underlying project issues are reviewed and understood by the stakeholders at an individual level. Similarly, minimum expectation levels of the stakeholders on each of the project issues should also be ascertained for setting up the overall expectation threshold for the project. Usually the Likert scale is quite effective for measuring these attributes. One of the key preceding tasks of collecting the mapping data is making stakeholders aware of all the infrastructure project concerns and involving them in and educate them about all relevant aspects. This is usually done using a range of media outlets including town hall meetings as discussed earlier.

6.5.4 Evaluation of social performance

By way of collecting the data on interests, impacts and expectations, network models can be constructed for performing social network analysis (SNA). The details of the SNA models, underlying network characteristics and algorithms for converting individual stakeholders' responses into a single social performance indicator at the project level, are discussed in Doloi (2012, 2018). A brief discussion on the SNA methodological aspect is included in the following section. As the stakeholders are recruited from all four categories as shown in Figure 6.1, the amount of data gathered in this process could be quite overwhelming and analysis of the networks could be computationally challenging as well. The time of data being collected from respondents would reflect the current state of the project. Thus, data needs to be collected from the relevant stakeholders as the infrastructure project proceeds over its life cycle so that social performance can reflect the overall health of the infrastructure project and encompass both short- and long-term objectives.

6.5.5 Monitor, control and maintain

As seen in Figure 6.2, continuous assessment and evaluation is the key to monitoring and maintaining social performance as projects proceed over the life cycle. There are two aspects to the assessment and evaluation of social performance: real or perceived performance being collected at a particular time in the project and the highest level of standard or performance expected at the same time. The expected level of performance usually sets the infrastructure project's performance threshold, which is instrumental in understanding how the infrastructure project is performing at the time of consideration. Thus, the social performance indicator needs to be compared with the project's social threshold for evaluating "Go" or "No Go" decisions. If the performance is equal or higher, strict monitoring and controlling processes are put in place for maintaining the performance at the expected levels. If the performance level is found to be below the acceptable threshold, project options including design configurations are altered or adjusted to repeat the same process until a favourable outcome is achieved.

Stakeholder engagement in major infrastructure projects

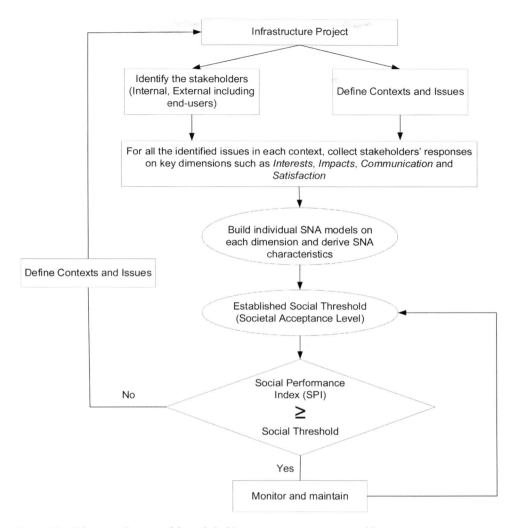

Figure 6.2 Schematic diagram of the stakeholder engagement process in public projects

6.6 Assessment of stakeholders impacts and their inter-relationships

Drawing from the author's own research, this section focuses on the methodological underpinnings of social performance assessments with respect to extended stakeholders' contributions leading to the social sustainability outcomes. Central to the evaluation framework is that organisations can best achieve social objectives in projects by integrating the stakes of extended stakeholders into their core organisational strategies and business operations. Figure 6.3 shows the conceptual measurement of social value performance against the threshold value over the project life cycle. As mentioned earlier, the social performance threshold is a function of the highest level of expected performance in the project informed by the collective responses being collected from wider stakeholders at the stipulated time phase within the project life cycle. Social performance threshold is not necessarily a fixed value but it's a value up to 100 percent depending on how the stakeholders at large have perceived the benefits of the project being developed. The change

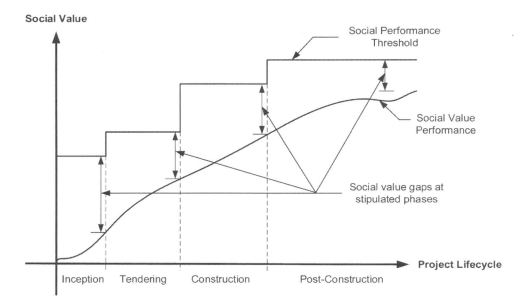

Figure 6.3 Conceptual measurement of social value performance against the threshold value over the project life cycle

in social performance value of projects is shown as a stepped graph in Figure 6.3. For instance, an infrastructure project at the middle of a busy urban precinct may create a massive operational disturbance to the community during the construction phase. Without much knowledge, the stakeholders generally perceive such a project to be not quite useful and thus provide poor rating, reflecting a lower social threshold value at the particular state of interaction. Similarly, social value performance at the time of the data being collected is the function of collective perceptions of the stakeholders in relation to the interests, impacts, communication, understanding and satisfaction on all the key project issues. With the same example project, suffering from the longer operational disturbances, a section of the community could completely disregard any potential benefits and rate the social value performance at its lowest.

While the measurement of social performance threshold is relatively easy, assessment of social value performance requires a great deal of computations and network analysis using SNA. The difference between the social performance threshold and social value performance is the social value gap at the stipulated project phases. The social value gap provides a clear indication about the opportunities for improving the social bottom line in projects from effective stakeholder engagement. A brief discussion of the methodology is included in the following section.

6.7 Social network analysis and network characteristics

Based on the stakeholder theory and behavioural science principles, criticality and legitimacy and power of stakeholders can be analysed and considered for characterising the web of interests and relative trade-offs within the project. As the social performance is primarily focused on the attitude of the society towards the project, SNA is a powerful tool to analyse the influence of stakeholders and the impact of the project on society. In recent years, SNA has been increasingly

Stakeholder engagement in major infrastructure projects

used by researchers to investigate complex patterns of interactions among stakeholders (Lyles, 2014; Mandarano, 2009). The network perspective examines the actors with respect to their relative positions, degree of connectivity and power of influence in the relationships shared directly or indirectly in projects (Pryke, 2015). Based on two case studies conducted in the United States, Lyles (2014) illustrated applicability of the SNA concepts for examining the planners' involvement and their roles in the environmentally oriented planning processes. The research asserted that greater integration of planners in stronger networks results in greater contributions in practice led by other professions (Lyles, 2014). Diversity of stakeholders with interdependent interests can exercise significant power and influence on one another in the network. Evaluation of the formation of new relationships and underlying structures is highly significant for building social capital and achieving optimal social benefits in the collaborative planning processes (Mandarano, 2009).

The application of social network analysis requires comprehensive understanding of the actors (stakeholder groups), their roles and relationships to other stakeholders or issues associated with the project. SNA analysis provides meaningful insights in terms of stakeholder influence in project development contexts. Thus the key contribution of this chapter is a robust sustainability assessment framework, especially with a particular focus on social dimension by employing SNA as the research method. The framework allows comprehensive social performance evaluation to be integrated into the complex interactions with the multitude of stakeholders; the results of the model allow a precise and deeper understanding in social outcomes within the construction projects. With reference to the network characteristics that result from the SNA process, both centrality and density measures are considered as good indicators of social capital outcomes at both individual and community levels. The next sections will discuss the framework focusing on the processes, modelling, applications, results and interpretations.

6.7.1 Evaluation of social value performance

The SNA models are built with the data collected from stakeholders' responses either in binary or in weighted scales. If binary, respondents are required to only indicate whether they have any interest in or impact from a particular infrastructure project issue being presented. Most of the time, such responses are based on one's own perception gathered from the available infrastructure project information and other indirect sources such as social media. For weighted scale, respondents are asked to provide not only their level of understanding but also degree of significance in relation to the specific project issues.

While SNA yields numerous network characteristics, centrality measures provide significant information about how a particular actor can function within the network. With the information on how actors (i.e., stakeholders) interconnect with each other, identifying the most influential stakeholders is an essential step in stakeholder analysis. Centrality in the network identifies prominent actors who have high involvement in many relationssships, namely highly influential stakeholders in the network (Freeman, 1984). The degree centrality (DC) is calculated by the sum of direct ties to other actors. The eigenvector centrality (EV) signifies the influence of one actor over the others, which is essentially the measure of extent of actors' nodes being connected within the network structure. Strong and extensive ties to other nodes indicate that the stakeholder is more likely to influence others and is thus more important (central) in the network (Knoke and Yang, 2008).

Referring to Doloi (2018), the mathematical expression of DC and EV are shown in Equations 1 and 2 below.

DC is expressed in Equation 1 as below:

$$C_{DC}(N_i) = \frac{\sum_{j=1}^{g} x_{ij}}{(g-1)}, \; i \neq j \qquad \qquad \textbf{Equation 1}$$

where

$C_{DC}(N_i)$ denotes the degree centrality for node i; $\sum_{j=1}^{g} x_{ij}$ sums up the intensity of directties

that node i has to the $g-1$ but excluding j nodes (exclude node i itself)

EV is expressed in Equation 2 as below:

$$C_{EV}(N_i) = \frac{\lambda \sum_{j=1}^{g} x_{ij} e_j \lambda}{(g-1)}, \; i \neq j \qquad \qquad \textbf{Equation 2}$$

where

$C_{EV}(N_i)$ denotes the eigenvector centrality for node i, e_j is the eigenvector centrality of node

j and λ is the eigen value. $\sum_{j=1}^{g} x_{ij} e_j$ sums up the influence of direct ties that node i has to the

g-1 but excluding j nodes (exclude node i itself). The social value from an individual stakeholder's perspective is derived from the assessment of their interests in the project (Doloi, 2012, 2018). The interest of the stakeholders is converted into the weight for the criteria, whereas the individual evaluation of social performance is the weighted average of the satisfaction level. The mathematical expression of evaluating social value is shown in Equation 3 below:

$$\text{Social value } \hat{S} = \text{norm} \sum_{i=1}^{n} \frac{\sum L_i I_i}{\sum I_i} \qquad \qquad \textbf{Equation 3}$$

where L_i is the satisfaction level in terms of the criterion i, and I_i is the stakeholder's interest of criterion i. 'Norm' denotes the normalisation to obtain a large coefficient of each of the network topology measurements (e.g., value 0.000 and 1.000).

Social value threshold is expressed in Equation 4 below:

$$\text{Social value threshold (target) } \hat{\xi} = \text{norm} \sum_{i=1}^{n} \frac{\sum E_i I_i R_i}{\sum n_i} \qquad \qquad \textbf{Equation 4}$$

where E_i, I_i, and R_i are the expectation, interest and impact levels in terms of the criterion i. 'Norm' denotes the normalisation to obtain a large coefficient of each of the network topology measurements (e.g., value 0.000 and 1.000). The overall social value of the project can be derived according to the individual social value and the respective stakeholder's influence in the network. The overall social value of a project is represented by an index known as SPI (Doloi, 2012). Thus, SPI is a measure of social performance of the project and is defined as the sum of social value

Stakeholder engagement in major infrastructure projects

from individual stakeholders multiplied by the influence of the stakeholder, i.e., degree centrality and the mathematical expression is shown in Equations 1 and 2.

$$SPI_{DC} = \text{norm} \sum_{i=1}^{n} \sum_{i=1}^{n} C_{DC}(N_i) \star S_i \Bigg/ \sum_{i=1}^{n} C_{DC}(N_i)$$

Equation 5

where

SPI_{DC} is the social performance index based on the network being characterised by the degree centrality, $C_{DC}(N_i)$ is the degree centrality of node i, and S_i is the social value with regard to stakeholder i. A similar representation of ths social performance index based on the network being characterised by the eigenvector centrality (SPI_{EV}) of can be expressed in Equation 4 below:

$$SPI_{EV} = \text{norm} \sum_{i=1}^{n} \sum_{i=1}^{n} C_{EV}(N_i) \star S_i \Bigg/ \sum_{i=1}^{n} C_{EV}(N_i)$$

Equation 6

As seen above, SNA models result in most of the required parameters for calculating the social performance indicators (SPI) shown in Equations 5 and 6. As the eigenvector is a measure of power or influence, the determination of SPI using eigenvector (e.g., Equation 6) is considered to be more reliable than the SPI using degree centrality values.

6.8 Chapter discussion questions

Having discussed the SNA-based model for encompassing wider stakeholders in the form of associated network structures and thereby evaluating social performance in infrastructure projects, some questions emerge while applying the model in practice. Following are a few questions that require answers when applying the model in practice.

1. What are the advantages of aligning stakeholder interest with project objectives?
2. How do stakeholder engagement processes on infrastructure projects apply?
3. What benefits of social network analysis can be used to enhance infrastructure project delivery?

As discussed above, by the virtue of infrastructure projects being complex, alignment of the large stakeholder base with varied stakes and interests is a difficult task. Yet, objective integration of their interests and mitigation of adverse impacts provide an aided advantage in achieving the true value outcomes in projects. Lack of adequate alignment of stakeholder interest with project objectives during the front-end planning phase could potentially result in community backlashes during the construction stage of the projects. In such a situation, financial risks at the operation phase of the project eventually make the projects fail across multiple fronts. In the event of stakeholders being misaligned with project objectives, perceived failure, especially in the social contexts, is one of the most significant risks in capital intensive infrastructure projects. Composition of stakeholders in relation to infrastructure projects is highly complex. Thus, appropriate methodology and processes for engaging the stakeholders to ensure inclusivity and partnerships is fundamentally important for achieving success in infrastructure projects. Among numerous processes being applied for stakeholder engagement, network structure is found to be efficient for dealing with complexity of interaction and behavioural dynamics leading to deriving tangible outcomes. The following section demonstrates the application of SNA and its efficacy in enhancing the delivery of infrastructure projects.

6.9 Case study

In this section, a new research-based stakeholder engagement practice as an alternative to the traditional practice is being demonstrated by applying it to an infrastructure project. Based on a rigorous stakeholder engagement practice, assessments of the social value-based performance were undertaken. As seen above, SNA models resulted in most of the required parameters for calculating the SPI values shown in Equations 5 and 6. For demonstration, degree centrality measures are used to determine SPI values (e.g., Equation 6), which are then compared against the project threshold with respect to the specific project phase. The significance of the findings is highlighted in the context of the stakeholder engagement process.

6.9.1 Case study: A foreshore development project (Project A)

An inner city local council in Melbourne has been working for almost 10 years on an aspirational and inclusive design of a foreshore development project. Due to commercial confidentiality, the project has been anonymised as Project A. Once developed, the site for the project will have the potential to add significant social and economic value with increased liveability and high visitor traffic. Due to an active local community and sensitive heritage overlay in the location, the effort of finding an acceptable design solution has not been successful.

Following the traditional stakeholder engagement practices, numerous co-design workshops have been conducted by both the council and the project task force. A design consultant was engaged to develop design options by incorporating responses collected from co-design workshops and community consultations. Focusing on one of the most preferred designs at the time (Design D), research was conducted by presenting the design to the community and evaluating its general acceptance.

Following the university's research ethics protocol, respondents were recruited over a two-month period and data was collected in a specified questionnaire template suitable for SNA modelling as a research method. While a vast number of project issues exist, for the sake of brevity, only eight key issues were selected for the research: 1) local park/public space, 2) walking and cycling paths, 3) complements the role and form of a heritage theatre, 4) adequate car parking, 5) serves the wider community, 6) accommodates a high volume of people, 7) a green, urban character, and 8) maintains views of the bay.

Upon presenting the above broad issues and appropriate descriptions to make the questions understandable for the respondents, data was collected to capture respondents' interests, impacts, communication and satisfaction on each issue on either a five-point Likert scale or on a binary Scale. Similarly, the value judgement was also collected using a 5 point Likert Scale in the context of how the respondents see or realise the value in the overall proposition of the project. This data was required for establishing the project's Social Threshold Value as shown in Figure 6.3.

Two-mode SNA models depicting the issues versus interests, impacts and satisfaction were developed to study the underlying network characteristics. Also, one-mode communication networks on each of the eight issues were developed for depicting the relative position of stakeholders in the network structure. Figure 6.4 shows the sociogram of a two-mode network of 54 respondents and eight issues; the nodes are based on the degree centrality

Stakeholder engagement in major infrastructure projects

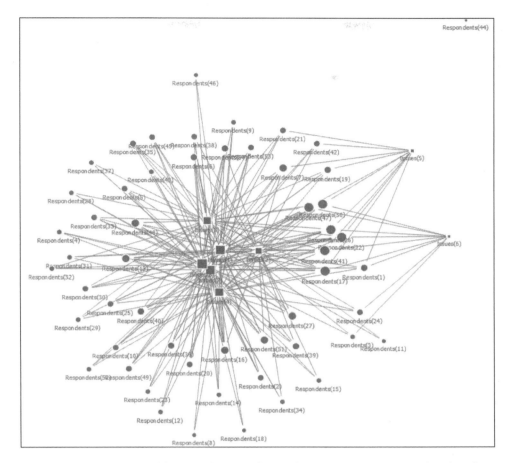

Figure 6.4 Two-mode network between the respondents and project issues in Project A (based on degree centrality in the impact network)

measures. Figure 6.5 shows a corresponding concentric map of the respondents depicting the relative spread of the nodes from center to periphery based on the degree centrality measures. Both Figures 6.4 and 6.5 clearly demonstrate the relative positioning of the respondents with respect to the impacts being exerted by the project issues, respectively. By studying all the key characteristics across all the relevant networks and performing the computations as per Equations 1–6 presented earlier, social value performance in the project was determined. Figure 6.6 depicts the comparison of the social value performance (SPI =19.35 percent) with the social value threshold (65.45 percent) in the project. As seen, there is a social value gap of 46.10 percent at the time the study is being undertaken (e.g., design stage). While the social gap reveals an immediate opportunity for improving the project design option and lifting the project's bottom line in the societal context, a threshold value gap of 34.55 percent (100 percent minus 65.44 percent) indicates a significant gap in a project's business case and perceived understanding of the benefits among the community at large. In an idealistic situation, the project design and the business case should yield a social value threshold close to 100 percent so that the opportunity for improving the social value performance is maximised.

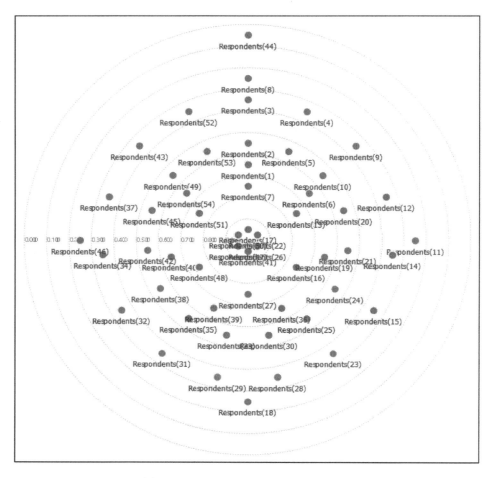

Figure 6.5 Concentric map of the respondents in Project A in the impact network

6.9.2 Case discussion questions

While the comparison of social value performance with social value threshold highlights the value gap in the current state of the project and thereby signifies the opportunity for improvement, the efficacy of the results depends on the quality of the data and authentication of the sources. Some of the obvious questions could well be how the study of social networks of stakeholders in relation to the measured phenomena (such as communication, interest, impacts, and satisfaction) supports and enhances the overarching objective of stakeholder engagement practice in infrastructure projects. The answer to this question lies within the efficacy of the SNA methodology where network study and underlying network characteristics not only have the potential for pinpointing particular stakeholders with respect to their relative stake but also strategies for taking targeted measures for managing the expectations well in the project. By the virtue of infrastructure projects being complex and large, a multitude of stakeholders gets involved in various stages. Traditional stakeholder management practice usually engages the stakeholders through common platforms such as social media, handouts and mass communications or even

Stakeholder engagement in major infrastructure projects

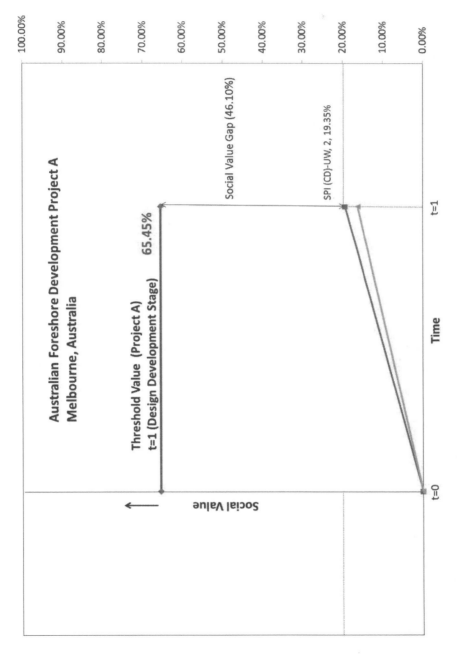

Figure 6.6 Social value performance in Project A (design stage)

drop-in information sessions. In such an approach, often customised stakeholders' response plans commensurate with the individualistic stake are ignored. Instead, both relevant and irrelevant information are communicated with all the stakeholders at large. As a result, stakeholders with specific interests or concerns on particular project issues do not get the necessary attention and the overall process is perceived to be poorly managed in the community. SNA-based methodology provides a rational approach where stakeholders with relative stakes are easy is identify where a targeted response plan can be devised to manage the expectation at an individual level.

Referring to Figures 6.4 and 6.5, the connectivity of all the respondents' nodes with the issues are not quite uniform. For instance, in Figure 6.4, while 56 of 57 respondents are somehow connected to at least one issue in the network, respondent 44 is completely isolated. This means that respondent 44 has either no impact from any issues or is completely unaware about the project. In such a situation, the underlying attributes or properties of the respondent are important to review, investigate and establish the root cause of disconnect. If the respondent is found to be left unconnected, an attempt must be made to engage the stakeholder and manage the expectations commensurate with their stake in the project. In regard to Figure 6.5, scattered distribution of the stakeholders' location in the concentric circle depicts the differences in impacts of the issues in relative terms. When combined with interest and communication networks, the resultant position of the nodes provides a clear understanding of relative positioning of the stakeholders with respect to the project.

While studying stakeholder networks in infrastructure projects, apparently questions such as how to connect and engage the stakeholders in a meaningful manner remain unanswered. By bringing the stakeholders closer to the project, closer or dense network structure is possible to ascertain. Well-connected stakeholder networks are usually the result of exhaustive and targeted communication which potentially leads to increased interest and high degree of satisfaction. By studying the key characteristics of all relevant networks, the strategy should be devised in such a way that the resulting SNA measures are improved and balanced networks are achieved with respect to the stipulated project issues. As these quantitative or structured analysis of the social networks of the stakeholders is closely linked to the social performance outcomes, a stringent process for not only engaging but also on-boarding all the key stakeholders is highly crucial for improving the social bottom line in projects.

6.10 Chapter summary

This chapter puts forward a research-based new approach for engaging stakeholders in infrastructure projects. Involvement of a multitude of stakeholders, especially in large infrastructure projects, cannot be viewed as a single one-off phenomenon at the front end of the project. Rather, understanding the evolution of different stakeholder communities, identifying their unique needs and requirements and effectively managing the project as it proceeds over the life cycle, is crucial for ensuring inclusivity of the public-funded projects in societal context. The traditional boundary of the stakeholder base is often limited to the business community and general public. While the business community is considered to be an insider with direct involvement in the project as a business, the general public is usually the people living in the vicinity of the project. In such a broad categorisation of the stakeholders, targeted communication and response plans are difficult to implement and often mass communication results in spreading incorrect perceptions, undermining the strategic intents of the project. Moreover, the practice of community consultation including co-design workshops as in case project A does not allow tracking and tracing of the community responses in progressive improvement of design options leading to the an optimal and socially inclusive project solution.

The chapter puts forward an alternative framework for engaging the broad stakeholder base with respect to the key project issues rather than the entire project. Considering the intents, objectives and types, stakeholders have been divided into four categories: industry community, neighbourhood community, end-user community and governance community.

The process of stakeholders' identification and mapping in relation to the key project issues has been demonstrated in both theoretical and practice perspectives. The significance of the social network of stakeholders and network characteristics in determining the social value performance in the project is discussed. Having applied the SNA-based research methodology at the design stage of the case study project, the importance of continuous stakeholders' engagement spanning the project life cycle is clearly highlighted. The social value-based performance assessment framework, especially in the public infrastructure projects, is a clear shift to ensure social inclusivity and improve the traditional front-end stakeholder engagement practice. The quantitative assessment of social performance coupled with the existing practices of environmental and economic assessments adds a significant new opportunity for lifting the TBL performance of the infrastructure projects. This framework is expected to enhance the implementation of the social procurement policies being rolled out by the public organisations with structured or quantitative measures of benefits in infrastructure projects.

References

Almahmoud, E.S. and Doloi, H. (2015). Assessment of social sustainability in construction projects using social network analysis. *Facilities*, 33(3/4), pp. 152–176.

Baker, E. (2012). Planning effective stakeholder management strategies to do the same thing! Paper presented at PMI® Global Congress 2012—North America, Vancouver, British Columbia, Canada. Newtown Square, PA: Project Management Institute.

Bahadorestani, A., Karlsen, J.T., and Farimani, N.M. (2019). A comprehensive stakeholder-typology model based on salience attributes in construction projects. *Journal of Construction Engineering and Management*, 145(9), 04019048.

Collinge, W. (2020). Stakeholder engagement in construction: Exploring corporate social responsibility, ethical behaviors, and practices. *Journal of Construction Engineering and Management*, 146(3), 04020003.

Delgado, L. (2007). Education for Sustainability in Local Government: Handbook. Canberra: Australian Government Department of the Environment, Water, Heritage and the Arts, and Australian Research Institute in Education for Sustainability.

Doloi, H. (2011). Understanding stakeholders' perspective of cost estimation in Project Management. *International Journal of Project Management*, 29(5), pp. 622–636.

Doloi, H. (2012). Assessing stakeholders' influence on social performance of infrastructure projects. *Facilities*, 30(11), pp. 531–550.

Doloi, H. (2013). Cost overruns and failure in project management: Understanding the roles of key stakeholders in construction projects. *Journal of Construction Engineering and Management*, 139(3), pp. 267–279.

Doloi, H. (2018). Community-Centric Model for Evaluating Social Value in Projects. *Journal of Construction Engineering and Management*, 144(5), *DOI:* 10.1061/(ASCE)CO.1943-7862.0001473

Elkington, J. (1998). Partnerships from cannibals with forks: The triple bottom line of 21st-century business. *Environmental Quality Management*, 8, pp. 37–51.

Freeman, E.R. (1984). *Strategic management: A stakeholder approach*. Boston, MA: Pitman.

Jing, Y., Shen, G.Q., Manfong, H., Drew, D.S., and Xiaolong X. (2011). Stakeholder management in construction: An empirical study to address research gaps in previous studies. *International Journal of Project Management*, 29, pp. 900–910.

Knoke, D. and Yang, S. (2008). *Social Network Analysis: Quantitative Applications in the Social Sciences*, Sage Publications, Thousand Oaks, CA.

Leshinsky, R. (2008). Knowing the social in urban planning law decision making. *Urban Policy and Research*, 26(4), pp. 415–427.

Lockie, S. (2001). SIA in review: Setting the agenda for impact assessment in the 21st century. *Impact Assessment and Project Appraisal*, 19(4), pp. 277–287.

Lyles, W. (2014). Using social network analysis to examine planner involvement in environmentally oriented planning processes led by non-planning professions. *Journal of Environment Planning and Management*, 58(11), pp. 1961–1987.

Mandarano, L.A. (2009). Social network analysis of social capital in collaborative planning. *Society and Natural Resources*, Issue 22, pp. 245–260.

PMBOK (2017). A guide to project management body of knowledge, Project Management Institution (PMI), Sixth Edition, USA.

Pryke, S. (2015). Analysing construction project coalitions: Exploring the application of social network analysis. *Construction Management and Economics*, 22(8), pp. 787–797.

Victorian Government (2020). Social procurement—Victorian Government approach, https://www.buyingfor.vic.gov.au/social-procurement-victorian-government-approach [Accessed on 1 June 2020].

Wittig, R. (2013). Mega project development: Optimising current practices and strategies. *13th Coal Operators' Conference. The Australian Institute of Mining and Metallurgy Association of Australia*, pp. 392–398.

Wu Lufeng, Guangshe Jia and Nikhaphone Mackhaphonh (2019), Case Study on Improving the Effectiveness of Public Participation in Public Infrastructure Megaprojects, *Journal of Construction Engineering and Management*, 145(4), DOI: 10.1061/(ASCE)CO.1943-7862.0001623

Yu, J. and Leung, M-Y. (2018). Structural stakeholder model in public engagement for construction development projects. *Journal of Construction Engineering and Management*, 2018, 144(6), 04018046. https://doi.org/10.1061/(ASCE)CO.1943-7862.0001462.

7

MANAGING MULTI-PARTNER COLLABORATIONS ON MAJOR INFRASTRUCTURE PROJECTS

Diana Ominde and Edward Ochieng

7.1 Introduction

The delivery of complex major infrastructure projects requires multi-partner collaborations (Chew, 2004). Each phase of a major infrastructure project life cycle presents multiple opportunities for collaborative practices, and in many cases, these practices have a meaningful impact on delivery times as well as costs and can improve project performance on other metrics (Chew, 2004). To achieve these benefits, infrastructure owners must be open to incorporate contractors' input early in the process, select the right contractors, clearly articulate the potential incentives and then work collaboratively with those contractors to develop, apply and standardise best practices. This chapter will examine how to formulate a comprehensive multi-partner model for infrastructure systems delivery.

7.1.1 Chapter aim and objectives

This chapter aimed to evaluate some of the management practices in collaborative infrastructure projects and assess the modalities through which these management practices influence the outcome of these projects.

- Multi-partner collaborations come with immense challenges both to the implementing teams and the project as a whole. What this means is that the adoption of a multi-partner project collaboration model faces certain management challenges that ought to be examined in detail if these project outcomes are to be enhanced. In this regard, this chapter will be discussing some of the challenges that have been experienced in these multi-partner collaborations.

- At the same time, the chapter shall evaluate some of the perceived barriers to these multi-partner collaborations. The ultimate focus of establishing the barriers as well as challenges is to help develop a "best practice" framework in managing these multi-partnered projects.
- Within the context of monitoring and evaluation, the problem is that there is no established standard of evaluating these multi-collaboration project initiatives. This chapter will, therefore, offer a detailed discussion regarding the standardisation process and the benefits of having a standardized model of evaluating the multi-partner collaborations on major infrastructure projects.
- Are there specific defined benefits of multi-partner collaborations on major infrastructures? The significance of developing a management model for multi-partner collaborations will be evaluated in detail. The chapter will assess the avenues through which project stakeholders in any major infrastructure projects can accrue the relevant benefits of this initiative. In the conclusive sections of the chapter, a detailed case scenario will be offered to give more insight into the concept of management of the multi-partner collaborations in infrastructure projects.

7.1.2 Learning outcomes

The following learning outcomes have been identified for this chapter. Readers will be able to:

- Formulate a comprehensive multi-partner framework for infrastructure systems;
- Identify key challenges of multi-partner collaborations and propose solutions to those challenges; and
- Depict how multi-collaborations can be aligned and adopted to infrastructure systems.

7.2 Planning for a comprehensive multi-partner framework for infrastructure systems delivery

The complexity of infrastructure projects today has emboldened research into the modalities of partnering amongst the various stakeholders to enhance the outcome of these projects. Indeed, the many value creation processes, such as major investments in infrastructures, have become too complex in terms of scope to the extent that single organisations are becoming increasingly strained in managing these infrastructures (Chew, 2004). There are several inherent challenges as well as complexities that are related to the multi-partner project collaborations. Despite these challenges, a comprehensive multi-partner framework for the delivery of infrastructures has not been fully developed. A bid to counter these inherent challenges related to multi-partner projects is based on the suggestions that there is very little known regarding the management practices that are required in the enhancement of the collaborative framework required for the multi-partner infrastructure projects. Moreover, this chapter will seek to develop a broad framework that applies to manage multi-partner collaborations in infrastructure projects. Based on the notion that an increasingly bigger number of infrastructure projects are aiming to strengthen various collaborative models to be able to enhance project outcomes. This has been largely achieved through the collocation of the project team members from diverse implementing entities within the same physical space as a way of advancing the project outcomes.

The conceptualisation of an effective multi-partner collaboration framework must be founded within a well-designed stakeholder integration model Demirkesen and Ozorhon (2017) conventionally; multi-partner projects come with a heightened challenge of stakeholder management in the project. Multi-partner projects encompass an extra group of stakeholders whose interest in the projects has to be adequately taken care of during the project execution phase. Naturally, this will further complicate the project stakeholder management schema because a cluster of these stakeholders has been grafted within the project execution phase. In the development of an operational framework for these projects, therefore, the question of stakeholder integration and management ought to dominate these discussions regarding the framework of multi-partner collaborations in these infrastructure projects (Sperry and Jetter, 2019).

We are therefore staring at a restructured stakeholder integration model in the project when designing a multi-partner collaboration strategy for infrastructure systems. The project implementing teams are mandated to configure conceptual scaffolds that would be applied to ensure that the newly included stakeholders within the project are well integrated within the project implementation matrix. What emerges, therefore, is that the multi-partner collaboration approach on major infrastructure systems has to go beyond the mere fact that a new form of collaboration has been established and has to interrogate the modalities through which the new crop of project implementers will be adequately included within the main project execution schedule. One could therefore suggest that one of the basic weaves of thread that are important in laying down foundations for a strong multi-partner collaboration could be based on the notion that a well-defined stakeholder management framework is mandatory. By talking of the stakeholder integration process, what comes out is the extent to which the project is likely to engage the two clusters of stakeholders and enable them to engage systematically to align their expectations in the process (Doloi *et al.*, 2016). What is evident is the fact that the construction of a comprehensive multi-partner framework for infrastructure projects ought to, create a platform onto which these partners can pillar their operations. The initial phase in designing this multi-partner framework of project operations ought to be a scaffold within a well-defined stakeholder integration framework.

One of the drivers of a well-designed stakeholder integration model in these multi-partner projects is based on the notion that the project implementing teams do share spaces (Eberlein, 2008). Often these more integrated ways of working include the utilisation of a shared workspace where project team member from different companies are physically co-located. These shared locations are referred to as "big rooms' or 'co-locations"; they are physical spaces that accommodate all these classes of stakeholders implementing the project (Xia *et al.*, 2018). But more deeply, the concept of co-location or a big room does not limit these stakeholders into a specific physical space. However, it equally refers to the information as well as the formal practices of collaboration that do occur within these spaces. These co-locations are meant to enhance the face-to-face practices of collaboration amongst these partners in the project (Kokkonen and Vaagaasar 2018).

What is evident, therefore, is that developing a framework for co-partnerships in advancing infrastructure projects needs to have a platform onto which this interaction ought to be developed. Even though progress in research studies has not been able to conclusively suggest the modalities of organising the space or the "big room" there have been insinuations by several systematic reviews that the "outlook" of these spaces can be well defined within a specific stakeholder integration model. In essence, this further emboldens earlier suggestions that the very basis of constructing the multi-partner framework for the growth of infrastructure ought to be the establishment of the dimensions of the "big room," and this is well achieved through the stakeholder integration and engagement model adopted in the project.

7.2.1 Collaboration mediators

A conceptual analysis of extant literature has effectively identified several collaboration antecedents. At the same time, several studies have proposed that there are several collaboration outcome elements in any infrastructure project. In essence, therefore, the design of any multi-partner framework for infrastructure systems thus ought to consider these antecedents for collaborations and tailor them towards realising the anticipated project outcomes. What is important, however, and has been postulated by several publications is that there ought to be a focused discussion on the mediators of project collaborations. Developing a design to manage multi-partner collaborations in any project ought to be cognizant of these mediators as key drivers of the project delivery (Mortensen and Haas 2018).

7.2.2 Knowledge integration capacity

The ability of the project to effectively apply theory into practice or knowledge into an actionable event is the basis of knowledge integration (Wang *et al.,* 2018). The integration of knowledge is one of the biggest challenges faced in the configuring of a stakeholder integration model in multi-partner projects. Concerning the difficulties experienced regarding knowledge integration, individuals who are from varying areas of competencies are paramount. Those who have been socialised differently in terms of their key competencies must find it very hard to operate within a common plank that is acceptable for all of them. The individual knowledge within the collaboration locus must be well integrated as a way of advancing the project progress. As a mediator in the multi-partner collaboration projects, knowledge integration is required within the project execution matrix to be able to harness the key skills and competencies in an area and apply them to advance the delivery of these projects (Kokkonen and Vaagaasar 2018).

The translation of knowledge into action means that the teams that are collaborating, regardless of the methodological variations in the manner in which these skills are applied, have to find some kind of modality to apply their operations. There is a suggestion from a number of publications that multi-partner frameworks for the advancement of the delivery of infrastructure projects ought to be developed from the perspectives of knowledge integration. The suggestions herein proposed by several research studies means that in drafting an operational manual for the multi-partner framework for infrastructure development, the modalities into which the existing theoretical knowledge by the various teams can be well integrated within a standard and acceptable means to advance the delivery of the projects has to be considered. Knowledge integration capacities are required to ensure that collaboration and specialist knowledge are aligned with the project benefits (Kokkonen and Vaagaasar 2018).

The framework for operationalising multi-partner projects thus relies greatly on the aspect of knowledge transfer to be effective. Having a platform on which these teams can operate is indeed a great place to advance these projects. Various other aspects of knowledge development have been proposed as critical in the transformation of theory into practice. In essence, this means that the framework for operationalising multi-partner projects ought to be very concerned about the modalities by which these teams are able to operate within the same plane of philosophy as a way of enhancing the delivery objectives of the infrastructure project. An aspect of knowledge capability seems to be quite salient in the framing of a template for these multi-partner projects. The project management teams, ideally, must construct the avenues through which theory is translated into actions. It is not sufficient to have a wider skill base in these collaborative infrastructure projects. Suffice to say, no meaningful progress can be made regarding the research if there are no proper operational procedures that are not keen on establishing how theory or skills can be transformed into actions within the project management strategy.

It is notable, based on the definition of knowledge integration capability in the project management schema, that there are several aspects of the conditions of the organisations implementing these projects that influence knowledge integration capability. To be precise, certain factors, such as the project organisation conditions, can be taken to be very crucial in complementing the modalities through which knowledge integration is undertaken within the infrastructure project. Moreover, the conditions of the organisation have a direct influence on the process progress. There is a need for knowledge to be easily accessible and this can only be achieved if the concept of knowledge integration is synthesised and "externalised" in such a way that the relevant recipients can access it. Most notably, though, the underlying activities that can be applied to enhance the accessibility of knowledge provide an even more important framework for translating theory into knowledge, yet these are highly dependent on organisational conditions. This supports the earlier suggestion that in the establishment of a template for the multi-partner framework for infrastructure systems, the conditions of the organisations must be considered within the broader context of knowledge transfer capability (Kokkonen and Vaagaasar, 2018).

Commonly, the process of "knowledge integration" ought to be viewed from the perspective of a continuous process. It is equally a collective process where the stakeholders in the project implementation schedule are engaged in the "process of constructing, articulating, and redefining shared beliefs through the social interaction of organisational members" (Kokkonen and Vaagaasar, 2018). This infers that it is a crucial ingredient in the delivery of these projects so that high performing project teams in these projects are capable of managing these complex terms within a developed framework. This suggests that within the context of knowledge integration transfer as a mediator in collaborative projects, the implementing agencies must find a modality of exploring the dynamics through which knowledge integration is undertaken either beyond the precincts of high-performing teams within the projects, or the teams as well (Chew, 2004).

7.2.3 Project collaboration quality

The accomplishment of any infrastructure project goal involving multiple partners is dependent on the established collaborative strategies that have been adopted by these organisations. In essence, collaboration between different organisations and organisational parts is often critical for the accomplishment of the common goal. Project collaboration quality is therefore considered to be an important factor in explaining the outcomes of any organisational input as well as the performance of these organisations. Current empirical research studies have been able to have assessed the performance quality of collaborations in consideration of the outcomes of these collaborative initiatives. Past research publications have tried to identify some of the factors that are responsible for the enhancement of the modalities of collaboration within the partners in the project, suggesting that the design of these collaborations ought to be as clear as possible to enhance the outcomes of these projects (Xia *et al.,* 2018). In a sense, these publications aid in the production of very valuable information for stakeholders involved in the multi-partnership programs. Specifically, they have been able to provide a broader understanding of the modalities that are applicable by the project management teams to be able to understand precisely how they can effectively design better collaborative relationships amongst them. Despite their relevance in this, these studies have failed to address, albeit directly, how effective collaborations occur, rather preferring to point out the consequences of these collaborations.

What comes out is that within the context of management of multi-partnership collaborations on infrastructure projects, the question of the quality of these collaborative approaches has to be taken into consideration. The partners in these projects are not only seeking to deliver these projects, but they are equally keen to deliver them within the specified standards. In designing the multi-partner framework for infrastructure systems, it therefore makes sense if the quality of these collaborations has been taken into details. The partners in these projects have to find a locus of engagement to be able to elevate the quality of these collaborations as a way of enhancing the delivery of these projects.

Within the context of project collaboration quality, it is important to describe the concept of quality from the perspective of the model of interaction amongst the stakeholders: as discussed earlier, "the basis of the fluency of interactional activities" that occur between the actors involved in the project. In essence, developing a framework for multi-partner collaborations ought to be cognizant of the type or quality of collaborations amongst the various collaborative actors in these projects (Demirkesen and Ozorhon, 2017). In the evaluation of the collaboration models in infrastructure projects involving several partners, five elements of collaborative quality have been proposed (see Table 7.1). These elements are considered to be important points of emphasis in designing the multi-partner collaborative framework.

Communication refers to a process through which individuals or stakeholders within the infrastructure project can effectively exchange ideas. Effective exchange of these ideas would naturally occur well if the mode of exchange of these ideas (communication) is improved). To be precise, it then follows that the communication quality within a setup of multi-partnership collaborations ought to indicate that the actors are capable of sharing their ideas. The consequence of such a finding means that the framing of a multi-partner template in the infrastructure systems ought to be undertaken with a clear grasp of the communication processes involved in the project. It is important to develop the framework with communication in view since, through communication, the partners within the infrastructure project can share their ideas regarding the project (Weiss *et al.*, 2017).

Coordination, as an element of quality collaboration, infers the "shared mutual understanding of the goals, related activities, interdependencies between the activities, and the status of member contributions" (Weiss *et al.*, 2017). As an aspect of collaboration, coordination is concerned with how the activities regarding the projects are planned in such a way that the partners in the project remain in total agreement to these plants. High-quality coordination remains a prerequisite for multi-partner projects to reinforce the "fluency of interactions in collaborative settings and ensures harmonized and synchronized co-action," which is an essential aspect of high-quality collaboration amongst the partners involved in the project. High-quality collaboration is considered synonymous with the existence of mutual support emanating from those actors participating in the project. This means that high-quality coordination aims to make the participants in the project able to ensure that they deliver on the goals that were agreed upon as well as remain flexible

Table 7.1 Elements of collaboration quality modified

Element	Code	High-quality characteristics
Communication	**COM**	Sufficient, open, and efficient information exchange between collaborative actors.
Coordination	**COR**	Shared mutual understanding of goals, necessary activities, and contributions needed to be performed by collaborating actors.
Mutual support	**MS**	Willingness of collaborating actors to help each other in achieving commonly agreed upon goals. Existence of mutual flexibility in case of unforeseen incidents and changes.
Aligned efforts	**ALEF**	Alignment of contributions provided by collaborating actors with the expectations of the contributions. The correspondence between actors' priorities in collaboration (e.g., resource usage) and commonly agreed upon priorities.
Cohesion	**COHE**	Existence of the collaborative spirit between actors.

Source: Hoegl and Gemuenden, (2001)

when certain projects risks arise. Other than being able to render mutual support in infrastructure projects, high-quality coordination aims to instill shared expectations amongst the actors in the project. Therefore, what is notable is that there ought to be an alignment between the various expectations in terms of the efforts of the project actors as well as the realized efforts in the multi-partnership projects. Within the context of multi-partnership projects, it then follows that high-quality collaboration necessitates that each participant involved in the collaborative project not only respects but equally accepts the norms that concern the efforts required in the project.

7.3 Challenges and barriers to multi-partner collaborations

The appeal to partnerships as an effective means of handling large infrastructure projects and enhancing project delivery has long been found to be a well-established mantra in project sustainability. Developing a collaborative framework for the delivery of infrastructure projects appears to be a very effective strategy in improving project outcomes, yet it comes with immense challenges. The challenge of integrating all the partners involved in infrastructure projects within the project execution matrix still inhibit the complete realisation of the defined benefits associated with the projects. Below is a summary of the challenges in multi-partner collaborations.

7.3.1 *Organisational culture*

The question of variation in organisational culture has hindered the implementation of an operational framework for multi-partnership collaborations in infrastructural projects. While the complex challenges in large infrastructure projects can easily be addressed through these collaborative initiatives, the problem of having a homogenous work model seems to be a real issue to the collaborating actors. Usually, these collaborators come into a setting where the divergence of their perspectives regarding the project is evident; their expectations regarding these projects differ immensely, an incidence that is likely to impair the smooth delivery of these projects. Variations in cultural subscriptions by partners in a collaborative environment have been discussed within the larger context of transnational partnerships. In evaluating the modalities of developing a harmonious engagement plan amongst project collaborators, the initial step for the actors is to learn of these cultural variations and accept that, indeed, they have different strategies of approaching the various aspects of these projects. This admission in itself, amongst the project collaborators, forms a very important foundation of the development of an engagement program amongst the project collaborators (Hoda and Murugesan, 2016).

Within any partnership arrangement, the project collaborators are keen to apply what they have been accustomed to undertaking. This has been the salient theme in several recent studies focusing on stakeholder engagement models based on the notion that variations in cultures of the collaborators within the project tend to lead to conflicts within the project management model. Some studies admit that collaborative templates have been fundamentally hindered through the theory of applying for knowledge transfer in these projects since every actor acts differently. In concept, therefore, cultural difference thus provides a very firm foundation onto which emergent project conflicts can arise. Developing an effective teamwork operational model requires that the team members involved in the execution of the project have a clear view of the goals of the project and the modalities of achieving these goals. Putting this into perspective, it is important for infrastructure operators to harness the skills of the teams involved in the collaboration to be able to advance the objectives of the infrastructure project. But while these recent studies admit that forming a reasonable team management strategy is essential in the advancement of the project delivery outcomes, finding a consensus amongst team members who have been cultured differently remains a stumbling block in weaving an effective collaboration model (Niazi *et al.*, 2016). This suggestion has been discussed in detail in several research platforms that have proposed some of the strategies essential in the development of an effective collaborative strategy. The studies suggest that the project management teams selected to advance the project goals must develop a focused work schedule. When talking of focus, the research confines itself to vision, yet vision is never achieved with variations in perspectives.

An emergent factor, therefore, is that multi-project teams have to be undertaken within a frame of cultural subscription that applies to the members of the teams. This is to say that all the team members in the project delivery team have to somewhat subscribe to an agreeable operational culture to be able to drive the objectives of the project. Perhaps these variations are some of the driving factors behind the focus on stakeholder engagement theories that have recently formed the crux of research studies in project management. A review of the current publications on project management suggests that the concept of stakeholder engagement and integration is a dominant discussion point; in the context of multi-partnership project planning models, it is an admission that cultural variations are indeed a challenge in handling these projects. In conceiving stakeholder management theories, it is notable that the essence of these stakeholder management theories is to come up with a homogenous cultural framework that is acceptable to all the teams implementing these projects. Therefore, we are not just looking at culture in terms of socialisation, but rather within the context of knowledge transfer as earlier discussed. The variation in which knowledge as theory is being transformed into meaningful actions is indeed an issue that requires the attention of the project management teams. In this regard, it is imperative to frame a strategy of operationalising a knowledge transfer system in the project.

7.3.2 *Management approaches*

The whole debate on multi-partnership infrastructure projects can be consummated through the concept of management. Different theories of project management have been applied in different projects to achieve the same objectives. This means that project implementing teams have entirely different styles of managing their projects even though the outcome of these projects is the same. In discussing management styles, we look at the entire frame of operations of handling issues in the project from financial management principles through to human resource administration to the very detail of project execution models. Management approaches differ from actor to actor. Moreover, management problems vary from context to context; traditionally, the onus is on the implementing firms to adopt the management style that is workable to them to advance the project goals, but these have to be homogenous if the collaborative model adopted ought to be working. From a management perspective, there is sufficient evidence to suggest that variations in management have profound implications on the partnerships undertaken amongst the stakeholders in the project. From a broader perspective, the system of managing the project generally influences how the teams operate (Binder, 2016). Since management has to be contextualised within the project setting, there is a need for the project implementing teams to come up with a management strategy that befits the infrastructure project; again, has to be an agreeable position by the implementing teams.

7.3.3 *Relational challenges*

Team partnerships, as a model of project management, equally experience relational challenges as well. The concept of relational competencies is related to the willingness of the partnering firms to be able to incorporate other firms in the advancement of the infrastructure project objectives. Moreover, in discussing the relational challenges in partnerships, the question of the ability of both the parties in the collaboration to form mutually beneficial relations based on trust equally arises. In a sense, the collaboration foundation has to be based not only on the ability of these forms to deliver the objectives of these projects, but the question of trust becomes key to the debate of collaborative projects (Eberlein, 2008). As discussed earlier, whenever these relational competencies don't match in any partnership engagement, there is a high probability of a compromised relationship.

Major research studies have been undertaken to evaluate whether having some form of engagement or relationship beyond contractual obligations can help insulate the project against the risk of collaborators falling out. A significant number of firms involved in multicultural partnership infrastructure projects have proposed that despite contractual obligations being binding within any project management collaboration, they prefer working with entities they trust; they are consistently seeking for collaborations with other teams they can form meaningful engagements with based on mutual trust. In essence, it is evident from the disclosures of several research studies that the bounds of engagement in any collaboration go beyond the legal obligations that are expected of them. It is accurate, in the context of infrastructure project collaborations and execution, how these firms that are involved in some sort of partnership in a project have a direct influence on the project delivery strategy or outcomes. It follows that when these relational competencies between the firms are absent, the implementation of these infrastructure projects is threatened because of some level of compromise at play.

What comes out is that collaborative efforts concerning project implementation can be adequately facilitated through what studies refer to as "the willingness to collaborate." In this regard, there ought to be some level of compatibility that gradually builds trust that is ultimately required in multi-partnership engagements. What follows is ensuring the performance of the partners as well as their commitment to the achievement of the infrastructure project objectives; one would argue that contractual agreements become very instrumental in developing a firm foundation of collaboration amongst the partners involved in the project (Zulch, 2014). It was suggested earlier that it is not sufficient for project collaborators to entirely rely on the binding nature of these contracts but rather operate entirely on trust. Firms occasionally would want to operate from a locus of trust as well as repetitive collaborations in a bid to govern their partners (Henttonen and Blomqvist, 2005). This is an affirmation that relational challenges have very profound effects on collaborations and partnerships in the development of infrastructure projects. Despite the nature of the contract or the clauses that bind the partners together, there are still a lot of relational issues that determine the quality of these partnerships.

Given the various dispositions regarding relational challenges, one would want to inquire about the precise manner in which these relational challenges have become an impediment to meaningful collaborations in project management. One notable issue in the debate regarding relational issues in commercial engagements is that the binding nature of these engagements hinges on trust and has very little to do with the contractual obligations. As earlier intimated, it follows that a firm that ideally ought to base its relationships or rather obligations to the other firm in a partnership engagement will be more concerned about the level of trust that they have with the other firm other than just establishing their engagement only on a firm's legal basis. What follows herein is that two firms that have never dealt with each other in any project are likely not to be involved in a partnership based on the feeling that they have not reached a level where they trust each other deeply.

In discussing relational challenges, there is a sense in which firms that have not yet been able to engage in any project activity tend to be wary of how they deal with each other as compared to two firms that at one point engaged in a project. While this is a factual issue on the ground, there are several problems with the argument since firms are likely to forego meaningful and strategic partnerships with other firms that are not considered to be "friendly" but get involved with firms that are lackadaisical in their operations simply because they have once been involved in a business process (Nguyen *et al.*, 2009). Several studies have concluded that issues of relational challenges are indeed very detrimental in the conceptualization of a partnership model amongst project collaborators. One notable issue, as already discussed, is that project collaborators would want to engage only those firms that they feel are 'trustworthy.' This means that they limit their engagements to firms they either have previously dealt with or use some sort of metrics that indicate the "trustworthiness" of these firms. Naturally, this would confine these firms to a narrow locus of operations as far as issues of partnerships are concerned. It follow that these firms will not be willing to open themselves up to engagement with newer firms in any partnership engagement and would choose to deal with only those whose trust they feel they have. This is an obvious hurdle to the formation of any meaningful partnership amongst the project collaborators (Van Os *et al.*, 2015).

7.3.4 *Communication*

The basis of any project success is the communication model adopted to manage the project's progress. Impediments to communication strategies are likely to compromise the project's progress profoundly. Currently, the research progress focusing on stakeholder engagement has particularly dealt with the issues of cross-cultural projects, citing major challenges arising from partnerships in projects that cut across multiple cultures. Evidence in research has emphatically suggested that a major concern for the project implementing teams where infrastructure projects cut across multiple cultures is the aspect of communication. For instance, the majority of Chinese infrastructure projects being implemented in Africa suggest there is a real problem with the communication strategies that are being used in these projects. Similarly, it is deducible that the element of communication is indeed a very huge challenge in developing meaningful partnerships in infrastructural projects.

Developing a framework for engaging all the parties involved in the project advancement relies entirely on communication systems and strategies that have been developed by the project actors. Every sphere of the project implementation phases, ideally, relies on a very strategic communication model to advance. Within the context of stakeholder management, this has been an even more advanced discussion. The stakeholders are held together through a well-defined communication model; in a way, the project success is entirely intertwined with the model of communication applied in the delivery schema of the project (Van Os *et al.*, 2015).

Nowhere is the communication problem in project delivery as pronounced as instances where the project implementing teams have barriers to a common communication tool. This goes ahead to affirm the significance of communication in collaborations and partnership projects. In a way, the partners in the infrastructure project ought to have a well-defined model of communicating amongst themselves and even with the other stakeholders, yet in a collaborative model where communication is a challenge, any form of objectivity in the project's progress remains a mirage. Based on these suggestions, it would be very accurate to suggest that ineffective communication models, which are advanced through a number of both internal and external factors, can compromise the quality of the infrastructure project involving multiple partners. In essence, therefore, communication is the basic thread that weaves the project collaborators together. Proper communication has to be based on not only shared ideas but also the language or rather the tool of communication (Zulch, 2014). Collaboration and partnerships are thus limited when the basic framework of communication is missing from the partnership arrangement. The significance of communication as a tool for enhancing project collaboration is so important that systematic reviews of research publications have proposed communication as the major prescription for infrastructure projects across multiple cultures. The significance of communication in advancing the project objectives is the reason why it remains the obvious prescription for problems that may involve multiple partners or projects that run across diverse cultures.

7.3.5 Benefits of multi-collaborations

Despite the perceived project challenges faced through a multi-partnership approach, there are indications based on trends today that multi-collaborations remain key factors that drive efficient delivery of infrastructural projects. Indeed, the advances in research studies on project partnerships have unequivocally stated that infrastructure projects that are undertaken within the framework of collaborations and partnerships have a very solid foundation of sustainability. This, in essence, elevates project partnerships and collaborations as the very critical determinants of project outputs; they are essential in improving the project outcomes in cases where the collaboration strategy is well pursued.

7.3.6 Faster innovation

Project teams leverage collaborations and partnerships to develop more innovative ideas in advancing their project goals. Any form of collaboration promotes the transfer of knowledge and skills. It advances each of the team members willing to absorb knowledge from the other partners in the collaboration continuum. Earlier on, the discussion of knowledge transfer as a mediator in any collaborative model suggested that the variations in applying skills into action may at one moment deter the project from proceeding smoothly, yet at the same time, the avenues or modalities of knowledge transfer equally come with several benefits.

Firms that are engaged in project collaboration put themselves along the path of innovation and innovation transfer. They are likely to build up their skills portfolio by strategically integrating newer models of operations that have been used elsewhere and applying the same in their operational manual leading to improved innovations. In the context of project delivery, it is becoming more important that a well-structured engagement strategy of collaboration leads to skill transfer. The project actors have to be capable of borrowing the modalities through which certain initiatives are undertaken in relation to developing a project collaboration initiative. To achieve this, one has to establish a foundation onto which the actors involved in the project are likely to have a platform onto which they can transfer skills (Cheung *et al.*, 2013). It gives the teams an angle through which they can studiously explore the modalities in which certain roles are undertaken.

This is so true that certain research studies have suggested that one of the issues they have against collaborations and partnerships in project execution is based on the idea that they are likely to lose some of their innovative strategies of undertaking projects to their commercial rivals. This refers to the fact that as the project progresses, there is a transfer of learning to their competitors in the market, an issue that would ultimately limit their business operational ability. Conclusively, project partnerships expose the project implementing teams to new modalities of executing their objectives. It is an important phase of learning that leads to the improvement of the strategies that are traditionally applied in delivering infrastructure projects, to modern and innovative strategies. Most profoundly, though, it acts as a platform where innovative ideas and skills are applied to new knowledge frontiers to aid in the advancement of the project objectives.

The collaborators are likely more inclined to learn from each other regarding various aspects of the project implementation.

7.3.7 Efficiency improved

Firms are looking at modalities through which they can optimise various aspects of operations. This is why skill transfer is rife in project delivery and management. There are deliberate strategies by project implementing firms to enhance their operations through the collaborations being implemented and adopted in project delivery for infrastructure projects. In an evaluation of the significance of project partnerships, some level of efficiency is instilled in the delivery teams. One fundamental implication of a collaborative approach in infrastructure project execution is that the actors in the project get to engage in more of a skill transfer program as earlier suggested. But other than transfer, the firms come together in the pursuance of specific aims; firms are offered a platform through which they can share skills in the name of consultations and find a solid platform of engaging each other (Rooney, 2009). Critically, this means that the level of efficiency in these firms where knowledge has been combined is excellent. The collaborators are likely to have a higher level of operations based on the sense that the teams are always willing to be part of the infrastructure project through sharing ideas. Such an arrangement leads to improved strategies of undertaking roles and engagements regarding the project.

Efficiency is equally looked at in terms of accountability. One would imagine that any collaborative environment for project managers comes with some sense of accountability. The project actors, who are involved in any collaboration initiative, in the conventional sense, are usually accountable to each other. They are concerned about the application of very optimal project delivery strategies to prove to the other partners that they are equal to the task. In concept, through, it follows that having a well-planned, collaborative project structure would come with an improved level of efficiency at the workplace (Binder, 2016). From a project management concept, the adoption of a multi-partnership framework in managing and executing infrastructure projects comes with an elevated understanding of the modalities of achieving the project variables. The collaborators have a better conception of the project's deliverables based on the notion that critical ideas regarding the infrastructure project are shared. At the same time, new approaches to understanding project delivery methods are obtained during the collaborations.

7.4 Standardising multi-collaborations best practices

There have been several suggestions in research regarding the development of an assessment criterion for collaborations in projects. This means that project implementation and management practitioners ought to come up with standard measures that are geared towards developing better assessment criteria for multi-collaboration projects. The need to standardise the best practices in collaborative projects becomes imperative based on the notion that this would lead to an adoption of a standard assessment procedure.

7.4.1 Stakeholder engagement

The complexity of project delivery initiatives becomes reality when infrastructure projects adopt a collaborative approach. Indeed, this has been the major focus of major research studies. Accordingly, stakeholder engagement has emerged as a major focus area in addressing the emerging problem of stakeholder issues in a collaborative project (Crawford and Helm, 2009). There is a need to further develop an effective platform for which stakeholder engagement can be optimised within the project management process. Stakeholder engagement is a best practice activity in multi-collaborations projects but unless strategies are being developed to measure the stakeholder engagement modalities, there may never be a framework for assessing the stakeholder engagement and integration in infrastructure projects (Demirkesen and Ozorhon, 2017). This has formed a critical theme in several reviewed papers, which have proposed that infrastructure project policymakers and practitioners should focus on developing a scorecard for stakeholder engagement in projects as a best practice approach in project management and execution.

The suggestions herein affirm what has been the focus of the current studies on stakeholder engagement theories. It has since been established that project management schedules ought to have a well-defined operational scorecard or assessment criteria regarding the manner and level in which the project stakeholders are engaged. Development of such a framework helps in laying out a basis for assessing and monitoring these projects within the context of stakeholder theory (Zulch, 2014). But coming back to the question of standardising collaboration strategies, it is notable that the conceptualisation of an effective assessment platform for stakeholders in projects helps to lay a very firm foundation about the modalities of evaluating the progress of collaborative infrastructure projects.

Standardisation of collaboration best practices ought to equally be cognizant of the need to build a homogenous approach of working amongst the stakeholders or the collaborators in the project. Initially, it was noted that the project collaborators are likely to have issues regarding the varying management styles. Rather, the project collaborators have no uniform card through which they can compare their project execution milestones. The development of a standard procedure for collaboration is a crucial process that aids in harmonising these initiatives. It develops a common platform onto which the progress of the infrastructure project can be monitored regardless of the number of collaborators involved in the project delivery.

7.4.2 Communication standards and tools

Earlier on, we proposed that communication is the basic framework for the operationalisation of any project execution model. This would mean that any hitch in the communication tools and model developed in any project would invariably compromise the progress of the project. To the project collaborators, the relevance of communication in the project execution continuum means that very drastic initiatives have to be taken to ensure that the progress of the partnership in these infrastructural projects is not interfered with in any way. But perhaps the most important strategy that ought to be deployed would be to standardise the communication model in the projects to be able to enhance the delivery of these projects. Before attempting to evaluate the relevance of the communication standardization initiatives in a collaborative model, it would be prudent to have a glimpse into the aspect of communication as a "best practice" in project collaborations. Through communication, various project execution deliverables are expressed to the teams. The expectations, risks and associated risk management procedures, as well as the integration strategies of the stakeholders, are well executed within a well-defined communication model. In this sense, what follows is that the conceptualisation of a well-developed model of communication amongst the project collaborators would enhance the delivery of the infrastructure projects.

In the context of standardisation of communication, it's important to have a uniform model and tool through which the communication amongst the various project stakeholders can be well executed. The need to standardise communication templates is crafted on the idea that despite the variation in the modalities of operations of the implementing teams, and regardless of differing management styles and cultures in the project, there is a point of congruence through which the project actors can interact effectively and advance the project objectives (Strawbridge, 2005). It is the prerogative of the team members involved in the infrastructure project that notwithstanding the variance in their perspectives of project delivery, a middle ground ought to be reached for the sake of the advancement of these projects (Lawer, 2019). When talking about the "middle ground," it is applied in the context of a workable approach. Therefore, one would view standardisation procedures in infrastructure projects involving multiple partners not as an opportunity of creating a platform for assessment as earlier intimated but rather a moment of reflection on what is best for the infrastructure project despite the existing differences. It is a moment when the collaborators engage each other and establish a proper tangent of operationalising their activities despite their differences without really compromising the quality of the project progress (Brady, 2011).

Best practices in project delivery involving multiple teams have to be encapsulated within standard operational measures. This could be hinged on two critical issues. One is the need to develop a standard measurement template for the development of a scorecard in the project delivery matrix. It can be used to assess or monitor the progression of the infrastructure project. However, what is more important is that the standardisation process enables collaborators with different perspectives and modalities of executing their duties to have a platform that is sufficiently uniform to be able to advance the project objectives. It can be seen as an opportunity for the multiple partners in the project to find a locus onto which they can operationalize their activities. One could thus suggest that despite the challenges in the delivery of infrastructure projects through a collaborative affair, the conceptualisation of a standard operational model would offer the implementing team a very concrete and uniform operational domain that would aid in streamlining their operations and thus limit the various project risks associated with collaborator or partner conflicts.

7.5 Chapter summary

As shown in this chapter, the complexities of current infrastructural projects require that a rational approach is used in their management. Currently, there seems to be a growing shift towards partnerships in the execution of projected deliverables to circumvent these complexities in project management. However, this has come with the challenges entailed in the management of multiple partnership collaborations in infrastructure projects. In designing a framework for the operationalisation of multi-partner projects, there is a need to take into consideration the stakeholder integration procedures that are relevant to the project. At the same time, other mediators in multi-partnership-modeled infrastructure projects such as knowledge integration and transfer capability ought to be taken into consideration as well. They are critical considerations in the formulation of a framework for optimising the delivery of outcomes of multi-partnership infrastructure project collaborations.

While there are several benefits related to collaborative approaches in infrastructural projects, there are several challenges that are likely to impair an effective collaboration model in infrastructure projects. Some of these challenges include the problem of relational experience and managerial and cultural variations. These critical issues have to be taken into consideration when formulating a partnership model for infrastructure projects. On the contrary, this chapter has addressed several benefits of partnerships such as knowledge transfer and improved efficiency. These, however, have to be standardised to offer the partners a uniform operational platform relevant for advancing project deliverables.

Managing multi-partner collaborations on major infrastructure projects

7.5.1 Chapter discussion questions

1. What are the key best practices for successful multi-collaborations on infrastructure systems?
2. Identify and describe the key challenges to multi-collaborations on infrastructure systems.
3. Propose a comprehensive multi-partner framework and align the proposed framework with your local infrastructure models. What are the key differences between your proposed model and your local infrastructure models?

7.6 Case study

As an example of a case scenario of multi-partner collaborations on infrastructure projects and the associated challenges, benefits, and need for standardisation, the Standard Gauge Railway (SGR) project in Kenya presents a quintessential description of the modalities of managing these types of projects. The SGR further gives important insight into the considerations that ought to be observed in designing a comprehensive multi-partner framework for infrastructure systems. The benefits, as well as the challenges associated with these collaborations' models, shall be reviewed through the case of SGR in Kenya. The SGR project – Standard Gauge Railway, a Chinese-funded railway infrastructure project in Kenya, was begun in 2008 and completed in 2017 (Ministry of Industrialisation, Trade and Enterprise Development, 2017). It was an upgrade of the old railway line from the traditional dimensions to a more standard model in line with the advances in the global railway transport systems.

While the details of the project in terms of the contractual obligations of the Kenyan government remain scanty, the stakeholder management model adopted in the project has been constantly questioned. For instance, a recent court ruling showed that the procurement of the contractors was done unilaterally (Okiya Omtata et al., (2020)). This points to the larger problem of stakeholder engagement in Kenya considering that the petitioner was a citizen of Kenya and a prospective beneficiary of the project. The approach offered to the SGR project within the context of project management provides very insightful information regarding the modalities of managing multi-partner collaborations on infrastructure projects. The construction of the rail was undertaken by China Road and Bridges Company for US dollars3.6 billion with a sizeable percentage of the money being advanced by Exim Bank of China. The Kenyan government paid the remaining 10 percent of the total expenses of the railway project. The construction was undertaken under the Chinese belt, and the road initiative undertaken by the Chinese government. The first section, completed in 2017 between Mombasa and Nairobi, the capital, measures 609 kilometres. The railway officially began passenger services on the first day of June 2017.

7.6.1 Nairobi-Naivasha section

The second phase of the project connects Nairobi and Naivasha, an inland city located about 120 kilometres away from the capital. The total projected cost of the project was US dollars 150 million, a sum that was also advanced by Exim Bank to fund the project. In December 2019, the project was formally launched. The Nairobi–Kisumu section, which was expected to cost US dollars 3.8 billion, is currently in the formative stages undergoing feasibility tests. The SGR, in effect, was meant to ease the pressure off the transport of cargo from Mombasa to the hinterland. There was a need, based on the state of the roads, to develop an alternative transport corridor for transit goods into the neighboring countries and the hinterland. At the same time, the focus was to develop an inland container terminal in Naivasha to ease the pressure on the port services in

Mombasa, an issue that generated equally immense controversy in the project. The annual capacity for the SGR was to oversee the transport of 20 million tonnes of cargo.

7.6.2 Relationship with the community

The SGR project in Kenya generated both controversy and excitement in equal measure. In terms of the relationship with the community members, there have been several stakeholders and public consultations over the project since its initiation from Mombasa to Nairobi. In 2015, there were questions about land compensation by those whose lands had been hived off for the project, which built up tension with the local communities over delays and the amount of compensation. The compensation was mired in so much controversy that at one point the progress was halted. At the same time, other stakeholders in the project raised an alarm over the Chinese dominance of the project, citing out that the locals had been denied an equal share of the project. This has been the major issue levelled against the project. In other instances, the project has had run-ins with activities over the blockage of the wildlife corridors in Nairobi. Despite these oppositions, the first two phases of the SGR project went smoothly.

7.6.3 Case discussion questions

1. Would the controversies facing land compensation issues during the implementation of the SGR project have been circumvented if adequate stakeholder engagement was adopted to guide the implementation process? Explain your answer within the context of stakeholder engagement theories.
2. What is the significance of the project implementing teams encouraging dissenting voices to be more actively engaged in the project? Was this approach applied in the case of Kenya's SGR project, and to what extent did it achieve the desired objectives?
3. To what extent should the national infrastructure projects rely on supplies from foreign entities? Discuss this in the context of the SGR project.

References

Binder, J. (2016). *Global project management: Communication, collaboration and management across borders*. New York: Routledge.

Brady, T. (2011). Creating and sustaining a supply network to deliver routine and complex one-off airport infrastructure projects. *International Journal of Innovation and Technology Management*, 8(3), pp. 469–481.

Cheung, S.O., Yiu, T.W., and Lam, M.C. (2013). Interweaving trust and communication with project performance. *Journal of Construction Engineering and Management*, 139(8), pp. 941–950.

Chew, A. (2004). Alliancing in delivery of major infrastructure projects and outsourcing services in Australia-An overview of legal issues. *International Law Review*, 21, pp. 319–355.

Crawford, L.H. and Helm, J. (2009). Government and governance: The value of project management in the public sector. *Project Management Journal*, 40(1), pp. 73–87.

Demirkesen, S. and Ozorhon, B. (2017). Impact of integration management on construction project management performance. *International Journal of Project Management*, 35(8), pp. 1639–1654.

Doloi, H., Pryke, S., and Badi, S.M. (2016). The practice of stakeholder engagement in infrastructure projects: A comparative study of two major projects in Australia and the UK. Royal Institution of Chartered Surveyors (RICS): London, UK.

Eberlein, M. (2008). Culture as a critical success factor for successful global project management in multinational IT service projects. *Journal of Information Technology Management*, 19(3), pp. 27–42.

Hoda, R. and Murugesan, L.K. (2016). Multi-level agile project management challenges: A self-organizing team perspective. *Journal of Systems and Software*, 117, pp. 245–257.

Hoegl, M., and Gemuenden, H. G. (2001). Teamwork quality and the success of innovative projects: A theoretical concept and empirical evidence. *Organization Science*, 12(4), pp. 435–449.

Henttonen, K. and Blomqvist, K. (2005). Managing distance in a global virtual team: The evolution of trust through technology-mediated relational communication. *Strategic Change*, 14(2), pp. 107–119.

Kokkonen, A. and Vaagaasar, A.L. (2018). Managing collaborative space in multi-partner projects. *Construction Management and Economics*, 36(2), pp. 83–95.

Lawer, E.T. (2019). Examining stakeholder participation and conflicts associated with large scale infrastructure projects: The case of Tema port expansion project, Ghana. *Maritime Policy and Management*, 46(6), pp. 735–756.

Ministry of Industrialization, Trade and Enterprise Development (2017). Available from: http://industrialization.go.ke/index.php/media-center/blog/358-high-speed-rail-will-be-an-economic-boost-to-kenya [cited 22 June 2020].

Mortensen, M. and Haas, M.R. (2018). Perspective—rethinking teams: From bounded membership to dynamic participation. *Organisation Science*, 29(2), pp. 341–355.

Niazi, M., Mahmood, S., Alshayeb, M., Riaz, M.R., Faisal, K., Cerpa, N., Khan, S.U., and Richardson, I. (2016). Challenges of project management in global software development: A client-vendor analysis. *Information and Software Technology*, 80, pp. 1–19.

Nguyen, N.H., Skitmore, M., and Wong, J.K.W. (2009). Stakeholder impact analysis of infrastructure project management in developing countries: A study of perception of project managers in state-owned engineering firms in Vietnam. *Construction Management and Economics*, 27(11), pp. 1129–1140.

Okiya Omtata and 2 Others v Attorney General and 4 Others (2020). Available from: http://kenyalaw.org/caselaw/cases/view/103808/ [cited 22 June 2020].

Rooney, G. (2009). Project alliancing–the process architecture of a relationship-based project delivery system for complex infrastructure projects. Available from: https://papers.ssrn.com/sol3/papers.cfm?abstract_id=1809267 [cited 7 May 2020].

Strawbridge, C. (2005). September. Project management in large collaborations: SNS lessons learned for ITER. In 21st IEEE/NPS Symposium on Fusion Engineering SOFE 05 (pp. 1–5). IEEE.

Sperry, R.C. and Jetter, A.J. (2019). A systems approach to project stakeholder management: Fuzzy cognitive map modeling. *Project Management Journal*, 50(6), pp. 699–715.

Van Os, A., Van Berkel, F., De Gilder, D., Van Dyck, C., and Groenewegen, P. (2015). Project risk as identity threat: Explaining the development and consequences of risk discourse in an infrastructure project. *International Journal of Project Management*, 33(4), pp. 877–888.

Wang, M.C., Chen, P.C., and Fang, S.C. (2018). A critical view of knowledge networks and innovation performance: The mediation role of firms' knowledge integration capability. *Journal of Business Research*, 88, pp. 222–233.

Weiss, M., Hoegl, M., and Gibbert, M. (2017). How does material resource adequacy affect innovation project performance? A meta-analysis. *Journal of Product Innovation Management*, 34(6), pp. 842–863.

Xia, N., Zou, P.X., Griffin, M.A., Wang, X., and Zhong, R. (2018). Towards integrating construction risk management and stakeholder management: A systematic literature review and future research agendas. *International Journal of Project Management*, 36(5), pp. 701–715.

Zulch, B. (2014). Leadership communication in project management. *Procedia-Social and Behavioral Sciences*, 119, pp. 172–181.

PART II

Enabling planning and execution

174

8

DELIVERING MAJOR INFRASTRUCTURE PROJECTS EFFECTIVELY AND EFFICIENTLY

Ambisisi Ambituuni

8.1 Introduction

The delivery of major infrastructure projects needs to consider and respond to the various factors within the project business environment for the project to be delivered effectively. Effective delivery should consider the time, cost and quality constraints, as well as ensuring that the project is fit for purpose. Some of the purposes to be considered should include meeting and resolving the national infrastructure deficit, the sustainable development goals (SDG) of the county, and value for money. Given the complex, multi-stakeholder and political changing environment within which major infrastructure projects are delivered, it is always important to ensure that such a project effectively delivers impact to socioeconomic, environmental, cultural and sustainability of a nation. Indeed, to achieve such an impact, the project needs to be aligned with developmental priorities, vision and goals of the country. The effective delivery and running of well-considered major infrastructure projects will in turn provide competitive economic development and sustainable advantages to the country. In this chapter, the author provides a critical account of key considerations that should be considered to ensure the effective delivery of major infrastructure projects that are fit for purpose and capable of delivering impactful gains to a country. The chapter sets out to achieve the following aim and objectives.

8.2 Chapter aim and objectives

This chapter aims to explore some of the challenges, issues and processes used by decision-makers and governments in the quest to deliver major infrastructure projects that are effective and efficient. The main objectives are to:

- Examine the approach used by various countries and decision-makers to appraise their current infrastructure situation and government needs.
- Explore the factors and key considerations for creating an infrastructure vision and goals for the future.
- Examine how major infrastructure projects are delivered effectively and efficiently, including the considerations for finalising plans and moving from planning to action.

8.2.1 Learning outcomes

The following learning outcomes have been identified for this chapter. Readers will be able to:

- Articulate the process of prioritising major infrastructure projects;
- Appreciate the value of prioritisation and cost-benefit analysis (CBA);
- Gain insight on current infrastructure situations and government needs;
- Comprehend how to create a vision and goals for the future infrastructure assets;
- Apply appropriate principles of ensuring major infrastructure projects are delivered effectively and efficiently;
- Demonstrate how to finalise a major infrastructure plan; and
- Gain insight on managing delays and cost overrun.

8.3 Prioritising major infrastructure projects

It is important to plan and prioritise infrastructure project investments in order to attain optimal productivity and as a means of engendering infrastructure contributions that will have long-term sustainable and economic growth. Indeed, the economic stage and progression of a country will affect the planning and prioritisation of infrastructure projects and the impact of investments on economic growth. This considerably varies the extent of attainment of the economic benefit from infrastructure investment and the way in which resources are allocated to infrastructure projects. For instance, developing countries will tend to invest in the delivery of capital projects with the aim of developing infrastructure to support its growth and global competitiveness. Conversely, for mature countries and ones moving towards developed status, capital functional infrastructure such as transportation, communication, sewage, water, and electric infrastructures are more likely to be already in place. The focus, therefore, tend to move more towards providing resources to sustain, improve, expand and maintain these functional infrastructures. Governmental preparedness on national economic infrastructure will help optimise the portfolio of infrastructure investments and allow governments to identify links between sectors in order to identify and increase the opportunities to expand the provision of infrastructure funded from users. The effective delivery of infrastructure will also create the greatest impact in terms of economic growth, social uplift and sustainability. It is generally assumed that for every dollar spent on public infrastructure investment the gross domestic product of the country will increase by approximately US \$0.05 to US \$0.25 (WEF, 2012).

Impacts from investing in infrastructure are wide and varied and often form the basis for prioritising economic major infrastructure projects. Impacts may, for example, relate to (i) costs, (ii) convenience, (iii) environmental aspects, (iv) strategic factors, (v) political factors and national prestige, (vi) alignment with policy goals and objectives, and (vii) income distribution considerations (Schutte and Brits, 2012). The prioritisation of infrastructure projects hinge on economic, technical, institutional, financial, social, commercial and benefit cases. A review of literature shows various methodological approaches for infrastructure investment prioritisation. First is the literature that focuses on criteria that play a major role in mitigating the infrastructure gap (e.g., Berechman and Paaswell, 2005, Karydas and Gifun, 2006). The criteria tend to focus on the identification and measurement of both direct and indirect benefits of the projects mainly focused on economic benefit and cost benefit analysis. In Andres, Biller, and Dappe (2016) a stepwise methodology was developed with a particular focus on infrastructure prioritisation in developing countries where demand for investment is huge and financial resources are limited. The methodological framework consists of three main steps: (i) identifying factors that affect infrastructure investment decisions, (ii) quantifying identified factors and (iii) ranking the infrastructure projects. The authors categorised the factors affecting infrastructure decision-making into project-level factors, economy-wide impacts, project-related market failures and a country's institutional system. In their report (PwC, 2016) they further drew on their experience of working on infrastructure projects and set out a set of principles to help the decision-making process. This included the objective identification of current and future needs to ensure that infrastructure decision meets a need (Flyvbjerg, 2008), while also fitting the broader policy agenda for government. They also emphasised the need to assess the financial viability of the project including assessing the availability of funds as well as developing an in-depth understanding of the link to wider economic benefits.

8.4 Major infrastructure prioritisation and cost-benefit analysis

One of the most commonly used method for prioritising major infrastructure projects is cost-benefit analysis (CBA). CBA is a systematic method used to evaluate the strengths and weaknesses of alternatives in order to determine options that provide the best approach to achieving benefits while ensuring savings (Boardman *et al.,* 2017). It assesses and totals up the equivalent money value of the benefits and costs to the community of projects to establish whether they are worthwhile. CBA is used for improving government decision-making on project prioritisation. However, several problems undermine its use: the difficulties of capturing benefits; unrealistic cost estimations and lack of consistency in the approach used. These problems risk projects being approved incorrectly or turned down or delayed. For instance, Atkins *et al.* (2017) presented the CBA framework used by the United Kingdom's government to judge the net present value of an investment – whether benefits outweigh costs, discounted over time. In this framework, CBA forms an important part of government's project prioritisation, approval, and allocation of limited resources, with guidance provided by the Treasury's Green Book. It also forms part of the accountability framework that civil servants and ministers operate within. Within the frameworks, CBA is mainly used in three separate stages (see Figure 8.1).

In the first instance, departments use CBA to value projects. Then the Treasury uses CBA results to help allocate spending under fiscal constraints. Finally, upon allocating the capital budget by the Treasury, departments then revisit their project selection and use CBA to reprioritise the optimal mix of projects. In all business cases, CBA is referred to as the "economic case," and is only one of five cases sitting alongside the strategic, commercial, financial and management cases. At both analytical and decision-making stages of infrastructure project prioritisation, the results of CBA, in particular the benefit–cost ratio, does not form the sole basis for an infrastructure project approval. Other factors such as the project's deliverability, affordability and the prevailing politics of the day also influence the prioritisation and decision-making process. The framework ensures that alignment with the government's policy priorities, outlined in the strategic case for a project, is usually the main reason for going ahead with a project. Nonetheless, decision-makers would require significant justification to proceed with a project if analysts thought that the costs

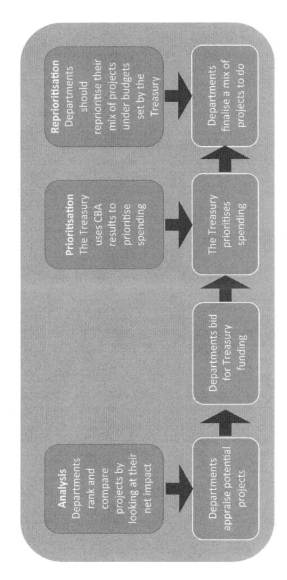

Figure 8.1 Cost-benefit analysis used by the UK government
Source: Original adapted from Atkins et al. (2017)

exceeded the benefits. CBAs should be thorough in order to influence decision-making. So, what does a good CBA in government infrastructure prioritisation look like?

- It must include all applicable and relevant impacts. Some major projects have dynamic effects. They have transformational objectives in the form of increasing economic growth and provide job opportunities that a static analysis will not account for. Wherever possible, these dynamic effects need to form an integral part of the economic case so that they are subject to a high level of internal scrutiny.
- The cost estimation and analysis must be realistic. Estimates that are over-optimistic can lead ministers to undeliverable targets and erode public faith in government when these are not met. Government must try to mitigate this tendency wherever possible, using data from past projects.
- CBA needs to be applied in a consistent way across all projects within the prioritisation framework to ensure comparability when prioritising.
- It must be transparent, devoid of bias and well communicated. Any CBA analysis must represent the forecasting risks and be transparent enough to allow decision-makers to debate on key assumptions made.

8.5 Appraising the current infrastructure situation and government needs

The role of government in planning major infrastructure projects varies from appraising the national infrastructure situation, and directing investment decisions and their coordination, to creating a framework to attract private investment. To ensure that infrastructure investments make an impactful contribution to the developmental objectives of a country, politicians and officials need to understand which infrastructure project investments are able to address these objectives in order to invest in it. However, it is important to acknowledge that new investments are not always the answer. Decision-makers need to be able to identify many other aspects of their infrastructure package such as upgrades of existing infrastructure, developing robust and improved maintenance regimes or setting a demand management system. If decision-makers decide to invest in new infrastructure project, a needs assessment will be required and this should include:

- An assessment of the performances of the current national infrastructure system vis-a-vis the vision and developmental objectives of the government. This should include a gap analysis of the performance of the current infrastructure and its effects on different regions, social groups and economic sectors.
- Consideration should be given to the changing demand on the new investments, including population growth, demography, economic growth, climate mitigation and adaptation, and technological change.
- A strategic evaluation of the impact of the infrastructure investment options on the national economy. This will provide a long-term perspective to projects with high upfront costs and the economic impact such projects can make in the long-term, such that decision-makers are able to make informed decisions about the long-term benefits of such projects.
- An assessment of the threats, opportunities and uncertainties related to the different options.

To better understand the condition of infrastructure assets and the barriers to further investment, the World Economic Forum (WEF, 2012) has developed a framework to help governments familiarise themselves with the main drivers of infrastructure investment – the Strategic Infrastructure Planner Framework. The framework in Table 8.1 identified 14 parameters within the four drivers

Table 8.1 Strategic Infrastructure Planner Framework: 14 economic infrastructure readiness parameters

Drivers of infrastructure readiness	Key parameters
Infrastructure quality	• Quality of land transport (road and rail) • Quality of ports and air transport • Availability and reliability of energy grids and power supplies • Availability and reliability of telecommunications networks • Quality of waste and water infrastructure
Government readiness	• Rule of law and effectiveness of law-making bodies • Government's openness and impartiality • Government's track record of infrastructure projects government's willingness to engage with private sector
Societal readiness	• Maturity of civil society • Government or public willingness to pay
Market readiness	• Competitiveness of construction industry and the supply chain – access to labour and materials • Access to finance

Source: Original adapted from WEF (2012)

of infrastructure readiness: the condition of infrastructure assets in the country; whether government policies and actions are conducive to infrastructure investment; whether there is support from wider society to invest in infrastructure and engage in debate; and whether there is a competitive construction industry that can easily access labour, building materials and finance.

These parameters serve as the basis for information collection about the condition of both privately and publicly owned economic infrastructure and evaluating a country's infrastructure readiness. Governments can obtain a comprehensive overview of the current state of infrastructure readiness in the country, which can then be used to assess and plan future requirements. This can be achieved by engaging different government departments and representative stakeholders drawn from public-sector organisations, private companies, financiers and investors, civil society, academia and non-governmental organisations (NGOs). The engagement activities are placed within the institutions that form a framework within which individuals, firms and governments interact to generate infrastructure project ideas, income, investments, etc. The framework also has a strong bearing on competitiveness and growth ambitions of a country and plays a central role in the ways in which societies distribute the benefits and bear the costs of infrastructural project strategies and policies, and it influences investment decisions and the organisation of production (Itani *et al.,* 2014). Government attitudes towards markets, freedoms and the efficiency of its operations are also very important when it comes to appraising infrastructure situations and developing a strategic understanding of government needs. Excessive bureaucracy, overregulation, corruption, dishonesty, lack of transparency and trustworthiness impose significant constraints on the appraisal process (Soto, 2001).

The appraisal of infrastructure situation and development of any plan needs to withstand the effects of changing political and governance landscapes such as elections and changes in government. The goal, therefore, is to develop an infrastructure plan that enjoys strong consensus by the government agencies involved, helps to set priorities, and identifies programmatic initiatives that go beyond elections such that it provides some level of certainties to stakeholders such as financiers and investors. A progressive and stable policy framework, therefore, supports the development of an agile and supportive infrastructure project and programme ecosystem, ensuring that contractors, developers, and investors remain positively responsive to opportunities.

Government's appraisal of infrastructure needs must provide plans that set directional momentum to address infrastructure deficits. Such a plan must also signal priority reforms and the institutional activities that are vital to eliminate obstacles to infrastructure investment. They force a holistic and integrated view of infrastructure needs beyond the boundaries of line departments and potentially help the resolution of overlaps and gaps in policies, institutions and programs to tackle infrastructure deficits.

8.6 Creating a vision and goals for the future infrastructure assets

In many instances, infrastructure assets are delivered to last for many decades. Therefore, the infrastructure vision over a long term should be the starting point for governments that are planning their infrastructure strategies. This vision must identify and develop an understanding of what they are trying to deliver, to whom and why. This process of establishing a national infrastructure project vision is important because it provides a point of reference on what needs to be achieved. Without it, there is a real risk that national infrastructure plans are little more than a collection of unrelated projects, all pursuing different goals and making little systematic impact on a country's infrastructure systems. Furthermore, a national infrastructure vision ensures that infrastructure decisions are aligned with sustainability goals, including the provision of inclusive development, mitigation of carbon dioxide emissions and adaptation to the impacts of climate change (Hall *et al.*, 2017). Before discussing the specifics of creating infrastructure vision, let us review the interlink between infrastructure and SDGs.

8.6.1 Creating an SDG-based vision

Infrastructure includes networked systems that deliver services such as energy, water, waste management, transport and telecommunications. In a broader context it also includes social infrastructure, such as social protection systems, healthcare systems (including public health), financial and insurance systems, education systems and law enforcement and justice (Thacker *et al.*, 2019). All these systems interact with the socioeconomic systems of society and are imbedded dynamics of human and environmental sustainability. For instance, a highways project will affect preferences for modes of transport and lock in patterns of urban development (Erickson *et al.*, 2015). Most infrastructure has a positive impact on society and, hence, an intertwined relationship with the SDGs. The access to transportation, energy, water, sewage systems, healthcare, education, etc., is underpinned by the provision of infrastructure to support the delivery of these services. Infrastructure also delivers vital factors of production (energy, water and access to labour markets).

Conversely, unreliable infrastructure systems limit the productivity of businesses and public services (Steinbuks and Foster, 2010). Indeed, whilst there is a positive side to infrastructure projects intertwined with SDGs, the construction and operation of infrastructure can also be profoundly harmful (Thomson and Koehler, 2016). The construction lifetime operation and maintenance of an airport, for example, can displace people. Even though it offers employment, it can expose people to hazardous working conditions. Fossil-fuel power used by airplanes is responsible for harmful air quality and greenhouse gas emissions. Furthermore, the construction of the airport infrastructure (runways, terminals, access roads, etc.) can destroy and fragment habitats and provides access that enables the exploitation of natural resources. However, there is increasing interest in sustainability infrastructure development aimed at substituting "grey infrastructure" with "green infrastructure" (Thacker et al., 2019), for instance, by utilising ponds and reed beds to treat sewage, wetlands to help recharge groundwater aquifers, and substituting afforestation for flood protection (Dadson *et al.*, 2017; Scholz and Lee, 2005; van der Kamp and Hayashi, 1998).

Figure 8.2 shows the UN SDGs. Thacker *et al.* (2019) analysed the extent to which infrastructure systems influence sustainable development outcomes, as defined by the targets of the SDGs. They found that infrastructure either directly or indirectly influences all 17 of the SDGs, including 121 of the 169 targets (72 percent). For 5 of the 17 SDG goals (SDGs 3, 6, 7, 9 and 11), all of the targets are influenced by infrastructure, whereas for 15 of the SDGs more than half of the targets are influenced by infrastructure. The water and energy infrastructure sectors were found to have the largest direct influence on individual SDGs: 6 (clean water and sanitation) and 7 (affordable and clean energy). Transport infrastructure enables access and participation in society and the economy, and has, therefore, a wide indirect influence. The increasing role of digital communications in enabling delivery of a wide range of services, from hazard warnings to remittance transfers, is demonstrated by this sector having the largest overall influence across the SDGs when also considering indirect effects. Thacker *et al.* (2019), however, observed that badly planned tendencies will have negative effects on the SDGs – for instance, when badly planned infrastructure leads to pollution, water and soil contamination, or disease transmission (SDG 3) or affects aquatic and terrestrial ecosystems (SDGs 14 and 15) and destroys culturally important sites (SDG11).

8.6.2 Understanding the tensions of infrastructure vision and goals

With every delivery of any type of infrastructure project we can recognise potential "winners." These winners could be, for example, a community that benefits from the expansion of a new road network; improved healthcare of a society from a new hospital; a business that developed enhanced capabilities because of the installation of a 5G network; and a local community with access to a new freshwater supply system. These winners are individuals, organisations and communities whose positions, activities, work quality or general well-being are enhanced by the delivery of infrastructure(s). Similarly, there are potential "losers" – persons, organisations or communities who miss out or are negatively affected by the emergence of an infrastructure. Examples are towns or communities bypassed by the expansion of a network, the loss of jobs from automation strategies, and the loss of farmland from the construction of residential homes. Emergent infrastructures function as redistribution mechanisms by reorganising resource flows across scales ranging from the local workplace to the global economy (Edwards *et al.,* 2007).

The perceptions of benefits and loss will influence the support (or otherwise) of any infrastructure's vision and goals and, hence, shape the environment within which the infrastructures emerge from. A receptive environment is such that allies (e.g., communities, financiers and funders) provide their support and innovation to extend the reach, quality and fit of infrastructure while a hostile environment is such that important user groups and audiences fail to be cooperative, undermine the vision, and refuse or oppose rival projects. "Failing to think proactively about the distributional consequences of infrastructure is not only bad politics, but bad business" (Edwards *et al.,* 2007, p. 22). A distinctive kind of tension emerges with groups that feel entitled by the emergence of infrastructure. These individuals, groups, and communities are stakeholders in advancing the development of infrastructure such that its emergence supports and extends their strengths. The historical constitution of powerful classes of infrastructural users, both individual and collective, may constitute a powerful conservative force confronting and constraining new infrastructural development. In this instance the tension exists when infrastructure is subject to "capture," in which powerful established constituencies exert their interest such that it overwhelms or constrains the potential development of favourable vision. This takes form when interests are exerted to effectively grasp infrastructure goals, resisting infrastructure initiatives and developments that favour new, less organised or less favourably placed actors, limiting the scope and vision of new infrastructural possibilities (Bowker *et al.,* 2010; Karasti *et al.,* 2010).

Delivering major infrastructure projects effectively and efficiently

Figure 8.2 SDGs

8.6.3 Drafting an infrastructure vision

Through the work of Thacker *et al.* (2019), we have seen a balanced illustration of the interlink between infrastructure and SDGs. Additionally, Edwards *et al.* (2007) provided some insight into the tension that exists in the context of setting and developing infrastructures. Therefore, it is essential for policymakers to establish long-term visions for sustainable national infrastructure systems but also factor in the tensions in order to ensure that the right infrastructure is built. Such a vision needs to be informed by the SDGs, and any infrastructure plan emanating from this vision must consider both the positive and negative impact of infrastructure projects (and operations) on the SDGs, as well as "winners" and "losers." Additionally, decision-makers must understand what infrastructure they are trying to deliver, to whom and why. Although final prioritisation decisions rest firmly with government, there is the need to invite key stakeholders to give their views and perceptions of the long-term vision and requirements. With the infrastructure vision drafted, the subsets of outcome-based, i.e., short-term, medium-term and long-term infrastructure goals, can then be prepared. This involves a methodological assessment of critical infrastructure gaps in order to identify critical priorities needed to drive socioeconomic transformation, create alignment with SDGs, set actionable goals around these priorities and identify infrastructure projects to aid the realisation of the goals.

Indeed, every country has specific infrastructure needs and priorities that are related to its history, level and ambition of economic development, socioeconomic and geographical considerations and domestic political choices. For instance, Norway and many oil-rich Gulf States are currently prioritising the diversification of their economies. Also, post-apartheid South Africa has focused on integrating rural, non-white communities into the national economy. Setting infrastructure vision, therefore, needs to be context specific to the country. In some instances this can also be linked to regional visions, as seen in the European Union (EU) and the Economic Community of West African States (ECOWAS). Objectives also change over time as a country moves through stages of development. Hence, vision needs to be reviewed consistently for relevance and benefits. For instance, over the post-war period, Singapore and Hong Kong both evolved from the provision of basic services like water and sanitation to a more strategic importance of provision of good quality of life, equality of opportunity and social cohesion.

8.7 Ensuring major infrastructure projects are delivered effectively and efficiently

For a national major infrastructure vision to be achieved, government and investors must ensure that the major infrastructure projects are delivered effectively and efficiently. This entails developing a strategic plan by converting the needs assessment into a viable plan aimed at improving the national infrastructure system and realising the set vision. A strategic plan will typically include a prioritised list of investment infrastructure projects and a package of related interventions aimed at improving the performance of the infrastructure system. The interventions could include the development of policy and regulatory changes. A strategic plan should ideally envisage the uncertainties inherent in a multi-decade timeframe and develop an approach that ensures the sustainable implementation of the infrastructure vision. It should also develop a framework for coordination with any regional, subnational, and sectoral plans, while also considering the funding and financing options. A process should be established to measure the progress of delivery against milestones and evaluate the success of the strategic plan in delivering the desired strategic effects (ICE, 2020); consideration must also be given to the opportunity of efficiently utilising scarce resources. The work of Ochieng *et al.* (2017) illustrates how efficiency can be achieved. This was expanded to include:

i. **Transparent and effective governance:** The creation of effective governance structure, staffed by skilled professionals, will aid the successful delivery of major infrastructure projects. Such a structure must include a clear division of responsibilities between the political executive, the national legislature, permanent officials and any expert independent infrastructure commission or body.

ii. **Institutions and institutional capabilities:** Institutions form the backbone of the effective delivery of major infrastructure projects. They provide the processes and skills, implement policies, and regulate and govern the infrastructure delivery environment. Their input does not only affect the delivery of infrastructure projects but also impacts the operations and the life cycle of the delivered asset. Ambituuni *et al.* (2019), for example, showed the need to assess institutional capabilities with respect to owning, operating and regulating infrastructure, and the need to align the capabilities of both the regulator and operator to ensure robust conceptualisation, delivery and operation of infrastructures. Indeed, when countries invest in infrastructure and this is accompanied by reforms to strengthen institutions and regulation, they experience relatively stronger impacts on productivity and economic growth. Institutional actions should be aimed at removing barriers to infrastructure investments, such as bureaucratic bottlenecks, corruption, conflicting priorities, etc. (Ambituuni *et al.,* 2014). This should ideally be anchored centrally by an institution as seen in countries like Rwanda, Sweden and the UK. In the UK, for instance, the National Infrastructure Commission (NIC) provides expert, independent analysis on pressing infrastructure issues, and is charged with preparing the National Infrastructure Assessment (NIA) to set an overarching, long-term vision and recommendations taking a 30-year perspective, while the Infrastructure and Projects Authority (IPA) prepares medium-term plans for a five-year period and also manages and provides regular updates to the National Infrastructure Pipeline. Such an approach provides a central converging point for infrastructure planning and can potentially help resolve overlaps and gaps in policies and programs to tackle infrastructure deficits.

iii. **Process:** Optimised infrastructures are delivered with optimised project decision-making, financing conception, and a planning and execution framework. Furthermore, processes that are standardised provide consistency in delivery and therefore save time and resources. They ensure consistency in performance matrices which allow for comparison across different projects. Optimised processes include effective adaptation of technologies and project management methodologies to reduce the risk of cost overruns, scope creep and delays. Moreover, funding processes further offer opportunities for optimised project delivery. For instance, for projects funded by taxpayers, a way to manage overall costs is to secure efficient sources of finance. Governments have two main options to pay for the construction and operating costs of the project: (i) using their own resources, either by spending existing cash reserves, selling government assets, raising taxes or issuing government bonds to fund the investment, or (ii) with PPP approaches, using project finance solutions and paying for the operational assets over years (WEF, 2012).

iv. **Public involvement:** The views of people and organisations likely to be affected by a major infrastructure project has well been established as critical in the delivery of the projects (UNECE, 1998). Such involvement is aimed at ensuring both fairness in decision-making and also to divulge information sources and viewpoints that technical analysts may have overlooked or misinterpreted. For certain elements of major infrastructure project such as environmental decision-making, public participation is a requirement of the 1998 Aarhus Convention 12. Public participation can help to improve the technical quality of the project processes. However, it is important to recognise that for public participation to be effective and enhance project delivery, objectives of engaging the public must be clear, and the method adopted for the engagement should be specifically designed to meet the set objectives.

v. **Risk management:** Major infrastructure projects are subject to both positive and negative risks. Negative risks can lead to inefficient and ineffective delivery of infrastructure projects due to resulting in their cancellation, serious project delays and cost overruns. The risks to major infrastructure projects can take the form of:

- **Inaccurate identification of the need for infrastructure:** For example, forecasters may overestimate demand, in which case benefits are lower than expected and poor value for money.
- **Policy uncertainty:** This could result in project sponsors, lenders and contractors deferring or abandoning projects in favour of opportunities elsewhere. Financing charges for projects may rise as investors and lenders perceive policy uncertainty as a risk.
- **Failure to assess the cumulative impact on consumers of funding infrastructure through user charges:** This increases the risk of financial hardship for consumers or the need for unplanned taxpayer support.
- **Taxpayer exposure to losses:** This could happen if the government guarantees to bear or share project risks and a risk subsequently materializes (e.g., cost overruns).
- **Delivery costs are higher than they should be:** This could result in higher costs for taxpayers and consumers and fewer projects going ahead than planned.

Assessing risk is at the heart of delivering effective infrastructure. Risk assessment should be underpinned by the financial analysis of a project or opportunity because each risk should be allocated a theoretical cost. In reality, however, this cost is likely to be a range of estimates within a probability function rather than a point estimate. The simple calculation is shown in the following equation:

$$Expected\ cost\ of\ risk = probability\ of\ risk\ occurring \times cost\ if\ risk\ occurs$$

Indeed, further consideration should be given to the impact of uncertainties. This is because while it is easier to put a price on risk, it can be very difficult, if not impossible, to put a price on uncertainty. Investors and lenders will consider the risks they choose to accept based on historical performance and specialist advice. But they will struggle to accept some particular events that may be regarded as uncertainties. This is because these events are beyond their control or management abilities. Consequently, it is possible, and perhaps more suitable, for the government to "own" and manage these uncertainties, especially where there is a joint public–private partnership in the delivery of major infrastructure projects (Lam, 1999). In managing the risk events, a number of options could be exploited including contract, finance, insurance and portfolio options.

8.8 Finalising the plan for major infrastructure projects

Once major infrastructure projects are prioritised and approved, governments can begin to address the practicalities of project delivery with the assurance that (i) the projects are aligned to the infrastructure vision, and (ii) the project will deliver the expected benefits and therefore proceed. A delivery plan can be developed in order to have visual insights on construction programmes and steps can be taken to start securing land use planning approvals. Some projects may also be subject to finalisation of statutory processes. Figure 8.3 illustrates how the UK government moves from vision to delivery of infrastructure projects and finalises project plans in Step 2. At Step 2, the portfolio choice and master plan are prepared. Drawing from the CBA earlier discussed in this chapter, the government reviews each CBA report to confirm that the

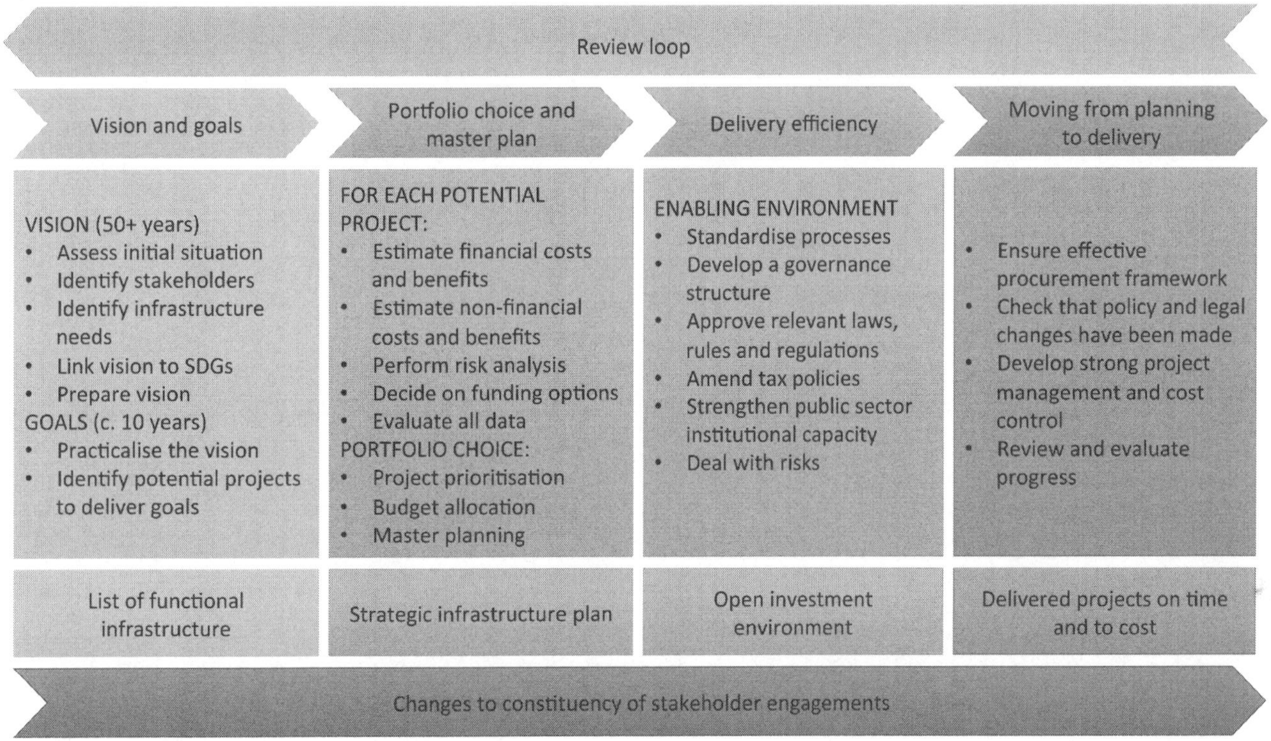

Delivering major infrastructure projects effectively and efficiently

Review loop

Vision and goals

VISION (50+ years)
- Assess initial situation
- Identify stakeholders
- Identify infrastructure needs
- Link vision to SDGs
- Prepare vision

GOALS (c. 10 years)
- Practicalise the vision
- Identify potential projects to deliver goals

List of functional infrastructure

Portfolio choice and master plan

FOR EACH POTENTIAL PROJECT:
- Estimate financial costs and benefits
- Estimate non-financial costs and benefits
- Perform risk analysis
- Decide on funding options
- Evaluate all data

PORTFOLIO CHOICE:
- Project prioritisation
- Budget allocation
- Master planning

Strategic infrastructure plan

Delivery efficiency

ENABLING ENVIRONMENT
- Standardise processes
- Develop a governance structure
- Approve relevant laws, rules and regulations
- Amend tax policies
- Strengthen public sector institutional capacity
- Deal with risks

Open investment environment

Moving from planning to delivery

- Ensure effective procurement framework
- Check that policy and legal changes have been made
- Develop strong project management and cost control
- Review and evaluate progress

Delivered projects on time and to cost

Changes to constituency of stakeholder engagements

Figure 8.3 Moving from vision to delivery of major infrastructure projects

Source: Original adapted from (WEF 2012)

recommended project represents best value for money. The Step 2 process will further determine which projects are the most appropriate to address social, economic and soft infrastructure deficiencies. These projects will usually have a cost-benefit ratio of more than 1. If the government is certain about which deficiencies to address first, the planning process in step 2 will result in fine-tuning the selected projects and considering the funding options.

The planning also needs to align with the budgetary cycles of government; ideally, the funding of projects within the budget should be for a longer period of time, beyond the tradition of annual budgetary allocations. Such an approach will reduce bureaucratic bottlenecks and enhance the procurement processes of the projects. It will also boost the confidence of investors and contractors. Any strategic infrastructure plan developed by governments and planners should provide a chronological visualisation for how infrastructure is delivered and developed over time. A master plans can, therefore, be used in this instance to guide national infrastructure decisions as well as other matters, such as the provision of social infrastructure (WEF, 2012). Such a plan also needs to be flexible and adaptive to the economic and sociocultural environment. For instance, a power plant designed to generate 100 megawatts based on a projected 20 years' demand can be planned to be delivered in stages. If the current demand is 30 megawatts, a phased delivery of the project will allow the project to meet the immediate needs of national demand while allowing capacity for expansion as the needs arise.

8.9 Moving from planning to delivery

Prioritised major infrastructure project plans will need to be published and marketed to stakeholders and potential investors. To further ensure that the projects are delivered effectively, new government policies may need to be developed to support the project context. For example, new public-private partnership (PPP) laws can be established to provide a legal funding framework that supports the project in a clear, reliable and assured way in order to boost investor confidence. The Lekki Toll Road in Nigeria is an example of how government established a PPP framework to support the delivery of a major infrastructure project. The Lekki Concession Company Limited was a special purpose vehicle set up specifically to execute the Eti-Osa Lekki Toll Road Concession Project. The project was designed to deliver essential road infrastructure on the Lekki Peninsular of Lagos state. Lekki Toll Road Concession was conceptualized in 2007 as a PPP scheme and uses the build-operate-transfer (BOT) model of infrastructure delivery. The concession was initially for a period of 30 years; after that, the assets will be transferred to Lagos' state government. In addition to establishing a suitable legal and policy framework for the project, detailed project preparation work should be commissioned to get projects ready for tendering. This should be aimed at making the procurement system very open and transparent.

8.10 Managing delays and cost overrun

Delays and cost overruns are an inherent part of many major infrastructure projects despite the multitude of research and case examples to learn from. In moving from planning to the actual delivery of these projects, one must be aware of the potential causes for delays and cost overruns, plus the possible mitigating actions to be taken. One factor that has been identified as a reason for cost overruns in many major infrastructure projects is design errors (Adam *et al.,* 2017; Narayanan *et al.,* 2019; Pinto and Slevin, 1987). It is important to note that proper representation of project requirements and the blueprint for achieving good technical input to project execution are usually mapped out based on project designs; thus, designs with errors involve incorrect or insufficient representations of project deliverables. They will lead to the wrong application of techniques in achieving results, andlead to delays and cost overruns. Another way design errors could lead to cost overruns and delays is the intersection between design and project estimations such that having errors in design in a form of omission or misrepresentation will mean that the estimation for the project costing also includes these omissions, thereby leading to extra work and change orders (Singh, 2010). Similarly, designs that are done without extensive investigation of a site could contain potential errors that emerge at the construction phase of the project. This could lead to additional work, revision of the scope of work, and contract revisions, which all affect the overall project delivery time and cost. Bordat *et al.* (2004) cited the typologies of design errors in projects as errors from inadequate field investigation, errors in specifications, planning errors and errors in design changes. In controlling project delays and cost overruns due to design errors, key consideration should be given to the involvement of professional and technical skills and application of competent tools in the project. Achieving error free design entails good communication with the entire design team and integrating a design process that is properly planned, giving enough time for corrections, extensive investigation, reviews and utilising incremental project implementation methodologies (Fathi *et al.,* 2020). Similarly, effective project planning, controlling and monitoring should be established to enhance project performance throughout the project life cycle. Proper site investigation should be done to ensure that all site conditions are noted in the design.

Delay and cost overrun in major infrastructure projects could also be as a result of scope change. Scope is the term that defines the entire deliverables that is expected at the end of a project. Therefore, logically, it can be said that all project plans, estimation, schedule, quality and baseline are usually designed base in the initial project scope. Thus, any change in the project scope during execution will mean a change in the initial project plan leading to budget and schedule reviews. With each scope change, precious project resources are diverted to activities that were not identified in the original project scope, leading to pressure on the project schedule and budget (Amadi, 2019; Morris, 1990; Preuß *et al.,* 2019). Project scope changes can result from an incorrect initial scope definition, inherent risks and uncertainties, sudden change of interest, or project funding changes. They can lead to change requests that result in changes in project deliverables, budget or even the entire project team. Poor scope change management can lead to disputes that may require spending time and money on arbitration and litigation between the contractor and the client. Hence, to achieve a proper control for scope change, it is important to first recognise that change is inevitable and could be beneficial to the entire project success. Therefore, major infrastructure projects should have an integrated and adequate change management process in place, taking a proactive approach to change, involving key stakeholders and incorporating their needs throughout the project life cycle. Similarly, to avoid disputes, it is important to always seek approval for changes from relevant project authorities and communicate changes in a timely way. For highly evolving projects, the scope could be frozen so as to concentrate on the expected deliverables.

The delivery of major infrastructure projects can take a long time and also be complex. This makes such projects susceptible to inappropriate or inadequate procurement and a faulty contractual management system (Flyvbjerg, 2014; Flyvbjerg *et al.,* 2004). Contracts spells out every aspect of project agreement, including payment terms, pricing, service levels, and discounts. Therefore, a contract that has not highlighted the entire project scenario may lead to dispute in the contract system. For instance, if the initial contract does not completely specify every relevant aspect of the project work, this may lead to long chains of negotiations, arbitration and/or mitigation due to work change orders and reviews of contractual agreements with new budgets and schedules. Similarly, ambiguous contractual agreements with unclear clauses can be of potential dispute, thereby generating project delays and cost overruns. Also, delays and cost overruns can result from poor contractor selection and unethical contract behaviours. Since the majority of major infrastructure projects are executed using contractors, it is important to note that the procurement process and contract management are critical to the successful completion of the projects. Thus, poor selection of contractors due to low bids, with no technical capability to handle the project, will lead to cost overruns, schedule delays, poor quality, and an unacceptable final result (PMI, 2018). Also, a poor contract management system with clients on a slow payment schedule can lead to the slowing down of project activities. To solve these problems, the ethical thing to do is identify the most qualified contractor and draft out the most suitable contract type as applicable to the conditions of the project and also explicitly define the terms and conditions that govern the contract in clear clauses. These clauses should spell out the penalties for delays and the party to bear risk associated with these events. Similarly, all important potential dispute (and dispute resolution) contract clauses should be stated in clear, unambiguous terms. The use of generic contract templates should be avoided and careful consideration should be given when forming the contracts.

Many major infrastructure projects tend to have a relatively long implementation period when compared to small projects. Consequently, the delivery of major projects are affected by inflation, a change in material prices and changes in exchange rates, so the initial budget may need to be supplemented to achieve completion. The result can be cost overruns and long chains of negotiation leading to delays. Similarly, projects with a high degree of complexity usually result in complex plans, schedules and estimates, and are therefore susceptible to omissions that can later lead to increased costs. Project complexity can also be defined in terms of the diversity of the stakeholders, all with different interests and an extensive communication channel (Cerpa and Verner, 2009). Capturing and considering stakeholder interests can take a lot of time and resources, which when overlooked can also result in conflicts and disputes. To eliminate or reduce the effect of delays and cost overruns due to project complexity, vigorous planning should be done that incorporates every important aspect of the project scope, milestones, detailed Work Breakdown Structure (WBS), delivery time, stakeholders and methodology to be used. Managing complex projects needs experience, expertise and exposure. It is therefore important to build an experienced team with the project's best interests and chance for success at heart.

The post-execution phase (closure) of a major infrastructure project is often ignored. Slow closure involves dragging the various handover activities caused by unresolved disputes. Disputes can be linked to contracts and procurement complexities, change order issues not resolved, final change orders not issued, a poor closure of the final account, poor documentation of project success and lessons learnt, and slow client acceptance and failing to close the work order, which can result in unexpected delays and stray charges made to the project. For instance, if the project team is not decommissioned on time after the project work has been completed, there is a tendency to run an idle team that may incur extra project expenses due to overhead; this may overrun the project costs. Similarly, delays in payments to contractors and suppliers after project completion can lead to disputes and a delay in signing the certificate of final completion of the project. The following suggested actions should be considered during a project closure phase:

- **Completion:** Ensure that the project is 100 percent completed to avoid disputes and payment delays.
- **Documentation:** Detailed documentation will ensure that future changes are made with little extraordinary effort.
- **Project system closure:** This includes closing the financial systems, including all payments, work termination, etc.

- **Project review:** This step can help the transfer of tangible knowledge of time and cost, know-how and know-why.
- **Project Team Management** Disband project team as soon as possible to avoid cost overruns due to extra overhead.
- **Stakeholder satisfaction:** Provide all the necessary information required by stakeholders to avoid conflicts. This information can include a timeline showing the progress of the project from the beginning until the end, the milestones that were met or missed, the problems encountered and a brief financial presentation.

8.11 Chapter summary

This chapter examined some of the challenges, issues, processes and tools used by infrastructure decision-makers and governments to effectively and efficiently deliver major infrastructure projects in alignment with a country's developmental objectives. The chapter began with an insight into how the prioritisation of major infrastructure projects is achieved by showing the different approaches used, especially cost-benefit analysis. The chapter then discussed the issues and challenges of appraising major infrastructure projects in recognition of the need to ensure that infrastructure investments make an impactful contribution to the developmental objectives of a country. Furthermore, the chapter provided a review of the issues and factors to consider in the development of a national vision for major infrastructure projects. This focused mainly on the need to align national infrastructure vision with the SDGs and needs of stakeholders and beneficiaries. As shown in this chapter, a good major project infrastructure vision must be informed by the SDGs, align with the short- and long-term developmental objectives of the country and consider both the positive and negative impact of infrastructure projects (and operations) on the "winners" and "losers" in the context of the proposed infrastructure projects.

The key factors that will ensure major infrastructure projects are delivered effectively and efficiently were discussed. They include ensuring that a transparent and effective governance structure is in place, enhancing institutional capabilities, optimising the project procurement and delivery processes and methodologies, involving the public and effectively managing the risks associated with the project. Once major infrastructure projects are prioritised and approved, processes need to be put in place to ensure effective delivery. Governments can begin to address the practicalities of project delivery with the assurance that the projects are aligned to the infrastructure vision and the project will deliver the expected benefits. The project delivery plan can then be put in place, the budget allocated and the project initiated. The project will at this point proceed from planning to delivery. The delivery needs to consider the procurement approach to be used as well as the methods of project delivery to avoid delays and cost overruns and ensure that the project is delivered efficiently.

8.12 Chapter discussion questions

1. What other factors beyond the economic case can influence the prioritisation of infrastructure projects?
2. A CBA should be thorough and credible in order to influence decision-making. So, what does a good CBA in government infrastructure prioritisation look like?
3. What are the drivers of infrastructure readiness and how do they influence infrastructure prioritisation?

4. What are the SDGs? And how can the SDGs influence the creation of effective infrastructure?
5. What tension exists in the development of infrastructure vision and goals?
6. How can efficiency be achieved in the delivery of major infrastructure projects?
7. What are the causes and effects of delays and cost overruns in major infrastructure project delivery?

8.13 Case studies

8.13.1 Case 1: Country-based comparison of project appraisal processes in decision support

In this case study, the appraisal processes used in decision-making across Germany, Sweden, and the US are presented as a basis for comparison (Mackie and Worsley, 2013). The discussion focuses on the appraisal practice of transport-related major infrastructure projects.

Germany: Germany provides an example of a federal system with some differences between the methods of cost-benefit analysis used by the different tiers of government. The German approach is based on ranking schemes according to their benefit-cost ratios (BCRs) after taking full account of nonquantifiable impacts on habitats and on the environment. The appraisal of these impacts serves to establish what mitigation measures or alternatives will be implemented in order to protect natural resources and whether this is feasible and affordable. No analysis of the impacts of a scheme on the economic performance of the state or the region is carried out, although additional "points" are attributed to schemes which serve low income regions. Projects are ranked by their BCRs; projects with BCRs below 1 are not proceeded with.

Sweden: The principles of cost-benefit analysis are widely accepted as a means of delivering transport policy objectives in Sweden. Appraisal takes place in the context of a Ten-Year Transport Plan that is updated every four years. Schemes included in the plan generally have BCRs in excess of unity. The understanding that schemes with BCRs below unity are unlikely to be included in the plan influences the choice, design and specification of projects put forward for inclusion and therefore serves as a valuable tool for sifting out weak options. Evidence on how the ranking of schemes in the plan on the basis of their BCR influences decision-makers when they decide on which schemes to fund is more mixed. Decisions delegated to officials generally show that ranking by BCR is the norm. However, where the decision is made by politicians, other criteria, primarily those related to their perception of the local, regional or national economic impacts, tended to influence the decision. Road schemes approved by ministers tend to show higher BCRs than rail schemes funded. Analysis of the decisions made on transport schemes shows that the BCR has become more dominant over the past 20 years in the decision-making process and that the appraisal process is better suited to highway schemes than rail projects.

United States: The US, a country with a federal government, has adopted processes for making decisions that differ according to whether the source of funding is through a discretionary federal grant or through state funding supplemented by a formula-based federal contribution. The US Department of Transportation requires projects it funds to be appraised using a traditional cost-benefit analysis, with most environmental impacts valued in monetary terms. Interestingly, there is no guidance about those environmental capital impacts that European countries tend to measure on a qualitative scale, since, under US law, heritage is protected against any incursions. Objectives related to gross domestic product have recently been taken into account the guidelines for applications for funding under the Transportation Investment Generating Economic Recovery. There is no mandatory appraisal method required for projects funded by individual states. Each US state uses an appraisal process, but the information provided to decision-makers differs between states. Some states use multi-criteria analysis, identifying factors of particular importance to that state and its transport users, effects on productivity, and the degree of public support, and then these criteria are weighted to provide a summary table and score. Others use cost-benefit analysis supplemented by an analysis of the impact on the local economy, while other states focus most on the impact on the local economy.

The three countries discussed above were aware of two limitations of the cost-benefit analysis approach and attempted to ensure that decision-makers were provided with information that helped to ensure a more holistic process (Mackie and Worsley, 2013). The first limitation occurs because of the extent to which cost-benefit analysis, as practised there, is restricted to the impacts whose effects can be measured and valued in monetary terms. Most countries had adopted a means of scoring other significant impacts against a qualitative scale to ensure that the welfare economic framework that underpins cost-benefit analysis was more comprehensive than a process that omitted all non-monetised impacts and that these impacts were therefore drawn to the attention of decision-makers. However, the process for assessing the weights that were given to these impacts was largely judgemental and not documented. A second limitation is the policy priority given to the potential impacts of transport schemes that fall outside the welfare-based economic cost-benefit framework. This has resulted in public investment being targeted on productivity and growth. Decision-makers need to know how far the investment in transport schemes that they approve will contribute to increased productivity and to redressing the regional imbalance in output.

8.13.1.1 Case 1 discussion questions:

1. The countries that employed cost-benefit analysis were aware of two limitations of the approach.
2. What are the limitations?
3. How can they ensure that infrastructure decision-makers are provided with data that will create a holistic decision-making model?
4. Identify key differences between the German and US approaches.

8.13.2 Case 2: Delivering an SDG-based major infrastructure projects: The Indonesian case

Financing and delivering sustainable infrastructure often require cross-sector collaboration. This is the case in Indonesia, where a financing facility is bringing together a number of global public- and private-sector stakeholders to foster investments in renewable energy and improved management of forests, biodiversity and ecosystem restoration services throughout the country. The Tropical Landscapes Finance Facility (TLFF) was launched in October 2016 by the Indonesian government and is a partnership between the UN Environment Programme, World Agroforestry, ADM Capital and BNP Paribas. With two sources of capital—a lending platform run by ADM Capital and BNP Paribas and a grant fund run by UN Environment and World Agroforestry— the TLFF provides technical assistance and co-funds early-stage development costs, enabling donors and foundations to harness private-sector funding.

TLFF funds South East Asia's first corporate sustainability bond, a multi-tranche, long-dated sustainability bond arranged by BNP Paribas and issued by TLFF Pte Ltd for the Royal Lestari Utama (RLU), a joint venture between Indonesia's Barito Pacific Group and France's Michelin. ADM Capital acts as facility and ESG manager for TLFF. It offers funding for climate smart, wildlife friendly, socially inclusive production of natural rubber in Jambi, Sumatra, and East Kalimantan provinces of Indonesia. Out of a concession area of 88,000 hectares, 34,000 will be planted with commercial rubber, while more than half will be set aside for community livelihoods

Table 8.2 Project performance metrics

Core objectives	Output and impact indicators	Value (as provided by RLU or source)
Forest Retention	Hectares of actively managed forest	Jambi – 2,000 hectares East Kalimantan – 6,500 hectares
Improved Rural Livelihoods	Number of smallholders rubber farmers engaged as part of the community partnership program	Jambi: 18 Kalimantan: 300 Training: 266 Total: 584
	Number of smallholder households impacted by the project	584 out of estimated 10,000 families living in and around the concessions
	Number of farmers selling into the RLU supply chain	300 in East Kalimantan
	Number of direct jobs created	Jambi: 3,579 East Kalimantan: 851
Reduced Emissions	Number of trees planted	18,622
	Carbon footprint (in tCO_2e)	12,836★
	Greenhouse gas emissions absorbed by protected forest and planted trees (in tCO_2e)	Plantations: 365,106 Forests: 123,286
Biodiversity Protection	Conservation programmes implemented	Protection forest Wildlife conservation area Human–wildlife conflict Wildlife monitoring
	Species protected in the concessions	Critically endangered species: elephants, tigers, orangutans, mitred monkeys, Malayan tapirs

and conservation. The project will also include 9,700 hectares for a wildlife conservation area. At maturity, the plantation will provide 16,000 jobs for local communities, while at the same time the commercial plantations in Jambi will serve as a buffer zone to protect the 143,000 hectares of Bukit Tigapuluh National Park, replete with important biodiversity and endangered species such as Sumatran elephants and tigers. Annual project monitoring includes an assessment against an environmental social action plan. Beyond direct employment, community livelihoods will be supported by the establishment of 7,000 hectares of smallholder plantations with smallholder financing and other livelihood programmes falling under a community partnership program. This also includes training in best practice in rubber production and the purchase of rubber from community program participants at a slight premium. As of September 2019 the project reported the following performance metrics (see Table 8.2):

8.13.2.1 Case discussion questions

1. Drawing on the outlined SDGs in this chapter, discuss the SDGs that the RLU joint venture forest infrastructure project will impact.
2. What lessons can be learned from the RLU joint venture forest infrastructure project in terms of stakeholder engagement?

Delivering major infrastructure projects effectively and efficiently

References

Adam, A., Josephson, P.-E.B. and Lindahl, G. (2017). Aggregation of factors causing cost overruns and time delays in large public construction projects: Trends and implications. *Engineering, Construction and Architectural Management* 24(3), pp. 393–406. DOI: 10.1108/ECAM-09-2015-0135.

Amadi, A. (2019). A cross-sectional snapshot of the insider view of highway infrastructure delivery in the developing world. *International Journal of Construction Management* 19(6), pp. 472–491. DOI: 10.1080/15623599.2018.1452097.

Ambituuni, A., Amezaga, J. and Emeseh, E. (2014). Analysis of safety and environmental regulations for downstream petroleum industry operations in Nigeria: Problems and prospects. *Environmental Development* 9(Supplement C), pp. 43–60. DOI: 10.1016/j.envdev.2013.12.002.

Ambituuni, A., Ochieng, E. and Amezaga, J.M. (2019). Optimising the integrity of safety critical petroleum assets: A project conceptualization approach. *IEEE Transactions on Engineering Management* 66(2), pp. 208–223. DOI: 10.1109/TEM.2018.2839518.

Andres, L., Biller, D. and Dappe, M.H. (2016). A methodological framework for prioritising infrastructure investment. *Journal of Infrastructure Development* 8(2), pp. 111–127. DOI: 10.1177/0974930616667886.

Atkins, G., Bishop, T.K. and Davies, N. (2017). How to value infrastructure: Improving cost benefit analysis. Available at: https://www.instituteforgovernment.org.uk/publications/value-infrastructure-september-2017 [cited 23 May 2020].

Berechman, J. and Paaswell, R.E. (2005). Evaluation, prioritisation and selection of transportation investment projects in New York City. *Transportation* 32(3), pp. 223–249. DOI: 10.1007/s11116-004-7271-x.

Boardman, A.E., Greenberg, D.H. and Vining, A.R. (2017). *Cost-Benefit Analysis: Concepts and Practice*. Cambridge: Cambridge University Press.

Bordat, C., McCullouch, B., Labi. S. and Sinha, K.C. (2004). An analysis of cost overruns and time delays of INDOT projects. *JTRP Technical Reports*. Available from: https://docs.lib.purdue.edu/cgi/viewcontent.cgi?article=1482&context=jtrp [cited 23 May 2020].

Bowker, G.C., Baker, K., Millerand, F. and Ribes, D. (2010). Toward information infrastructure studies: Ways of knowing in a networked environment. In: Hunsinger J., Klastrup L., and Allen M. (Eds.), *International Handbook of Internet Research*. Dordrecht: Springer Netherlands, pp. 97–117. DOI: 10.1007/978-1-4020-9789-8_5.

Cerpa, N. and Verner, J.M. (2009). Why did your project fail? *Communications of the ACM* 52(12), pp. 130–134. DOI: 10.1145/1610252.1610286.

Dadson, S.J., Hall, J.W., Murgatroyd, A., Acreman, M., Bates, P., Beven, K., Heathwaite, L., Holden, J., Holman, I.P., Lane, S.N., O'Connell, E., Penning-Roswell, E., Reynard, N., Sear, D., Thorne, C. and Wilby, R. (2017). A restatement of the natural science evidence concerning catchment-based 'natural' flood management in the UK. *Proceedings. Mathematical, Physical, and Engineering Sciences* 473(2199), p. 20160706. DOI: 10.1098/rspa.2016.0706.

Edwards, P.N., Jackson, S.J., Bowker, G.C. and Knobel, C.P. (2007) Understanding infrastructure: Dynamics, tensions and design. Available from: https://deepblue.lib.umich.edu/bitstream/handle/2027.42/49353/UnderstandingInfrastructure2007.pdf?sequence=3&isAllowed=y [cited 23 May 2020].

Erickson, P., Kartha, S., Lazarus, M. and Tempest, K. (2015). Assessing carbon lock-in. *Environmental Research Letters* 10(8). IOP Publishing: 084023. DOI: 10.1088/1748-9326/10/8/084023.

Fathi, M., Shrestha, P.P. and Shakya, B. (2020). Change orders and schedule performance of design-build infrastructure projects: Comparison between highway and water and wastewater projects. *Journal of Legal Affairs and Dispute Resolution in Engineering and Construction* 12(1). American Society of Civil Engineers: February 2020. DOI: 10.1061/(ASCE)LA.1943-4170.0000353.

Flyvbjerg, B. (2008). Curbing optimism bias and strategic misrepresentation in planning: Reference class forecasting in practice. *European Planning Studies* 16(1), pp. 3–21. DOI: 10.1080/09654310701747936.

Flyvbjerg, B. (2014). What you should know about megaprojects and why: An overview. *Project Management Journal* 45(2), pp. 6–19.

Flyvbjerg, B., Holm, M.K.S. and Buhl, S.L. (2004). What causes cost overrun in transport infrastructure projects? *Transport Reviews* 24(1), pp. 3–18.

Hall, J.W., Scott, T., Cao, Y., Chaudry, M., Blainey, S.P. and Oughton, E.J. (2017). Strategic analysis of the future of national infrastructure. *Proceedings of the Institution of Civil Engineers - Civil Engineering* 170(1), pp. 39–47.

ICE (2020) *Enabling Better Infrastructure: 12 guiding principles for prioritising and planning infrastructure*. Institute of Civil Engineering. Available from: https://www.ice.org.uk/ICEDevelopmentWebPortal/media/Documents/Media/ice-enabling-better-infrastructure-report.pdf [cited 1 June 2020].

Itani, N., O'Connell, J.F. and Mason, K. (2014). A macro-environment approach to civil aviation strategic planning. *Transport Policy* 33, pp. 125–135. DOI: 10.1016/j.tranpol.2014.02.024.

Karasti, H., Baker, K.S. and Millerand, F. (2010). Infrastructure time: Long-term matters in collaborative development. *Computer Supported Cooperative Work (CSCW)* 19(3), pp. 377–415. DOI: 10.1007/s10606-010-9113-z.

Karydas, D.M. and Gifun, J.F. (2006). A method for the efficient prioritization of infrastructure renewal projects. *Reliability Engineering and System Safety* 91(1), pp. 84–99. DOI: 10.1016/j.ress.2004.11.016.

Lam, P. (1999). A sectoral review of risks associated with major infrastructure projects. *International Journal of Project Management* 17(2), pp. 77–87. DOI: 10.1016/S0263-7863(98)00017-9.

Mackie, P and Worsley, T., (2013) International Comparisons of Transport Appraisal Practice. Overview Report. Institute for Transport Studies, Faculty Of Environment, Leeds University: Available from: https://assets.publishing.service.gov.uk/government/uploads/system/uploads/attachment_data/file/209530/final-overview-report.pdf [cited 7 June 2020)

Morris, S. (1990). Cost and time overruns in public sector projects. *Economic and Political Weekly* 25(47). Available from: https://www.epw.in/journal/1990/47/review-industry-and-management-uncategorised/cost-and-time-overruns-public-sector [cited 2 June 2020].

Narayanan, S., Kure, A.M. and Palaniappan, S. (2019). Study on time and cost overruns in mega infrastructure projects in India. *Journal of The Institution of Engineers (India): Series A* 100(1), pp. 139–145. DOI: 10.1007/s40030-018-0328-1.

Ochieng, E., Price. A.D.F. and Moore, D. (2017) *Major Infrastructure Projects: Planning for Delivery*. London: Palgrave Macmillan.

Pinto, J.K. and Slevin, D.P. (1987). Critical factors in successful project implementation. *IEEE Transactions on Engineering Management* 34(1), pp. 22–27. DOI: 10.1109/TEM.1987.6498856.

PMI. (2018). *Guide to the Project Management Body of Knowledge (PMBOK)*. 6th ed. Project Management Institute.

Preuß, H., Andreff, W. and Weitzmann, M. (2019). *Cost and Revenue Overruns of the Olympic Games 2000–2018*. Springer Nature. DOI: 10.1007/978-3-658-24996-0. Available from: https://www.springer.com/gp/book/9783658249953 [cited 5 June 2020].

PwC. (2016). How to prioritise public infrastructure investments. Available from: https://www.pwc.com/gx/en/issues/economy/global-economy-watch/prioritise-public-infrastructure-investments.html [cited 4 March 2020].

Scholz, M. and Lee, B. (2005). Constructed wetlands: A review. *International Journal of Environmental Studies* 62(4), pp. 421–447. DOI: 10.1080/00207230500119783.

Schutte, I.C. and Brits, A. (2012). Prioritising transport infrastructure projects: Towards a multi-criterion analysis. 16(3), pp. 97–117.

Singh, R. (2010). Delays and cost overruns in infrastructure projects: Extent, causes and remedies. *Economic and Political Weekly* 45(21), 43–54.

Soto, H.D. (2001). *The Mystery of Capital*. New ed. London: Black Swan.

Steinbuks, J. and Foster, V. (2010). When do firms generate? Evidence on in-house electricity supply in Africa. *Energy Economics* 32(3), pp. 505–514. DOI: 10.1016/j.eneco.2009.10.012.

Thacker, S., Adshead, D., Fay, M., Hallegate, S., Harvey, M., Meller, H., O'Regan, N., Rozenberg, J., Watkins, G. and Hall, J.W. (2019). Infrastructure for sustainable development. *Nature Sustainability* 2(4), pp. 324–331. DOI: 10.1038/s41893-019-0256-8.

Thomson, P., and Koehler, J. (2016). Performance-oriented monitoring for the water SDG – challenges, tensions and opportunities. *Aquatic Procedia* 6, pp. 87–95.

UNECE. (1998). *Convention on access to information, public participation in decision-making and access to justice in environmental matters*. Done at Aarhus, Denmark: The United Nations Economic Commission for Europe (UNECE). Available from: https://ec.europa.eu/environment/aarhus/ [cited 1 June 2020].

van der Kamp, G. and Hayashi, M. (1998). The groundwater recharge function of small wetlands in the semi-arid Northern Prairies. *Great Plains Research: A Journal of Natural and Social Sciences*. Available from: https://digitalcommons.unl.edu/greatplainsresearch/366 [cited 6 June 2020].

WEF. (2012). Strategic infrastructure: Steps to prioritise and deliver infrastructure effectively and efficiently. *World Economic Forum Report. Geneva Switzerland: World Economic Forum*. Available from: https://www.weforum.org/reports/strategic-infrastructure-steps-prioritize-and-deliver-infrastructure-effectively-and-efficiently [cited 7 June 2020].

9

ENHANCING THE WHOLE LIFE-CYCLE PERFORMANCE OF MAJOR INFRASTRUCTURE PROJECTS

Edward Ochieng

9.1 Introduction

As shown in previous chapters, major infrastructure projects have wider benefits to society, and as such these benefits should be well assessed. Whole life-cycle performance infrastructure metrics can be applied to infrastructure systems and can adopt a range of basic approaches depending on the maturity of the infrastructure project. However, their application may be more beneficial to those interdependent and interconnected projects that have the greatest value, require considerable funding, are high risk and/or are seen as critical. In some cases, complex approaches may be applied and in these circumstances, higher quality data and predictive modelling techniques will often be needed. Achieving whole life cycle value is a step beyond whole life cost and entails making infrastructure decisions based on broader criteria, while also considering the needs of a wider range of stakeholders instead of just those typically involved in the immediate decision-making process. Understanding the whole life of an infrastructure project can pay dividends when it comes to costing, construction, maintenance and operation decisions. As expounded in this chapter, the ultimate aim for infrastructure clients and contractors should not only be to construct good infrastructure projects, but also ensure that the right infrastructure projects are constructed to meet the requirements during delivery and completion of an infrastructure project. Moreover, there is a need to ensure infrastructure projects meet the requirements of all stakeholders, particularly the end users.

9.1.1 Chapter aim and objectives

As a consequence of the above, this chapter examines how the whole life-cycle performance of infrastructure projects can be achieved and sustained. The main objectives are to:

- Examine the process of whole life costing of infrastructure investment;
- Depict how owners and contractors can integrate circular economy principles into infrastructure whole life costing;
- Appraise different types of collaborative procurement models;
- Establish the value of fostering early client involvement and usage of BIM; and
- Ascertain how the client team and contractors can reduce costs at the design and pre-construction phases.

9.1.2 Learning outcomes

The following learning outcomes have been identified for this chapter. Readers will be able to:

- Articulate the process of whole life costing of infrastructure investment;
- Comprehend the value of embedding circular economy principles into the infrastructure of whole life costing;
- Differentiate different types of collaborative procurement models;
- Appreciate the value of fostering early client involvement;
- Gain an insight into the use of building information modelling (BIM); and
- Describe how to reduce costs at both the preliminary design phase and pre-construction phase.

9.2 Whole life costing of infrastructure investment

According to Opoku (2013), there are several definitions of whole life costing. According to Construction Research and Innovation Strategy Panel (CBPP, 1998), "the systematic consideration of all relevant costs and revenues associated with the acquisition and ownership of an asset," and according to the British Standard (2011), "a technique which enables comparative cost assessment to be made over a specified period of time, taking into account all relevant economic factors both in terms of initial capital costs and future operational cost" – they are vague definitions. Opoku (2013) noted that both definitions agree on one theme: the inclusions of all relevant costs. Clift and Bourke (1999) affirmed that whole life costing should take into account the initial construction or major refurbishment cost plus the recurring or occupancy cost such as cleaning, maintenance and repair. Olanrewaju (2013) showed that the origin of whole life-cycle costing and the time it was first applied to construction projects is not available, and it can be safely concluded that it preceded value management techniques. Despite the nomenclature, the main aim is to consider future costs in the determination of the true cost of projects. In other words, whole life cycle costing can be viewed as a technique that can be used to relate the initial cost with future costs like running, operation, maintenance, replacement and alternation costs (Olanrewaju, 2013).

It is worth noting that whole life costing allows a practical economic comparison of alternative options, in terms of both the present and future costs. This is to establish, in the final appraisal, how much additional capital outlay is warranted today in order to achieve future benefit over the entire life of the infrastructure project. It can therefore be taken as the relationship between the initial cost and other future cost. The primary aim of infrastructure whole life costing is to

Enhancing the whole life-cycle performance of major infrastructure projects

provide a means to aid the decision-making process in the lifespan of an infrastructure asset. It is a method that can be used to appraise various options at an early stage in the life of an infrastructure project to enable the client and contractors to make informed decisions. Below is a summary of whole life costing benefits:

1. The contractor can use whole life costing to demonstrate to the client how they will assess and attain sustainability during the delivery of an infrastructure project;
2. The client team and contractors can use the whole life costing model to project future capital expenditure to be applied to long-term costing assessment; and
3. If formulated well, the client team and contractors can use the whole life costing model to assess risks and propose mitigation strategies that can be adopted on infrastructure projects.

The decision regarding the life cost of an infrastructure project has to be ascertained right from the infrastructure project's conceptual phase as to whether to reduce the initial cost at the detriment of the maintenance and running costs. This depends on the infrastructure owner's value system on the infrastructure projects; however, in order to ensure meaningful selection, effective balance must be realised. According to Olanrewaju (2013) the initial construction costs which are projected costs and other unpredictable cost should be considered. It is worth emphasising that issues of whole life costing are more important to infrastructure clients than to the contractors, who only construct the infrastructure project. In this case, clients and end users are left to bear the maintenance costs. Olanrewaju (2013) acknowledged that the contemporary procurement model (design, building and operation) is a possibly good channel to consider building the life cycle. Moreover, whole life costing is a tool often utilised by the clients and management team to procure value for investments. There are numerous costs associated with constructing, operating and maintaining infrastructure projects (see-**Chapter 10**). Infrastructure, just like other types of capital expenditure, will become outdated after years of continuous usage and exposure to various climate conditions. In order to continue the operations, ageing infrastructure systems (such as roads, ports, rail, highways, airports, power and water) will require periodic maintenance, modernisation and replacement in whole or parts. Ramachandran (2019) asserted that the annual spending on maintenance and modernisation would typically range from two percent to 20 percent or more depending on numerous factors, not just the size, nature and capacity but also the purpose, design, materials, quality of construction and most importantly the way it operates and is maintained by the client team.

9.2.1 Life cycle cost analysis

According to Ramachandran (2019) operational expenditure (OPEX) variables are expenditures over a period of 20 to 30 years or more. It is highly possible that the capital spending on an infrastructure project throughout its life cycle will be much higher than the initial capital expenditure to construct it. Interestingly, from the reviewed literature it was found that there is not enough analysis being done when constructing assets, cities or major infrastructure projects. As illustrated in Figure 9.1, there is a critical need to examine this and optimise the expenditures on the infrastructure throughout its life cycle. In order to ensure the cost-effectiveness of an infrastructure project, Ramachandran (2019) recommended the use of life cycle cost analysis (LCCA).

Life cycle cost analysis is a tool that can be used to determine the most cost-effective option among different competing options to purchase, own, operate, maintain and dispose of an infrastructure asset when each is equally appropriate to be implemented on technical grounds. It is worth noting the LCCA takes into account all costs incurred in the life cycle of the infrastructure.

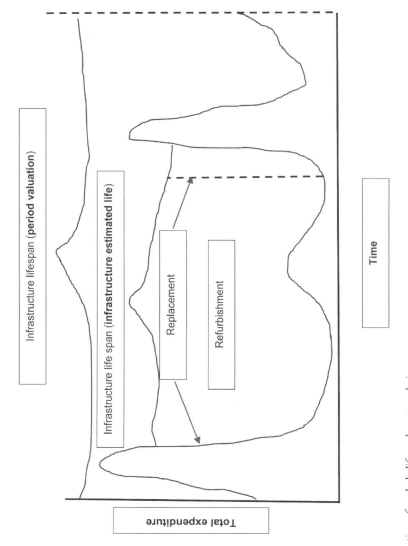

Figure 9.1 An illustration of a whole life cycle cost analysis

This signifies to the most cost-effective option to build, operate, maintain and decommission the infrastructure. For instance, to provide more clarity for a highway pavement, LCCA will consider the initial construction cost, user costs (e.g., reduced capacity at work zones), and agency costs associated with future activities, including future periodic maintenance and rehabilitation. All the costs are usually reduced and tailored to present-day value known as net present value (NPV) (Ramachandran, 2019). In order to provide an all-inclusive and comprehensive model, one has to incorporate a life cycle model with the total cost of ownership. To provide certainty in asset performance, it is vital to replace infrastructure assets at the right time and in line with good practice, organisation policy and contractors' recommendations. Thus, well-thought-out capital planning before incurring actual costs can enhance the efficiency in the usage of time and resources. Cost savings can also be achieved through standardising the approach, and reducing costs and obsolescence through supply chain partners. The aim is to improve stakeholders' confidence as reduced uncertainty in cost effectiveness can lead to improved asset performance and minimise the disruption and risks. With the context of this book, the life cycle can be viewed as the asset's estimated life before the next replacement occurs, whereas the lifespan can consist of more than one life cycle. A typical infrastructure life cycle costing appraisal should entail capital expenditure (CAPEX), OPEX, replacement expenditure and cost of disposal.

A meticulous life cycle cost (LCC) modelling can enable the design to be value engineered so that the cost profile of an infrastructure project can effectively minimise the expenditure throughout the life cycle by a considerable amount. The LCC modelling can be achieved by conducting a series of sensitivity tests on the cost components. By utilising the LCC modelling to conduct extensive testing, comparison and analysis, infrastructure operators can determine the optimised development combination for a particular infrastructure project that is likely to have the most preferred cost profile and financial exposure. Ramachandran (2019) further emphasises that life cycle analysis can be used to provide an overall framework for considering total incremental costs over the lifespan of the infrastructure project. Moreover, LCCA can be used to provide "win-win" strategies in terms of identifying suitable technologies and services that are economically, socially and environmentally sustainable. Consideration of life cycle costs should include infrastructure evaluation, construction, operation, maintenance, regeneration and rehabilitation. As shown in this section, it enhances better decision-making from a more accurate and realistic assessment of revenues and costs.

9.2.2 Embedding circular economy principles into infrastructure whole life costing

Current construction business models miss opportunities to create value by considering the whole life of buildings and infrastructure (Boyd, 2017). Boyd (2017) suggests that circular economy principles provide a new approach to the planning, design, delivery and operational of infrastructure assets, offering value-added and productivity spending, high levels of fragmentation and inadequate risk-sharing. Ellen Macarthur Foundation (n.d) highlighted a circular economy aims to refine growth, focusing on positive society-wide benefits. It progressively decouples economic activity from the consumption of finite resources, and designs waste out of the model. Underpinned by a transition to renewable energy sources, the circular framework builds economic, natural and social capital. As observed from the reviewed literature, a circular economy is based on three themes (Ellen Macarthur Foundation, n.d):

- Design out waste pollution;
- Keep products and materials in use; and
- Regenerate natural systems.

The concept recognises the significance of the economy needing to work effectively at all levels for large and small businesses, for organisations and individuals, and both globally and locally. Adopting a circular economy does not only amount to amendments targeted at minimising the negative impacts of the linear economy. Rather, it represents a systematic shift that fosters long-term resilience, generates business and economic opportunities and provides environmental and social gains (Ellen Macarthur Foundation, n.d). The model discerns between technical and biological cycles. Consumption occurs only in biological cycles, where food and biologically-based resources (such as cotton or wood) are designed to feed back into the system through processes like composting and anaerobic digestion. These cycles reproduce living systems, such as soil, which make available renewable resources for the economy. Technical cycles recover and reinstate products, components and materials through strategies like reuse, repair, remanufacture or recycling. According to Arup (2018), adopting the circular economy model in a high growth, high waste sector like the built environment presents huge opportunities for businesses, governments and cities to minimise structural waste and realise greater value from built environment assets. Within the context of a circular economy, renewable materials are used where possible, energy is provided from renewable sources, natural systems are conserved and improved, and waste and negative impacts are designed out. Materials, products and components are instead managed in loops, sustaining them at their highest possible intrinsic value.

Integrating the three principles of the circular economy in infrastructure projects through the use of new technology, business models and partnerships could lower the construction and infrastructure delivery costs, reduce negative environmental impacts and make urban areas more liveable, productive and convenient. As highlighted by Arup (2018), the adoption of circular economy principles requires the application of systems thinking and new approaches to the way the industry designs, operates and maintains assets. Such thinking can be seen as a natural extension of the all-inclusive approaches already being applied by contractors. It is worth noting that the utmost opportunities are realised when the circular economy thinking model is used for strategic decision-making. For instance:

- Urban planning and construction could incorporate contemporary technologies and nature-based solutions to create more resilience to the impacts of climate change;
- Removing harmful resources from buildings can enhance occupants' health and productivity.

MI-ROG (2016) affirmed that procurement is vital in embedding circular economy principles in infrastructure design and operation. It is worth noting that recognition of the need for greater resilience in infrastructure supply chains is driving a growing interest in the circular economy. MI-ROG (2016) showed that a whole life cycle approach to the development of critical infrastructure, with assets and materials kept at their highest value as long as possible, can lead to potential capital effectiveness and net positive environmental impacts. Aligning circular economy principles such as designing for disassembly and remanufacture and innovative procurement that encourages service over product, requires effective partnership. Infrastructure operators need to consider circular economy principles as a natural progression from pre-existing infrastructure initiatives, commitments and practices, such as resource efficiency, revenue generation through materials sales and reducing landfill waste disposal.

Green Alliance (2019) noted the utmost potential for improving resource efficiency and contributing to the circular economy in infrastructure delivery occurs during the feasibility, design and selection phases. By integrating procurement criterion focused on circular economy principles from the onset of the infrastructure development, operators can achieve cost efficiencies. Boyd (2017) accentuated that the current whole life costing model has not kept pace with the technological change or client expectations and its profligate utilisation of resources is increasingly at odds with efforts to mitigate climate change. As established from the reviewed literature, the current approach to infrastructure is based on a traditional linear model which disconnects the planning, construction, operation and maintenance of infrastructure assets into separate activities (Boyd, 2017; Green Alliance, 2019; MI-ROG, 2016). The circular economy model challenges the existing life costing model by requiring the whole of an asset to be considered from the onset. Moreover, it utilises a systems-thinking approach to identify the most cost-effective solutions that can cut across traditional sector boundaries, implementing life cycle appraisal, digital technology and new business models to achieve capital effectiveness, amplify asset utilisation and augment efficiency across a range of sectors, such as integrating circular economy principles in infrastructure delivery changes where value is realised in the supply chain and offers an opportunity for new, more capital gain at all phases of an infrastructure project. Strategically adopting a circular economy approach changes the question: rather than considering sector by sector what infrastructure is required, a circular economy approach takes a step back and considers whether the infrastructure is required at all (Boyd, 2017). For instance, congestion of a specific infrastructure route represents demand for connectivity. A traditional approach would be to provide more capacity whereas a circular economy approach may meet demand in a different way using virtualised services (that is, it provides data that facilitates users to shift to alternative under-utilised assets on the same route) (Boyd, 2017).

MI-ROG (2016) established that taking a circular economy approach requires a degree of innovation in procurement and technical specifications. Existing technical standards can be an impediment to innovation, which may result in procurement guidelines restricting innovation by being too prescriptive, leading to tenders based on familiar methods. MI-ROG (2016) further stressed that innovation must offer enhancements over industry best practice, but also be implementable, which will require a good track record to substantiate any benefits gained. Different methods of procurement may support innovation by asking for "services" rather than established specifications, particularly at the contract level. The potential risk in service delivery to both client and contractor can be managed by an appraisal of new solutions through demonstration or pilot projects before widespread adoption (MI-ROG, 2016). Boyd (2017) suggested that the utilisation of design-build-operate-maintain (DBOM) contracts will transform infrastructure from a product into service. Placing the responsibility for the whole life of an infrastructure into the hands of one contractor will allow value to be achieved across the whole life of an infrastructure project. The returns on investment in research into new durable, replaceable and recyclable resources can be recouped if the main contractor and operator are one entity. Additionally, alternative approaches for risk mitigation may include risk-sharing opportunities through mechanisms such as collaborative performance models or design-build-finance-operate (DBFO) procurement models. The procurement approaches appraised in this section can more readily entail desired circular economy outcomes. Note that value chain plans will be vital, too, as operations, teams and technology will change over the infrastructure lifespan. From an asset management viewpoint, the application of circular economy guidelines will allow operators to use new digital services to manage demand in real time, collect data on how assets are used and to augment the user experience. Moreover, smart sensors and intelligent monitoring will minimise disruption by allowing maintenance to take place only when necessary.

9.3 Collaborative procurement models

As construction and infrastructure clients, suppliers and consultants all come under ever-increasing pressure to deliver infrastructure-based works and services more efficiently and effectively, so the drive to eliminate waste and duplication intensifies. According to Construction

Excellence (2009) this cost and value for money pressure are compounded by the fact that capital and people resources have become more difficult to obtain and retain, which has encouraged infrastructure operators to identify new ways of ensuring that their short- and long-term strategic objectives can be achieved. As suggested by Construction Excellence (2009), several clients are now partnering and using different forms of contract models that encourage collaboration, greater certainty of cost and time, dispute resolution processes together with effective schemes of performance measurement that are aligned to financial schemes that reward cost reduction, continuous improvement and innovation. The successes gained have shown that the construction and infrastructure sector has achieved higher levels of success when teams collaborate, whether it be for a single contract or more importantly over a long period where a guaranteed flow of work can be assured, as with framework agreements. As specified by Construction Excellence (2009), within the public sector, collaborative procurement has managed to deliver considerable capital gains, improved working practices and cashable gains that have provided the much-required financial resources for enhancing public services or works.

At this juncture, it is worth noting that the term "collaborative procurement" can mean something to both individuals and organisations depending on their needs and the knowledge and experience gained. According to Construction Excellence (2009), collaborative procurement is considered as two or more organisations that agree to work together and having identified key benefits that can be gained by aligning their purchasing power and resources to deliver financial savings, efficiencies and effectiveness without any detriment to the project. Collaborative working can also be viewed as two or more contractors working together in a collaborative team-based environment to deliver contract or service objectives efficiently, effectively and that demonstrate value for money. This reflects the "partnering" culture and associated processes that have been widely accepted in the construction and infrastructure sector. Collaborative procurement can be undertaken as part of new or existing collaborative relationships. As asserted by PwC (2018), contemporary procurement models (see Table 9.1) have for quite some time been preferred by infrastructure project owners for their simplicity and for the certainty and risk transfer they provide to owners. Contemporary models create commercial incentives for non–owner participants to act in a manner contrary to the interests of the infrastructure project owner and vice-versa. As stated by PwC (2018), this misalignment of commercial interests has discouraged the collaboration between project stakeholders that is needed to improve construction productivity. It was from a desire to address this misalignment of interests that "collaborative contracting" was introduced.

Collaborative contracting entails a wide and flexible range of models of managing the relationship between infrastructure project owners and other project stakeholders and is based on the recognition that there can be mutual benefit in a more collaborative and cooperative relationship between them. If the infrastructure project owner contractually agrees to share the gains it receives from outstanding performance by the non-owner participants, the contract can financially motivate the non–owner stakeholders to achieve such outcomes, even if they need to expend more effort and capital to do so. This can be expressed as the establishment of a win–win scenario. Collaborative contracts entail features specifically designed to overcome the misalignment of commercial incentives associated with conventional fixed-price contracts. These features can be summarised this way (PwC, 2018):

- Contractual commitments to co-operate and act in "good faith";
- Early warning mechanisms, formulated to alert other stakeholders to emerging issues, so that solutions can be created and agreed before the issues escalates;

Enhancing the whole life-cycle performance of major infrastructure projects

Table 9.1 An overview of the main collaborative contracting models

Collaborative models	Overview
Partnering model	The main aim of partnering is to create an environment of trust and co-operation, to minimise disputes and facilitate the completion of a successful infrastructure project. The partnering process commences with a workshop at which stakeholders seek to identify common objectives, establish communication channels and discuss guidelines for dealing and avoiding disputes before the commencement of the contract. At the end of the workshop, stakeholders involved sign a partnering charter that will be part of the conventional contract, create detailed mission and common objectives of the stakeholders and demonstrate the commitment of the key people involved.
Integrated project delivery model	Five features differentiate IPD from conventional construction procurement: a The remuneration regime: IPD alters the renumeration arrangements and risk allocation found in conventional fixed-price contracts by replacing the fixed price with a performance based renumeration regime. b Creation of virtual organization: The integrated project team or alliance is comprised of individual team members from the infrastructure project owner and each non-owner participant. c Continuous involvement of all non-owner participants from the moment the contractual relationship is developed. d A requirement for decisions regarding the infrastructure project to be made by way of unanimous agreement between the infrastructure client and other participants in the integrated project team. e No blame, no dispute clause: Here each party agrees that it will have no right to bring legal claims against any other stakeholders in the integrated project team, except in the very limited circumstance of willful default by another stakeholder. It is worth noting that some IPD agreements do not fully embrace all of the features listed above. The IDP model is best suited to infrastructure projects with the following traits: • Complex risks, interfaces, and stakeholder issues that are difficult to allocate and price. • The scope of work is not sufficiently formulated to enable sensible pricing, or scope changes are likely to be significant. • Considerable scope value-adding through innovation. • Tight schedules require scope definition, design and construction to occur concurrently. • The scale of the infrastructure project, and the benefits which can be derived from utilising the IPD framework, are sufficient to justify the additional procurement and contract establishment costs associated with the framework.
Managing contractor model	According to PwC (2018), the managing contractor model is an innovative arrangement that shares some of its features with design and construct (DC) or engineering, procurement, construction and management (EPCM) models. It is worth noting the managing contractor is responsible for the design and construction of the project from feasibility right through to the commissioning phase. The agreement usually entails the client entering into one contract with the managing contractor, who then subcontracts out all of its design and construction responsibilities.

Table 9.1 An overview of the main collaborative contracting models *continued*

Collaborative models	Overview
Engineering, procurement and construction management (EPCM)	The primary role of an EPCM contractor is very similar to the role of a managing contractor. The EPCM contractor is appointed by the client early in the infrastructure development process to oversee the feasibility phase of the infrastructure project before going on to manage the design/engineering, procurement and construction phases of the infrastructure project. The characteristic that distinguishes EPCM from the managing contractor framework is the lower level of risk that an EPCM contractor is exposed to in terms of the quality of project tasks. The EPCM framework typically differs from the managing contractor model in terms of how it assigns the risk of design and construction defects, whereas a managing contractor framework accepts responsibility for ensuring that the design is fit for purpose and the project tasks are constructed from any defects.
Delivery partner model	The delivery model comprises managing contractor, IPD and engineering, procurement and construction management models. The delivery model, if implemented well, allows the project owner to supplement its internal project management capacities by including one or more delivery partners to assist the project owner with planning, programming, design management and construction management services. By incorporating this expertise, the client is able, with the help of its delivery partners, to integrate a sophisticated client procurement strategy involving direct engagement of suppliers and subcontractors, as opposed to inviting a major contractor to coordinate this process. This can lead to huge financial savings and other gains for the client. As affirmed by PwC (2018), the delivery partner model has been utilised successfully in the context of public major infrastructure projects. For instance, the model was used by the UK government during the construction of infrastructure for the London Olympics Games. A more traditional delivery model was found unsuitable, due to the complexity of the project and time-critical date for completion. The utilisation of the delivery partner model allowed the UK Olympic Delivery Authority (ODA) to acquire the required expertise where the ODA did not have the time to source out and engage personnel of the required skill to meet the timeline for the project. It is vital to note that the delivery partner model is best suited to major infrastructure projects where the client wishes to achieve time and capital effectiveness and is willing to embrace and manage the integration.

- Early involvement of the primary contractor and key specialist sub-contractors in the design phase;
- Governance arrangements that will lead to collective problem-solving and decision-making;
- Payment arrangements that financially motivate stakeholders to act in a manner that will best suit the project, rather than best for individual participants;
- The agreement of each stakeholder to waive its right to institute legal proceedings against any other participant for errors, breach or negligence by another stakeholder (except in the case of willful default).

Within the context of infrastructure project delivery, collaborative contracts take different forms. For instance, a number of conventional contracts try to facilitate greater collaboration by integrating contractual promises to work together, as well as early warning mechanisms. However, these contracts have been found not to address the real challenges to greater collaboration that are inherent in conventional fixed price contracts. To address these issues, other forms of collaborative contracting such as the delivery partner model and the Australian Department of Defence's managing contractor model have been introduced. More recently, the project alliance model was introduced by the American construction sector, where it has been called integrated project delivery (IPD) and has increasingly being used with great success (PwC, 2018). Table 9.1 provides an overview of the main collaborative contracting models (Bakker *et al.*, 2008; PwC, 2018; Wondimu *et al.*, 2019):

From the above, it could be suggested that collaborative contracting models are generally considered unsuitable if project owners wish to raise capital on a project finance basis; that is where investors may only look to cash flows and assets of the project to secure repayment and not to the balance sheet of the owner. It is not impossible to raise capital for a project delivered under a collaborative contract model. To mitigate greater risks assumed by an infrastructure project owner under collaborative models, project financiers may need the following (PwC, 2018; Deloitte, 2006):

1. The equity investors in the special project vehicle/borrower to provide more equity upfront, together with binding commitments to provide additional equity in the event of delays or cost overruns.
2. The creation of a standalone cost overrun facilities with higher margins.
3. The contract to include certain characteristics such as well-formulated gainshare/pain share agreement, as well as a prescriptive subcontracting system.
4. More thoroughness in relation to infrastructure project risks, technical issues and the capabilities of the contractors.
5. Tailored insurance policies.

In addition to the above, it is worth emphasising that there is no "one-size-fits-all" when it comes to deciding which contract strategy to go with. The contract model that will best suit a particular infrastructure project will depend upon several variables including the project owner's objectives, the features of the project and the state of the construction market. What is important is that the client and project team have to fully understand the features of the contracting models and how they can be refined to meet the infrastructure project owner objectives.

9.4 Fostering early contractor involvement

The preparation work for large major infrastructure projects requires an extraordinary amount of time, capital and human resources and is not particularly cost-effective. It could be suggested that some of this inefficiency is caused by traditional procurement methods which bring contractors into the process after several decisions have been deliberated. The client and contractors are required to make design decisions with sufficient data and know-how as to available technology, equipment and innovative solutions. It is worth noting that the early contractor involvement (ECI) model provides an efficient means of designing and planning infrastructure projects in a cost-effective, more efficient and less adversarial structure. The application of ECI with a well-structured contract that reflects a partnering relationship should maximise transparency, reduce risk, enhance shared responsibility and limit the reasons for litigation (IADC, 2012). As stated by Wondimu *et al.* (2019), ECI facilitates implementation of innovative, efficient and value-adding solutions through building trust-based cooperation between clients and contractors. The main aim of ECI is to bring the contractor's construction knowledge and experience into pre-construction phases of infrastructure projects. Of particular interest is the enhancement in value for money and project delivery time by comparison with traditional project delivery models.

IPD, alliance and partnering are the three relational project delivery methods that have been widely applied internationally. IPD's project delivery method assimilates people, systems, business structures and practices by using contracts (Gokhale, 2011). Early involvement of contractors is at the centre of IPD. Even though ECI does not require the application of technological tools, the amalgamation of BIM with IPD has greatly enhanced the efficiency of collaboration in all phases of infrastructure projects. BIM is considered an important tool that enables ECI (Rowlinson, 2017). BIM allows for collaboration among clients, designers, contractors, users and other key stakeholders throughout the life cycle of an infrastructure project with the application of 3D models, specifically encouraging ECI (Ferme *et al.*, 2018). The reviewed literature suggests that BIM is an essential tool that enables ECI but is not an ECI approach itself. Partnering is an integral part of ECI that entails mutual commitment (Walker and Lloyd-Walker, 2012). Within the context of IPD, partnering requires a long commitment between the client and contractor for the purpose of achieving specific business objectives. In addition, it requires that each participant enhances the effectiveness of their resources (Chan *et al.*, 2004). According to Wondimu *et al.* (2019), an alliance is a relational project delivery arrangement whereby the client and contractor work together as an integrated, collaborative team and make common decisions. Walker and Lloyd-Walker (2012) showed that alliancing is an ECI approach. In this approach, the project risks are managed cooperatively, and the outcome of the infrastructure project is shared (Lahdenpera, 2012). Wondimu *et al.* (2019) suggested that the approaches used to implement ECI in Australia's infrastructure have been divided into three activities:

- Selection of one contractor based on non-price aspects;
- The alliance contract for the design development; and
- The design-build (DB) contract in the design and construction phase.

In New Zealand, three individual sections are used to implement ECI in an infrastructure project. The feasibility study is performed in the first segment. The second segment entails the preparation of detailed design, negotiation of commercial terms (fixed price negotiation) and contract duration. The third segment includes completion of the detailed design and physical works based on the design-build contract. It is worth noting that the second segment of this approach shares similarities with target cost contracting (TCC). As stated by Chan *et al.* (2010b), in TCC, a fixed target cost is set based on given parameters at the start of an infrastructure project by the client and the contractor. Any cost savings or overruns between target cost and actual cost are shared between the contracting parties based on a pre-determined share ratio specified in the contract. Interestingly, project alliances similar in form of those utilised in Australia are being used in Europe (Laan *et al.*, 2011). Furthermore, the emergence of competitive dialogue (DC) has simplified the use of project alliances in Europe (Walker and Lloyd-Walker, 2012). Alliances can be sub-divided:

- **Pure alliance:** This method is based on a single target cost contract. In this method, the client is required to select only one contractor primarily based on experience, capability and attitude. The target cost is determined after the client and contractor have formulated the project together.

Enhancing the whole life-cycle performance of major infrastructure projects

- **Competitive alliance:** In this method, the procurement procedure is either a negotiated procedure or a competitive dialogue. The contract is awarded to the contractor with the most economically merited tender (Lahdenperä, 2009). This means price is one of the criteria, together with technical qualifications, former experiences, etc. Competitive alliance can be performed either through the competitive single TCC or the multiple/dual TCC.

According to Wondimu *et al.* (2019), a two-phase procurement process has been supported as an ECI approach that sustains the degree of competition for the contractor selection. The first phase of the selection process is typically based on price and qualitative criteria. The price-based criteria entail the following submissions of the contractor:

- Profit margin;
- Overheads;
- Pre-construction stage fee; and
- Approach to risk pricing and other cost components.

The qualitative benchmark typically entails:

- The proposed construction method;
- Ability to deal with unanticipated issues;
- Ability to deliver a similar type of projects on schedule;
- Experience with similar projects (track record); and
- Familiarity with local sub-contractors and contractors (Wondimu *et al.,* 2019).

ECI requires clear infrastructure objectives, careful forward planning and commitment from the client and contractors to adhere to best project benchmark standards for the life of the infrastructure project, not just focusing on the ECI phase. If poorly formulated and managed, an ECI process can lead to additional challenges and its benefits may not be ultimately realised. As Alden and Gordon (2018) asserted, ECI can minimise conflict and contract variations throughout an infrastructure project as the contractor can develop and revise the design or scope of the infrastructure project with the principal before the contract is awarded. A better understanding of project risks prior to awarding a contract allows the contractor and client to better assign those risks, which ultimately leads to better price optimisation. The question of whether an ECI is suitable for a particular infrastructure project should be considered early in the procurement phase based on the following components (Alden and Gordon, 2018):

1. **Cost and time versus overall value:** An ECI process necessitates an up-front commitment in both time and cost which needs to be aligned with the overall value that the ECI process will provide to the infrastructure project. If the project design or scope of services is well defined, such that specialist knowledge from the market is not needed, a traditional procurement is likely to be more suitable. Conversely, if the infrastructure project design or service requirements are incomplete or complex, and would benefit from specialist market knowledge or innovations, the time and cost needed in the ECI phase will likely be compensated by the overall value of a well-tested and resolved design or specification resulting in fewer alterations during the contact delivery phase.
2. **Availability of right resources:** To enhance the value of an ECI process, both the client and contractors must commit the right resources for not only the ECI workshop period but

also the whole of the infrastructure project. The resources committed to the ECI workshop need to have an appropriate decision-making authority, project knowledge and expertise.

3. **Expertise of resources:** An ECI process necessitates a thorough understanding of the process and a commitment and belief in the value of the process. The stakeholders involved need to have the required knowledge of an ECI process and understand what is needed of them to enhance the project outcomes.

4. **Loss of competitive tension:** An ECI process can lead to the loss of competitive tension in the procurement. This can be mitigated by running a double or triple ECI process, which is needed for any public infrastructure project. Participants need to ensure that the loss of competitive tension in a single ECI process is weighed against the increased resource commitment for a double or triple ECI process.

5. **Management of the ECI process:** For an infrastructure ECI process to be successful, the whole process must be well formulated, planned and managed. This should entail consideration of:

   a. whether the process should be a single, double or triple ECI process with regard to legislative procurement requirements (if any);
   b. the length of the ECI workshop period and number of ECI workshops each proponent will need to attend;
   c. ensuring the ECI process preserves advocates confidentiality and security of data;
   d. ensuring the client has clearly specified the objectives of the ECI process and how the ECI workshop will achieve that objective;
   e. ensuring that the ECI workshops are planned well in advance to safeguard all required primary stakeholders and external secondary stakeholders.

9.5 Mandating the use of building information modelling

BIM has taken hold in every aspect of construction and infrastructure project development. It is not new to the construction and infrastructure sector; it's been around for a while. Bleasby (2019) affirmed that its international global acceptance is confirmed by the fact that nations such as Brazil, Chile, Denmark, Finland, Norway, the United Kingdom, South Korea, Singapore and Vietnam have started using BIM in their public infrastructure projects. Lee and Borrmann (2020) showed that the advancements in software and hardware technologies have primarily boosted the deployment and development of BIM. Examples of such BIM technologies entail object-based and parametric modelling, automated design quality and compliance checking, inter-operability, scan-BIM, BIM-fabrication, filed BIM management, BIM facility-management, digital twins and mobile and cloud-based BIM collaboration (Lee and Borrmann, 2020). Nonetheless, several BIM projects have confirmed that the mere integration of BIM does not guarantee the success of projects and that outcomes vary depending on how the technologies are applied and by whom (Kang *et al.,* 2013). In the adoption of a new technology, the people and the process are often recognised as key features in addition to the technology itself. Lee and Borrmann (2020) referred to this concept as the people, process and technology (PPT) model. From the reviewed literature, it was established that the PPT model is predominant, but the definite origin is unknown. It is also vital to note that BIM as a technology also requires users (people) and best practices (processes) to be integrated by the construction and infrastructure sector, as well as a policy to facilitate the adoption process.

The drive by governments, the major client of public construction and infrastructure projects, is important for BIM adoption by the sector. BIM policies can range from a strict mandate of BIM in all publicly procured infrastructure projects through legislation changes, where necessary, to provide financial and organisational aid down to low-level encouragement and support. Lee and Borrmann (2020) noted that Singapore, Finland, Korea, the US, the UK and Australia form the international face of BIM policy making. These six nations' subsidary authorities have played a major role in fostering the adoption of BIM. In 2004, Singapore made it mandatory to submit construction documents for public construction projects via an internet platform (Khemlani, 2006). This entailed the submission of BIM frameworks in the software-neutral format. In Finland, public authorities have made the use of digital models a requirement for all public projects since 2007 with projected budgets over one million euros (Senate Properties, 2007). In the US, major governmental building clients, such as General Service Administration (GSA) and the US Army Corps Engineers (USACE), require the use of BIM applications for project execution (GSA, 2007). In 2011, the UK government decided to introduce legislation to reduce costs and lower the carbon footprint of construction projects through the use of BIM methods and technologies.

The key aspect of the UK government strategy has been to demand fully collaborative 3D-BIM, which aligns with BIM Level 2, for all centrally procured construction projects from 2016. Among the most advanced nations are Finland, Sweden, Norway and the Netherlands (Lee and Borrmann, 2020). In 2014, France made significant investments to support the adoption of digital technologies (PTNB, Delcambre, 2014). In Germany, the Ministry of Transport issued a BIM roadmap in 2015 that affirmed the use of BIM methods for all federal infrastructure projects (BMVI, 2015). Interestingly, China started to develop BIM standards and guidelines in 2001 (Liu *et al.,* 2017). The Ministry of Housing and Urban-Rural Development issued "Outline of Development of Construction Industry Information-2011-2015," which stressed BIM as a core technology to support and enhance the construction sector. In Japan, the term BIM is used for building projects, and the term construction or civil information modelling (CIM) is used for civil engineering projects; it is also called construction information modelling for both building and civil engineering sectors (Tateyama, 2017; Teh, 2019).

BIM is an attractive tool for infrastructure operators because its potential extends beyond the planning and design. If implemented effectively, BIM can generate cost savings over the entire infrastructure project life cycle. By facilitating collaboration and shared knowledge between stakeholders, it can be used to generate more accurate cost estimates before an infrastructure project commences. During the construction of an infrastructure project, real-time communication can minimise wasted materials and unnecessary labour. As noted by Imperiale (2018), BIM uses three-dimensional modelling and a common data environment to access and share data. Digital modelling allows the representation of a design in three dimensions and from different perspectives, facilitating identification of conflicts, thereby reducing design errors and resolving constructability issues much earlier. Within the context of construction and infrastructure, advances in digital technology will rapidly change the way contractors, infrastructure owners and engineers exchange data and streamline efficiency. A key element of this transformation will be BIM, the processes and tools that, among other things, digitally represent the physical and functional characteristics of an infrastructure project. A global survey showed that 75 percent of highly BIM engaged contactors acknowledged that BIM has the ability to capture detailed comprehensive data of a building project, in comparison to 41 percent of contractors admitting a reduction of design errors using clash detection as the top-ranking benefit of BIM adoption. Imperiale (2018) emphasised that BIM's data sharing is the driving force behind greater efficiency, but the philosophical underpinnings of BIM create a tension with the traditional allocation of liability among project stakeholders. Imperiale (2018) further identified six potential issues that should be addressed in any BIM realisation:

1. **Requirements:** Specific BIM protocols must be developed at the outset of each infrastructure project and tailored to each level of BIM so that clients and contractors have clarity as to what they are buying and providing.
2. **Design responsibility:** At country level, it is vital to establish the rights, responsibilities and liabilities of the contractor in charge, and other designers, and provide guiding principles on when the level or state of design is sufficiently mature to be relied upon, and for what purposes and by whom. This may entail a well delineated method for granting access and unlocking data, as these guidelines will also have the implications in liability disputes.
3. **Standard of care:** As shown in previous sections, BIM has the potential to fundamentally change how construction professionals think about formulating designs, producing construction data and constructing infrastructure projects. BIM encourages and facilitates the flow of diverse data between client, contractors and other parties. How contributions affect the legal liability of all the infrastructure project stakeholders is an issue that will need to be first delineated in contracts and laws. Unanticipated issues will no doubt arise and present themselves as the technology is now widely used and continues to develop.
4. **Intellectual property rights:** Infrastructure clients and contractors must work with legal counsel to create and negotiate contract clauses that entail ownership of intellectual property and downstream uses of BIM data. Imperiale (2018) affirmed that when BIM is used as envisioned, the contributions by various contractors and subcontractors all come together to formulate a final model that can be adopted. This raises legal questions regarding who owns and who has the licence to use the final model. While clients will insist upon ownership of the final model, the intellectual property rights of the model's contributors also must be taken into account. Given that a designer's expertise and know-how is advanced over time, across several projects, and often under proprietary processes and procedures, it is advisable that designers retain intellectual property use for future use.
5. **Confidentiality:** The usage of BIM requires data sharing, both before and during the delivery of an infrastructure project. A model BIM protocol should embed meaningful limits on what data is subject to sharing and how shared data is protected so that contractors can sustain their competitive advantage.
6. **Security:** A well-developed BIM protocol should provide on guide how contractors will allocate costs linked with data exchange and should outline the roles, responsibilities and procedures for cybersecurity. These procedures should be sufficiently uniform so that contractors can factor the costs/risks into their pricing

Many are hopeful that BIM's technology will minimise errors and omissions and promote greater collaboration in the construction and infrastructure sector. While at country level industry groups have promoted the use of BIM and provided guidelines, a more state-of-the-art protocol that affirmatively considers the above six issues would benefit the industry. Additionally, Poljansek (2018) asserted that regularisation consists of building a society around a standard with an implied script that brings people and things together in a world already made of competing conventions and standards. As shown in Figure 9.2, standards related to ICT are usually divided into three themes, data models and processed. Common concepts and classification are required to ensure that contractors speak the same language. Neural formats for data models are needed for systems and contractors to exchange data clearly. A uniform process for data delivery and a

Enhancing the whole life-cycle performance of major infrastructure projects

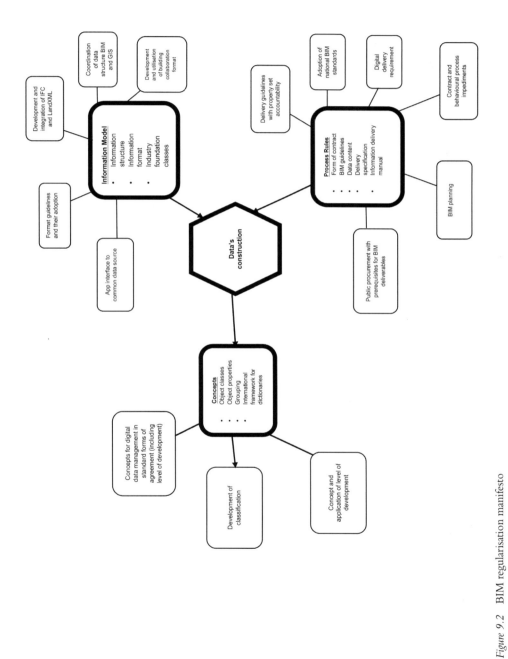

Figure 9.2 BIM regularisation manifesto
Source: Original adapted from (Poljansek, 2018)

common working approach is necessary. Around these three themes the construction and infrastructure sector can achieve the standardisation of BIM.

9.6 Reducing the costs at preliminary design phase and pre-construction phase

To support the delivery of efficient whole life infrastructure outcomes, there needs to be informed client leadership from the outset of the infrastructure project. This should begin with a well-defined brief for the infrastructure project that should be aligned with services that the investment is to deliver. Scottish Future Trust (2016) highlighted that a clear brief offers the design team a platform to develop efficient and creative solutions that can be used to boost the achievement of the project outcomes. The need to procure suitably skilled and resourceful contractors is also key to achieving efficient whole life outcomes. A commitment to design-led thinking and creation of an environment that supports whole life outcomes should, therefore, be established early in the infrastructure development process. This commitment needs to be sustained throughout the infrastructure project process in order to deliver the outcomes and benefits identified in the feasibility phase. Thus, there is a need to balance issues such as cost, environmental impact, functionality, social impacts, economic impacts, flexibility and other relevant issues. Using a collaborative and high-quality preliminary design and pre-construction process to evaluate these issues can help to address this balance and optimise infrastructure project outcomes. Every infrastructure project is unique and should be treated as such. Close attention to detail in every step of the infrastructure project is critical to give the client a quality product that is tailored to the client's needs. The design phase of an infrastructure project can be sub-divided into as many steps as necessary to resolve all design problems and to incorporate the concepts into functional infrastructure plan. It is primarily recommended that the following phases are used (Alarcon and Mardones, 1998):

- Preliminary design
- Final design

In addition to the above, the remaining components in the design phase are:

- Cost re-evaluation or value engineering (which is optional)
- Construction documents

During the preliminary design phase, emphasis should be placed on civil, mechanical and architectural and engineering drawings, accurate construction costs, estimated operational costs and schedules of construction. There are also engineering specifications of materials, machinery and equipment for each phase. A preliminary report should specify all the structural components and mechanical processes of the infrastructure project, and how they integrate. Moreover, the report should provide an outline of materials and equipment specifications that can be used as a basis for revising initial construction costs estimates. Costs and scheduling revisions can be compared with the original financial objectives and constraints to ensure that the infrastructure project remains financially feasible. It is vital to validate at the end of the final design phase that the infrastructure project remains economically viable. If by some chance it does not, the client team and contractor must agree on changes that can be made and design solutions that can be adopted. It is worth noting that any changes at this phase are most costly; hence, the final design is the most expensive of the planning processes. One of the most frequent approaches of sustaining economic viability at this phase is to develop and construct the infrastructure project in stages and to

Enhancing the whole life-cycle performance of major infrastructure projects

ensure that each phase of the construction programme is financially viable. Cost re-evaluation or value engineering is an optional element of the design process. For large infrastructure projects, it can be useful to incorporate a value engineering phase. The value engineering process can be used to re-evaluate all the major costs and determine if the most cost-effective design approaches have been adopted. Value engineering is not a costly or time-consuming process (Selim *et al.*, 2017). If used well, it can result in huge cost savings to the infrastructure project by highlighting inconsistencies between operational and required design solutions. Any recommendations made by the value engineering team should be incorporated into the final design report.

The pre-construction phase can be used to specify the key requirements of an infrastructure project and give the client a clear picture of what their project task is going to look like, how it is going to get done, and when every aspect of the project task will be completed. The pre-construction can start in broad terms with an overall project evaluation. This should emerge from the initial meeting between the client and contractor. The initial evaluation should lay the foundation for the pre-construction services and the actual construction work that will be carried out. It is at the pre-construction phase the client team has to ensure that the needs and expectations for the project have been clearly defined and understood by all those set to participate in the delivery of an infrastructure project. There are a number of components that go into estimating the budget of an infrastructure project. These may entail the cost of materials, subcontractors, suppliers, equipment and more. As highlighted by Xenidis and Stavrakas (2013), cost estimations and determining how to budget infrastructure projects are part of an iterative process that takes place in the other phases of an infrastructure project life cycle. The project progresses, and cost estimations become more accurate, since they are supported by a large amount real-time data.

As asserted by Xenidis and Stavrakas (2013), the need for reliable and updated information is essential to appraise realistically the infrastructure project's costs for performance and budgeting needs. As established by Koo *et al.* (2010), this can be achieved by ensuring that the team of experts possesses the required cost estimation knowledge. Common practice shows that the use of cost estimation aligned with pre-construction will lead to cost savings. To prevent escalating construction cost growth due to corrections for errors and omissions, pre-construction services should be well defined. Accurate pre-construction services budgets should be well developed to appropriately quantify the infrastructure project scope (that is, the technical scope and also the site-specific determinants of the infrastructure project). Puerto *et al.* (2016) emphasised that top-down estimating is the most appropriate estimating method to estimate pre-construction services costs for projects at the early phases of an infrastructure project. This approach comprises the determination of pre-construction services costs based usually on the contractor's experience. For infrastructure projects with limited data available in terms of requirements and scope, contractors should analyse and compare the available historical information of pre-construction services for similar infrastructure projects to calculate and develop the pre-construction service estimate.

The bottom-up estimating approach requires a greater level of detailed data about an infrastructure project (Puerto *et al.*, 2016). It comprises data from the aggregation at each functional level of estimates generated from a work breakdown structure. Each functional level will contain the estimate of the pre-construction services costs for specific areas, which will be eventually integrated to determine the total costs for the infrastructure project. The structure of the method enables decision-making at different functional levels of budgeting, scheduling and the allocation of resources. The two estimating methods are related. The method selected will depend on the available data at every phase of the infrastructure project. The estimates are required to make strategic financial decisions at the planning and pre-construction phases. The appropriate utilisation of these two approaches at the right time can result in final accurate estimates and well-informed decisions.

The pre-construction and pre-design phases allow the client team and contractors to evaluate the entire infrastructure project and formulate a plan that will take it all the way through its conclusion.

9.7 Chapter summary

As shown in this chapter, the decisions regarding the life cost of an infrastructure project have to be ascertained right from the infrastructure project's conceptual phase as to whether to reduce the initial cost at the detriment of the maintenance and running costs. Life cycle cost analysis is a tool that can be used to determine the most cost-effective option among different competing options to purchase, own, operate, maintain and dispose of an object or process when each is equally appropriate to be implemented on technical grounds. Meticulous life cycle cost modelling can enable the design to be value-engineered so that the cost profile of an infrastructure asset can effectively minimise the expenditure throughout the life cycle by a significant amount. Whole life can be used to appraise design decisions on infrastructure projects. Infrastructure clients instruct the design and construction of their infrastructure assets with consideration for the whole life cycle, facilitated by design-build-operate-maintain contracts. The operation of these contracts can be aligned with whole-life economic, environmental and social indicators. Infrastructure clients can show leadership by creating new criteria for circular procurement and clearly communicating these to the supply chain. As asserted in this chapter, infrastructure operators need to consider circular economy principles as a natural progression from pre-existing infrastructure initiatives, commitments and practices, such as resource efficiency, revenue generation through materials sales and reducing landfill use. Collaborative contracting models are generally considered unsuitable if project owners wish to raise capital on a project finance basis; that is where investors may only look to cash flows and assets of the project to secure repayment and not to the balance sheet to the owner. ECI provides an efficient means of designing and planning infrastructure projects in a cost-effective, more efficient and less adversarial structure. Moreover, BIM has created a solid connection between design and construction that had never been witnessed before. Instead of asking how we can get architects, engineers, and construction managers to collaborate efficiently, we are now asking how we can use BIM to cut costs and make the process even faster and more efficient. Well documented pre-design and pre-construction phases will provide a solid foundation for any infrastructure project.

9.7.1 Chapter discussion questions

1. Identify an infrastructure of your choice and critically discuss how you would incorporate the process of whole life costing in the project. Also, identify some of key practical benefits of whole life costing.
2. Critically discuss the value of embedding circular economy principles into infrastructure projects.
3. Having gone through the collaborative procurement model section, differentiate the models appraised with respect to infrastructure projects.
4. Within the context of infrastructure project delivery, what are some of the benefits of early client involvement?
5. What are some of the advantages and disadvantages of using BIM on infrastructure projects?
6. Identify a high-risk infrastructure project and critically describe how you would reduce costs at the pre-design and pre-construction phases.

Enhancing the whole life-cycle performance of major infrastructure projects

9.8 Case study: E39 coastal highway route infrastructure project (Norway)

The E39 Coastal Highway Route project is the largest infrastructure project in modern Norwegian history and quite possibly the largest ongoing project globally (Vegvesen, n.d.). The rugged west coast of Norway, home to a large populace, is a challenge to travel by car. The harsh weather conditions also make it a precarious route with closed roads and ferries often having to cancel their departures due to heavy snow or winds or high waves. As noted by Vegvesen (n.d), a continuous E39 that is always accessible, with fixed links between the islands and the mainland, will make the western coast of Norway more reachable for the populace who inhabit the coast but also for tourists and the transportation of goods. About one third of Norway's populace of 5.3 million live in the western region. As observed from the reviewed project data, the Norwegian government has had a long-term goal to construct the E39 as an improved and continuous coastal route between the cities of Kristiansand in the south and Trondheim in the north by 2050 (Vegvesen, n.d). The Norwegian Public Roads Administration is the main contractor for the project, which will stretch 1,100 kilometres. The route will run through six counties through the cities of Stavanger, Bergen, Alseund and Molde. The total travel time of the existing roadway is 21 hours, and users require seven ferry connections. The primary objective of the project is to create an enhanced E39 without ferries, which will cut the travel time in half. The details of the route are as follows:

- It will be almost 50 kilometres shorter;
- The reduction of travel time will be achieved by replacing ferries with bridges and tunnels;
- The government aims to upgrade a number of road sections;
- Sixty percent of Norway's export goods are produced on the west coast, so an efficient and accessible transport infrastructure system will boost the national economy;
- It will create new patterns of habitation;
- The new road will shorten the path to reach healthcare facilities, jobs and academic institutions; and
- The preliminary project data indicate that the required investments and enhancements will cost approximately Norwegian krone 340 billion.

9.8.1 Technical aspects of the project

To replace the present seven ferrie,s the primary client and primary contractor are:

- Planning to construct alternative constructions, either bridges or tunnels;
- Considering a new structure – a submerged floating tube bridge (SFTB) – for some of the deepest and longest fjords exposed to harsh weather conditions, where suspension floating bridges will be a challenge to construct. When a fjord is deeper than a few hundred metres or wider than two to three kilometres, existing engineering solutions are required;
- Performing tests to assess what a driver's experience will be; of course, safety is a key priority for the government whenever they plan any type of construction project;
- Fires and explosions are some of the loads which have been included in the design, since the early phases;
- As can be seen in Figure 9.3, the submerged floating tube bridge is certainly an engineering phenomenon. As specified by Vegvesen (n.d), the first known project proposal was in 1886 by UK naval architect Sir James Edward Reed.

217

Figure 9.3 An illustration of a submerged floating tube bridge

- In Norway, the idea was initiated in 1923 and, after the feasibility studies were conducted by the Norwegian Public Road Administration for the E39 project, the Norwegian government realised that several other nations are considering constructing the same type of structure.

9.8.2 *Case discussion questions*

1. As a sub-contractor, you have been tasked by the Norwegian Public Roads Administration to plan, design, construct and maintain the project. Propose whole life costing guidelines that can be used by the main contractor to achieve sustainable whole life outcomes. In addition, justify how you will reduce costs at the pre-design and pre-construction phases.
2. In modern infrastructure planning, it is important to recognise the value of circular economy principles. Propose a circular economy model that can be adopted by the main contractor.
3. The choice of your procurement method will be vitally important to the success of your construction project. As shown in this chapter, a number of construction procurement models exist. Identify a procurement method that will be best suited for this infrastructure project and provide the essential ingredients for your method, as well as its relative advantages and disadvantages.
4. Within the context of this project, identify the key benefits of early client involvement and BIM.

References

Alarcon, L. and Mardones, D. (1998). Improving the design–construction interface. Paper presented at the 6th Conference of the International Group for Lean Construction, Guaruja, Brazil, 13–15 August, Available from: https://www.researchgate.net/profile/Luis_Alarcon2/publication/228707074_Improving_the_design-construction_interface/links/02e7e51db431d509a8000000.pdf [cited 25 July 2020].

Alden, S. and Gordon, V. (2018). Australia: Unlocking the value-applying early contractor involvement (ECI) to service contracts. Available from: https://www.mondaq.com/australia/construction-planning/731566/unlocking-the-value-applying-early-contractor-involvement-eci-to-service-contracts [cited 20 July 2020].

Arup (2018). From principles to practices: First steps towards a circular built environment. Available from: https://www.ellenmacarthurfoundation.org/assets/downloads/First-steps-towards-a-circular-built-environment-2018.pdf [cited 26 July 2020].

Bakker, E., Walker, H., Schotanus, F., and Harland, C. (2008). Choosing an organisational form: The case of collaborative procurement initiatives. *International Journal of Procurement Management*, 1(3), pp. 297–317.

BMVI (2015). Roadmap for digital design and construction, federal ministry of transport and digital infrastructure. Berlin, Germany. Available from: https://www.bmvi.de/SharedDocs/EN/publications/road-map-for-digital-design-and-construction.html [cited 23 July 2020].

Boyd, R. (2017). Why we need to grasp the whole life cycle. Available from: https://www.ice.org.uk/news-and-insight/ice-thinks/infrastructure-transformation/why-we-need-to-grasp-the-whole-life-cycle [cited 26 July 2020].

BSI (2011). Buildings and constructed assets: Service life planning, general principles and framework, BS ISO 15686-1:2011. British Standard Institution, London, United Kingdom.

CBPP (1998). Introduction to Whole Life Costing: Fact sheets. Construction Best Practice Programme, Department of the Environment, Transport and the Regions (DETR), London, United Kingdom.

Chan, A.P., Chan, D.W., Chiang, Y., Tang, B., Chan, E.H., and Ho, K.S. (2004). Exploring critical success factors for partnering in construction projects. *Journal of Construction Engineering and Management*, 130(2), pp. 188–198.

Chan, D.W., Lam, P.T., Chan, A.P., and Wong, J.M. (2010b). Achieving better performance through target cost contracts: The tale of an underground railway station modification project. *Facilities*, 28(5/6), pp. 261–277.

Clift, M. and Bourke, K. (1999). *Study of whole life costing*. London, United Kingdom: BRE Press.

Construction Excellence (2009). Collaborative procurement. Available from: https://constructingexcellence.org.uk/wp-content/uploads/2015/01/Collaborative_Procurement_Guide.pdf [20 July 2020].

Del Puerto, C.L., Agosto, L.G.C., and Gransberg, D.D. (2016). A case for incorporating preconstruction cost estimating in construction engineering and management programs. ASEE's 123rd Annual Conference and Exposition. New Orleans, LA, June 26-29. Available from: https://peer.asee.org/a-case-for-incorporating-pre-construction-cost-estimating-in-construction-engineering-and-management-programs.pdf[cited 24 July 2020].

Deloitte (2006). Building flexibility: New delivery models for public infrastructure projects. Available from: https://www2.deloitte.com/content/dam/Deloitte/uk/Documents/infrastructure-and-capital-projects/deloitte-uk-building-flexibility-report.pdf [cited 26th July 2020].

Ellen Macarthur Foundation (n.d.). What is circular economy? A framework for an economy that is restorative and regenerative by design. Available from: https://www.ellenmacarthurfoundation.org/circular-economy/concept [cited 26 June 2020].

Ferme, L., Zuo, J. and Rameezdeen, R. (2018). Improving collaboration among stakeholders in green building projects: role of early contractor involvement, *Journal of Legal Affairs and Dispute Resolution in Engineering and Construction*, Vol. 10 No. 4, p. 4518020.

Green Alliance (2019). Building a circular economy: How a new approach to infrastructure can put an end to waste: Available from: https://circulareconomy.europa.eu/platform/sites/default/files/building_a_circular_economy.pdf [cited 25 July 2020].

Gokhale, S. (2011). Integrated project delivery method for trenchless projects. International Conference on Pipelines and Trenchless Technology *2011: Sustainable Solutions for Water, Sewer, Gas, and Oil Pipelines*, pp. 604–614. https://doi.org/10.1061/41202(423)66. Available from: https://ascelibrary.org/doi/10.1061/41202%28423%2966 [cited 20 July 2020].

GSA (2007). GSA BIM Guide 01 - Overview. GSA BIM Guide Series 01. Washington DC, Public Buildings Service, U.S. General Services Administration (GSA).

IADC (2012). Early contractor involvement in infrastructure projects. Available from: https://www.porttechnology.org/technical-papers/early_contractor_involvement_in_infrastructure_projects/ [cited 20 July 2020].

Imperiale, J.T. (2018). Standardisation of BIM implementation in the United States. https://www.troutman.com/insights/standardization-of-bim-implementation-in-the-united-states.html [cited 21 July 2020].

Kang, Y., O'Brien, W.J., and Mulva, S.P. (2013). Value of IT: Indirect impact of IT on construction project performance via best practices. *Automation in Construction*, 35, pp. 383–396.

Khemlani, L. (2006). 2006 2nd annual BIM awards, Part 1. AECBytes [Available from: http://www.aecbytes.com/buildingthefuture/2006/BIM_ Awards.html [cited 23 July 2020].

Koo, C., Hong, T., Hyun, C., and Koo, K. (2010). A CBR-based hybrid model for predicting a construction duration and cost based on project characteristics in multi-family housing projects. *Canadian Journal of Civil Engineering*, 37(5), pp. 739–752.

Laan, A., Voordijk, H., and Dewulf, G. (2011). Reducing opportunistic behaviour through a project alliance. *International Journal of Managing Projects in Business*, 4(4), pp. 660–679.

Lahdenperä, P. (2009). Project Alliance the Competitive Single Target-Cost Approach, Finland VTT Technical Research Centre of Finland, Vuorimiehentie.

Lee, G. and Borrmann, A. (2020). BIM policy and management. *Construction Management and Economics*, 38(5), pp. 413–419. DOI: 10.1080/01446193.2020.1726979

Liu, B., Wang, M., Zhang, Y., Liu, R., and Wang, A. (2017). Review and prospect of BIM policy in China. IOP Conference Series: Materials Science and Engineering. Available from: https://iopscience.iop.org/article/10.1088/1757-899X/245/2/022021 [cited 20 July 2020].

MI-ROG (2016). Embedding circular economy principles into infrastructure operator procurement activities. Available from: https://constructingexcellence.org.uk/wp-content/uploads/2016/11/Embedding-Circular-Economy-into-Procurement-MI-ROG-White-paper_October-2016.pdf [cited 26 July 2020].

Olanrewaju, L.A. (2013). A critical review of value management and whole-life costing on construction projects. Available from: https://fmlink.com/articles/a-critical-view-of-value-management-and-whole-life-costing-on-construction-projects/ [cited 27 July 2020].

Opoku, A. (2013). The application of whole life costing in the UK construction sector: Benefits and barriers. *International Journal of Architecture, Engineering and Construction*, 2(1), pp. 35–42.

Poljansek, M. (2018). Building information modelling (BIM) standardisation. Available from: https://ec.europa.eu/jrc/en/publication/building-information-modelling-bim-standardization#:~:text=BIM%2C%20short%20for%20Building%20Information,asset%20planning%20running%20and%20cooperation [cited 26 July 2020].

PwC (2018). Collaborative contracting. Available from: https://www.pwc.com.au/legal/assets/collaborative-contracting-mar18.pdf [cited 15 July 2020].

Ramachandran, G. (2019). Analysing infrastructure life-cycle. Available from: https://theaseanpost.com/article/analysing-infrastructures-life-cycle [cited 27 July 2020].

Rowlinson, S. (2017). Building information modelling, integrated project delivery and all that, *Construction Innovation*, Vol. 17 No. 1, pp. 45–49.

Scottish future trust (2016). Review of public sector procurement for construction: Whole life appraisal tool for the built environment. Available from: https://www.scottishfuturestrust.org.uk/files/publications/Whole_Life_Appraisal_Tool_For_Construction.pdf [cited 23 July 2020].

Senate Properties (2007). BIM requirements for architectural design. Helsinki, Finland, Senate Properties.

Selim, A.M., Meetkees, O.A., and Hagag, M.R. (2017). Value engineering application in infrastructure projects by public private partnerships (PPPs). *International Journal of Applied Engineering Research*, 12(20), pp. 10376–10373.

Vegvesen, S. (n.d). Norway takes on its largest infrastructure project in modern history. Available from: https://www.vegvesen.no/en/roads/Roads+and+bridges/Road+projects/e39coastalhighwayroute/news/norway-takes-on-its-largest-infrastructure-project-in-modern-history [cited 23 July 2020].

Tateyama, K. (2017). A new stage of construction in Japan, i-Construction. *IPA Newsletter*, 2(2), pp. 2–11.

Teh, N.J. (2019). *Japan transforming construction 2019*. London, UK: Innovate UK.

Walker, D.H. and Lloyd-Walker, B. (2012). Understanding early contractor involvement (ECI) procurement forms. 28th Annual ARCOM Conference, 3-5 September 2012 Edinburgh, UK. Association of Researchers in Construction Management, pp. 877–887.

Wondimu, P.A., Klakegg, O.J., and Laedre, O. (2019). Early contractor involvement: Ways to do it in public projects. *Journal of Public Procurement*, 20(1), pp. 62–87.

Xenidis, Y. and Stavrakas, E. (2013). Risk based budgeting of infrastructure projects. Procedia–Social and Behavioural Sciences, 74, pp. 478–487. 26th IPMA World Congress, Crete, Greece.

10

THE ROLE OF OPERATIONS AND MAINTENANCE IN INFRASTRUCTURE MANAGEMENT

Edward Ochieng and Diana Ominde

10.1 Introduction

A complementary and potentially more cost-effective approach is to improve the utilisation, efficiency and longevity of the built and existing infrastructure projects. Many governments in both developed and developing countries neglect their existing assets, and current operations and maintenance practices are often seriously deficient. In operations, they fail to maximise asset utilisation and to meet adequate user quality standards while incurring needlessly high costs as well as environmental and social externalities. Maintenance is all too often neglected since political bias is towards funding new assets. Similarly, resilience to natural disasters tends to be ignored, although such hazards are becoming more common and more destructive because of climate change. As a result of the maintenance backlog and the lack of resilience measures, existing assets deteriorate much faster than necessary, shortening their useful life. It is worth noting that satisfactory lifetime performance of transport infrastructure is critically important to sustained economic growth and social development in both developed and developing economies.

In recent years, governments have had to reduce the operations and maintenance budgets, creating a situation of inadequate maintenance planning, analysis and tasks. As shown in this chapter, inadequate maintenance naturally leads to underperforming infrastructure systems that cause increased risks, potential service disruptions, and premature infrastructure breakdown (BC, 2019). The net outcome is a strain on operators and either higher life cycle budgets of managing and delivering the expected level of service, or minimal level of service. Making decisions about enhancing maintenance and the timing of infrastructure upgrades, replacements or decommissioning requires tracking maintenance data – including costs – individually from operations data. Interestingly, most governments in developed and developing countries integrate operations and maintenance budgets. Tools and systems for tracking costs are used separately on large infrastructure programme projects. In addition, the utilisation of tools and systems for enhancing operations and management, along with increased collaboration among internal government departments, are used to enhance project outcomes.

As suggested by Frangopol and Liu (2007), a deteriorating infrastructure system leads to increased costs for business and ultimately for users. Catastrophic failure of infrastructure systems can cause extensive social and economic consequences. Well-preserved transport infrastructure systems can extensively maximise a nation's competitiveness in a global economy and enhance resilience to adverse conditions such as natural hazards (e.g., earthquakes, hurricanes and floods) and man-made disasters (e.g., vehicular collisions). According to Frangopol and Liu (2007), researchers and engineers have been developing and implementing a variety of maintenance management programs to achieve desirable solutions that can be used to sustain satisfactory infrastructure performance from a long-term economical viewpoint (Frangopol and Liu, 2007). A number of existing methodologies for the maintenance and management of the transport infrastructure systems have been based on least life-cycle cost while enforcing relevant performance levels. Kong and Frangopol (2003) found that there exist practical challenges with these treatments. For instance, if the budget allocated is more than the computed minimum life-cycle cost, the infrastructure system performance can be sustained at a higher level than what was initially prescribed for deriving the minimum life-cycle cost solution. Alternatively, if the available capital is not enough to meet the computed minimum life-cycle cost, operators have to come up with a solution that can be used to enhance the structure of performance to the highest possible level under budget constraints. In order to ensure that transport infrastructure systems are well preserved at the highest level, operators need to adopt maintenance and operation strategies that can be used sustain infrastructure systems. Consequently, the predominant aim of this chapter was to appraise premises and the application of operations and maintenance strategies.

10.1.1 Chapter aim and objectives

This chapter will examine how the operations and maintenance strategies can be used to maintain infrastructure systems efficiently. The main objectives are to:

- Demonstrate the significance in operations and management in the life cycle of the infrastructure and the process of asset management;
- Evaluate how operations and maintenance can be used to enhance service delivery and reduce life cycle costs;
- Provide guidance that can be used by practitioners to implement best practices and optimise existing infrastructure systems;
- Highlight the importance of measuring infrastructure spending and performance for infrastructure systems; and
- Evaluate the alignment of infrastructure spending with economic growth.

10.1.2 Learning outcomes

The following learning outcomes have been identified for this chapter. Readers will be able to:

- Describe the significance in operations and management in the life cycle of the infrastructure and the process of asset management;

- Depict how operations and maintenance can be used to enhance service delivery and reduce life cycle costs;
- Formulate operations and maintenance guidance and best practices that can be used to optimise existing infrastructure systems;
- Highlight the importance of measuring infrastructure spending and performance for transport infrastructure; and
- Appreciate the alignment of infrastructure spending with economic growth.

10.2 Definition of operations and maintenance

As suggested by Castro *et al.* (2009), operations and maintenance signifies all of the activities required to run and sustain an infrastructure project. The main aim of operations and the maintenance phase is to ensure efficiency, effectiveness and sustainability of an infrastructure project. The activities that fall under operations and maintenance are very different in nature. Operations entails the direct access of the infrastructure system by the user, to the activities of any operational staff, and to the rules or by-laws, which may be planned to govern who may access to the infrastructure system, when, and under what conditions. Moreover, operations can also be defined as the design or implementation of programs, services, policies or systems and related guidelines of a city (BC, 2019). Thus, operations refer to the daily tasks needed to provide service delivery to the citizens, businesses, schools and other users. In comparison, maintenance entails technical activities, planned or retrieved, which are required to sustain the working system. Maintenance requires skills, tools and spare parts. Maintenance necessitates functional checks, monitoring, testing, measuring, servicing, repairing or replacing of necessary tools and supporting utilities so that the infrastructure system can perform the required functions and deliver the set-out service delivery objectives throughout the life cycle of the infrastructure system (Carter, 2009). BC (2019) asserted that operations and maintenance are often considered together, and most government budgets do separate operations and maintenance. They serve different functions in the life cycle of an infrastructure system and without good understanding of how much is invested in each, it can be challenging to track the efficacy of individual operations and maintenance tasks in enhancing the service of infrastructure systems. There are three major categories of maintenance (BC, 2019; Carter, 2009):

1. **Proactive maintenance**

   - *Preventive maintenance:* Comprises work that is planned and carried out on a daily basis to sustain and keep the infrastructure system in a perfect condition. The activities can be in the form of materials, equipment or facilities which have to be inspected, maintained and protected before they collapse, or other failures occur.
   - *Predictive maintenance:* Entails the use of sensor information that can be used to monitor an infrastructure system continuously or appraised against historical trends to predict any breakdown.

   Proactive maintenance can be used by operators to ensure that the infrastructure assets reach full-service life potential or source out regular information about the condition and function of assets. This makes it easier for operators to anticipate investments in infrastructure and avoid surprises. In addition, operators can be equipped with vital data about the assets and how they are maintained to better communicate with the users about how potential decisions are made.

2. Reactive maintenance

♦ *Corrective maintenance:* Includes replacing or repairing any task that was not done correctly. Corrective maintenance will take place when there is a failure, malfunctioning or breakdown of an infrastructure system. The key to ensuring effective maintenance of an infrastructure system is to ensure that responsibilities are well specified, and the maintenance team has the tools and skills to deal with any failures which might arise.

Preventive and predictive maintenance are proactive and are usually applied to prevent failure, minimise wear, enhance efficiency and maximise the life of infrastructure components. Applying proactive strategies can be more cost-effective than relying on corrective maintenance. While corrective maintenance is sometimes necessary, it should be kept at a low level if possible because it can lead to unpredictable spikes in costs and can disrupt service delivery (SIP, 2017). Devoting capital to proactive maintenance can help lessen the need for costly reinvestment by enhancing the service life of an infrastructure system. The ease of operations and maintenance of an infrastructure system is central to its sustainability; hence, it must be carefully considered at the design phase. Some operations and maintenance issues can be aligned to the location, but it is worth appreciating that urban and countryside infrastructure projects differ significantly in the complexity of the technologies applied. The standardisation of parts, designs, construction methods, etc. can be different. The planning of operations and maintenance do not represent a huge task; however, they require a certain level of methodical planning, commitment and monitoring. The facets to be planned are (Castro *et al.,* 2009):

- **What:** The task to be carried out.
- **When:** The frequency of the task.
- **Who:** The human resources required for the task.
- **With what:** What materials, tools and equipment will be required.

The following case study shows how Central Japan Company applied maintenance and preventive measures.

Vignette 10.1 Maintenance management of Tokaido Shinkansen case study

As noted by Nakamura *et al.* (2019), maintenance on railway infrastructure systems can be divided into periodic routine maintenance work for tracks and medium- and long-term maintenance work against deterioration of materials and natural disasters. As specified below, for Shinkansen maintenance a special train for dynamic inspection (called Doctor Yellow) runs all sections at the operating speed (see Figure 10.1).

The Shinkansen test train, a brightly coloured train that speeds around Japan's high-speed rail network, is used to assess the condition of rail tracks and overhead wires, helping to sustain the enviable safety record of the rail service. According to Nakamura *et al.* (2019), Doctor Yellow operates every 10 days and inspects the tracks and electrical and signal facilities by checking rail distortion, overhead lines and signal current. The generated information is instantly analysed and necessary repairs required are revealed. Visual inspections of civil structures are performed every two years. The aim is to identify repair sites that need more detailed inspection and carry out special inspections for checking fatigue and deterioration. It is worth noting that the interior of the Yellow Doctor comprises specialised

The role of operations and maintenance in infrastructure management

Figure 10.1 Doctor Yellow

monitoring equipment in each of the seven cars (Medhurst, 2016). A crew of nine includes two drivers, three technicians responsible for the tracks, and four others to monitor power-related issues.

Two methods for civil engineering structure maintenance are utilised is maintaining the rail infrastructure (Nakamura *et al.* (2019):

- The first method is mainly used for repair and reinforcement in advance. If the crew finds any kind of problem in the structure, they perform maintenance work in advance for other structures in which problems have not yet appeared.
- The second method is the preventive maintenance management. With this method, the crew carries out preventive maintenance measures for the parts of steel bridges or reinforced concrete structures that would have hitches in order to keep the facilities available. For instance, the neutralisation depth of concrete is created based on the inspection data forecast and surface protection is scheduled as preventive maintenance for supressing the neutralisation.

Interestingly, since starting its operations, Shinkansen has had no serious accidents and provides such punctual service that the average delay time is less than 30 seconds (Nakamura *et al.*, 2019). The high standards of safety and stability are assisted by the successive and elaborate maintenance work on railway facilities.

Case discussion questions

1. Within the context of your country:

   ♦ Discuss how you would apply proactive and reactive maintenance strategies in your railway network.
   ♦ Identify key benefits of repair and reinforcement in advance, as well as preventive maintenance management.

2. What are some of the short- and long-term benefits the Japanese government has gained from the utilisation of Doctor Yellow.

10.3 Infrastructure maintenance and operations research

According to SIP (2017), transport infrastructure systems such as roads, railways, harbours and airports support everyday life and social economic activities. However, a number of infrastructure systems were constructed during periods of the high economic growth. As they get older, an increase in maintenance and repair expenditures, along with the possibility of fatal accidents occurring during the service, becomes a serious social issue. In order to allow for an efficient and effective preventive maintenance management of transport infrastructures and establish a safe and secure infrastructure system, it is vital to have technologies that can be used to precisely diagnose and take appropriate measures by closely appraising a number of infrastructures individually on-site. It is also important to reduce the hazard risks that are linked with manual handling in the workplace. For instance, in Japan, the Cross-Ministerial Strategic Innovation Promotion Program (SIP) (2017) examined the standard and maintenance of infrastructure systems by using low-cost preventive maintenance while stressing the necessity to match the needs of infrastructure maintenance with the seeds of technical development and development of new technologies into more attractive technologies that could be used on-site. The aim was to contribute to regional revitalisation and maintenance of vital internal infrastructures to high standard. Within the context of infrastructure maintenance and operations, it was found that the research community has broadly clustered their focus on the following five research themes listed below:

- **Inspection, monitoring and diagnosis technologies:** They have been developed to fully estimate damages of civil infrastructures. Research and development themes include: an internal defect inspection technology using supersensitive magnetic non-destructive testing, an integrated diagnostic system using high-speed traveling noncontact radar, remote diagnostic technology using supersensitive near infra-red spectroscopy, a pavement inspection system, floor slab deterioration detection using onboard underground probe radar, and the displacement monitoring technique for infrastructures using satellite SAR (synthetic aperture radar).
- **Structural materials, deterioration mechanisms, repairs and reinforcement technologies:** Within this theme, extensive R&D activities have extensively progressed. There are quite a number of simulation models for the deterioration mechanisms of structural materials and innovative estimation systems for deterioration progress of infrastructures to organise a core base for R&D of structural materials to develop effective maintenance technologies, and to support the commercialisation and wider application of precast members using highly-durable concrete for society.
- **Robotics technologies:** As found from the reviewed literature (SIP, 2017), a number of robots are currently being used to inspect transport infrastructures. At the same time, a wide range of R&D tasks are ongoing for a study of applicable structures for the introduction of robotics technologies, and establishing an integrated database to centrally manage the data for the effective usage of robotics technologies. With these efforts, the implementation of robots in society for infrastructure maintenance is being highly utilised.
- **Information and communication technologies:** Within this theme, the main R&D has been to create data management systems fully utilising information and communication technologies (ICT) to take advantage of the huge amount of data on maintenance, repair and renewal of civil infrastructures for contributing to the real application of advanced ICT for society. Specific R&D fields are in data screening based on integrated large-scale sensor data for pavements and bridges. Data management enables comprehensive control of a variety of information and data analysis and visualisation technologies for making the stored data effectively applicable for real operation on-site.
- **Asset management technologies:** Within this category, the advanced technologies entail non-destructive test methods and innovative mathematical tools have been created for maintenance of road structures.

10.4 Implementing operations and maintenance best practices

Understanding how major infrastructure projects evolve over their life cycle will provide an essential viewpoint in developing an efficient operations and maintenance program that can be adopted to sustain safety and efficiency and reduce operations costs in the long term. WEF (2014) projected the global infrastructure demand at about US$4 trillion in annual expenditure with a gap – or missed opportunity – of at least US$1 trillion every year. Globally, one of the essential areas for investment is indeed infrastructure. This specific asset class should be considered when appraising the needs of the fastest growing populations in Africa and Asia. Despite the growing gap in constructing new infrastructure projects, it should be noted that the global stock of existing infrastructure is worth about US$50 trillion, which is of the same order magnitude as the global stock market capitalisation (US$55 trillion) and comparable to a certain extent to the global GDP (US$72 trillion). As shown below in Table 10.1, the American Society of Civil Engineers (ASCE, 2017) appraised the status of the following infrastructures (bridges, roads, railways, water and sewage, energy, dams, ports and airports) and projected the maintenance and replacement costs. As can be observed, a number of the infrastructure systems were evaluated as fourth (D, or "risky") out of the five measures, and highlighted national understanding of the need for infrastructure maintenance and replacement (ASCE, 2017).

In addition, ASCE (2017) projected the investment budget for sustaining the infrastructure level in the US as "B" or good out of the five points on the scale. Nakamura *et al.* (2019), stated that the cost has increased on a yearly basis since 2013 and that in 2020 the cost would be $3.6 trillion, $1.6 trillion short due to lack of financial resources. This highlights the significance of

Table 10.1 Current status of infrastructure in the United States as appraised by the American Society of Civil Engineers

Infrastructure type	1988	1998	2001	2005	2009	2013
Aviation	B–	C–	D	D+	D	D
Bridges	–	C–	C	C	C	C+
Dams	–	D	D	D+	D	D
Drinking water	B–	D	D	D+	D	D
Energy	–	–	D+	D	D+	D+
Hazardous water	D	D–	D+	D	D	D
Inland waterways	B–	–	D+	D–	D–	D–
Levees	–	–	–	–	D–	D–
Roads	C+	D–	D+	D	D–	D
Schools	D	F	D–	D	D	D
Solid waste	C–	C–	C+	C+	C+	B–
Transit	C–	C–	C–	D+	D	D
Wastewater	C	D+	D	D–	D–	D
Port	–	–	–	–	–	C
Average value of America's infrastructure cost	C	D	D+	D	D	D+
Improvement cost	–	–	**$1.3 trillion**	**$1.6 trillion**	**$2.2 trillion**	**$3.6 trillion**

A = Exceptional: Fit for the future, B = Good: Adequate for now, C = Mediocre: Requires attention, D = Poor: At risk, F= Falling/Critical: Unfit for purpose.

Source: Original adapted from American Society of Civil Engineers. *2013 Report Card for America's Infrastructure.* Reston.

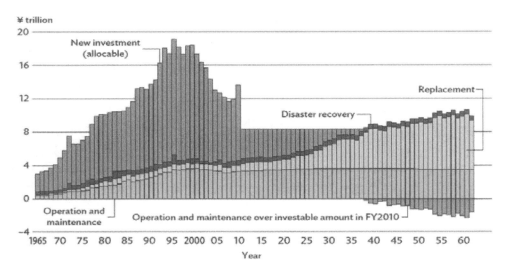

Figure 10.2 Maintenance and replacement cost of infrastructure evaluated
Source: Nakamura *et al.* (2019)

infrastructure maintenance and the scale of maintenance cost to society. In Japan, Nakamura *et al.* (2019) found that the project cost for maintenance and replacement including roads, ports, airports, public spaces, airports, public rentals, housing, sewage, city parks, rivers and coastal management facilities between 2011 and 2060 would be about $1.9 trillion, about $300 billion short (see Figure 10.2). This projection offers a reference for infrastructure maintenance and replacement in Japan.

Infrastructure maintenance requires a sound plan that should be progressively executed. Asset management can be used as a methodology to achieve this. The key concept of asset management is preventive maintenance management. Conventional maintenance methodology requires maintenance engineers to collect data on damage and deterioration of the infrastructure through periodic appraisals and repair them palliatively. Conversely, predictive maintenance requires maintenance engineers to forecast the future status of the infrastructure based on damage factors and deterioration mechanisms of facilities and to repair them in advance by making good use of the newest innovation (Ochieng *et al.*, 2017). As specified by the International Organisation for Standardisation (ISO), the process of asset management can be outlined as plan-do-check-act (PDCA) (Nakamura *et al.*, 2019). The phases of PDCA are summarised below:

Phase 1: Plan

- Set management goals and financial conditions
- Examine, judge, and monitor
- Forecast life cycle cost (LCC) and decide on management level
- Calculate necessary maintenance and repair costs
- Propose priority for repairs
- Make budget
- Make maintenance and repair plan

Phase 2: Do

- Execute maintenance and repairs

Phase 3: Check

- Carry out ex post evaluation of maintenance and repair plan
- Update asset information

Step 4: Act

- Review and revise maintenance and repair policy

As shown by WEF (2014), asset management is an all-inclusive holistic endeavour; its aim goes well beyond considering a single outcome (such as minimising an asset total cost of ownership). Instead, it seeks to enhance the asset over its life cycle, across functions and tasks, in social, economic and environmental aspects of the infrastructure. BC (2019) established that asset management involves the process of making informed decisions and considers cost, risk and service. The task of continuous enhancement of operations and maintenance is a key determinant of the asset management life cycle as operations and maintenance practices can considerably impact asset life cycle costs, management of risks and service delivery performance of an asset. Well-organised and executed operations can allow operators to deliver their services effectively, efficiently and economically (BC, 2019). The application of asset management requires operators to set service delivery priorities through infrastructure planning, budgeting and investment processes (BC, 2019). It also means ensuring decisions are set to achieve "value for money" and appropriate quality and quantity of resources are dedicated to meet the intended objective. Regular and proactive maintenance are key to enhancing service life and minimising service disruptions. As found by Ochieng *et al.* (2017), neglecting the maintenance of an infrastructure by deferring it to future years can be an easy option in the short term; however, this creates a false economy. Over time, unplanned maintenance can be more costly than planned maintenance, and it can lead to reduced service life or compromise the functions of the infrastructure system. Continuous enhancement of operations and management tasks, and consideration of operations and management throughout all phases of the asset life cycle, can lead to (BC, 2019):

- Reduction of long-term operating costs through decisions made during design and construction, such as reduced annual energy costs due to strategic tools selection, or reduced maintenance requirements due to efficient design;
- Deferment of capital costs by extending the life of existing infrastructure;
- Ensuring service levels are achieved, assets are functioning as expected and unexpected service breakdowns are lessened;
- Protection of the environment;
- Adaptation of the impact of climate change on assets through modified operations and management tasks; and
- Reduction of risks to main services delivered by natural assets.

Distinctive characteristics of a well-developed asset management model, in contrast to traditional approaches, include the following (Prieto, 2013; Uddin *et al.,* 2013):

- **Value perspective:** In a number of traditional models, the emphasis is on cutting down the total cost of owning and operating fixed capital assets (while providing the desired level of service). A well-constructed infrastructure asset can provide the greatest cost saving, possible life cycle revenues and user benefits. This can lead to socioeconomic returns.
- **Triple bottom line objectives:** All-inclusive asset management not only focuses on operational and financial goals but also includes the environmental and social dimensions (externalised costs and benefits).
- **Risk recognition:** As with financial asset management, practitioners have to ensure that infrastructure asset management optimises risk-adjusted returns, that is, secures a maximum (e.g., amount, user quality) return at a given appropriate level of risk (e.g., downtime, congestion, resilience).
- **Systematic scope:** A sound asset management is concerned with system-level performance, not just with performance at asset level.
- **Integrated tasks:** A well-created asset model should entail more than effective maintenance. It should comprise operating, managing and optimisation phases, and be built in an innovative way.
- **Complete measures:** –Stakeholders have to put some effort into other prerequisites, such as getting more out of the allocated capital though operational excellence, which in turn reduces costs. It is worth noting that advanced asset management not only entails operational levers to enhance efficiency, but also focuses on other strategic business objectives of the asset owner to improve effectiveness. Rather than just focusing on short-term fixes, stakeholders should also focus on measures that will facilitate the operations and maintenance long-term benefits, that is, creating appropriate funding capabilities and governance for the infrastructure asset.
- **Organisational integration:** A well-constructed asset management model should conscientiously integrate and interconnect maintenance with design, engineering and construction, and amalgamate maintenance planning with operations planning.
- **Proactive culture:** Asset management should be proactive and innovative, rather than reactive, and take a formal approach. It should be based on fact-based reviews of the asset base, propose innovations and dynamically adjust its processes and organisation to new issues.

With increasing cost pressures, service demands and unsustainable funding methods, operators will have to change the way they think about managing their assets, recovering revenues and delivering services to the society. As shown below, the implementation of an asset management model has a number of benefits for operators (BC, 2019):

- Robust data to support the decision-making process;
- Effective and reliable delivery of key services, both presently and into the future;
- Minimised life cycle budgets of service delivery;
- Improved value to the society's investment in assets over the life cycle;
- Defensible prioritisation of minimal resources using a consistent and repeatable system;
- Enhanced financial planning and better management of unfunded liability aligned with renewing or replacing aging engineered assets;
- Alignment of departmental and society objectives with technical and financial decisions and actions;
- Provision of stewardship that can lead confidence with shareholders, users and other stakeholders.

The following section provides a set of selected approaches that can be used to optimise existing infrastructure assets.

10.4.1 Maximise asset utilisation

Globally, a number of existing infrastructure assets suffer from congestion because demand has risen well beyond the level projected. Since their capacity is limited, and it might not be possible to build new assets, it is vital for operators to make the most of their existing infrastructure asset capacity. As found by WEF (2014), even a slight increase in effective amount can make a large difference, since congestion is a strongly non-linear occurrence. To ease congestion and enhance the usage, operators can adopt a three-part best practice strategy: improve peak capacity and effective amount, utilise demand management to minimise peak demand by shifting some of it off-peak to optimise availability and reduce downtime. In order to maximise asset utilisation, infrastructure operators can apply the following strategies: enhance peak capacity and effective output, apply demand management and optimise availability to reduce downtime (see Figure 10.3):

10.4.2 Enhance quality for users

It could be argued that, historically, users of infrastructure assets have not been precedence consideration for a number of operators. As observed from the reviewed literature(WEF, 2014), infrastructure facilities tend to enjoy natural control, so there is minimal competitive pressure to provide high quality customer service. In its place, a public service mentality tends to predominate the dutiful provision of predefined, uniform service for users rather than an integrated effort to monitor their changing requirements and satisfy their needs. Globally, service quality is a constant issue and is becoming more severe because of the deteriorating infrastructure systems. Poor quality service is not only demoralising for users but is also costly for the society. To resolve this issue, infrastructure operators need to introduce strategies that would advance their service levels. The following are some strategies that could be introduced (Akinyemi and Zuidgeest, 2002; Gharehbaghi and Georgy, 2015; Harvey, 2012; Nakamura *et al.*, 2019; Ochieng *et al.*, 2017; Prieto, 2013; Uddin *et al.*, 2013):

1. **Adopt a user–centric operating framework:** The construction of infrastructure assets has always been an engineering-driven venture, which puts greater emphasis on the infrastructure asset than on the user. Traditionally, the operators have always adhered to predefined requirements derived from stable, well-set-out standardised regulatory prerequisites, technical requirements and pay less attention to users' diverse needs. Providers and policymakers need to put greater effort in enhancing the technical topographies and integrate the requirements of the users. In order to embrace user orientation, providers and policymakers should seek to optimise user insight and adopt an operating framework that allows continuous customer focus and service.
2. **Enhance the end–to–user experience:** Operators need to create a positive end-to-end user experience in two key areas: the operators should excel in ancillary services by holistically incorporating all user needs, such as ease of use, availability and reliability, performance, safety, affordability and accessibility. They also need to enhance the whole network performance by partnering with other stakeholders and head-to-head assets. In order to optimise the value that users will derive from using the infrastructure, operators need to consider all head-to-head service topographies that might be beneficial to the user.

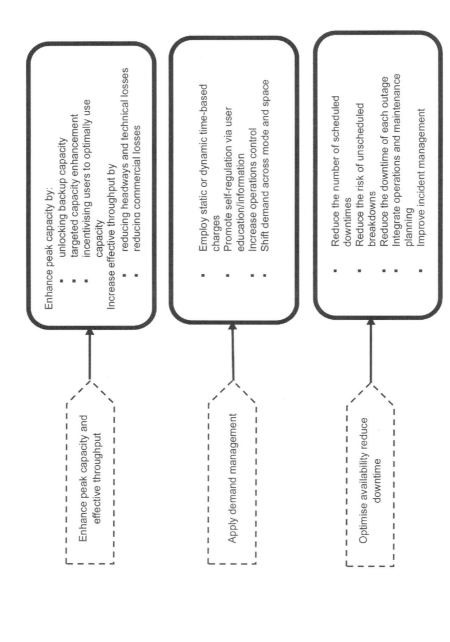

Figure 10.3 A summary of asset utilisation strategies

3. **Use smart technologies to refine user performance:** As digitisation has become a key driving force for infrastructure, better and wider application is needed. By employing new innovative technologies, infrastructure providers can ease a number of the traditionally quality–cost–trade-offs and attain win–win solutions. Smart innovative technologies can be harnessed to enhance three users' needs: better user data and interaction, enhanced system performance and safety, and faster billing and payment.

4. **Reduce operational and maintenance costs:** Despite receiving far less attention in the past, operational efficiency has become a watchword in most governments. As witnessed in most economies, public budgets are constrained and users are increasingly demanding lower prices. Operators can continuously respond to the challenge by making processes more efficient by implementing lean and automated processes, cutting unit costs by enhancing procurement costs and outsourcing and enhancing overheads by rightsizing management and support.

   ♦ Implementation of lean and automated processes strategy comprises two main strands: adjusting processes in line with lean principles and leveraging automation.

   ♦ Rightsizing management support and functions: A number of governments regard their infrastructure operators as oversized and users perceive their operations as being of low quality. To address this problem, WEF (2014) proposed the following remedies for operators: redesign the organisations, since a high performance organisation should have a right structure, clearly defined employees, partnership procedures and a well-structured knowledge management system; undertake delayering, that is, operators need to cut down on duplication and minimise the distance between top management and the user; and adopt a shared service model.

   ♦ Optimisation of procurement costs and outsourcing strategy should entail a total-cost-of ownership analysis and a whole life-cycle review of any high value system, equipment and component being purchased. Operators need to ensure that an all-inclusive procurement policy is aligned to the various commercial levers such as supplier management, bundling and best-cost country sourcing. Additionally, it should seek to optimise procurement processes and technical procurement levers such as standardisation and make-or-buy where possible. Table 10.2 provides a summary of procurement levers that can be considered by operators.

5. **Mitigate externalities:** Progressively, infrastructure projects are being scrutinised for their environmental and social sustainability. Against a background of resource limitations – scarcity of water, energy and raw materials and climate change challenges – infrastructure operators are now being pressed to meet tighter environmental requirements set by government policymakers but also have to deal with the secondary stakeholder criticism. Three broad strategies can be adopted: planning a comprehensive programme of sustainability/health, safety and environment interventions; integrating sustainable principles into everyday operations; and cooperating with other key stakeholders.

   ♦ **Setting out comprehensive sustainability plans:** It is important for the operators to take action and set clear KPIs which are aligned to environmental and social themes – in particular, improving resource efficiency; minimising the impact on biodiversity and the ecosystem; cutting down noise, dust and emissions; and committing to ethical business operations. As can be seen in Table 10.3, the French highways authority proposed a comprehensive sustainability framework which incorporates key indicators of sustainability.

Table 10.2 Procurement levels that can be considered by operators

	Lever	Typical actions	Typical actions
Commercial	1. Supplier management	• Review and develop supplier base • Facilitate supplier switch/ quality new suppliers • Renegotiate contracts and terms/conditions	• Develop framework agreements • Adjust frequency of negotiation • Optimise negotiations: multi round, e-auctions
	2. Bundling	• Centralise purchases for scale economies • Use global tenders and master agreements • Coordinate purchases across departments	• Compare prices and terms with benchmarks • Optimise orders over time and consolidate orders • Optimise single-order quantity and quality
Process	3. Best-cost country sourcing	• Purchase products from best-cost countries • Ask suppliers from best-cost countries • Buffer cost uncertainty via indexing/hedging	• Leverage global economic price trends • Introduce global procurement organisation • Optimise taxation
	4. Demand management	• Assess buying needs regularly • Plan volume by purchasing department • Control unplanned maverick buying	• Check compliance with supplier agreements • Simplify order process and streamline logistics • Coordinate and balance workload of contractors
Technical	5. Process optimisation	• Simplify bidding process • Standardise bidding process/documents • Use e-tools in bidding and negotiation	• Implement standard terms and conditions • Transmit orders and invoices electronically • Optimise supplier stocks to share risk
	6. Standardisation/ redesign	• Standardise specs to reduce variants in use • Modularise or use market standard products • Simplify specs via design-to-cost	• Get supplier input for specifications • Redesign services without compromising quality • Enforce stability of specifications/limit changes
	7. Make-or-buy	• Assess make-or-buy opportunities regularly • Optimise ratio of own vs external production • Enforce transparency on cost/margin structure	• Purchase inputs for suppliers • Understand "should" cost and cost-drivers • Assess supplier performance permanently

Source: Original adapted from WEF (2014)

Table 10.3 French highways sustainability framework

Emission reduction	Carbon footprint reduction of traffic
	• Reassurance of car-sharing via park-and-ride schemes • Electronic toll collection, avoiding stops at toll booths • Lively traffic control systems and speed regulations
Noise protection	**Improvement to residents' quality of life through noise protection**
	• Noise reflection/absorption barriers and facade isolation on 271km • Noise reduced by 2–10 decibels (equivalent to 35% less traffic)
Waste reduction	**Environmentally friendly renovation of 311 stations**
	• Normalisation of waste separation and installation of sewage treatment • Energy-efficiency measures (e.g., energy-saving lamps)
Biodiversity protection	**Preservation of flora and fauna in 120 projects**
	• Construction of 686 wildlife crossings and ponds, insect hotels, fish ladders • Installation of biotopes and ornithological observatories with walk-throughs
Water protection	**Prevention of water pollution**
	• Water retention via basins and drainage; treatment facilities • Use of hydraulic construction and rainwater disposal on >115 sites

Source: Original adapted from WEF (2014)

♦ **Integrate sustainability into routine operations:** In order to ensure that a sustainability strategy will last, operators need to ensure that it is fully integrated into the processes and the organisational structure.

♦ **Work with key stakeholders:** Operators need to have an active programme of communicating with key stakeholders. In addition, they need to work in partnership with fellow operators across the infrastructure project life cycle and embrace a multi-stakeholder approach.

6. **Extend asset life:** The utilisation of infrastructure systems as long as possible provides clear-cut benefits. As most of the cost of delivering infrastructure systems comprises fixed past investments, and given relatively low existing operating costs, any additional year of service provides high value as the asset is repaid. To enhance infrastructure system life, one can pursue three broad strategies: invest in preventive and predictive maintenance; avoid and control excessive asset consumption and stress; and optimise the infrastructure system resilience against disaster. As noted in the literature (Nakamura, *et al.,* 2019), for politically minded policymakers, little incentive exists to invest in preventive maintenance and resilience. The immediate benefits are usually hardly visible, and the long-term positive impact is difficult to assess and validate.

It is important to customise the maintenance strategy for each infrastructure asset. Within the context of infrastructure vulnerability and criticality, one could implement the following four broad strategies. A corrective/reactive or failure-based maintenance strategy will activate repair work in the event of a breakdown and should be designed to bring a failed system back to its operational state. This strategy can be adopted if the risk of failure is very low and if the rate of failure is fairly low. Scheduled maintenance should be carried out at defined intervals, either after a certain amount of usage (use-based). This strategy can be used when failures are costly, or safety is significant. Scheduled maintenance can also be used when the failure rate is fairly low. Condition-based strategy or predictive maintenance strategy can be used to trigger maintenance activities when the infrastructure system condition falls below a certain level. The aim is to ensure that maintenance work is timed optimally and performed in moderation. Risk-based or reliability maintenance focuses on the infrastructure system's current condition and the likely consequences of failure. The main aim of this approach is to minimise the overall risk and impact of unexpected failures. This approach entails inspecting and monitoring high-risk components of the asset. Figure 10.4 provides a summary of the four-maintenance described above.

Infrastructure systems recurrently face natural hazards, and at times can suffer major devastation. The economic losses caused by disasters over the past 30 years have been estimated at US $3.5 trillion globally (Francis and Robert, 2012). Major disasters can cost about six percent of GDP, such as the 2011 tsunami in Japan, and more than 100 percent of GDP in low-income countries (Francis and Robert, 2012). Although these disasters are infrequent, infrastructure systems with a lifetime of several decades remain at risk. Additionally, evidence suggests that their occurrence and severity are likely to increase (Shabnam, 2014). In the future, it has been suggested that climate change could intensify storms, droughts, flooding, landslides, extreme temperatures and forest fires (WEF, 2014). To reinforce the infrastructure systems' ability to resist, cope with and recover from severe weather conditions, operators will have to develop master plans for resilience, enhance risk identification and assessment, combine structural with non-structural measures, prepare for managing residual risk, reconstruct for resilience and make financial and institutional arrangements that sustain resilience.

7. **Reinvest with a life cycle view:** One could suggest that globally a number of infrastructure assets are approaching the end of their useful life. For instance, GLA (2003) suggested that about half of the main water pipes in London are more than 100 years old, and one-third could be older than 150 years. In Western economies, it was found that a number of social infrastructure systems are ageing (WEF, 2014). Faced with an ageing asset base, it will be vital for governments to embrace a three-part reinvestment strategy: ensuring the prioritisation of project options by using a whole life-cycle cost benefit analysis, selecting the contracting method that offers a best value for money, and ensuring a comprehensive preparation for the efficient delivery of infrastructure projects. The prioritisation of project options can be achieved by conducting diligent baselining; this assessment should be used to establish exactly how the infrastructure system in its current form falls short of users for both present and future needs; it should identify all potential solutions so that operators can assess the various options by looking into how they can improve, expand or replace the existing infrastructure systems. Once different project options have been appraised, a whole life-cycle benefit analysis can be performed. The selection of contracting methods should entail an appraisal of all the procurement and contracting options (e.g., design-bid-build; design-build; build-operate-transfer; design-build-rperate). Public-private partnership (PPPs) have been found to be well-suited for infrastructure upgrades, replacements and rehabilitation. The preparation for infrastructure project delivery depends on efficient and high-quality project preparation, including feasibility studies on the technical, financial, legal and environment aspects of the project (Ochieng et al., 2017).

It is worth noting that infrastructure enters the operations and management phase after construction starts to offer service (Nakamura et al., 2019). Operations information varies depending on the type of infrastructure; operations and management matters common to type are (i) facility operations, (ii) demand creation, (iii) risk management and (iv) business management. As specified

The role of operations and maintenance in infrastructure management

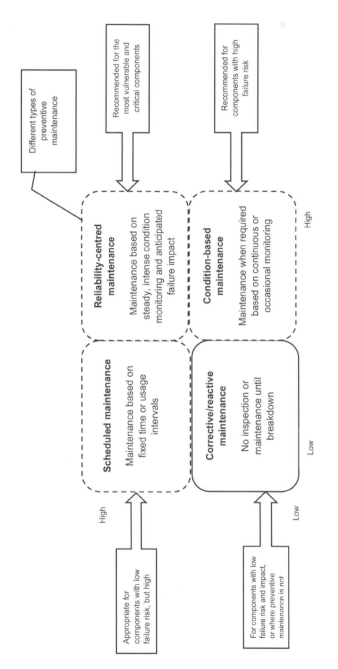

Figure 10.4 A summary of the four maintenance strategies
Source: Original adapted from WEF (2014)

by Nakamura *et al.* (2019), facility operations refers to a compilation of tasks for which the main function is to provide good service to facility users. An airport, for instance, does not affect its function until procedures such as control and logistics operations are carried out. Demand creation can be used to support the use of infrastructure systems. The social determinant, economic determinant and revenue of the operating entities of transportation infrastructure projects will increase as the percentage of users grows. Within the context of operations management, risk management can be used to mitigate a number of risks that will negatively impact the operations of the infrastructure system and prevent or reduce degradation of the infrastructure system when a major disaster occurs. Operators can use business management to highlight the sustainable services in public–private mixed and private business categories.

10.5 Alignment of infrastructure spending with economic growth

Universally, very few economists would suggest that good infrastructure investment does not maximise a country's economic capacity. As demonstrated by KPMG (2016), a good transport infrastructure project can augment labour mobility; for instance, good energy infrastructure can deliver low-cost fuel to the energy sector. In the case of macroeconomic health, a number of legislators and economists have highlighted that a lot more state-led infrastructure projects are needed to boost growth (McDonnell, 2016 and Wren-Lewis, 2013). There could be two separate opinions for this, which could be conflated but need to be analysed distinctly. The first is that infrastructure spending is a significant tool of Keynesian economics. That is, when an economy is in a recession or slowing down, and unemployment is significantly high, government spending to finance and/or construct infrastructure projects can help in lessening unemployment directly and have a strong multiplier effect on the economy more generally. The second is that infrastructure spending can assist in boosting the productive potential of the economy by enhancing its supply side. This argument suggest that greater state-financed infrastructure investment can maximise the productive potential of the economy by greasing the wheels of the economic activity in future (Bourne and Zuluaga, 2016). Skidelsky and Miller (2013) found that these two rationales are often made together; however, they are really discrete. If policymakers were purely interested in short-term demand management, it does not matter which type of infrastructure projects are executed. What matters is cutting down the level of unemployment directly and injecting resources back to work. If one has multiple objectives, that is, demand management and long-run growth maximisation, this can create variation. It is worth noting that good macro negates what we know about good micro (Eichner and Kregel, 1975).

There are a number of implied assumptions supporting Keynesian demand management generally, not least surrounding the interaction with monetary policy. For open economies with floating exchange rates and high debt levels, minimal evidence suggests that discretionary stimulus has a huge impact on GDP (Ilzetzki *et al.,* 2010); however, in theory, government spending could be used to smooth the business cycle. Constant revisions to GDP data make it difficult for policymakers to forecast downturns and even track the health of the economy over time, especially during and after recessions when economic change tends to be at its most rapid. This is significant because infrastructure projects usually have long lead times, due to the complexity and government-imposed constraints such as planning laws and meeting environmental appraisals. In addition, provisional government infrastructure investments can also create permanent expectations; even as the economy recovers there can be resistance to cutting down state capital spending.

Infrastructure spending can also be used to boost the supply side. For instance, with low interest rates policymakers can push investment in roads, rail, energy, housing and ports. In the provision of public goods and on infrastructure projects where the return rate will be higher than those delivered on private sector projects, well-targeted infrastructure projects undertaken based on disciplined cost benefit analysis can be used to enhance the economic well-being (Ochieng *et al.*, 2017). A low interest rate environment can make investment in both the public and private sector more cost effective. On project selection, legislators often prefer to spend on infrastructure in areas of political advantage or for regeneration purposes rather than growth. While a number of research studies have highlighted the significant positive impacts of infrastructure investment, it is always important to take a cautious approach when adopting some of these studies (Calderon *et al.*, 2011; Glaeser, 2016). At a global level, some examples suggest that perhaps for many of the observations appraised above, large amounts of infrastructure spending are neither essential nor sufficient condition for healthy growth. For instance, the US, which ranks worse that the UK on the quality of its infrastructure, economically grows more robustly than the UK. Despite Japan being ranked highly by the World Economic Forum, the country is arguably less vibrant after spending US \$6.3 trillion on heavy engineering construction projects (Glaeser, 2016). Ansar *et al.* (2016) showed that a number of Chinese infrastructure projects were plagued by cost overruns and overestimated benefits, meaning that 55 percent of the infrastructure projects had a benefit-to-cost ratio below 1. The majority of the country's government infrastructure projects require a full feasibility study. Economic appraisals are usually conducted as part of the project appraisal; the aim is to go beyond demonstrating the financial feasibility. The economic appraisal of the infrastructure project can focus on key growth indicators such as job creation, GDP growth, tax revenues, environmental impact, deliverability, governance and asset management.

10.6 Measuring infrastructure spending

As highlighted by Gibson *et al.* (2011), energy consumption, greenhouse gas emissions (GHG) and costs of road transport infrastructure fall broadly into three stages: (i) construction, (ii) maintenance, and (iii) operations. The construction and maintenance costs of a road transport infrastructure will vary according to country, location and availability of materials. During its operation, costs, energy consumption and GHG emissions mainly result from electricity use – lighting, signals and signage – and so will vary significantly depending on a country's local conditions. Considerable savings in costs and environmental impacts can be gained during road operation by using specific materials and design methods to improve energy efficiency. The impact of road transport infrastructure from the energy and environmental perspective can be linked to the construction, associated materials and services. However, the importance of the whole life cycle of a road transport infrastructure, including the use of vehicles and their own life cycle, has been acknowledged in design, planning and decision processes. For instance, promoters of road building have highlighted the congestion relieving merits of constructing additional lanes for busy routes. However, Gibson *et al.* (2011) and Treloar *et al.* (2004) suggested that, considering the increased demand induced by a bigger capacity, this approach may not be the most suitable option to address road transport issues due to the increased GHG emissions over the life cycle of additional routes. Figure 10.5 displays an abridged life-cycle energy framework for a road infrastructure that entails vehicles and allied maintenance. As can be seen in Figure 10.5, the life cycle evaluation entails energy consumption and GHG emissions at each phase.

Principles of best practice can be integrated into the design and construction phases of road transport infrastructure projects. Best practice for road design can include perpetual pavements, which can be built in thin layers to enhance durability and prevent surface stresses from penetrating through to lower layers, permitting top layer repairs/maintenance only (Acott, 2009). Specific design needs can lead to the utilisation of different construction techniques such as the

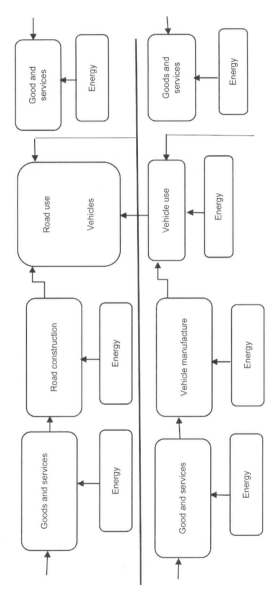

Figure 10.5 Life cycle energy framework for road transport infrastructure
Source: Original adapted from Treloar *et al.* (2004)

application of porous draining asphalt paving that allows water to filter through the road surface and prevents vehicles from sliding in wet conditions. Best practices can also entail considering existing road infrastructure and materials as a substitute to virgin materials (Gibson *et al.*, 2011). Gibson *et al.* (2011), further notes that other best practices can include obtaining resources locally and/or procuring materials that have been manufactured using best practice methods (e.g., energy efficient manufacturing or energy obtained from alternatives to fossil fuels). Design and engineering considerations still have an essential role to play in the reduction of GHG emissions from the road transport infrastructure. Design considerations should include road operational emissions and costs. For instance, a Canadian research study showed that heavy trucks that run on concrete (rather than asphalt) are more efficient in terms in fuel use. It was found that for a high-volume roadway, asphalt produced 738 tonnes of carbon dioxide equivalent/kilometre (tCO_2e/km) compared to 674 tCO_2e/km produced by concrete over a lifetime cycle (Gibson *et al.*, 2011).

10.6.1 Measuring performance of infrastructure

According to Schutte (2008), road management entails the following tools:

- Policy formulation: Definition of standards and policies for the road sector;
- Monitoring: Knowledge of the network extent, conditions and traffic characteristics;
- Needs assessment: Determination of the required expenditures for management and operations;
- Capital budgeting: Appraisal and ranking of investment options;
- Monitoring maintenance: Monitoring of maintenance projects; and
- Monitoring performance: Obtaining performance measures of operations.

The above listed stages, if combined, can be used to measure the performance of road networks and planning their maintenance, comparing maintenance strategies against operational requirements, and for programming future maintenance and improvement activities based on available resources. In order to enhance the performance, these procedures can be coupled with the need to select pre-defined performance standards. In this sense, positioning performance objectives and connected indicators is a key ingredient of road management tools. Performance has to be assessed in a methodical manner and must be compared against indicators set by road operators. Karlaftis and Kepapsoglou (2012) noted that, despite the fact that the application of performance indicators and thresholds (condition ratings and levels of service as examples) have been at the centre of road management for quite a number of years, performance based management of roads infrastructure was introduced relatively not long ago. Zietlow (2004) found that, in the past, maintenance and operations were allocated or contracted on the basis of design and material requirements, and were linked to the amount of work carried out as part of a maintenance project. Sultana *et al.* (2012) asserted that not long ago the concept of performance-based management was utilised by road agency authorities. Under performance-based contracting, minimum performance standards (or targets) are set and reimbursement of contractors can be based on how well the contractors adhere to the standards and not the amount of accomplished work.

Contemporary road management performance is based on programming, implementation of maintenance and operational activities which are driven by appropriately well-specified performance indicators. As found by Shaw (2003), performance assessment is usually encountered in a number of tasks and processes which are aligned to engineering, economics, health and many more disciplines. Based on the above, one could suggest that performance measurement is a task needed for assessing and enhancing features and operations of a system, process or an infrastructure. Shaw (2003), provided an all-inclusive definition of performance definition by suggesting that performance management is a process of appraising progress toward achieving predetermined objectives, including data on the effectiveness with which resources are transformed into goods and services (outputs), the quality of those outputs (how well they are delivered to operators and the extent to which operators are satisfied) and outcomes (the results of a project activity compared to its intended objective), and the efficiency of government operations in terms of their specific contributors to the project objectives.

As summarised below, Haas *et al.* (2009), asserted that transport infrastructure can be assessed from a number of viewpoints and reasons:

- To assess current and future conditions of transport infrastructures;
- To evaluate the road authority's efficiency with respect to provided services, productivity, protection of the environment, capital effectiveness and many more.

Moreover, transport infrastructure entails a number of stakeholders, often with a number of interests and expectations. This can lead to the need for appraising and measuring a number of dimensions of performance within this domain. Figure 10.6 summarises these viewpoints.

10.6.2 Performance indicators for transport infrastructure

Transport infrastructure operators collect and retain extensive data sets that are related to their operations and the life cycle of their infrastructure systems. The proper collection, analysis, refinement and presentation of that information is a requirement for using them for proper reporting to key stakeholders. As such, the integration of appropriate performance indicators is needed for linking transportation and infrastructure information for transport infrastructure management. According to OECD (1997), performance indicators can be considered as a tool that will enable the effectiveness of an operation or an organisation to be measured or an achieved result to be assessed in relation to a set of objectives. As summarised below, objectives aligned to introducing performance indicators may include the following (Humplick and Paterson, 1994; Haas *et al.*, 2009):

- Physical condition appraisal, with respect to level of service offered, structural integrity and safety provision of transport infrastructures;
- Support in management of the transport network in terms of decision-making for investments, expenditures and operations;
- Diagnosis of critical transport infrastructure network determinants with respect to deterioration and remedial action aligned to decision-making;
- Tracking and monitoring of policies with respect to their effectiveness and compliance with associated objectives;
- Data provision to infrastructure transport users and service operators;
- Optimal allocation of resources through the quantification of the efficiency of transport infrastructure investments and other administrative activities; and
- Cost (and relevant information) tracking with respect to construction and maintenance of transport infrastructures and equipment.

Performance indicators can be utilised in a number of ways, such as in-house decision-making and better communication between key transport network stakeholders. As suggested by Humplick and Paterson (1994), performance indicators can further be used to assess:

- Compliance with operational and policy objectives;

The role of operations and maintenance in infrastructure management

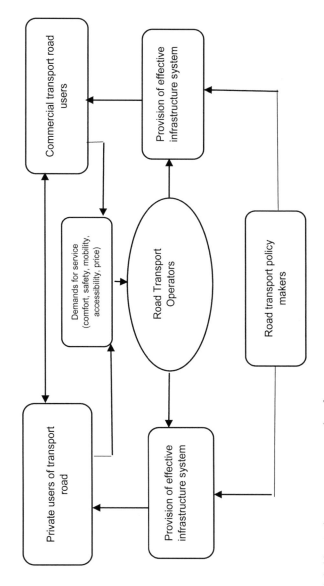

Figure 10.6 Stakeholders in the transport sector and performance measurement

Source: Original adapted from Humplick and Paterson (1994)

- Satisfaction of the transport infrastructure users with respect to network services;
- Efficiency of transportation network service providers; and
- The relationship between policymakers and the road administrator.

Cambridge Systematics (2000) set out the following conditions for performance selection. First, in order to assess future conditions and characteristics of alternative road management programs, performance indicators should be best suited for forecasting purposes. Second, it is vital to ensure that the indicators are uncomplicated and aligned to the requirements of stakeholders. Third, the performance indicators should be of great use; that is, the measures should appropriately reflect the objectives and capture cause and effect conditions between the actions of operators and the objectives. The operators need to also ensure that the performance indicators can be used to diagnose problems and most importantly reflect the actions that affect it. Last, the performance indicator selection process should entail the temporal effects of the measure and compatibility to programming of actions. A summary of key determinants is provided in Table 10.4:

All transport infrastructure require maintenance to keep them safe, serviceable and reliable. The key to sustaining and providing value for money is executing a timely and appropriate maintenance program that will (NAO, 2014):

- Limit the effect on road users;
- Prevent further deterioration; and
- Minimise the whole life-cycle cost of the infrastructure assets.

Planning maintenance at suitable phase during an infrastructure life can often lead to a good condition and extend its use. Infrastructure operators need to emphasise the significance of intervening at the right time for transport repairs. Particularly, when carrying out preventive maintenance checks. Given the long lives of some transport infrastructure, and the varying cyclic intervention points, operators need to utilise long-term management plan for each type of infrastructure system to allow them to schedule maintenance at the optimal time and minimise whole-life costs (NAO, 2014). Infrastructure operators can use data to optimise road performance. To achieve this, operators must understand the requirements of their infrastructure systems and the most cost-effective approach to sustaining and maintaining it. This will require the following:

- Current, all-inclusive and accurate data on what assets exist, their location, performance and condition;
- A good understanding of how different infrastructure systems are likely to deteriorate over time; and

Table 10.4 Key determinants of performance indicators

Property	Description
Relevance	The determinants should be relevant to the purpose it was developed for.
Clarity	The determinant should be well defined.
Reliability	Measurements for obtaining the determinants should not be affected by the process or individual performing them.
Precision	As precise measures as possible should be expected.
Availability	A determinant should be readily available as long as the cost of obtaining it does not exceed its usefulness and should still be useful and up-to-date when available to the road administrator.

Source: Karlaftis and Kepaptsoglou (2012)

- It is vital for operators to apply knowledge acquired of the network to support good decision-making.

10.7 Operations and maintenance challenges

Climate change will have an impact on built and natural assets in a number of countries because the conditions in which infrastructure systems operate are changing. Depending on the impacts, assets may face wear and tear through changes of rainfall patterns, wind, intense storms and changes in temperature highs and lows (BC, 2019; WEF, 2014). Recognising these impacts, climate change may reduce infrastructure service life if operations and maintenance tasks are not adjusted to changing conditions. Thus, as highlighted in previous sections, it is important for operators to review operations and maintenance principles from time to time. Operators can apply proactive maintenance in ensuring that the infrastructure systems function as intended. For instance, climate change will result in augmented intensity of storms in many countries, which can also result in more debris in storm drains and creeks (BC, 2019). Regular freeing of stormwater drainage can help in minimising the risk of flooding due to blockage. In addition, the maintaining the operations of rivers and streams and keeping them clear of debris can minimise reliance on the built stormwater system.

The management of operations activities use considerable workforce and capital, and are most of the time prioritised because they have a direct and immediate impact on the intended services. For instance, the speed and scale of snow freeing has an immediate impact on the level of service of transport infrastructure systems (e.g., roads, rail and airports). If operations are reduced, the users usually notice the impacts immediately. Most of the time, there is often sufficient resources or time devoted to regular review and improvement of operations activities to enhance the service life of the infrastructure system. The necessities of the operations are impacted by a number of factors, such as land use planning, demand management, design, availability of natural infrastructure systems, environmental conditions and the design of the infrastructure system (BC, 2019; WEF, 2014). The following are scenarios which might impact the operations and maintenance of infrastructure systems:

- **Population growth and a high percentage of development:** This scenario is likely to increase the demand on infrastructure systems. Depending on the infrastructure system, this may lead to wear and tear of the service life that was intended.
- **Land utilisation and urban pattern:** These may create pressure on natural assets (or loss of infrastructure systems) that offer vital service to the society, reducing the capacity to function optimally.
- **Planning assessments on land use:** This scenario is likely to dictate the type of infrastructure system that is needed, which will form the types of operations and maintenance activities that will be required by operators.
- **Technology advances:** Within this theme, it is envisaged the specific tools, technologies and skill sets that will be needed to operate and maintain infrastructure systems.
- **Insufficient resourcing maintenance of an infrastructure system:** This can increase the resources required to operate it and create a cascading negative impact on other infrastructure systems.

10.8 Chapter summary

This chapter has provided guidelines for enhancing operations and management on major infrastructure projects. As shown, operations and management are vital components of delivering desired levels of service, mitigating risk and ensuring long-term affordability. Operations and maintenance practices can be enhanced continuously to increase the efficiency, effectiveness and reliability of infrastructure systems. Operators with a good understanding of their infrastructure systems can incorporate operations and maintenance principles into their projects incrementally. There is no one-size-fits-all operations and maintenance strategy. The sequencing and scheduling of approaches being applied on infrastructure systems will vary. The integration of operations and maintenance approaches in infrastructure management can generate questions and decisions such as:

- Does the current level of maintenance enhance the infrastructure life?
- Are current operational levels sufficient to achieve users' objectives?
- Should an infrastructure be maintained, replaced or retired?
- Are new infrastructure projects required? What are the implications of new infrastructure systems and how can operators and policymakers plan those needs?
- How can we augment operations and maintenance through design, construction and procurement decisions?
- Do we continue to provide the current level of service or does the level of service need to be amended?
- What are the costs and benefits of running the infrastructure systems?
- How do operations and maintenance impact the infrastructure system?

10.8.1 Chapter discussion questions

1. What is the significance of operations and maintenance in the life cycle of the infrastructure and the process of asset management?
2. How can operations and maintenance be used to enhance service delivery and reduce life cycle costs?
3. Identify key operations and maintenance guidance and best practices that can be used to optimise existing infrastructure systems in your home country.
4. Why is it important to measure infrastructure spending and performance for transport infrastructure systems?

10.9 Case: Infrastructure systems in the United States

According to McBride (2018), the $18 trillion US economy relies on a vast network of infrastructure systems such as road, bridges, freight rail, ports, electrical grids and internet provision. The systems currently in place were constructed decades ago. Industry practitioners and economists have suggested that the delays and maintenance costs are holding back the country's economic performance (ASCE, 2017). Engineering practitioners have raised safety concerns, warning that a number of bridges are structurally deficient and that antiquated drinking water and wastewater systems pose risks to public health. As it has been claimed (McBride, 2018), investing in both new and current infrastructure would positively impact the US economy in a number of positive ways. Augmenting efficiency, reliability and lowering transportation costs would enhance long-term US competitiveness and insulate the economy from shocks. In addition, it would also directly add demand and employment.

As specified earlier in this chapter, the American Society of Civil Engineers (ASCE, 2017) found the nation's infrastructure systems average a "D," meaning that conditions are below the required standards and significantly deteriorating, which could lead to high-risk failure. The

The role of operations and maintenance in infrastructure management

ASCE (2017) estimated that there is a total infrastructure gap of nearly \$1.5 trillion required by 2025. Interestingly, the US Department of Transportation suggested that over \$800 billion will needed to shore up the country's roads and bridges, while McKinsey highlighted that \$150 billion per year will be needed before 2030 to sustain the country's infrastructure needs. ASCE (2017) further noted the following:

- Transportation will require the largest chunk of funding needs;
- The DOT established that one in four bridges are structurally deficient or not constructed for the traffic they support;
- Its aviation infrastructure is overburdened with some 20 percent of all arrivals and departures delayed;
- The country's rail systems are a mix of commercial and passenger; however, the passenger rail lines are in poor condition;
- The country's water and energy are under stress; the Environmental Protection Agency has suggested that a budget of \$632 billion will be required over the next decade;
- Ports and waterways handle over one-fourth of the nation's freight transport; however, they face mounting delays;
- The US electrical grid is struggling to integrate the required investments; and
- There is a huge broadband gap; rural and low-income communities suffer from a lack infrastructure to deliver reliable and fast internet.

10.9.1 How does the United States compare internationally?

If compared with other developed countries, the US generally lags behind (Schwab, 2017). As shown in the World Economic Forum Global Competitiveness Report (Schwab, 2017), in 2016 the US was ranked tenth in the world in a broad measure of infrastructure quality; it was in fifth place in 2002. That places the US behind Japan, Germany, Spain and France. According to McBride (2018), the US infrastructure systems' performance mainly suffers from its moderately low quality, with consequences for businesses and society as a whole. Much of the inconsistency between the US and other nations can be found in very different funding levels. As stated by McBride (2018), on average, European nations spend about five percent of GDP on constructing, improving and sustaining their infrastructure systems, while the US spends 2.4 percent. Other nations, including Australia, Canada, France and the United Kingdom, also have well-thought-out infrastructure models that allow the government to allocate and prioritise infrastructure projects in a way that the US's more decentralised model has struggled to do.

10.9.2 Case discussion questions

1. What proposals can be adopted by policymakers to improve the overarching infrastructure system in the US?
2. America's approach to planning, financing, building, maintaining and operating its infrastructure systems is fragmented and inefficient. What mechanisms would you adopt to improve them?
3. The US infrastructure systems average a "D." What strategies would you recommend the government adopt to improve the standards and deterioration of the infrastructure systems specified by the American Society of Civil Engineers and the US Department of Transport?

4. Propose a revised allocation and prioritisation model to be used by US infrastructure policymakers.
5. From an operation and maintenance viewpoint, what are some of key lessons you have learnt?

References

Acott, M. (2009). Hearing on the role of research in addressing climate change in transportation infrastructure before the subcommittee on technology and innovation. Available from: https://trid.trb.org/view/899378 [cited 28 April 2020].

Akinyemi, E.O. and Zuidgeest, M.H.P. (2002). Managing transportation infrastructure for sustainable development. *Computer Aided Civil and Infrastructure Engineering*. 17, pp. 148–161.

Ansar, A., Flyybjerg, B., Budzier, A., and Lunn, D. (2016). Does infrastructure investment lead to economic growth or economic fragility? Evidence from China. *Oxford Review of Economic Policy*, 32(3), pp. 360–390.

ASCE (2017). 2013 Report card for America's infrastructure. Available from: https://www.infrastructurereportcard.org/making-the-grade/report-card-history/2013-report-card/ [cited 16 April 2020].

BC (2019). A companion document to asset management for sustainable service delivery: A BC framework. Available from: https://waterbucket.ca/cfa/2019/10/11/guidance-document-primer-on-integrating-natural-assets-into-asset-management-released-by-asset-management-bc-september-2019/ [cited 16 May 2020].

Bourne, R. and Zuluaga, D. (2016). Infrastructure spending and economic growth. Available from: https://iea.org.uk/wp-content/uploads/2016/11/IEA-Infrastructure-spending-briefing.pdf [cited 27 April 2020].

Calderón, C., Moral-Benito, E., and Servén, L. (2011). Is Infrastructure capital productive? A dynamic heterogeneous approach. *Journal of Applied Economics*, 30(2), pp. 177–198.

Cambridge Systematics (2000). A guidebook for performance-based transportation planning. National Cooperative Highway Research Program (NCHRP) Report 446. National Academy Press, Washington, DC.

Carter, R.C. (2009). Operation and maintenance of rural water supplies. In: Perspectives No. 2. St. Gallen: Rural water supply network. Available from: https://www.rural-water-supply.net/en/resources/details/207 [cited 7 May 2020].

Castro, V., Msuya, N., and Makoye, C. (2009). Sustainable community management of urban water and sanitation schemes (A Training Manual). Nairobi: Water and sanitation program-Africa. Available from: http://www.wsp.org/wsp/sites/wsp.org/files/publications/africa_training_manual.pdf [cited 7 May 2020].

Eichner, A.S. and Kregel, J.A. (1975). An essay of post-Keynesian theory: A new paradigm in economics. *Journal of Economic Literature*, 13(4), pp. 1293–1314.

Francis, G. and Robert, R. (2012). The Sendai report: Managing disaster risks for a resilient future (English). Washington DC; World Bank. Available from: http://documents.worldbank.org/curated/en/851321468339912993/The-Sendai-report-managing-disaster-risks-for-a-resilient-future [cited 25 April 2020].

Frangopol, D.M. and Liu, M. (2007). Maintenance and management of civil infrastructure based on condition, safety, optimisation, and life-cycle cost*. *Structure and Infrastructure Engineering*, 3(1), pp. 29–41.

Gharehbaghi, K. and Georgy, M. (2015). Utilisation of infrastructure gateway system (IGS) as a transportation infrastructure optimisation tool. *International Journal of Traffic and Transportation Engineering*, 4(1), pp. 8–15.

Gibson, G., Milnes, R., Morris, M., and Hill, N. (2011). Road transport infrastructure. Available from: https://iea-etsap.org/E-TechDS/PDF/T14_Road%20Transport%20Infrastructure_v4_Final.pdf [cited 28 April 2020].

Glaeser, E. (2016). If you build It...myths and realities about America's infrastructure spending. *City Journal Magazine*. Summer 2016. Avilable from: https://www.city-journal.org/html/if-you-build-it-14606.html [cited 27 April 2020].

The role of operations and maintenance in infrastructure management

Greater London Authority (2003). London's water supply: A report by the London Assembly's Public Service Committee. Available from: https://www.london.gov.uk/sites/default/files/gla_migrate_files_destination/archives/assembly-reports-pubserv-water.pdf [cited 25 April 2020].

Haas, R., Felio, G., Lounis, Z., Cowe Falls, L. (2009). Measurable performance indicators for roads: Canadian and international practice. Proceedings of the 2009 Annual Conference of the Transportation Association of Canada, Vancouver, British Columbia.

Harvey, M.O. (2012). Optimising road maintenance. Bureao of international transport forum. Available from: https://www.oecd-ilibrary.org/transport/optimising-road-maintenance_5k8zvv39tt9s-en [cited 15 April 2020].

Humplick, F. and Paterson, W.D.O. (1994). Framework of performance indicators for managing road infrastructure and pavement. Proceedings of the 3rd International Conference on Managing Pavements, St. Antonio, TX, pp. 123–133. Available from: http://citeseerx.ist.psu.edu/viewdoc/download?doi=10.1.1.655.3713&rep=rep1&type=pdf [cited 16 May 2020].

Ilzetzki, E., Mendoza, E., and Végh, C. (2010). How big (small?) are fiscal multipliers? NBER Working Paper No. 16479. Available from: https://www.imf.org/en/Publications/WP/Issues/2016/12/31/How-Big-Small-are-Fiscal-Multipliers-24699 [cited 27 April 2020].

McBride, J. (2018). The state of United States infrastructure. Available from: https://www.cfr.org/backgrounder/state-us-infrastructure [cited 18 May 2020].

McDonnell, J. (2016). Labour Party conference speech by the Shadow Chancellor. 26 September 2016. Available from: https://www.taxjournal.com/articles/labour-party-conference-shadow-chancellors-speech-04102016 [cited 23rd April 2020].

Medhurst, R. (2016). Doctor yellow keeps Shinkansen network healthy. Available from: https://www.nippon.com/en/nipponblog/m00107/doctor-yellow-keeps-the-shinkansen-network-healthy.html [cited 20 May 2020].

Nakamura, H., Nagasawa, K., Hirashi, K., Hasegawa, A., Seetha Ram, K.E.S., Kim, C.J., and Xu, K. (2019). Principles of infrastructure: Case studies and best practices. Available from: https://www.adb.org/sites/default/files/publication/502801/adbi-principles-infrastructure-case-studies-best-practices.pdf [cited 23 April 2020].

NAO (2014). Maintaining strategic infrastructure: roads. Department of transport and highways agency. Available from: Available from: https://www.nao.org.uk/wp-content/uploads/2015/06/Maintaining-Strategic-Infrastructure-Roads.pdf [cited 6 May 2020].

Karlaftis, M. and Kepaptsglou, K. (2012). Performance measurement in the road sector: A cross-country review of experience. Discussion paper 10. Available from: https://www.itf-oecd.org/sites/default/files/docs/dp201210.pdf [cited 5 May 2020].

Kong, J.S. and Frangopol, D.M. (2003). Evaluation of expected life-cycle maintenance cost of deteriorating structures. *Journal of Structural Engineering*, 129(5), pp. 682–691.

KPMG (2016). Assessing the true value of infrastructure investment. Available from: https://assets.kpmg/content/dam/kpmg/pdf/2016/07/colombia-bt3-4inf-true-value-of-Infraestrucute-investment.pdf [cited 27 April 2020].

Ochieng, E.G., Price, A.D.F., and Moore, D. (2017). *Major infrastructure projects: Planning for delivery*. Basingstoke, UK: Palgrave Macmillan's Global Academic: https://he.palgrave.com/page/detail/major-infrastructure-projects-edward-ochieng/?sf1=barcode&st1=9781137515858.

Organisation for Economic Co-operation and Development (OECD). Performance indicators for the road sector. OECD, Paris, (1997). Available from: https://www.itf-oecd.org/performance-indicators-road-sector-summary-field-tests [cited 10 May 2020].

Prieto, R. (2013). Impediments for implementing a sound asset management system. Center for Advanced Infrastructure and Transportation State of Good Repair Summit. 27 March 2013. Available from: https://www.researchgate.net/publication/273001233_Impediments_for_Implementing_a_Sound_Asset_Management_System.

Schutte, I.G. (2008). A user guide to road management tools. Sub-Saharan Africa Transport Policy Program, The World Bank, Washington, DC.

Schwab, K. (2017). The global competitiveness report 2017-2018. Available from: http://www3.weforum.org/docs/GCR2017-2018/05FullReport/TheGlobalCompetitivenessReport2017%E2%80%932018.pdf [cited 17 May 2020].

Shabnam, N. (2014). Natural disasters and economic growth: A review. *International Journal of Disaster Risk Science*, 5, pp. 157–163.

Shaw, T. (2003). Performance measures of operational effectiveness for highway segments and systems. National Cooperative Highway Research Program (NCHRP) Synthesis of Highway Practice 311, National Academy Press, Washington, DC.

Skidelsky and Miller (2013). Supply matters—but so does demand. *Financial Times*. Available from: https://www.ft.com/content/f858826a-79c3-11e2-b377-00144feabdc0 [cited 27 April 2020].

SIP, (2017). Infrastructure maintenance renovation and management. Available from: https://www.jst.go.jp/sip/dl/k07/booklet_2017_en.pdf [cited 16 May 2020].

Sultana, M., Rahman, A., Chowdhury, S. (2012). An overview of issues to consider before introducing performance-based road maintenance contracting. *World Academy of Science, Engineering and Technology*, 62, pp. 350–355.

Trealoar, G.J., Love, P.E.D., and Crawford, R.H. (2004). Hybrid life-cycle inventory for road construction and use. *Journal of Construction Engineering and Management*, 130(1), pp. 43–49.

Uddin, W., Hudson, W.R., and Haas, R. (2013). *Public infrastructure asset management*, 2nd ed. New York: McGraw Hill Education.

WEF (2014). Strategic infrastructure steps to operate and maintain infrastructure efficiently and effectively. Available from: http://www3.weforum.org/docs/WEF_IU_StrategicInfrastructureSteps_Report_2014.pdf [cited 16 April 2020].

Wren-Lewis, S. (2013). Something we can all agree on? Mainly Macro blog post. 1 November 2013. Available from: https://mainlymacro.blogspot.com/2013/11/something-we-can-all-agree-on.html [cited 23rd April 2020].

Zietlow, G. (2004). Implementing performance-based road management and maintenance contracts in developing Countries - An Instrument of German Technical Cooperation. German Development Cooperation (GTZ), Eschborn, Germany.

PART III

Financing and contracting strategies for infrastructure systems

11

IMPROVING THE FINANCING AND DEVELOPMENT OF MAJOR INFRASTRUCTURE PROJECTS

Edward Ochieng and Maria Papadaki

11.1 Introduction

Globally, a number of nations are finding it difficult to finance the growing demand for essential infrastructure through funding alone. With massive infrastructure needs around the world, and the reality of constrained public–sector budgets, bold leadership is needed to prioritise public policy, harness private capital, and bring innovation to infrastructure funding and project delivery. While demand for infrastructure projects has been growing rapidly, supply has been confined by an inability to mobilise capital. The World Bank Group (2019), found that the estimated annual demand for infrastructure projects in emerging nations is about US $2.0 trillion per annum and current financing for infrastructure mobilised will be US $1.0 trillion, 50 percent of the amount that will be needed to meet investment needs (Bhattacharya and Romani, 2013). These projections are aligned to traditional sectors such as energy, transport and water. If social infrastructure projects was to be included, the investment capital needed could be much higher. For instance, currently spending in the health sector exceeds $4 trillion (nine percent of GDP globally). As shown in this chapter, governments are the primary source of finance for infrastructure projects, usually by raising sovereign debt from local and international markets that are serviced through tax financed budget allocations and retained earnings from national owned enterprises. To meet the increase demand of infrastructure projects, it will be essential to maximise the productivity of infrastructure budgets and tap into new private sources of funding (European Bank, 2018). As stated by the European Bank (2018), the traditional approach to infrastructure funding and finance has faced barriers. High country indebtedness in a number of emerging economies has deterred public debt driven delivery as a scalable alternative to construct urgently needed infrastructure projects. The fundamental shift of liquidity from banks towards institutional investors in the wake of the global financial crisis has led to further challenges to the traditional financing model for emerging-infrastructure markets. Thus, the overarching aim of this chapter is to explore the theory and practice of infrastructure financing and development.

11.1.1 Chapter aim and objectives

This chapter focuses on examining how the financing and development of infrastructure projects can be achieved. The main objectives are to:

- Appraise infrastructure project identification capacity, transaction costs, market and other investment decisions;
- Depict how investors approach the process of infrastructure investments;
- Examine sustainable governance in developing, developed and emerging economies;
- Evaluate key constraints of achieving sustainable infrastructure development in developed and developed economies;
- Assess the general principles of infrastructure cost benefit analysis;
- Establish how financial risk mitigation and deterrence strategies can be incorporated on infrastructure projects; and
- Evaluate investment financing mechanisms and assessment options that can be used for infrastructure projects.

11.1.2 Learning outcomes

The following learning outcomes have been identified for this chapter. Readers will be able to:

- Articulate the process of project identification capacity, transaction costs, markets and other disciplines of investment decisions;
- Gain an insight into how investors approach the process of infrastructure investments;
- Describe sustainable infrastructure governance within the context of developing, developed and emerging economies;
- Appreciate the benefits of cost benefit analysis within the context of infrastructure projects;
- Apply financial risk mitigation and deterrence strategies on infrastructure projects; and
- Differentiate various types of infrastructure investment financing and assessment methods.

11.2 Project identification capacity

According to Chan *et al.* (2009), the provision of public infrastructure entails the interrelated activities of funding, financing and investment; the three determinants have distinctive implications for government economic efficiency. As specified below:

Investment in infrastructure should add value to the society. Profitability alone is an appropriate criterion for infrastructure projects with significant spillover benefits that are not fully integrated in market price. It is worth noting that an investment is efficient in allocating resources if it delivers the highest value of benefits to costs compared to other alternatives. These alternatives can include options such as expenditure on other public services or returning the funds to taxpayers.

As suggested by Berssaneti and Carvalho (2015) and Esty (2015), funding sources should be aligned to users, with public funding making up the shortfall between the user charges and

Improving the financing and development of major infrastructure projects

the overall costs of infrastructure (construction and operation). These costs can include interest payments and principal repayments. Achieving efficiency requires a good balance between the effects of users and charges on demand (including the impact of additional users on taxpayers who may or may not use the service. If there is a need to fund the gap, this subsidy can be directly funded through budgetary processes to help ensure accountability and transparency of project funding decisions. Fletcher and Pendleton (2009) suggested that considerations of intergenerational equity can negatively impact infrastructure investment decisions as most infrastructure projects are long lived, and there is usually a trade-off in construction and maintenance costs. If not well appraised and used, funding decisions can have implications for future generations as commitments to subsidies can impose burdens on future generations.

It is vital for the project appraisers to ensure that the financing minimises the lifetime financing costs of the infrastructure project. Despite the fact that the main financing task is meeting upfront investment costs in an efficient way, the key efficiency issue is which financing vehicle will best mitigate the identified project risks. Chan *et al.* (2009) defined the financing vehicle as a method that can be used to raise capital to meet payments for construction and, in some situations, the operation of the infrastructure project. The financing vehicle can influence the funding gap through the incentives it generates for user charges and the disciplines it imposes on risk management. Financing vehicles may vary in their (Besner and Hobbs, 2012; OECD, 2015; Papke-Shields and Boyer-Wright, 2017):

- **Risk management:** The assignment on non-diversifiable infrastructure project risks and management of the overall project risk;
- **Transaction costs:** The cost of arranging and managing finance, and costs linked with delay or uncertainties with availability of finance; and
- **Exposure to market or other disciplines:** -The extent to which borrowers and lenders share, signal and act on data in the infrastructure project prospects and risks in the investment decision.

Financing vehicles that assign risks to the partner are best placed to manage different types of risk in a more efficient way, reducing the overall cost of the infrastructure project (Besner and Hobbs, 2012). There may also be capacity for financing vehicles to influence allocative efficiency by enacting greater discipline on investment and funding decisions. Figure 11.1 summarises the three considerations in public infrastructure investment decisions appraised in this section.

A basic principle of infrastructure theory is that an efficient investment is one in which an infrastructure project is expected to yield benefits that exceed risk-adjusted costs (Chan *et al.,* 2009). Where there are project constraints on the availability of capital, efficiency will require optimisation across projects to ensure the maximum overall returns. Infrastructure investments have been found to have numerous consequences and, therefore, nations have had to consider a number of macro and micro determinants when carrying out project appraisals and comparisons. The nature and size of an infrastructure project's benefit and costs will be influenced by policies that will underpin government decisions to provide infrastructure services. For instance, profitability alone is an inappropriate determinant for infrastructure projects with important spill-over benefits that are not fully integrated in market prices. In general, it is not straightforward to specifically identify and predict an infrastructure project's benefits and costs. Attaining precise cash flow forecasts, identifying an appropriate discount rate, and appraising risk evolution over the infrastructure project life cycle in a realistic manner are some of the challenges in infrastructure investment evaluation. These complications can be compounded by the presence of embedded contract opportunities in some infrastructure projects.

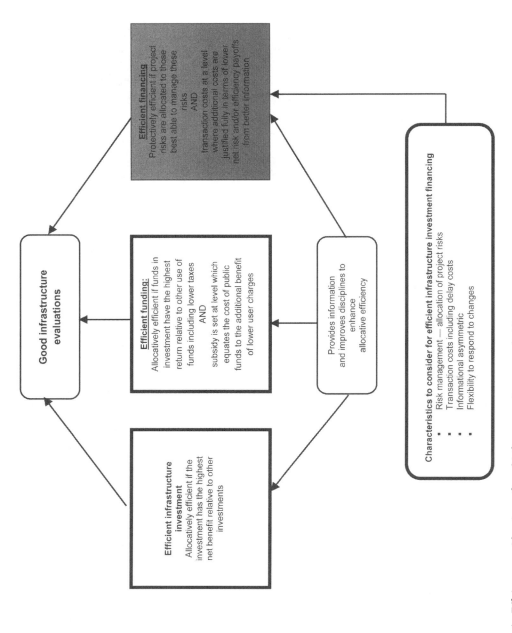

Figure 11.1 Efficiency considerations of good infrastructure investment decisions

Source: Original adapted from Chan et al. (2009)

Improving the financing and development of major infrastructure projects

The utilisation of capital funds should be sustainable and undertaken to ensure the solvency of the service provider. This means that the infrastructure project must produce more than it costs in net present value terms and thus maximise its net worth. Public infrastructure projects that require public funding should be evaluated by the market based on government commitment to the funding. This can be based on the nature of the financing vehicle as well as views on sovereign risk. As noted by Benjamin *et al.,* 2014 and OECD (2015), this may or may not be determined by the characteristics of the infrastructure project itself. As highlighted above, efficient financing vehicles are those that minimise the total cost of finance. This cost is usually made up of the return on the funds, the cost contingent liabilities and transactions costs. The main determinants that can affect the total cost of finance can include allocation of project risk, negotiation and management costs, other costs linked with adequacy of finance and its flexibility, and the disciplines brought to bear on investment decisions.

11.2.1 Project identification risks

If well-defined and aligned with responsibilities for managing project risks, financing vehicles reduce the overall cost of financing the infrastructure project. With full data, clients, whether private or public, need a premium to accommodate risks that they cannot mitigate. The lower the infrastructure risk, the lower the premium needed and hence the lower the total cost of financing the infrastructure project. Chan *et al.* (2009) identified a number of risks that can contribute to the variance of net returns from infrastructure investment (see Table 11.1).

When assessing the worthiness of an infrastructure project, clients need to take into account any non-diversifiable risks. In making infrastructure investment decisions, government policymakers, like any other investor, need to compare the risk-adjusted cost of capital, that is, taking all net benefit flows into account and the rate of return that government policymakers would otherwise

Table 11.1 Investment risk in infrastructure

The sources of investment risk associated with infrastructure include these types.

Construction risk arises from unexpected design problems, cost overruns and delays in construction works. This risk, which can be substantial for capital intensive infrastructure projects, exists during the construction and warranty phases of a project.

Operational risk arises because the planned level of service availability from an asset might not happen. This risk is commonly associated with unexpected problems in staff management, maintenance and other elements of operating the infrastructure. It is present from the commencement of operations.

Demand risk arises because the demand for infrastructure services and, hence, the project revenue might differ from expectations. This risk is present throughout the life of a project. For example, an unanticipated decline in demand could lead to a reduction in the value of the infrastructure asset.

Technological risk arises because purpose-built infrastructure assets might become obsolete or stranded when users switch to a new form of service delivery. This type of risk is present throughout the life of a project.

Financing risk arises because the expected availability and cost of finance might not materialise. This can occur as, for example, interest rates and exchange rates change over time. The financing risk is present throughout the life of a project.

Regulatory (sovereign) risk arises in infrastructure projects, either owned or managed by private entities, because government regulations might affect project profitability. Such a risk can be related to a change in planning and environmental requirements, pricing determinations, and regulatory conditions governing the entry of new service providers. In some cases, governments might expropriate privately owned infrastructure assets. Regulatory or sovereign risk is present throughout the life of a project.

Source: Original adapted from Chan *et al.,* 2009

257

be able to obtain from previous projects with the same risk level as the infrastructure project being currently undertaken. The return required by clients will increase with risk and uncertainty to the extent a financing vehicle can minimise risk and uncertainty and minimise the total cost of financing an infrastructure project. It is worth noting that the conditions needed for infrastructure project risk to invariant (*i.e. the addition of a sure amount to capital reduces the risk by the same amount*) to the allocation of risk are rarely met in practice (OECD, 2015). Thus, a financing vehicle that can apportion infrastructure project risks to those who have the means to better mitigate the risks can reduce the overall level of infrastructure project risk.

A number of infrastructure projects have characteristics that will influence against standardised methods to identify and apportion risk. Moreover, some "grey parts" exist in the project life cycle where neither private nor public clients have a clear advantage in project identification risk management (OECD, 2015). However, there has been emerging agreement on which party should bear what risk at a particular phase of the project life cycle. In addition to the alignment of incentives, risk can be transferred to those more willing to take on risks. Government's ability to pool risk can translate into low credit risk with government debt, but taxpayers will have to bear the contingent liabilities, that is, any cost consequences that might arise if particular infrastructure projects fail to be completed as planned. It is worth mentioning that government's ability to pool risk will not lead to lower project risk, just transfer it. There will only be gains if those taking on the project risk are able to positively gain from the risk sharing. Abridged risk premiums as a result of beneficial risk sharing or risk transfer suggest a reduction in total costs of financing the infrastructure project. Meticulousness is required, as imposing a pattern of risk bearing that is not linked with the ability to manage the risk could maximise the total cost of finance through the higher contingent liabilities. For instance, financial risks that have been restructured for the private sector by government policymakers in public-private partnership agreements through the inclusion of material adverse integrate clauses in project contracts (Benjamin *et al.*, 2014). Where the government might have an influence over the risks identified, such as regulated prices, such risk transfer may be appropriately mitigated. On the other hand, the risks identified have to be properly priced in the contract. This can lead to the reduction required by the private clients and offset the additional cost of contingent liabilities for the government, but this is not always the case. The danger lies in a government accommodating risks that they are not able to mitigate or not being reimbursed for assuming additional infrastructure project risk. It is worth mentioning that the capital market can send convincing signals about exposure to risk. Pertinent market signs include yield spreads, liquidity and availability of credit suppliers, and the terms and conditions of loans and investment. Chan *et al.* (2009) asserted that auxiliary credit data, such as credit rations and project reviews by financial organisations, strengthen and compliment these capital market as summarised in Table 11.2.

11.2.2 Project identification transaction costs

According to Hepburn *et al.* (1997) transaction costs comprise the cost of attaining data, including credit data, creating appropriate project contracts and monitoring borrowers' operational and financial performance. In a number of cases, these are far from insignificant and could noticeably offset efficiency gains from an improvement in risk allocation. Taxation can complicate the conditions for financing efficiency. If not well applied, it can affect the relative capital costs between infrastructure projects, depending on the capability of individual infrastructure projects to create tax revenues for government and tax offsets for business. The pressure to sustain fiscal farsightedness could minimise the capability of governments to take on and finance on-budget investment. This can negatively affect the sufficiency of pay-as-you-go as a financing vehicle as

Improving the financing and development of major infrastructure projects

Table 11.2 Credit information reinforces market signals

Financial organisations help reveal infrastructure project risks by gathering public or proprietary data for relevant infrastructure investment:

- **Credit ratings** are predictive views of a borrower's likelihood to repay debt in a timely manner. Their reliability alters from time to time as they are subject to the quality, completeness and veracity of data used in the rating process, and the skills of the agency analysts.
- **Bank lending** relies on financial institutions' ability to retrieve credit data through their capability in credit analysis and their close relationships with borrowers. Relationship banking facilitates screening and monitoring of borrowers' investment activities. It is also influential for developing flexibility and discretion in bank loans.
- **Bond issues** convey credit data through various bond features apart from their credit ratings. For instance,, the total principal amount of a bond issue vitally bears on its marketability and liquidity, as do its duration and structure, and its alignment with other bond issues to add depth to that particular part of the market. The yield spread (the difference between the yields of a government bond and any other bond) is thought to reflect the relative risk associated with a project.
- **The frequency** with which a borrower issues bonds reveals the borrower's trustworthiness and experience. Other informative bond characteristics include issue purpose, backing (specific revenue sources or general obligations) and sale method (negotiation or auction).
- **Bond insurance** conveys credit rating opinions of specialist insurance companies (known as "monolines") on particular bond issues. Such views are backed by monolines' financial obligations in the event of bond default. For insured bonds, the responsibility of credit appraisal and monitoring is shifted from bond investors to bond insurers.
- **Underlying credit ratings** of insured bonds are sometimes disclosed to communicate additional credit data to bond investors. One researcher has estimated that this has the effect of reducing coupon rates by 0.04 percentage point on average (Peng, 2002).
- **Performance contracts in public–private partnerships** (PPPs) convey data on the contractual ability of private-sector sponsors to exploit economies of scope in designing, building, operating and maintaining infrastructure. If fully achieved, these economies have the potential to reduce total project costs. Possibly, the plausible validation for PPPs rests on the comparative strength of the private sector in productivity or technical capacity — not on its strength from ownership or financing (Engel, Fischer and Galetovic, 2007; Martimort and Pouyet, 2006).

Source: Original adapted from Chan *et al.,* 2009

capital may not be available when it is required for efficient infrastructure project delivery. The formulation of government budgets and accounting conventions could also result in bias in the choice of financing vehicle.

11.2.3 Project identification market and other disciplines on investment decisions

One barrier to allocative efficiency is "asymmetric data" or the unequal distribution of data aligned to risk–return features of infrastructure projects to be financed (Claus and Grimes, 2003). This issue is notably predominant in large-scale infrastructure projects where the project proponent has more data concerning the financial viability of the project, such as the probability and consequences of potential design amalgamations, cost overruns and unrealised anticipated demand, than the financer. Due diligence required by market players or parliamentary scrutiny provides an incentive to invest in data. It is worth noting that the financing method may sway this incentive and can also better accommodate the interests of different parties to the infrastructure investment to minimise data asymmetry. Data asymmetry is likely to be stern when funding and financing decisions entail multiple stakeholders or a number of government departments. The reason is that not all stakeholders involved in the delivery of the infrastructure project will have an incentive to supply reliable, validated data sufficient to identify the true realisation costs of the infrastructure project.

259

A deviation of interest may occur in agency, intergovernmental or public-private relationships, which can create incentive issues that will result in inefficient investment or costly infrastructure investment. Finance investors or contributors who face data disadvantage tend to sensibly fund infrastructure projects on terms reflecting their views of an average risk-return profile, plus an additional premium for bearing the uncertainty linked with unequal data risk (European Union, 2018). This additional premium is likely to increase the cost of financing the infrastructure project. In such kinds of situations, an unfavourable selection problem can arise, which may result in infrastructure projects with a relatively high net benefit facing greater costs of financing than their actual risk-return profiles warrant. These infrastructure projects could be possibly crowded out by projects that will have a smaller net benefit but higher certainty completion. Chan *et al.* (2009) suggested that particular financing vehicles could be useful for transmitting data in an incentive-compatible way, that is, stakeholders involved can all positively gain by truthfully revealing public-private relationships. For instance, the provision of a loan guarantee will ensure that the guarantor better understands and has a stronger confidence in the outcomes of a project than the lender. It can also show the lender's capability to minimise and/or willingness to assume the risks. It is worth noting that the lender will rely on the guarantor's financial capability and commitment. Financing efficiency prevents the financing of high risk-projects; however, the lender requires returns to be proportionate with project risk levels. In order to attain financing efficiency, both explicit (require rate of return) and implicit (contingent liabilities) costs have to be taken into consideration.

11.2.4 Project identification analytical framework

In the context of developing and developed countries, a number of conditions might affect an investor's willingness and ability to partake in infrastructure investments. Three primary factors have generally had a significant impact on the success of infrastructure investments: the presence or absence of local requirements favourable to investment, the type of modality used and the application of risk mitigation tools. As illustrated in Figure 11.2, the proposed analytical framework shows how the three determinants can interact at a high level to shape a successful infrastructure structure.

As shown in Figure 11.2, local conditions dictate much of what can be successfully attained in a developed or developing country. The methodology used to create the project and the risk tools available must address challenges raised by the local conditions. A meticulous appraisal of such conditions, therefore, is the starting point for deciding which approach may be successful (Vives *et al.*, 2006). Risk mitigation instruments may then be used to expand the range of possible approaches by mitigating risks that are raised in the local environment. A number of failures in investments in infrastructure can be ascribed to the utilisation of financial structures mostly borrowed from other environments. As illustrated in Figure 11.2, the analytical infrastructure model considers the feasibility of a number of approaches, given the current or likely to exist local conditions, and categorises them as being practicable, non-achievable or achievable only with improvements or risk mitigation strategies (available tools).

The presented analytical infrastructure framework encourages senior infrastructure practitioners and policymakers to assess eight key local conditions (or determinants) that have proven a high degree of effect on the success of infrastructure projects in developing and developed countries. Other vital determinants may exist, and each infrastructure investment appraisal should consider those most relevant for the investor and the country, locality and project. The model presented here can be a practical tool that can be used by investors and nations as projects are being identified, designed, implemented and improved. As highlighted in the previous sections, the framework can be adapted to the conditions existing for a specific country, infrastructure project and investor. While it is not meant

Improving the financing and development of major infrastructure projects

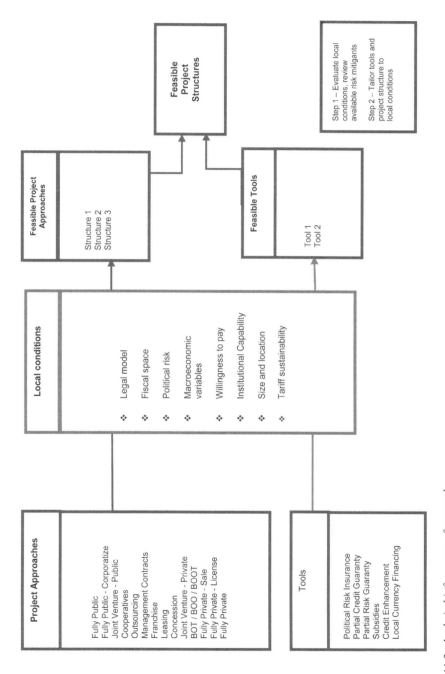

Figure 11.2 Analytical infrastructure framework

Source: Original adapted from Vives *et al.* (2006)

to substitute for thorough due diligence on the part of project sponsors or nations, it should provide vital recommendations and insight for appropriately conducted due diligence appraisals.

It is worth emphasizing that both private and public involvement are vital to meet the requirements of developed and developing governments. There is a need for a better a way to mobilise public and private investment, and to structure the way projects are being identified and appraised. According to Vives *et al.* (2006), the consequences of choosing a poor project methodology and structure were reflected in the high renegotiation rates of infrastructure contracts during the 1990s (40 percent of a sample of 796 infrastructure concessions were renegotiated; the average time to renegotiate was approximately 2.2 years). One of the main concerns highlighted by project sponsors was a change in bargaining power from the investment stage (which favours the firm) to the operational phase (which tilts the balance toward the government because the firm is attached to the investment). Under these conditions, project sponsors were easily expropriated where the local requirements had not been carefully examined and the project structure/appraisal did not match the existing shortcomings (Vives *et al.,* 2006).

11.3 Sustainable infrastructure governance in developing, developed and emerging economies

As noted by the International Monetary Fund (2020), public investment in developing and developed countries support the delivery of key public services through the construction of schools, hospitals, public housing and other social infrastructure. In addition, public investment also connects the society and organisations to economic opportunities through the provision of economic infrastructure centres such as airports and seaports and networks that support telecommunications, transport and electricity production and transmission. As shown in the preceding chapters, through the provision of economic and social infrastructure systems, public investment in developing and developed countries can serve as a significant catalyst for economic growth. A number of studies have underscored the positive relationship between investment in high-quality public infrastructure and economy-wide productivity. For instance, in the October 2014 World Economic Outlook, the IMF (2020) established that for a sample of developed economies, a one percentage point increase in GDP investment spending would boost the level of output by about 0.4 percent in the same year and by 1.5 percent after four years. Interestingly, after years of steady decline, public infrastructure investment has begun to recover in developing economies but remains at a historic low in developed economies (IMF, 2020). This largely highlights improvements that have been gained in the quality of and access to social infrastructure (e.g., schools and hospitals). However, large differences in economic infrastructure (e.g., roads and electricity) still remain.

Within the context of developed and developing economies, the economic and social impact of infrastructure investment will depend on its efficiency. A recent study by IMF (2015) examined a number of techniques used to measure public investment efficiency, which is defined as the relationship between the accrued public capital stock per capita and various determinants of the quality of and access to infrastructure. The findings of the study suggested that 30 percent of the potential benefits of public infrastructure investment were lost due to inadequacies in investment process on average. The size of the efficiency gap shrunk as income rose, with developing countries facing a gap of 40 percent, emerging economies facing a gap of 27 percent and developed economies facing a gap of 13 percent on average. The efficiency of public infrastructure investments will depend significantly on how governments manage their infrastructure investments. Governments that have a sturdy public infrastructure management governance structures and institutions, will have more predictable, credible, efficient and productive infrastructure investments (both social and economic infrastructure systems). In order to ensure that both developed and developing countries are able to critically assess their strength of the public and infrastructure investment management practices, the IMF (2020) proposed public investment management assessment (PIMA) framework that could be used to appraise 15 institutions of public investment decision-making at three key phases (see Figure 11.3):

- **Planning:** Sustainable infrastructure investment secured across the public sector;
- **Allocating:** Infrastructure investment given to the right sectors and projects; and
- **Implementing:** Infrastructure projects produced on time and on budget.

As illustrated in Figure 11.4, the PIMA model provides an all-inclusive methodical of the strengths and weaknesses of a nation's public infrastructure investment management system, allowing comparisons with similar categories and a nation's tailored guidelines.

As specified below (IMF, 2020), the PIMA framework can be applied at three key levels:

- **Nation agencies:** It can be used to provide a basis to create a selected reform plan that is tailored to their specific requirements and aligned with the nation's resources and capabilities;
- **Policymakers:** It can be used to foster dialogue with governments, including surveillance and fund-supported program design, which can result in better reflection of public infrastructure management issues in the fund's work programme; and
- **Project sponsors:** It can be used to assess needs, mobilise funding and enhance coordination among capability development providers.

Infrastructure investment in developed and developing economies can be used by policymakers to secure faster and more sustainable growth. To attain the most out of spending on infrastructure investment, the goal should be to improve the quality of spending and, where necessary, also the quantity. The following infrastructure policy-support initiative tools can be used by both developed and developing economies to enhance the efficiency of public infrastructure investment and explore ways to sustainably increase spending (IMF, 2020):

- **Public investment management assessment (PIMA):** As already highlighted above, policymakers can use PIMA to assess the efficiency of their country's public infrastructure investment and outline their relative institutions' strengths and weaknesses and provide practical guidelines that can be adopted to enhance efficiency and impact of public infrastructure investment.
- **PPP fiscal risk assessment model (PFRAM 2.0):** It can be used to appraise the potential costs and risks that might arise from PPP infrastructure projects. It is worth noting that, a number of infrastructure projects in emerging, developing and developed economies, have been procured as PPPs for efficiency reasons but also to avoid budget constraints and postpone recording the fiscal costs of creating infrastructure services.
- **Debt-investment-growth (DIG):** At the country level, policymakers can use this model to assess the macroeconomic implications and potential risks aligned with different financing methods (including a mix of external concessional, external commercial, and domestic debt) and different methods of public investment as well as the consequences of changes in infrastructure investment efficiency. This model can be modified to accommodate current economic issues (e.g., resource revenue management, building resilience to natural disasters, major outbreak of diseases).
- **Debt sustainability assessments (DSA):** Policymakers can use this tool to appraise the sustainability of fiscal policies, including increasing infrastructure investment. The DSA can also be used to appraise different scenarios, such as higher or lower expenditures.
- **Medium term debt management strategy (MTDS):** Policymakers in developing and developed countries can use this analytical model to provide a methodical and analytical approach

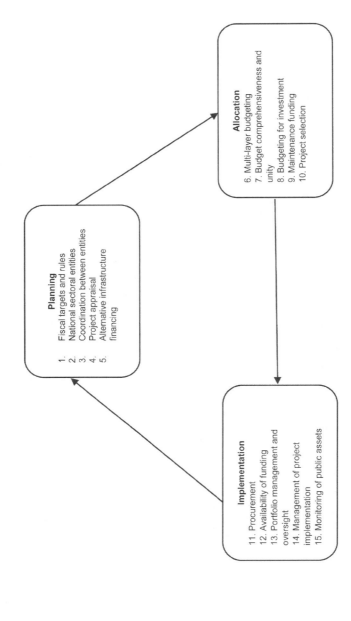

Figure 11.3 Public investment management assessment (PIMA) framework

Source: Original adapted from IMF (2020)

Improving the financing and development of major infrastructure projects

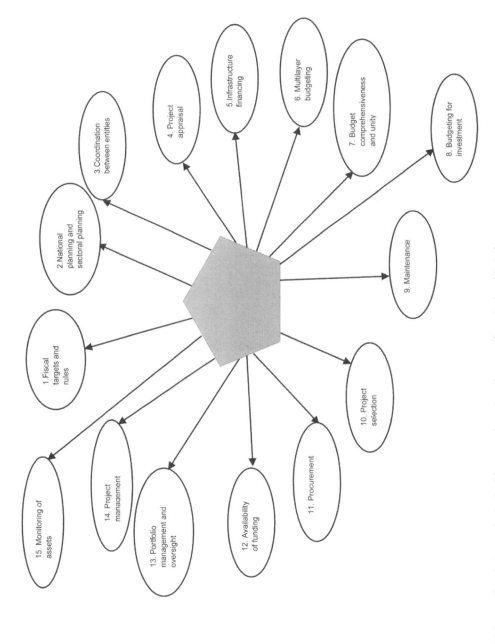

Figure 11.4 An all-inclusive methodical of the strengths and weaknesses of a nation's public infrastructure

Source: Original adapted from IMF (2020)

for debt management strategy at the country level. By creating a debt management strategy, policymakers can ensure that their country's financing needs and payment commitments are met at the lowest possible cost and are consistent with a sensible degree of risk.

11.3.1 Key constraints of achieving sustainable infrastructure development in developed and developing economies

In developing and developed economies, the planning and coordination of infrastructure projects fall within the competence of a number of government departments and ministries. The national and local governments will be involved if the infrastructure project falls within their area. This can be a heavy burden for advocates of unsolicited proposals if the coordination and communication are non-existent between the relevant government departments (Hodges and Dellacha, 2007). As summarised below, the delivery, management and financing of infrastructure projects faces a number of challenges. A set of solutions which can be adopted by policymakers in developed and developing countries are proposed below (WEF, 2015):

1. There is a need for international financial institutions to re-orientate their mandates to focus on activities which bring significant and unique value to infrastructure development. These can entail project planning, the navigation of complex and regulatory models, making positive contribution to the wider investment environment, the provision of financing and collaboration with construction organisations and governments to bring infrastructure projects to the operational phase. However, in some countries they do not yet have the financial-structuring skill-sets needed to meet investor needs; hence, there is a need to adopt an "originate-to-distribute" framework (as opposed to a hold-to-maturity framework) that could deliver greater financially additionality, especially in developing economies.
2. Both in developing and developed economies, there is a need to accelerate bankable projects which are already in the pipeline. There is a need for increasing successful project preparation facilities, especially those that bring together a broad group of skills from international financial institutions, private financial institutions and construction and legal experts. International best standards could be adopted in national infrastructure development, particularly when it comes to implementation of national infrastructure strategy and execution, supporting nations to engage more effectively with private finance, and introducing digital project preparation.
3. There is a need for utilising syndication and securitisation financing frameworks for infrastructure development. This would lead to more effective intermediation between international financial institution projects and the investment requirements of institutional investors. Syndication and securitisation can be used to facilitate the financial restricting of individual projects into portfolios of assets, which can be further be divided into tranches that meet the credit and liquidity needs of institutional investors. In addition, syndication and securitisation can be used to provide international financial institutions to recycle capital, reduce capital replenishment and allow greater engagement between private financial institutions and specialist funds.
4. There is a need for "fit-for-purpose" hedging tools for project sponsors, particularly to mitigate foreign exchange and political risk. International financial institutions have successfully seed-funded expert providers of such hedging, increasing the liquidity and availability of tools. However, the tools need to be affordable, more flexible and longer in duration. As suggested by Tyson (2018), innovative policy methods should be scaled up and consideration be given to whether this area might merit public subsidy to increase project sponsor intake. It is worth mentioning that severe fluctuations in exchange rates have negatively impacted

Improving the financing and development of major infrastructure projects

private infrastructure investment in Africa less than in other global economies. One main reason is the region's minimal ability to raise debt in international markets. The second is its bias toward infrastructure projects with minimal regulatory intervention (Sheppard *et al.*, 2006).

Mitigating regulatory risk aligned to changes in exchange rates could allow acceleration of foreign financing for infrastructure projects that earn mainly local currency revenues and are subject to tariff regulation. According to Sheppard *et al.* (2006), mitigation of such risks would protect infrastructure projects against arbitrary interference by regulatory authorities that would prevent tariff modifications commensurate with cost upsurges caused by exchange rate movements. Sheppard *et al.* (2006), argued that Africa escaped the currency challenges have affected a number of developing regions. However, African currencies have nonetheless seen exchange rate shifts large enough to negatively impact a project's ability to service foreign currency debt and meet project sponsors' prospects on rates of return. In Africa, as in other nations, investors have sought to maximise local markets' contributions to the debt funding of infrastructure projects that create mostly local currency incomes.

Gurara *et al.* (2018), noted that the way forward is not easy and the scope for increasing public investment in developed and developing countries is rather limited, even though the wide range of financing rations indicate that a number of countries may have some room for scaling up. Over recent years, public debt levels in developing and developed countries have risen, external investment conditions have tightened, and growth outlooks have weakened, especially in developing countries. Gurara *et al.* (2018) suggested that countries with fiscal space should seek infrastructure investments on the most concessional terms possible, with the support from the international community. The connection between the sum of public investment (the input) and the quantity and quality of infrastructure in a nation (the outputs and the outcomes) is not very tight and although a number of determinants may contribute to this variation, differences in investment effectiveness are likely one of them. Dabla *et al.* (2012) found that low–income nations have relatively weak public investment management institutions, and that enhancing those institutions could maximise the efficiency of public investment. As summarised below, Petrie (2010) recommended eight key strategies that could be used to strengthen public infrastructure investment management and increase the efficiency of public investment:

- Strategic guidance and preliminary screening: Strategy data should be specific enough and have sufficient coherence and authority to guide public investment. Sector strategies should be fully costed and closely amalgamated and consistent with medium term budgets.
- Appraisal: Infrastructure projects should be appraised using a full range of techniques as appropriate. There should be a comprehensive central recommendation on project appraisal, including specific detailed guidance on the appraisal of public private partnerships.
- Independent review of appraisal: Projects should be subjected to independent review.
- Project selection and budgeting: Only projects that have been subject to thorough appraisal and have been independently reviewed should be selected for funding in the budget.
- Project implementation: There should be a strong focus on managing the total project costs over the life of each project with regular reporting on financial and non-financial progress.
- Project adjustment: Specific mechanisms should be put in place to trigger a review of a project's continued justification if there are material changes to project costs, schedule or expected benefits.
- Facility operation: An all-inclusive and reliable asset register should be maintained and be subject to external audit.
- Project evaluation: An ex post evaluation model should be put in place.

11.4 General principles of infrastructure cost benefit analysis (CBA)

Cost-benefit analysis (CBA) is an analytical tool that can be used to identify the key economic merits or demerits of a major investment decision by assessing its costs and benefits in order to assess the welfare change attributable to it (European Union, 2015). The main aim of CBA is to accelerate a more effective allocation of resources, demonstrating the convenience for society of a particular intervention rather than possible alternatives. As specified by EU (2015), the analytical model of CBA refers to a number of underlying concepts:

- **Opportunity cost of an infrastructure investment:** The opportunity cost of an investment is defined as the potential gain from the best alternative forgone when a decision needs to be made between a number of mutually exclusive options. The rationale of CBA lies in the observation that an investment decision taken on the basis of profit motivations and price mechanisms might lead to some negative circumstances (e.g., market failures such as asymmetry of data, externalities, etc). Conversely, if input, output (including tangible ones) and external effects of an infrastructure project are valued at their social opportunity costs, the return estimated is a proper measure of the infrastructure contribution to the society.
- **Long-term viewpoint of an infrastructure investment:** A long-term perspective needs to be adopted, which should range from a minimum of 10 to a maximum of 30 years or more. Thus, there is a need to forecast future costs and benefits, integrate appropriate discount rates to calculate the present value of future costs and benefits and take into account uncertainty by appraising the infrastructure project risks.
- **Calculation of economic performance determinants expressed in monetary terms:** It is worth noting that CBA is based on a set of predestined project outcomes, giving a monetary value to all the positive outcomes and negative costs and welfare effects of the intervention. These costs are discounted and then equalled in order to calculate a net benefit. The infrastructure project overall performance can be measured by the following determinants: net present value (NPV), expressed in monetary values, and the economic rate of return (ERR), which allows for comparability and ranking of competing infrastructure projects or substitutes.
- **Macroeconomic approach:** From a microeconomic viewpoint, CBA is used to assess the infrastructure project's impact on society as a whole via the calculation of economic performance variables; this provides an assessment of predictable welfare changes. While direct employment or external environmental outcomes realised by the project are captured in the NPV, indirect determinants such as secondary markets and wider outcomes (public funds, employment, regional growth, etc.) should be excluded. As noted by Boardman *et al.* (2006), this is mainly because of:

  - A number of indirect or wider effects are usually transformed, redistributed or capitalized; hence, the need to minimise the potential for benefits double-counting;
  - There remains minimal practice on how to translate them into robust variables for infrastructure project appraisal; hence, the need to disregard the analysis relies on assumptions whose reliability is a challenge to check; and
  - It is advisable, however, to provide a qualitative account of the impacts to better highlight the contribution of the government's national/regional policies.

- **Incremental approach:** Policymakers and senior infrastructure project practitioners can use CBA to compare a scenario with the project with a counterfactual baseline scenario without the project. As suggested by EU (2015), the incremental approach requires that:

Improving the financing and development of major infrastructure projects

♦ A counterfactual situation is defined as what would occur in the non-existence of the project. For this scenario, projections made focus on all cash flow related to the operations in the project location for each year during the project life cycle. In instances where the project consists of a completely new asset, that is, no pre-existing infrastructure, the without-the-project scenario will be one with no operations. In instances where investments are targeted at enhancing facility, it should entail the costs and the revenues/benefits to operate and maintain the infrastructure a level that is still operable or business as usual (BAU) or even small adaptation investments that are programmed to take place anyway (doing the minimum) (Boardman *et al.,* 2006). In particular, it is advisable to carry out an analysis of the project sponsors; historical cash flows as a basis for the forecasts where relevant. The choice between BAU and doing the minimum as a counterfactual should be done on a case by case basis, on the basis of the evidence about the most feasible, and likely, scenario. If there is uncertainty, a BAU situation should be adopted as a general rule. If policymakers decide to do the minimum as a counterfactual the scenario should be both feasible and credible and not lead to undue and unrealistic additional benefits or costs. As illustrated in the example below, the choice adopted may have significant implications on the outcomes of the analysis;

♦ Projections of cash flows can be made for the situation with the proposed project. This should take into account all the investment, financial and economic costs and benefits resulting from the infrastructure project. In situations of pre-existing infrastructure projects, it is recommended to perform an analysis of historical costs and revenues of the beneficiary (at least the three previous years) as a basis for the financial forecasts of the with-project scenario and as a reference for the without-infrastructure project scenario; otherwise, the incremental analysis can be very susceptible to manipulation; and

♦ Lastly, the CBA only reflects the difference between the cash flows in the with-the-project and the counterfactual scenarios. The financial and economic performance determinants are calculated on the incremental cash flows only (Florio, 2006).

The example below highlights the issue of the project performance in relation to what scenario is selected as counterfactual.

As seen in Table 11.3, the proposed project consisted of rehabilitating and expanding existing infrastructure capacity and involved investing EUR 450 million. The project was to result in benefits growing by five percent per year. The "do-minimum" scenario consisted of only rehabilitating existing capacity and investing 30 million, followed by constant benefits. The BAU entailed no investment at all, which in turn was to affect the amount of output the facility can

Table 11.3 An example of the choice of the counterfactual scenario

	Scenarios	$ (m)	NPV	1	2	10	21
1	Proposed project	Net benefit	1,058	45	47	70	119
		Investment	435	450			
2	Do-minimum	Net benefit	661	45	45	45	45
		Investment	29	30			
3	Business as usual	Net benefit	442	45	43	28	16
		Investment	0				
	Results						
1-2	Proposed project net of do-minimum	Net benefit	-9	-420	2	25	74
		ERR	3%				
1-3	Proposed project net of business as usual	Net flows	182	-450	4	42	103
		ERR	6%				

Source: Original adapted from EIB (2013)

produce, leading to a fall in net benefits of five percent per year. As specified in Table 11.3, the results of the CBA significantly different scenarios were adopted as counterfactual. By comparing the proposed project with the "do-minimum" scenario, the ERR equalled three percent. If the BAU is taken as a reference, the ERR increases to six percent. Hence, any choice should be duly justified by the project sponsor on the basis of clear evidence about the most feasible situation that would occur in the absence of the project.

11.4.1 Infrastructure project appraisal through CBA

As summarised below, standard infrastructure CBA is structured in seven phases:

- Description of the context
- Definition of project objectives
- Identification of the infrastructure project
- Technical feasibility and environmental sustainability
- Financial analysis
- Sustainability analysis
- Economic analysis
- Risk assessment

11.4.1.1 Description of the context

The first step of the infrastructure project appraisal should aim to appraise the social, economic, political and institutional contexts in which the project will be implemented. The key determinants to be depicted should be:

- The socioeconomic conditions of the nation/region that are aligned to the project, including the demographic dynamics, expected GDP growth, labour market conditions, unemployment trend, etc.;
- The policy and institutional aspects, including existing economic policies and development plans, organisation management services to be provided/developed by the project, as well as the capacity and quality of the institutions involved;
- The current infrastructure endowment and service provision, including determinants/information on coverage and quality of services provided, current operating costs and tariffs/fees/charges paid by users, if any (Florio, 2006);
- Other data and statistics that are relevant to better justify the context, for example, existence of environmental issues, environmental authorities likely to be part of the project, etc.;
- The views and prospects of the population with relation to the service to be provided, including, when relevant, the positions implemented by civil society organisations.

11.4.1.2 Definition of project objectives

The second phase of the infrastructure project appraisal should aim to delineate the objectives of the project. As asserted by Ochieng *et al.* (2017), the infrastructure project objectives should be delineated in explicit relation to the needs. Moreover, the project needs assessment should be built upon the description of the context and provide the basis for the objective's definition. It is vital to ensure that the objectives are quantified through key indicators and targeted in line with the result orientation principle of the government policy. This can be related to improvement of the output quality, to better accessibility to service, to maximise existing capacity, etc. A well-defined set of objectives are required to identify the effects of the project to be further evaluated in the CBA and validate the project's significance.

Improving the financing and development of major infrastructure projects

11.4.1.3 Identification of the infrastructure project

Below are analytical themes that should be considered in infrastructure project identification:

- The physical components and activities that will be included provide a given good service and achieve a well-defined set of objectives which should consist of a self-sufficient unit of analysis;
- The department responsible for the implementation (often referred to as the project sponsor) needs to be identified and its technical, financial and institutional capacities appraised; and
- The impact location, the final beneficiaries and all key stakeholders need to be identified. All stakeholders (both public and private) affected by the project need to be described. Large infrastructure projects usually do not only affect the client (mostly the government) and the direct consumers of the service but will generate wider effects on partners, suppliers, competitors, public administrators, local communities, etc.

11.4.1.4 Technical feasibility and environmental sustainability

The technical feasibility and environmental sustainability are key components of data that need to be assessed in the justification of infrastructure projects. Although both determinants are not formally part of the CBA, their data must be precisely reported and used as a primary information source within the CBA framework. Detailed data should specify the following: demand analysis, options analysis, environment and climate change considerations, technical design, cost estimates and implementation timeline. The CBA should be taken as an ongoing, multidisciplinary exercise performed throughout the project planning in parallel with other technical and environmental considerations. Requirements for the CBA of the recommended project solution are, however, the finalisation of the detailed demand analysis and the availability of investment and operational and management (O and M) cost estimates, including costs for environmental mitigation and adaptation measures. These should be based on the preliminary project designs, which are centrepieces of the technical feasibility study and environmental impact assessment (EIA). From the early phase of the infrastructure project, project practitioners preparing the CBA should adopt a multidisciplinary approach to project planning. Their involvement in the planning of the demand analysis and option analysis is significant (and often decisive) in the successful completion of the project. A full-scale CBA should be carried out at the end of the preliminary design phase. The aim is to ensure that a verification of adequacy and economic convenience of the proposed solution to meet the pre-established project objectives is provided to project planners. A summary of the proposed solution should entail the following themes (EU, 2015):

- **Location:** Description of the location of the project.
- **Technical design:** Description of the main constituents, technology that will adopted, design standards and specifications.
- **Production plan:** Description of the infrastructure capacity and the expected utilisation rate. These determinants should highlight the service provision from the supply side, project scope and size in the context of the forecasted demand.
- **Cost estimates:** Estimation of the financial requirements for the project realisation. The data should clearly specify whether the cost estimations are investor estimates, tender prices or out-turn costs.
- **Implementation timing:** A realistic project timeline and implementation schedule should be provided.

11.4.1.5 Financial analysis

The financial analysis should be performed in order to:

- Assess the consolidated project profitability;
- Assess the project profitability for the project client and some key stakeholders;
- Verify the project financial sustainability and key feasibility condition for any typology of the infrastructure project;
- Specify the cash flows which underpin the calculation of the socioeconomic costs and benefits.

In order to ensure that the financial appraisal of the infrastructure project captures the key components of the project, the following guidelines can be adopted:

- Only cash inflows and outflows should be considered in the analysis; reserves, depreciation, price and technical contingencies and other accounting items which do not align to actual flows are overlooked.
- Financial appraisal should, as a general rule, be performed from the point of view of the infrastructure project sponsor. If, in the provision of a general interest service, the client and operator are not the same unit, a consolidated financial analysis, which excludes the cash flows between the client and the operator, should be performed to assess the actual profitability of the infrastructure investment.
- In order to calculate the present value of the future cash flows, a suitable financial discount rate (FDR) should be adopted. The financial discount rate should reflect the opportunity cost of capital.
- Project cash-flow forecasts should entail a period appropriate to the infrastructure project's economically useful life and its likely long-term benefits. It is vital for the project financial analysts to ensure that the number of years for which forecasts will be provided align to the project's time horizon (or reference period). The choice of time horizon can affect the appraisal outcome. It is therefore vital to integrate a standard benchmark, differentiated by sector and based on internationally accepted practice.
- To ensure successful delivery, the financial analysis should usually be performed in constant (real) prices, that is, with prices fixed at a base year. The use of current (nominal) prices, prices adjusted by the consumer price index (CPI), should entail a forecast of CPI that does not seem always necessary.
- The analysis should comprise net of VAT, both on purchase (cost) and sales (revenues), if this is recoverable or not recoverable by the project client.
- For the purpose of financial sustainability verification, direct taxes (on capital, income or other) should be considered, which should be calculated before tax deductions. The aim is to avoid capital income tax rules' complexity.

11.4.1.6 Financial profitability

Determination of infrastructure costs, operating costs, revenues and sources of financing allows the appraisal of project profitability, which can be computed by the following indicators:

- Financial net present value, FNPV(C), and the financial rate of return, FRR(C), on investment;
- Financial net present value, FNPV(K), and the financial rate of return, FRR(K), on national capital.

11.4.1.7 Return on investment

The financial net present value of investment, FNPV(C), and the financial rate of return of the investment, FRR(C), can be used to compare the infrastructure project investment costs to net revenues and measure the extent to which the infrastructure project's net revenues are able to repay the investment, regardless of the sources or methods of financing. The financial NPV on investment is delineated as the sum that results when the expected investment and operating costs of the project (discounted) are deducted from the discounted value of the expected revenues:

$$\text{FNPV(C)} = \sum_{t=0}^{n} a_t S_t = \frac{S_0}{(1+i)^0} + \frac{S_1}{(1+i)^1} + \dots + \frac{S_n}{(1+i)^n}$$

Where: S_t is the balance of cash flow at time t, a_t is the financial discount factor chosen for discounting at time t and i is the financial discount rate. The financial rate of return on investment is defined as the discount rate that produces a zero FNPV; FRR is given by the solution of the following equation:

$$0 = \sum \frac{St}{(1+FRR)^t}$$

The FNPV(C) is expressed in money terms (\$/£/EUR) and must be related to the scale of the infrastructure project. The FRR(C) is a pure number and is scale invariant. Mainly, the project appraiser uses the FRR(C) in order to judge the future performance of the investment in comparison to other infrastructure projects or to a benchmark required rate of return. This calculation also contributes to deciding if the project requires financial support from the project client: when the FRR(C) is lower than the applied discount rate (or the FNPV(C) is negative), the revenues generated will not cover the costs and the project needs assistance from the project owner. This is often the case for public infrastructures, partly because of the tariff structure of these sectors (Boardman *et al.*, 2006; EU, 2015).

The return on investment should consider:

- (incremental) investment costs and operating costs as outflows;
- (incremental) revenues and residual value as inflows.

As a result of this, cost of financing is not included in the calculation of the performance of the investment FNPV(C), but it is included in the table for the analysis of the return on capital FNPV(K) (see Table 11.4). Moreover, as mentioned, capital, income or other direct taxes are included only in the financial sustainability table and not considered for the calculation of the financial profitability, which is calculated before deductions.

11.4.1.8 Return on national capital

The aim of the return of national capital computation is to assess the infrastructure project performance from the viewpoint of the assisted public, and possibly the private. The return on national capital is computed as outflows: the operating costs and the national (public and private) capital contributions to the project; the financial resources from loans at the same time in which they are reimbursed; and the related interest on loans. As far as replacement costs are concerned, if they are self-financed with the project revenues, they can be taken as operating costs (see Table 11.5).

Table 11.4 Calculation of the return on investment ($/£/EUR thousands)

				Years				
	1	2	3	4	5–9	10	11–29	30
Total revenues				11,598	...	12,011	...	12,222
Residual value								4,265
Total inflows	0	0	0	11,598	...	12,011	...	16,487
Total operating costs				5,561	...	5,662	...	5,713
Initial investment	8,465	75,176	42,890					
Replacement costs						11,890	9,760	
Total outflows	8,465	75,176	42,890	5,561	...	17,552	...	5,713
Net cash flow	–8,465	–75,176	–42,890	6,037	...	–5,540	...	10,774
FNPV(C)				– 34,284				
FRR(C)				1.4%				

A financial discount rate of 4 % has been applied to calculate this value.

Source: Original adapted from EU (2015)

Table 11.5 Calculation of the return on national capital (£/$/EUR thousands)

				Years				
	1	2	3	4	5–9	10	11–29	30
Total revenues				11,598	...	12,011	...	12,222
Residual value								4,265
Total inflows	0	0	0	11,598	...	12,011	...	16,487
Public contribution	3,148	27,956	15,950					
Private equity	1,085	9,632	5,495					
Loan repayment (including interest)					1,789	1,789	1,789	
Total operating and replacement costs				5,561	...	17,552	...	5,713
Total outflows	4,233	37,588	21,445	5,561	...	19,341	...	5,713
Net cash flow	–4,233	–37,588	–21,445	6,037	...	–7,329	...	10,774
FNPV(K)								11,198
FRR(K)								5.4 %

The loan is here an outflow and is only included when reimbursed. In this example, it is assumed to be paid back in ten constant payments starting in year 5.

In this example, replacement costs are self-financed with the project revenues. Accordingly, they are treated as operating costs.

Source: Original adapted from EU (2015)

Improving the financing and development of major infrastructure projects

The financial net present value of capital, FNPV(K), is the amount of the net discounted cash flows that accumulate to the national beneficiaries (public and private combined) due to the execution of the infrastructure project. The subsequent financial rate of return on capital, FRR(K), of these flows will determine the return in percentage points. When calculating FNPV(K) and FRR(K), all sources of financing should be taken into consideration. The sources can be taken as outflows (they are inflows in the financial sustainability account) instead of investment costs (as they form part of the financial return on investment computation). While the FRR(C) can be very low, or negative for the public infrastructure investments, the FRR(K) will be higher and, in some instance, even positive. It is worth highlighting that, for public infrastructure, a negative FNPV(K) does not mean that the infrastructure project is not required from the operator or the public's viewpoint and should be terminated. It just indicates that it does not provide sufficient financial return on national capital utilised based on the benchmark applied (that is, four percent in real terms).

11.4.1.9 Financial sustainability

An infrastructure project will be financially sustainable when the risk of running out of cash in the future, both during the investment and the operational phases, is zero. Project owners need to explicitly demonstrate how the sources of financing available (both internal and external) will consistently match payments on a yearly basis. In the instance of non-generating income infrastructure projects or if negative cash flows are expected in the future (that is, years in which large capital is required for asset replacements), a well-defined commitment to cover negative cash flows must be specified. It is worth noting that the inflows and outflows will show the deficit or surplus that will be accrued on a yearly basis. Sustainability that arises from the cumulated generated cash flow is positive for the years considered (see Table 11.6). The inflows can include:

Table 11.6 Financial sustainability (£/$/EUR thousands)

					Years				
	1	*2*	*3*	*4*	*5–9*	*10*	*11–29*	*30*
Sources of financing	8,465	75,176	42,890					
Total revenues				11,598	...	12,011	...	12,222
Total inflows	**8,465**	**75,176**	**42,890**	**11,598**	**...**	**12,011**	**...**	**12,222**
Initial investment	8,465	75,176	42,890					
Replacement costs						11,890	9,760	
Loan repayment (including interest)					1,789	1,789	1,789	
Total operating costs				5,561	...	5,662	...	5,713
Taxes				604	...	–733	...	651
Total outflows	**8,465**	**75,176**	**42,890**	**5,561**	**...**	**19,341**		**5,713**
Net cash flow	**0**	**0**	**0**	**6,037**	**...**	**–7,329**	**...**	**6,509**
Cumulated net cash flow	**0**	**0**	**0**	**6,037**	**...**	**20,726**	**...**	**133,835**

The cumulated cash flow should be zero (or positive) during the construction phase.

Financial sustainability is verified if the cumulated net cash flow row is greater than zero for all the years considered.

Source: Original adapted from EU (2015)

- Sources of financing;
- Operating revenues from the provision of goods and services; and
- Transfer, subsidies and other financial gains not stemming from charges paid by users for the use of infrastructure.

The dynamics of the inflows can be assessed against the outflows; these can relate to the following determinants:

- Initial investment;
- Replacement costs;
- Operating costs;
- Refund of loans and interest payments;
- Taxes on capital/income and other direct taxes.

It is vital for the project practitioners to ensure that the infrastructure project does not risk suffering from a deficiency of funding. In particular, in the case of vital reinvestments/upgrades, proof of disposal of sufficient resources must be captured in the sustainability analysis. It is advisable to perform a risk analysis that takes into account the possibility of the key determinants in the analysis (usually construction costs and demand) being worse than expected.

11.4.1.10 Economic analysis

In order to assess an infrastructure project's contribution to society, an economic analysis must be performed. The key concept that can be adopted is the use of shadow prices to reflect the social opportunity cost of goods and services, rather than prices observed in the economic market, which may be inaccurate. Sources of market inaccuracy might include (EU, 2015):

- Administered tariffs for utilities which may reflect the opportunity cost of inputs due to affordability and equity reasons;
- Non-efficient markets where the public sector and/or operators exercise their power (e.g., subsidies for energy generation from renewable sources, prices including a mark-up over the marginal costs in the case of monopoly, etc.);
- Some prices can include fiscal requirements (e.g., duties on import, excises, VAT and other indirect taxes, income taxation on wages, etc.);
- For some effects no markets (and prices) are available (e.g., reduction of air pollution, time savings).

The standard guide proposed in this book is consistent global practice, which is to move from financial to economic analysis. Starting from the account for the return on investment computation, the following adjustments should be adopted (Boardman *et al.,* 2006; EU, 2015):

- **Fiscal corrections:** Taxes and subsidies are transfer costs which do not represent real economic costs or benefits for the society as they entail merely a transfer of control over certain materials from one group in society to another. Some general guidelines can be adopted to correct inaccuracies:

    - Costs for input and output must be considered net of VAT;
    - Charges for input should be reflected net of direct and indirect taxes;
    - Costs (e.g., tariffs) used as a proxy for the value of outputs should be reflected net of any subsidy and other transfer granted by a public entity.

- **Conversion from market to shadow prices:** When market costs do not align with the opportunity cost of inputs and outputs, it is best to convert them into shadow prices and include the items in the financial analysis.
- **Utilisation of conversion determinants to project inputs:** Transforming inputs into market costs into shadow rates can be achieved through the application of conversion determinants. These can be defined as the ratio between shadow prices and market prices. They signify the determinant at which market costs have to be multiplied to achieve inflows valued at shadow cost. Formally:

$$k_i = \frac{v_i}{p_i} \Leftrightarrow v_i = k_i \cdot p_i$$

Where p_i are market prices for the good i, v_i are shadow prices for the same good and k_i are the conversion factors. If the conversion variable for one item is higher than the other, the observed cost is lower than the shadow rate, meaning that the opportunity rate of that item is higher than that captured by the economic market. Equally, if the conversion determinant is lower than 1, the observed price is higher than the shadow rate due to taxes or other market inaccuracies, which add to the marginal social value of an item and determine a higher market price.

- **Assessment of non-market impacts and correction for externalities:** Impacts on project users due to the use of a new or enhanced improved item or service, which are beneficial to the society, but for which a market value is not available, should be integrated as project direct benefits in economic analysis of the infrastructure project appraisal. In principle, the willingness-to-pay (WTP) projected for the use of service should include these effects and ease its inclusion in the analysis. Examples of (positive) non-market factors might include savings of travel time, increased life expectancy or quality of life, prevention of fatalities, injuries or accidents, enhancement of landscape, noise reduction, increased resilience to present and future climate change and reduced uncertainty and risk.

Market prices modification and non-market impacts approximation, costs and benefits occurring at different levels must be reduced. The reduced rate in the economic analysis of infrastructure investment projects, the social discount rate (SDR), should reflect the social view on how future benefits and costs will be valued against present ones (EU, 2015).

- **Evaluation of greenhouse gas (GHG) emissions:** Within the context of economical assessment, climate change impacts will occupy a significant position in the externalities appraisal because:

  - GHGs, especially carbon dioxide (CO_2), nitrous oxide (N_2O) and methane (CH_4), have a long lifetime in the atmosphere so that present emissions will add to impacts in the distant future;
  - The long-term effects of continued emissions of GHGs are difficult to forecast but can be catastrophic;
  - Scientific data has suggested that causes and future paths of climate change will become combined (e.g., the probabilities of temperature outcomes and effects of the natural environment, which are associated with different levels of stabilisation of GHGs in the air).

- As suggested by (EU, 2015), the best approach to integrate climate change externalities into economic assessment will be based on the European Investment Bank (EIB) Carbon Footprint Methodology. This should consist of the following phases:

- Quantification of the volume of emissions additionally emitted, or saved, in the air because of the infrastructure project. Emissions should be computed on the basis of project-specific emission determinants factors (e.g., t-CO_2 per unit of fuel burnt, kg-CO_2 per kilometre travelled, etc.) and expressed in tonnes per year. In the absence of project-specific information, project appraisers can use default emissions determinants from the economic literature.
- Calculation of total CO_2 equivalent emissions using global warming potentials (GWP). GHGs other than CO_2 can be converted by multiplying the sum of emissions of the specific GHG with a factor equivalent to its GWP. For instance, set the GWP of CO_2 equal to unity (=1), and the GWP for CH_4 and N_2O to 25 and 298, respectively, indicating that their climate impact is 25 and 298 times larger than the impact of the same amount of CO_2 emissions (IPPC, 2007).
- Evaluation of externality using a unit cost of CO_2-equivalent total tonnes of CO_2e should be multiplied by a unit sum expressed in your home currency/tonne.

11.4.1.11 Risk assessment

When appraising infrastructure projects, it is important to include a risk assessment in the CBA. The risk assessment is required to deal with the uncertainty that always infuses infrastructure investment projects, including the risk that adverse impacts of climate change may have on it. The proposed phases for appraising the infrastructure project risks are as follows:

11.4.1.11.1 Sensitive analysis

Sensitive analysis will allow the identification of critical determinants of the infrastructure project. Such determinants are those whose variations, whether positive or negative, have the biggest impact on the infrastructure project's financial and/or economic performance. The analysis should be performed by varying one determinant at a time and determining the effect of change on NPV. As a guiding condition, the proposed recommendation is to consider "critical" those determinants for which a variation of ±1% of the value adopted in the base case gives rise to a variation of more than 1% in the value of the NPV. It is vital to ensure that the tested determinants are deterministically independent and as disaggregated as possible. Correlated determinants could lead to misrepresentations in the results and double counting. Hence, before progressing with sensitive analysis, the CBA framework should be appraised with the aim of separating the independent determinants and removing the deterministic interdependencies (e.g., splitting a determinant in its independent constituents). For instance, "revenue" is a compound determinant, which relies on two independent variables, "quantity" and "tariff," both of which should evaluated. Table 11.7 shows a descriptive example of sensitivity analysis:

It is worth highlighting a particularly relevant determinant of the sensitivity analysis: the computation of the switching values. It is the value that the appraised determinant would have to take in order for the NPV of the project to become zero, or more generally, for the result of the infrastructure project to fall below the level of acceptability (see Table 11.8).

The use of substituting numbers in sensitivity analysis allows users to make some decisions on the risk of the infrastructure and the opportunity of undertaking risk-preventing actions. For example, as can be seen in Table 11.8, one must evaluate if a 19 percent investment cost increase, which would make the ENPV equal to zero, means the infrastructure project is too risky. There is a need to further assess the reasons of this risk, the likelihood of occurrence and identify possible

Improving the financing and development of major infrastructure projects

Table 11.7 Sensitivity analysis

Variable	Variation of the FNPV due to a ± 1 % variation	Criticality judgement	Variation of the ENPV due to a ± 1 % variation	Criticality judgement
Yearly population growth	0.5%	Not critical	2.2%	Critical
Per capita consumption	3.8%	Critical	4.9%	Critical
Unit tariff	2.6%	Critical	N/A	N/A
Total investment cost	8.0%	Critical	8.2%	Critical
Yearly maintenance cost	0.7%	Not critical	0.6%	Not critical
Per capita willingness to pay	Not applicable	–	12.3%	Critical
Annual noise emissions	Not applicable	–	0.8%	Not critical

Source: Original adapted from EU (2015)

mitigation strategies. It is vital to ensure that the sensitivity analysis is complemented with a scenario analysis. The scenario analysis can be used to study the impact of combinations of values taken by the critical determinants. Specifically, combinations of "optimistic" and "pessimistic" values of the critical determinants could be used to build a number of realistic scenarios, which might hold under certain hypotheses. In order to outline the optimistic and pessimistic situations it is necessary to choose for each variable the extreme (lower and upper) values (within a range defined as realistic) (Boardman *et al.*, 2006). Incremental project performance variables can then be computed for each combination. In addition, some decisions on the infrastructure project risks can be made on the basis of the analysis findings. For instance, if the ENPV remains positive, even in the pessimistic scenario, the infrastructure project risk can be evaluated as low.

Table 11.8 Switching values

Variable	Switching values	
Benefits / revenues		
Yearly population growth	Minimum increase before the FNPV equals 0	104%
	Maximum decrease before the ENPV equals 0	47%
Per capita consumption	Minimum increase before the FNPV equals 0	41%
	Maximum decrease before the ENPV equals 0	33%
Tariff	Minimum increase before the FNPV equals 0	60%
	Maximum decrease before the ENPV equals 0	Not applicable
Per capita willingness to pay	Minimum increase before the FNPV equals 0	Not applicable
	Maximum decrease before the ENPV equals 0	55%
Costs		
Investment cost	Maximum decrease before the FNPV equals 0	82%
	Minimum increase before the ENPV equals 0	19%
Yearly maintenance cost	Maximum decrease before the FNPV equals 0	95%
	Minimum increase before the ENPV equals 0	132%
Annual noise emissions	Maximum decrease before the FNPV equals 0	Not applicable
	Minimum increase before the ENPV equals 0	221%

Source: Original adapted from EU (2015)

11.4.1.11.2 Qualitative risk analysis

The qualitative analysis of an infrastructure project should include the following components:

- A list of adverse events to which the infrastructure project is exposed;
- A risk matrix for each adverse event highlighting the

  - Possible causes of occurrence;
  - Link with the sensitivity analysis, where appropriate;
  - Negative effects created on the infrastructure project;
  - Categorised levels of probability of occurrence and of the severity of impact; and
  - Risk level.

- An explanation of the risk matrix including the assessment of acceptable levels of risk; and
- An account of mitigation and/or prevention strategies for the key risks, stating who is responsible for the applicable mitigation strategies to minimise risk exposure, when they are considered necessary.

To carry out a qualitative risk analysis on infrastructure projects, the first phase should entail the identification of adverse events that the infrastructure project may face. Compiling a list of potential adverse events is significant as it helps to understand the complexities of the infrastructure project. Once the potential adverse events have been categorised, the corresponding risk matrix may be created. To operationally create a risk matrix, one could adopt the following steps:

1. It is important to identify the possible causes of risk materializing. These are the primary hazards that could occur during the project life cycle of an infrastructure project. All causes of each adverse event must be detected and appraised by the project team. One has to take into account the weaknesses of forecasting and planning and/or the project team may have similar consequences over the infrastructure project. Risk detection can be based on ad hoc analysis or looking at similar problems that have been detailed in past projects. Examples of the possible risks can include low contractor capacity, inadequate design cost estimates, insufficient site investigation, low political commitment, insufficient market approach, etc.
2. When suitable, the link with the data of the sensitivity analysis should be specified by highlighting which critical determinants will be affected by the adverse events.
3. For each adverse event, the general effect(s) presented on the infrastructure project and the relative consequences on the cash flows should be explained. For instance, if there are delays in construction time, it will postpone the operational stage, which in turn could have a negative impact on the financial sustainability of the infrastructure project. It is advisable to depict the effects in terms of what the project sponsor (or the project manager) might experience in terms of functional or business impacts.
4. A probability or likelihood of occurrence should be credited to each adverse event. The following categorisation can be used:

   a. very likely (0–10 percent probability)
   b. unlikely (10–33 percent probability)
   c. about as a likely as not (33–66 percent probability)
   d. likely (66–90 percent probability)
   e. very likely (90–100 percent probability)

   The probability exercise should be carried out during the planning stage so that the project team can decide on the acceptable level and what mitigation strategies can be used.

Improving the financing and development of major infrastructure projects

During the risk analysis integrated in the CBA, the remaining risks in the final design phase of the infrastructure project are appraised. In principle, no undesirable risks should remain.
5. The proposed intensity mitigation strategies should equal the level of risk. A stronger response and high level of commitment to managing the infrastructure project should be aligned with risks with a high level of impact and probability. Low impact risks should be closely monitored and controlled.

11.4.1.11.3 Probability analysis

With probability analysis, the project team can assign a probability distribution to each of the critical determinants of the sensitivity analysis. In order to recalculate the expected values of financial and economic performance determinants, a precise range of values around the best estimate can be used as the base case. The probability distribution for individual determinants can be derived from a number of sources, such as consultation from experts, distributions found in the literature for similar projects or experimental data. Once the probability distribution for the critical determinants has been established, the project team can proceed with the computation of the probability distribution of the FRR or NPV of the infrastructure project. For this purpose, the Monte Carlo method is recommended. The method comprises repeated random extraction of a set of values for the critical determinants, taken within the respective defined intervals, and then the calculation of the performance indices for the infrastructure project (FRR or NPV) resulting from each set of obtained values. The values extracted will allow decision-makers to infer vital decisions about the level of risk of the infrastructure project. As can be seen in Table 11.9, **ENPV** can generate negative values (or ERR lower than the social discount rate, SDR) with a probability of 5.3 percent, divulging an infrastructure project with a low risk level.

It is worth mentioning that the data of the Monte Carlo drawings, expressed in terms of the probability distribution or cumulated probability of the internal rate of return (IRR) or the NPV in the subsequent interval of values, will stipulate more all-inclusive data about the risk profile of an infrastructure project. Figure 11.5 displays a graphical example.

The cumulated probability curve presented in Table 11.10 evaluates the project risk. For instance, it confirms whether the cumulative probability for a given value of NPV or the IRR is higher or lower than a reference value that is thought to be critical. In the example presented in Figure 11.3, the cumulative probability of an ENPV value is £18, 824,851, which is set at 50 percent of the base value, 0.225, a value high enough to advocate taking preventive and mitigation strategies against the identified infrastructure project risk.

Table 11.9 Results of Monte Carlo simulation

Expected values	ENPV	ERR
Base case	36,649,663	7.56%
Mean	41,267,454	7.70%
Median	37,746,137	7.64%
Standard deviation	28,647,933	1.41%
Minimum value	−25,895,645	3.65%
Central value	55,205,591	7.66%
Maximum value	136,306,827	11.66%
Probability of the ENPV being lower than zero or ERR being lower than the reference discount rate		

Source: Original adapted from EU (2015)

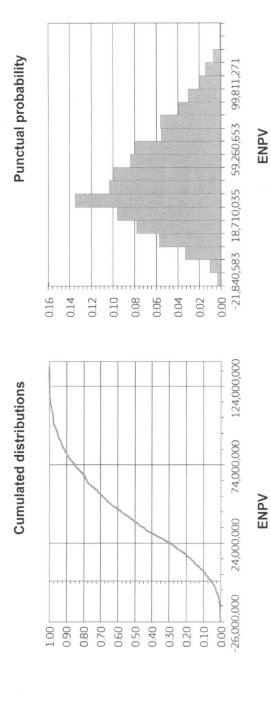

Figure 11.5 Example of cumulated and punctual probability distribution of the ENPV

Source: Original adapted from EU (2015)

Improving the financing and development of major infrastructure projects

Table 11.10 Infrastructure investment by sector

R$ billion	2012	2013	2014	2015	2016	BNDES forecast (period average) 2017-2020	Author forecast (period average) 2019-2024
Electric power	71.6	59.1	54.9	52.8	56.9	39.5	80.0
Highways	19.7	19.5	17.9	14.0	8.2	9.7	40.0
Railways	8.0	9.1	9.3	9.4	5.9	6.0	20.0
Ports	7.2	3.1	6.9	7.3	3.6	3.8	10.0
Airports	3.3	8.8	4.7	3.5	2.5	2.2	5.0
Sanitation	13.0	13.2	14.3	13.0	12.4	10.4	25.0
Solid waste	1.2	1.4	1.4	1.2	1.2	1.4	5.0
Urban mobility	3.5	7.3	9.8	17.0	6.5	4.8	20.0
Infrastructure– without telecom	127.5	121.5	119.2	118.2	97.2	77.8	205.0
Telecom	34.3	39.6	37.1	32.0	27.5	26.7	40.0
Infrastructure total	161.8	161.1	156.3	150.2	124.8	104.6	245.0
%GDP	**3.1%**	**3.0%**	**2.7%**	**2.5%**	**2.0%**	**1.5%**	**3.2%**

Source: WEF (2019)

11.4.1.11.4 Risk mitigation and deterrence strategies for infrastructure projects

The integration of the stages highlighted above can be adopted as deterrence and mitigation strategies for an infrastructure project. It is advisable to take a neutral approach towards risks because the public sector might be able to pool the risks of a large number of projects. In such instances, the appraisal of switching values and scenario analysis data, followed by a well-defined risk matrix (plus a probabilistic risk analysis), might be required. It is vital to ensure that the risk evaluation is the basis for risk assessment, which is the identification of strategies to minimise risks, including how to allocate them to key stakeholders involved and which risks to transfer to the contractors involved in the delivery of the infrastructure project. Within the context of infrastructure project delivery, if not well-defined, risk management can be a complex function. In order to ensure that the management of risks on infrastructure projects is smooth, the client, project manager and the team have to identify specific measures for mitigation/deterrence of the identified risks.

11.5 Investment financing methods: assessment options

As suggested by Ray (2015), traditionally infrastructure financing models have been over-reliant on a leverage structure complemented by the development of financial institutions, government institutions, multilateral institutions and export credit agencies, even while seeking to optimise private sector competences in project execution, cost optimisation and operational efficiencies, along with a steadfast, monopolistic revenue framework of such projects largely funded by governments and executed by competitively bid private sector counterparties. The entry of the private sector into developing and sponsoring infrastructure projects has led to a conscious focus on breaking monopolies. This has led to revenues being largely determined by market forces and the lease cost bidding model, making financing on high leverage a relatively risky scheme. Conversely, for infrastructure projects that are being delivered under regional monopolies, like airports, roads or ports, the financing and execution can be largely transferred to private-sector contractors whose own abilities to raise funds can be limited because of their reliance on the banking instructions.

11.5.1 Private activity bonds

Private activity bonds (PAB) are issued on behalf of governments and can be used to finance certain private (or partially private) infrastructure projects that will benefit the public (Dewar and Puttock, n.d; PwC, 2013; Ray, 2015). Generally, the income interest from the bonds issued by governments to finance government infrastructure projects is exempted from government income taxes, while the income generated from bonds issued for the benefit of private actors and mixed public-private businesses is subject to government income taxes. However, there is an exclusion for bonds issued to finance authorised infrastructure projects such as the construction of airports, docks, high-speed intercity rail, waste and flooding defences projects. In those specific projects, governments can issue bonds to finance infrastructure projects owned by private or mixed private-public businesses. From the reviewed literature (PwC 2013; UN 2017), it was found that no dominant project bond model has yet emerged, and at country level the conditions will always vary. While the specific bond model for each infrastructure project is likely to remain dynamic, the financing source for the infrastructure project will likely transition from bank debt to institutional investors. For a logical infrastructure debt, it would be sensible to use short-term bank debt for construction finance (which can be in the form of a supplier's credit with a take-out finance underwriting) and refinancing the same in the long-term institutional markets (Dewar and Puttock n.d). It is worth highlighting that the key risk with this framework is what refinancing risk arises in terms of infrastructure project operations, regulation, interest and exchange rate and who is likely to be the bearer of the risk. As recommended by Ray (2015), a deterrence strategy for such project-specific risks can be found in the securitised debt market; where financial institutions can package a bundle of project finance loans and sell them as securitized in the financial markets. In order to ensure that institutional markets have a sustained interest in the long term, it is vital for the infrastructure policymakers to ensure that they have (Dewar and Puttock, n.d; PwC, 2013; Ray, 2015;):

- Capital outside of the banking system;
- Sufficient governance and transparency in financial reporting;
- Balanced tax and commercial policies; and
- Project specific credit support/credit enhancement.

11.5.2 Public-private partnerships

Globally, public-private partnerships (PPP) have emerged as the preferred model for infrastructure development (Li *et al.*, 2017). They are a funding model whereby the public partner can be represented by the government and the private partner can be a privately-owned business, often with extensive capabilities that will add value and complement the overall goals of the infrastructure project. There are a number of potential benefits with this type of model, such as quicker delivery of infrastructure projects and minimization of debt that the government might have to take on. At the initial phase of the infrastructure project, if not well managed it can lead to disappointing experiences. These can arise as a result of inadequate pre-investment work, insufficient project planning, absence of proper feasibility studies, flawed project evaluations, absence of competitive tendering, poor contract design, complexities in land acquisition and inaccurate estimation of demand. It has been found that lack of see-through governance mechanisms can

Improving the financing and development of major infrastructure projects

further complicate the outcome of the project, which can lead to conflicted regulatory structure, arbitrary and populist government interference, lack of judicial independence and lack of a strong legal model specifying the rights and obligations of private investors (Kumari and Sharma, 2016). As established from the reviewed literature (Ray, 2015), there is emphasis on unbundling operational risks and assigning external risks to project entities, internal risks to project sponsors and residual risks to government stakeholders (see Figure 11.6).

For the model to have a better success rate, one has to ensure the following (UNESCAP, 2012):

a. To ensure transparency and accountability one has to integrate a global best practice model by fully disclosing bid criteria and making criteria easily available for public scrutiny.
b. Formulation of PPP units should be aligned with international best practices, such that the units should be designed to accelerate the PPP procurement and delivery process before contacts are finalized, enabling all linkages, permits and approvals, and having a transparent interface with the establishments that approve or deny projects.
c. There should be an independent, non-conflicted regulatory environment capable of monitoring the progress of the infrastructure project: its commissioning and operation. In addition, it should assist in the implementation of a reward and penalty structure through market mechanisms.
d. There should also be investment in human resources for PPP to enhance skills and knowledge across the broad spectrum of specialties, from institutional to technical to financial, by collaborating with experienced nations.

The main advantage of PPPs will be attained once project risks have been better managed through engaging one contractor to design, build and operate the infrastructure project. This can significantly reduce the contingent liabilities to the government and the investor. The trade-off will comprise of transactions costs, which at times can be high, in tendering and in negotiating and administering contracts that might efficiently shift the risk to the private partner (Chan *et al.*, 2009). Figure 11.7 provides a summary of the main considerations that need to be taken into account in comparing PPP to conventional government bond finance. It is worth noting that while this provides a guide to the considerations needed, the balance of cost savings in the total budget of financing with the additional costs needs to be considered on a case by case basis.

11.5.3 Cross-border public-private partnership

Financing cross-border projects through the PPP route presents even greater challenges as such infrastructure projects are designed for substantial spill-over benefits and nations involved may have a number of financial challenges in terms of financial capacity (Kumari and Sharma, 2016; Ray, 2015). Nations with minimal financial markets not only face funding gaps but even a gross deficiency in the institutional infrastructure supporting PPPs. Local politics can also hinder the creation of such infrastructure projects as the duration is often very long with immediate tangible benefits in the short term. It is worth stressing that cross-country infrastructure projects are more complex than national infrastructure projects and entail constructing infrastructure in less-developed border areas with benefits that are spread over a longer period and not easy to capture (Fay *et al.*, 2018). Such kinds of infrastructure projects do not lend themselves easily to the PPP framework. As a consequence, it is vital to identify and analyse potential infrastructure projects in order to decide which fit private investment and which can only be implemented using public capital. The viewpoints of nations involved often differ in terms of identification of budgets and benefits to each. In addition, the complexities in the development, approval, preparation,

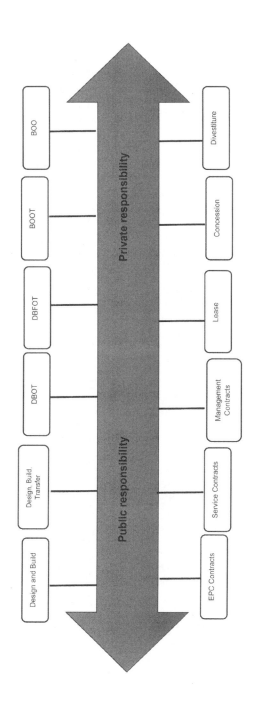

Figure 11.6 Public–private partnership model

Key: DBOT-Design Build Operate Transfer; DBFOT-Design Build Finance Operate Transfer; BOOT-Build Own Operate Transfer; BOO-Build Own Operate

Improving the financing and development of major infrastructure projects

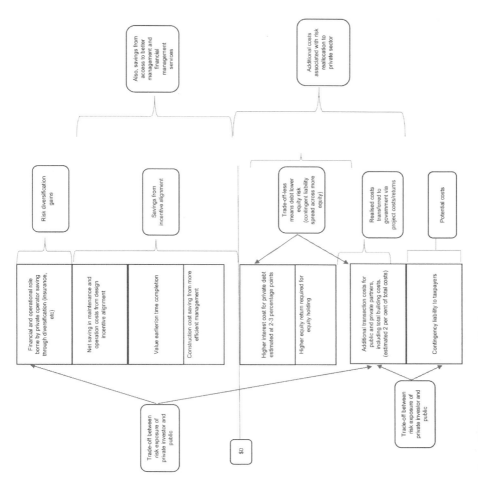

Figure 11.7 Cost of PPP financing, relative to government debt finance

Source: Original adapted from Chan et al. (2009)

assessment, implementation, management, operation and maintenance of cross-country infrastructure projects make their financing very difficult and at times not doable without some form of government guarantee. As specified below, some of the key challenges in delivering these infrastructure projects through cross-border PPP include (Ray, 2015):

- Incongruent cross-border economic regulations between nations;
- Lack of capital market coordination and variance in sovereign risk and rating of the participating nations minimise investor options for the whole infrastructure project;
- Lack of amalgamation between regional financial markets affects the ability to procure long-term infrastructure finance;
- Multiple currency incomes can lead to uncertainty in income and debt service estimation; and
- Lack of coordination between nations.

Some key processes that can be adopted to promote cross-border PPPs include (Fay *et al.*, 2018; Li *et al.*, 2017):

- Establishing a strong sovereign guarantee mechanism;
- Spending more time in identification and development of infrastructure projects to make them bankable;
- Establishing non-discriminatory measures for managing currency risk, e.g., innovative swap instruments;
- Creating strong legal, institutional and regulatory model;
- Establishing strong foreign direct investment to facilitate PPPs;
- Amalgamating national markets and establishing regional bond/equity markets; and
- Negotiating and procuring non-discriminatory investment protection treaties for greater private investment and foreign direct investment.

A number of developing countries have ill-defined PPP models that, because of their nebulousness, inhibit private involvement, while capital controls or unavailability of foreign exchange discourage foreign investors who worry that they may not be able to recoup their capital and profits. Weak regulatory or legal frameworks have intensified the risk, and shallow or illiquid capital markets complicate existing strategies.

11.5.4 Regional infrastructure funds

A regional infrastructure fund (RIF) can be used to facilitate the timely availability of capital to regional infrastructure projects, which will deliver significant benefits to the social/economic growth of regions (UN, 2017). RIFs can be efficient drivers for accelerating funds into regional infrastructure projects that cannot be sufficiently funded through traditional methods of private and public funding. In order to be successful, they need to be tailored to the specific prerequisites and priorities of a specific region. If well managed, RIFs can prove effective in fine tuning infrastructure projects from feasibility to customised solutions with robust financial and economic benefits. As highlighted below, RIFs can be used to add value to infrastructure development by (Ray, 2015; UN, 2017):

- Being flexible and providing development finance at the early phase of infrastructure development;
- Assisting countries to focus on local and cross-border priorities by making optimal and efficient use of public resources;

Improving the financing and development of major infrastructure projects

- Generating project level investment attractiveness for private sector infrastructure development and financing involvement;
- Generating infrastructure development projects as a commercial scheme, backed by a well-defined business plan and techno-commercial viability, while procuring cooperation and coordination at national, regional, sub-regional and local levels for infrastructure planning and delivery;
- Formulating a risk mitigation framework by cross-pooling sovereign support of the regional economies; and
- Integrating strategies for sovereign support, transparent formulation of user charges, and tariff escalation, leading to a pronounced impact through regional infrastructure funding.

Well-managed RIFs can assist the growth of regions by providing grants for infrastructure projects that will have the potential to stimulate regional economic activity and enhance the social competitiveness of regions.

11.5.5 *Project bonds*

In 2008, the global financial crisis resulted in stricter regulations on financial institutions and banks and their lending requirements, which meant that infrastructure projects could no longer be financed by traditional debt alone. Other sources of infrastructure financing, such as project bonds, were considered and implemented. Globally, project bonds opened up an alternative debt funding mechanism to source financing for infrastructure projects. As suggested by Dewar and Puttock (n.d), project bonds enable sponsors to reduce the cost of finance to the extent that tenors are often longer those found in the commercial loan market. According to Dewar and Puttock (n.d) and PwC (2013), for capital markets investors (mainly insurance organisations, bank treasuries, pension funds and asset managers seeking long term stable assets), infrastructure projects offer inflation, risk-adjusted capital returns with minimal correlation to an economic cycle, thus providing for predictable, steady returns. As illustrated in Figure 11.8, the project bond entails the following phases: permission to lead, pre-launch, launch and roadshow, pricing and signing/issue of final.

As asserted by PwC (2013), project bonds could provide a flow of suitable highly-rated assets direct to pension plans and life insurance. Some of the key considerations to be taken into account when deciding to raise finance for an infrastructure project are summarised below (Dewar and Puttock (n.d); PwC (2013):

- **Regulatory requirements:** Generally, issuers will want to configure their project bond offering so that they can make offers, sell into the market and ensure access to adequate investor demand, therefore offering competitive terms. Project bonds are subject to extensive and complex securities laws. In most countries, the legislation requires issuers to undergo extensive due diligence, disclosure and reporting obligations, both prior and after the offering.
- **Credit rating requirements:** A number of institutional investors, which make the project bond market, are required to meet the "minimum investment grade" regardless of the capability of the project sponsor or the project's risk mitigants. As found Dewar and Puttock (n.d) a number of project organisations located in emerging jurisdictions have lacked the capability to gain a robust credit rating as a result of poor sovereign rating of the host nation.
- **Consent and intercreditor issues:** Within the context of a project bond, the typical method for seeking consent through a trustee has been found to be complex and time consuming. Reconciling the interests of a large number of lenders (potentially commercial banks, export credit and development agencies and bond holders), often with deviating interests (capital market investors being particularly driven by short-term benefits from trading their project

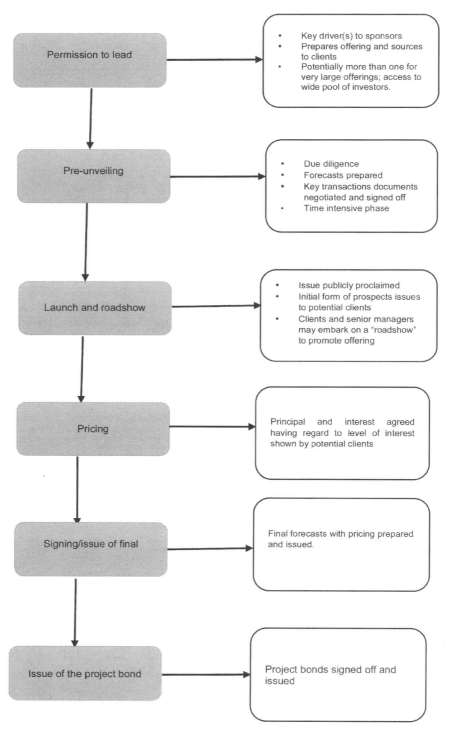

Figure 11.8 Principal stages of a project issuance

debt), has been found to be complex and will require meticulous handling by solicitors when creating the intercreditor mechanics.

- **Construction risk:** Construction risk has been found to be one of the key risks in an infrastructure project because of the project's reliance on a limited number of assets to create revenue. Possible strategies that could be used to mitigate the risks include a fixed-price "turnkey" construction consisting of appropriate performance incentives, an on-demand, unconditional letter of credit or performance issued by a creditworthy organisation, a financing model that allows payment or scheduled debt service under a downside construction scenario. The creation of project bonds for infrastructure projects at the construction phase requires considerable thought from stakeholders involved in the structuring of the deal.
- **Operating period risk:** From the construction phase, one could argue that no significant or unforeseeable costs are needed to be borne by the infrastructure project, which should reduce risk and allow a steady cash flow during the payback period of the bond. The de-risking of the infrastructure project should lead to a successful placement of the project bond; however, one should ensure that they have put together a set of operational mitigation strategies such as the use of recognised technology, obtaining all-inclusive insurance policies and the use of experienced contractors.

11.5.6 Innovative infrastructure financing

As established from the reviewed literature, innovative or alternative infrastructure financing entails an umbrella concept that comprises traditional infrastructure funding sources and financing methods (Chen, 2016a). It also encompasses any strategy involving new funding sources, new financing methods, and new financial arrangements in the provision of infrastructure:

- **Funding** refers to a revenue stream or capital that will pay for an infrastructure project. It may comprise a revenue source from local tax receipts or grants, or it may refer to proceeds of debt financing (Chan and Bartle, 2017). A major infrastructure project will consist of multiple sources of funding, including national, local and international sources.
- **Financing** refers to borrowing capital to pay for a major infrastructure project, typically through loans, bonds or other debt methods such as a line of credit (Chan and Bartle, 2017).

As suggested by Chen (2016a), alternative infrastructure financing can be categorised into three main types:

- **New funding sources:** These can be any new measures that generate additional revenue resources to pay for government infrastructure projects. They can include new taxes such as local option taxes that have been targeted for major infrastructure projects, or different value-capture mechanisms such as impact fees or development exactions, which can be charged to counterweigh the cost of constructing a new infrastructure improvement project during the construction phase.
- **New financing mechanisms:** These are comprised of new methods for borrowing capital in flexible and/or potentially cost-effective ways to pay for a major infrastructure project. The new mode of financing can include credit assistance tools such as loans, loan guarantees and lines of credit. These can mainly be offered by governments and alternative bond and debt financing tools such as green bonds, grant application revenue, national bond banks and social impact bonds.
- **New financial arrangements:** These can include new partners such as the private sector, the non-profit sector or the general public.

11.6 Chapter summary

As shown in this chapter, infrastructure projects may be financed and managed by either governments or private investors and, in the case of government ownership, may be developed and operated by the private sector or through public-private partnerships. Infrastructure financing can take many forms, including direct lending, commercial bank loans, project bonds issued via public capital markets or private placements, direct equity investment funds and capital raised from managed infrastructure bonds. While most of the financial instruments appraised in this chapter can be generally used to finance infrastructure projects, the selection must be undertaken on a case-by-case basis. The selection of sources of financing infrastructure projects requires the appraisal of risks. Infrastructure projects suffer from significant undermanagement of financial risk in practically all phases of the value chain and throughout the project life cycle. The long-term delivery of modern infrastructure projects will require well-thought-out financial strategies that reflect the uncertainty and huge number of risks they are exposed to during their delivery.

11.6.1 Chapter discussion questions

1. Identify some of the key cost benefit analysis determinants that you could use to appraise infrastructure projects.
2. Critically discuss how you would apply financial risk mitigation and deterrence strategies to infrastructure projects.
3. Identify the advantages and disadvantages of the following investment financing methods:

   a. Private activity bonds
   b. Public-private partnerships
   c. Cross border public-private partnerships
   d. Regional infrastructure funds
   e. Project bonds
   f. Innovative infrastructure financing

4. Describe how you would apply the process of project identification capacity, transaction costs, market and other investment decisions on infrastructure projects.
5. Within the context of developing, developed and emerging economies, discuss how you would apply the principles of sustainable governance on infrastructure projects.

11.7 Case: Improving infrastructure financing in Brazil

Brazil's infrastructure investment as a proportion of its GDP is estimated to be around two percent, while it is near seven percent in China and five and a half percent in India (WEF, 2019). This poses a noteworthy challenge to Brazil's growth potential. Higher investment in infrastructure would maximise export competitiveness and boost productivity growth. As found in the literature, Brazil has transitioned from a model in which public financing is the key source for its infrastructure requirements to a more balanced framework in which the National Bank for Economic and Social Development (BNDES) has catalysed domestic and foreign funding (WEF, 2019). This transition has required fundamental changes in financial instruments and rules to better manage risks and standardise the processes, contracts, financing instruments and insurance policies related to infrastructure projects and concessions required to attract institutional investors.

Since 2014, the infrastructure financing model that was being utilised by policymakers had long succeeded in Brazil and relied heavily on concessional BNDES lending, but it had

Improving the financing and development of major infrastructure projects

been challenged by a number of events ranging from corruption scandals, recessions, defaults affecting long-term and bridge loans from financial institutions, and fewer instruments available for covering project risks. Currently, Brazil's infrastructure is insufficient to meet the nation's development requirements after years of underinvestment and needs investment conventionally approximated at around 3.2 percent of GDP over 2019–2024 (WEF, 2019). The total capital required over the period is estimated at around R$205 billion yearly. It is approximated that more than half of that must be generated from private domestic sources. Enhancing access to private finance will require amending the role of BNDES from a mere financier to a catalyst capable of mobilising other sources of finance for infrastructure projects. This means that while BNDES will still remain a significant source of financing infrastructure projects, it will also need to "crowd in" other actors by mitigating risks, such as completion risk and currency risk. With reference to currency risk, other Latin American nations, such as Chile and Peru, have shown in the energy sector field that by propounding power purchase agreements (PPA) in US dollars they have been able to access cheaper and longer-term capital, particularly for renewables, which has led to lower costs for electricity generation.

In May 2016, the Brazilian government introduced efforts to further enhance infrastructure PPPs and concessions. Aside from introducing a dedicated unit (known as the PPI Special Secretariat) to better coordinate government ministries, regulatory agencies, financiers and concessionaires, the Brazilian government took important steps to maximise transparency in concession procurement processes and to increase investor confidence and financing settings. Some of the steps were the gradual removal of the Taxa de Juros de Longo Prazo (TJPL) and its replacement with the market referenced rate Taxa de Longo Prazo (TLP), or Long-Term Rate, the diminution of BNDES support to infrastructure projects along with an encouragement to utilise capital market instruments (debentures), and improved integration of capital expenses with revenues in project cash flows (WEF, 2019). Table 11.10a summarises the amounts in Brazilian reals of investment in each infrastructure area since 2012 and the BNDES estimates for 2017–2020, based on customers' queries for new loans. As can be seen in Table 11.10, telecommunication has been separated from other infrastructure systems because telecom investments in Brazil are usually prepared by private companies and financed by corporate debt and equity rather than depending on BNDES. As illustrated in Table 11.11, R$60 billion will be invested per year on the assumption that R$60 billion per year (30 percent would be financed via equity, some RS$30 billion a year would be financed by BNDES and R$115 billion by other sources of funding (that is, for 2019–2024).

As illustrated in Table 11.10b, R$60 billion will be invested per year on the assumption that R$60 billion per year (30 percent) would be financed via equity, RS$30 billion a year would be financed by BNDES and R$115 billion by other sources of funding (that is, for 2019–2024). Other sources will be generated from domestic and foreign capital markets (infrastructure debentures). Thus, determining how to enhance access to markets should be a cornerstone of government infrastructure policy. The increase in volumes specified by the shading can only be attained by attracting foreign sources of funding.

11.7.1 Case discussion questions

1. Critically discuss why higher investment in infrastructure would maximise export competitiveness and boost productivity growth in Brazil.
2. What are some reasons why Brazil has transitioned from a model in which public financing is the key source for its infrastructure requirements to a more balanced framework in which the National Bank for Economic and Social Development (BNDES) is.

Table 11.11 Sources of infrastructure financing

R$ billion	2012	2013	2014	2015	2016	BNDES forecast (period average)	Author forecast (period average)
						2017–2020	2019–2024
% Equity (estimated)	34%	20%	10%	23%	36%	37%	29%
% Debt (estimated)	66%	80%	90%	77%	64%	63%	71%
Infrastructure debt financing of which:	84.2	97.6	106.9	91.0	62.1	49.0	145.0
BNDES disbursements (without telecom)	48.2	60.0	64.3	52.3	23.8	15.0	30.0
Infrastructure debentures	2.5	4.1	5.5	5.0	4.5	5.0	**20.0**
FI–FGTS=Fundo de Investimento do Fundo de Garantia por Tempo de Serviço (Investment Fund of the Worker's Severance Guarantee Fund)	2.5	2.5	3.2	3.0	3.0	3.0	5.0
CEF = Caixa Econômica Federal (federal government-owned savings bank)	10.0	10.0	18.3	15.0	15.0	10.0	15.0
Corporate debentures	4.0	4.0	6.6	6.6	6.6	4.0	**40.0**
Multilaterals+other DFIs	2.0	2.0	4.1	4.1	4.1	2.0	5.0
Commercial banks+ others	15.0	15.0	5.0	5.0	5.0	10.0	**30.0**

Source: WEF (2019)

3. As a policymaker, how would you deal with the following challenges?

   a. Corruption scandals
   b. Recessions
   c. Defaults affecting long-term projects
   d. Bridge loans from financial institutions
   e. Fewer instruments available for covering project risks

4. Critically discuss why BNDES will remain a significant source of funding in Brazil.
5. What led to the removal of the Taxa de Juros de Longo Prazo (TJPL) and its replacement with the market referenced Taxa de Longo Prazo (TLP)?
6. From Tables 11.10a and b, propose next steps and action plans that can be adopted by the Brazilian policymakers to improve infrastructure financing.

References

Benjamin, C.E., Chavich, C., and Sesia, A. (2014). An overview of project finance and infrastructure finance. Harvard Business School Background Note 214-083. Available from: https://www.hbs.edu/faculty/Pages/item.aspx?num=47358 [cited 16 May 2020].

Berssaneti, F.T. and Carvalho, M.M. (2015). Identification of variables that impact success of Brazilian companies. *International Journal of Project Management*, 33(3), pp. 638–649.

Besner, C. and Hobbs, B. (2012). An empirical identification of project management toolsets and a comparison among project types. *Project Management Journal*, 43(5), pp. 24–46.

Bhattacharya, A. and Romani, M. (2013). Meeting the infrastructure challenge: The case for new development bank. A presentation to G-24 technical group meeting. Washington, DC, 21 March. Available from: http://g24.org/wp-content/uploads/2016/01/Session-4_2-1.pdf [cited 14 April 2020].

Boardman, A.E., Greenberg, D.H., Vining, A.R., and Weimer, D.L. (2006). *Cost-benefit analysis: concepts and practice*, 3rd ed.. Upper Saddle River, New Jersey: Pearson Prentice Hall.

Chan, C., Forwood, D., Roper, H., and Sayers, C. (2009). Public infrastructure financing: An international perspective, productivity commission. Staff working paper. Available from: https://core.ac.uk/download/pdf/30685486.pdf [cited 10 March 2020].

Chen, C. (2016a). Innovative infrastructure financing tools. In: Daniel L. Smith and Jonathan B. Justice (Eds.), *Encyclopedia of Public Administration and Policy*. New York: Taylor and Francis Group Press.

Chan, C. and Bartle, J.R. (2017). Infrastructure financing: A guide for local government managers. Public Administration Publications 77. Available from: https://digitalcommons.unomaha.edu/pubadfacpub/77/ [cited 14 April 2020].

Claus, I. and Grimes, A. (2003). Asymmetric information, financial intermediation and the monetary transmission mechanism: A critical review, Working Paper no. 03/19, New Zealand Treasury. Available from: https://motu.nz/our-work/wellbeing-and-macroeconomics/money-and-banking/asymmetric-information-financial-intermediation-and-the-monetary-transmission-mechanism-a-critical-review/ [cited 11 March 2020].

Dabla-Norris, E., Brumby, J., Kyobe, A., Mills, Z., and Papageorgiou C. (2012). Investing in public investment: An index of public investment management efficiency. *Journal of Economic Growth*, 17(3), pp. 235–266.

Dewar, J. and Puttock, C. (n.d). Project bonds: Growing liquidity for energy and infrastructure finance. Available from: https://www.milbank.com/images/content/2/5/v7/25919/Energy.pdf [cited 26 March 2020].

Engel, E., Fischer, R., and Galetovic, A. (2007). The basic public finance of public-private partnerships. Discussion Paper no. 1618, Cowles Foundation for Research in Economics, Yale University.

European Investment Bank (2013). The economic appraisal of investment projects at the EIB. Available at: http://www.eib.org/infocentre/publications/all/economic-appraisal-of-investment-projects.htm. [cited 26 March 2020].

European Bank (2018). The EBRD's Infrastructure project preparation facility. Available from: https://www.ebrd.com/infrastructure/infrastructure-IPPF.com [cited 14 April 2020].

Fay, M., Martimort, D., and Straub, S. (2018). Funding and financing infrastructure: The joint-use of public and private finance. Policy research working paper 8496. Available from: http://documents.worldbank.org/curated/en/176101530040441739/Funding-and-financing-infrastructure-the-joint-use-of-public-and-private-finance [cited 26 March 2020].

Fletcher, P. and Pendleton, A. (2014). Identifying and managing project finance risks: Overview (UK). Available from: https://www.milbank.com/images/content/1/6/v6/16376/5-564-5045-pl-milbank-updated.pdf [cited 25 March 2020]

Florio, M. (2006). Cost-benefit analysis and the European Union cohesion fund: On the social cost of capital and labour. *Regional Studies*, 40(2), pp. 211–224.

Hepburn, G., Pucar, M., Sayers, C., and Shields, D. (1997). Private investment in urban roads, industry commission staff research paper, AGPS, Canberra. Available from: https://www.pc.gov.au/research/supporting/urban-roads [cited 11 March 2020].

Hodges, J.T. and Dellacha, G. (2007). Unsolicited infrastructure proposals: How some countries introduce competition and transparency. Available from: http://documents.worldbank.org/curated/en/142981468777252745/Unsolicited-infrastructure-proposals-how-some-countries-introduce-competition-and-transparency [cited 12 March 2020.]

Gurara, D., Klyuev, V., Mwase, N., and Presbitero, A.F. (2018). Trends and challenges in infrastructure investment in developing countries. *International Development Policy*. Revue internationale de politique de développement. Available from: http://journals.openedition.org/poldev/2802; DOI: https://doi.org/10.4000/poldev.2802 [cited 18 May 2020].

IMF (2015). Making public investment more efficient. Available from: https://www.imf.org/en/Publications/Policy-Papers/Issues/2016/12/31/Making-Public-Investment-More-Efficient-PP4959 [cited 11 March 2020].

IMF (2020). The international monetary funding and infrastructure governance. Available from: https://www.imf.org/external/np/fad/publicinvestment/#1 [cited 11 March 2020].

IPCC, (2007) *Climate change 2007: synthesis report*. Contribution of Working Groups I, II and III to the fourth assessment report of the intergovernmental panel on climate *change* [Core Writing Team, Pachauri, R.K and Reisinger, A. (eds.)]. IPCC, Geneva, Switzerland, 104 pp. Available from: https://www.ipcc.ch/report/ar4/syr/ [cited 24 March 2020].

Kumari, A. and Sharma, A.K. (2016). Infrastructure financing and development: A bibliometric review. *International Journal of Critical Infrastructure Protection*, 16, pp. 49–65.

Li, S., Abraham, D., and Cai, H. (2017). Infrastructure financing with project bond and credit default swap under public–private partnerships. *International Journal of Project Management*, 35, pp. 406–419.

Martimort, D. and Pouyet, J. (2006). Build it or not: Normative and positive theories of public–private partnerships, Discussion Paper no. 5610, Centre for Economic Policy Research, Washington, DC.

Ochieng, E.G., Price, A.D.F., and Moore, D. (2017). *Major infrastructure projects: Planning for delivery*. Basingstoke, UK: Palgrave Macmillan's Global Academic. Available from: https://he.palgrave.com/page/detail/major-infrastructure-projects-edward-ochieng/?sf1=barcode&st1=9781137515858.

OECD (2015). Infrastructure financing instruments and incentives. Available from: http://www.oecd.org/finance/private-pensions/Infrastructure-Financing-Instruments-and-Incentives.pdf [cited 22 March 2020].

Papke-Shields, K.E. and Boyer-Wright, K.M. (2017). Strategic planning characteristics applied to project management. *International Journal of Project Management*, 33(3), pp. 638–649.

Peng, J. (2002). Do investors look beyond insured triple-A rating? An analysis of standard and poor's underlying ratings. *Public Budgeting and Finance*, 22(3), pp. 115–31.

Petrie, M. (2010). Promoting public investment efficiency: A synthesis country experience, paper presented for world bank preparatory workshop promoting public investment efficiency global lessons and resources for strengthening World Bank support for client countries (Washington, DC: World Bank).

PwC (2013). Capital markets: The rise of non-bank infrastructure project finance. Available from: https://www.pwc.com/gx/en/psrc/publications/assets/pwc-capital-markets-the-rise-of-non-bank-infrastructure-project-finance.pdf [cited 26 March 2020].

Ray, S. (2015). Infrastructure finance and financial sector development. ABDI working paper series. Available from: https://www.adb.org/publications/infrastructure-finance-and-financial-sector-development [cited 28 March 2020].

Sheppard, R., Klaudy, S.V. and Kumar, G. (2006). Financing infrastructure in Africa: How the region can attract more project finance. Available from: http://documents.worldbank.org/curated/en/122421468140367177/Financing-infrastructure-in-Africa-how-the-region-can-attract-more-project-finance [cited 11 March 2020].

Tyson, J.E. (2018). Private infrastructure financing in developing countries. Five challenges, five solutions. Available from: https://www.odi.org/publications/11168-private-infrastructure-finance-developing-countries-five-challenges-five-solutions [cited 10 March 2020].

UN (2017). Innovative ways to financing transport infrastructure. Available from: https://www.unece.org/fileadmin/DAM/trans/main/wp5/publications/ECE_TRANS_264_E_Web_Optimized.pdf [cited 25 March 2020].

United Nations Economic and Social Commission for Asia and the Pacific (UNESCAP). (2012). A new vision for public–private partnerships (PPP) in Asia-Pacific. Speech by Noeleen Heyzer at Ministerial Conference on Public Private Partnerships for Infrastructure Development, Third Session, Tehran, Iran. 14 November. http://www.unescap.org/speeches/new-vision-public-private-partnerships-ppp-asia-pacific [cited 26 March 2020]

Vives, A., Paris, A.M., Benavides, J., Raymond, P.D., Quiroga, D., and Marcus, J. (2006). *Financial structuring of infrastructure projects in public-private partnerships: An application to water projects*. Washington D.C.: Inter-American Development Bank. Available from: https://publications.iadb.org/en/publication/10977/financial-structuring-infrastructure-projects-public-private-partnerships [cited 11 March 2020].

WEF (2014). African strategic infrastructure initiative managing transnational initiative managing translational infrastructure programmes in Africa-challenges and best practices. Available from: http://www3.weforum.org/docs/WEF_AfricanStrategicInfrastructure_Report_2014.pdf [cited 10 March 2020].

WEF (2019) Improving infrastructure financing in Brazil. Available from: https://www.weforum.org/reports/improving-infrastructure-financing-in-brazil [cited 16 May 2020].

World Bank Group (2019). Public-private partnership for emerging market health. A briefing paper from the IFC public-private partnership (PPP) think tank discussion at the 2019 global private health care conference. Available from: https://www.ifc.org/wps/wcm/connect/industry_ext_content/ifc_external_corporate_site/health/publications/eiu+briefing+paper+-+ppps [cited 14 April 2020].

12

REGULATORY PROCESS FOR INFRASTRUCTURE SYSTEMS DEVELOPMENT

Nicholas Chileshe and Neema Kavishe

12.1 Introduction

12.1.1 Infrastructure regulatory frameworks

Despite the unprecedented level of infrastructure projects in developed and developing economics, there still remain a number of challenges affecting the delivery of infrastructure development and investment. In particular, the majority of developing countries in Africa have a number of their mega-projects funded by China. Some of the countries, particularly those in developing economies use the public–private partnerships (PPPs) strategy for the delivery and execution of the different types of infrastructure projects in sectors such as transportation, water, communications and social. More so, the success of the projects' delivery is heavily dependent on the supportive enabling environment in which these projects are executed and underpinned by strong regulatory and legal frameworks.

However, some developing countries have weak legal and regulatory frameworks of their contracts such as PPPs in infrastructure projects. Therefore, in order for the infrastructure to contribute to the economic output of a country, there is a need to review the regulatory frameworks and assess their effectiveness. Given the broader definition of infrastructure, this chapter focuses more on the sector of construction and transportation to not only illustrate and establish the mechanisms for recognising good and bad infrastructure in a number of selected sub-Saharan African countries, but also to identify and evaluate the prevailing political and regulatory risks and how they affect the overall success of the infrastructure projects. Some mitigating strategies in overcoming the identified risks are also suggested in this chapter. That allows the practitioners to draw upon what lessons can be learned, and tailor them to the local conditions. In addressing the weak legal frameworks, the chapter also reviews benchmarks for regulatory governance – key standards, the standards, procedures and tools – and explores how some of the robust infrastructure contracts have been applied in selected countries. Thus, the overarching aim of this chapter is to identify mechanisms for enhancing the effectiveness of the regulatory processes for infrastructure system development, through the auditing of the environment by evaluating political and regulatory risks, application of robust benchmarking evaluation systems, standards and contracts. The PPP contract is used as the lens for developing the infrastructure projects.

12.1.2 Chapter aim and objectives

The aim of this chapter is to examine how the regulatory process for infrastructure systems development can be streamlined. The main objectives are to:

- Explore the approaches to evaluating regulatory effectiveness;
- Establish mechanisms for recognising good and bad infrastructure regulations;
- Evaluate the political and regulatory risk;
- Articulate the benchmarks for key standards of regulatory governance;
- Review the standards, procedures and tools;
- Apply robust infrastructure and contracts; and
- Establish best practices in achieving international commitments.

12.1.3 Learning outcomes

The following learning outcomes have been identified for this chapter. Readers will be able to:

- Apply the approaches to evaluating regulatory effectiveness;
- Identify the major challenges to implementing regulations for infrastructure development;
- Identify mechanisms for benchmarking infrastructure regulations;
- Develop coping strategies for managing political and regulatory risks;
- Understand the critical success factors (CSFs) influencing the implementation of infrastructure regulations; and
- Appreciate different funding mechanisms for infrastructure projects in developing countries.

12.2 Approaches to evaluating regulatory effectiveness

According to one of major studies on this subject by Brown *et al.* (2006), evaluation of the regulatory effectiveness is an important attribute of monitoring and periodical review of their performance to ensure that there functional and on target in meeting the government set out objectives and aspirations. The evaluation also acts as a safeguard against potential corruptive activities which according to Githaiga *et al.* (2019) has the potential of increasing the long-term costs of infrastructure projects for countries involved. Studies have also shown that availability and effectiveness of proper regulatory and legal framework for PPPs are major critical success factors for infrastructure project delivery in the UAE (Al-Saadi and Abdou, 2016), Kenya (Chileshe *et al.*, 2020); Afghanistan and Turkey (Bayat *et al.*, 2019), and Ethiopia (Debela, 2019).

The chapter builds on the proposed three levels of evaluation, namely short-, mid-level, and in-depth, albeit from the electricity sector, and applies these to the housing and infrastructure projects such as ports and roads as undertaken in emerging economies, particularly African countries. The findings and discussions are based on data and information from literature review, the author's previous research findings mostly around PPPs within developing countries, and focused on the following areas: challenges, barriers, CSFs, regulatory issues, and capacity building initiatives. Most importantly, PPPs are used as the basis for the chapter as they are widely used in delivering infrastructure projects in both developed and developing economies (Cui *et al.*, 2018), and for promoting and accelerating public sustainable infrastructure in Africa (Cui *et al.*, 2018).

12.2.1 Regulatory effectiveness – the case of Tanzania

In order for the regulations to be effective, some government bodies must be constituted. From the PPPs perspective, the majority of developing countries have specific units designed to implement the regulations underpinning the PPPs. In Tanzania, according to the World Bank (2016), and Kavishe *et al.*, 2018), the PPP finance unit and PPP coordinating unit are responsible for the assessment and approval as well as the coordination of all PPP projects in Tanzania. These units have been involved in the formulation of PPP policy as well as the regulations. The PPP Coordination Unit was established by the *2010 PPP Act* within the Tanzania Investment Centre (TIC) to coordinate and oversee the mainland Tanzanian PPP projects and PPP Financing Unit within the Ministry of Finance with the duty of assessing and examining all PPP proposals in their financial aspects. The PPP units are generally accepted as institutional structures with oversights as those of an independent regulatory agency for infrastructure services (Mourgues and Kingombe, 2017). However, despite the existence of these PPP units, studies such as Kavishe *et al.* (2018) have reported that the PPP unit has little or no impact on the proposed infrastructure PPP projects despite their existence. This suggests a radical shift in how their regulatory effectiveness could be assessed. The following subsection identifies some best practices for assessing regulatory systems as undertaken in other strict industries such as Atomic and Space.

12.2.2 Components of regulatory effectiveness

According to Brown *et al.* (2006), the three most common forms of evaluation are cross country statistical studies, cross country descriptive analyses, and single country structured case studies. While the literature is replete with cross country statistical studies such as the World Bank (2016) report which highlights some of the countries with weak regulatory systems, the focus and approach of this chapter are rather based on the single country structured case study and findings as reported based on the authors' prior studies, and use a mixed approach evaluation which includes the following: 1) quick evaluation; 2) mid-level evaluation; and 3) in-depth evaluation. The rationale for selecting the single country structured case studies is that it's the most effective (Brown *et al.*, 2006). Therefore, following the recommendation suggested by Brown *et al.* (2006) the reported evaluation examines the following two basic dimensions of any regulatory system: *regulatory governance* and *regulatory substance*.

There also exists in literature a number of tools and a benchmarking index that assess the capacity of countries in Africa to carry out sustainable PPPs in infrastructure (EIU, 2015) While the focus of the "Infrascope analytical framework" as developed in 2009 was on capacity building, the Infrascope index addresses the following six categories and associated indicators with their weightings shown in parentheses: 1) legal and regulatory framework (weighted 25 percent); 2) institutional framework (weighted 20 percent); 3) operational maturity (weighted 15 percent); 4) investment climate (weighted 15 percent); 5) financial facilities (weighted 15 percent); and 6) subnational adjustment factor (weighted 10 percent). Based on the regulatory framework indicator, Tanzania was ranked sixth out of the 15 countries included in the EIU (2015) benchmarking exercise. This suggests that aspects related to the following are above average:

i. Consistency and quality of PPP regulations;
ii. Effective PPP selection and decision-making;
iii. Fairness/openness of bids, contract changes; and
iv. Dispute-resolution mechanisms.

The above discussion further reinforces the need for developing countries to have better cohesion amongst their PPP policies, guidelines, legal framework and procurement regulations which would be expected to act as a set of rules, with the power to manage and inform the actions of members of a society or organisation (i.e., PPP coordinating units, Tanzanian public and private organisations/stakeholders).

12.3 How to recognise good and bad infrastructure regulations

This sub-section builds on the approach as suggested by Brown *et al.* (2006) by discussing how to recognise "good" and "bad" elements of a regulatory system, with respect to both regulatory governance and regulatory substance. As with the previous sub-sections, the focus is on Tanzania based on the author's previous research, and reinforced with supporting literature.

12.3.1 Background to Tanzania

According to the World Bank (2016) and EIU (2015), Tanzania has a robust legal and institutional framework for the establishment and implementation of PPPs. The government approved a PPP policy in 2009, following which it promulgated the *PPP Act* in 2010 and the associated PPP Regulations in 2011. However, according to the EIU (2015), most African countries are faced with challenges in ensuring strong rules and regulations, as well as effective implementation. This is further exacerbated by the fact that the housing PPPs are still in an early stage in Tanzania primarily because of a lack of direct experience and inadequate new investment. To date, two public organisations, namely the National Housing Corporation (NHC) and the National Social Security Fund (NSSF), have used the PPP method for housing provision. Since the 1990s, NHC used the PPP approach in building development, but most of these partnership projects were not very successful. This was primarily due to the lack of an adequate PPP legal framework to guide the implementation of such projects, which delayed its progression (URT, 2009).

12.3.2 Good regulations

In order to understand whether the regulations are "good" or "bad," the terminology or meaning of "regulation" should be provided. This chapter draws on the earlier works of Brown *et al.* (2006) who stated that "regulation" means government-imposed controls on particular aspects of business activity. In Tanzania, the government has moved in enhancing the regulatory frameworks via PPP regulations in 2011 and the *Public Procurement Act 2011*. This is despite the existence of a number of challenges such as the relative infancy of rather complex PPPs and lack of experience across the stakeholder chain (Mboya, 2013).

In addition to following the best principles acknowledged in literature as being significant in the implementation of infrastructure projects such as PPPs, the Tanzanian government further included two important components to be established for enhancing the regulatory framework: a PPP unit to oversee the implementation activities and a legal framework (law, policies and regulations). However, despite the establishment and inclusion of the PPP unit, the Tanzania units are considered as newer start-ups (Mbaya, 2013). Some examples of good regulatory actions are as follows:

- To facilitate, improve and promote private sector participation in PPP projects in Tanzania, in 2009 the government of Tanzania issued a PPP policy, then in 2010 the act was passed, and in 2011 the PPP regulations were approved (Mboya, 2013).

The importance of good regulations is also acknowledged in a number of developing economies such as Malaysia (Abdul-Aziz and Kassim, 2011; Abdullahi and Aziz, 2011) and India (Sengupta, 2004; 2006). For example, within the Indian PPP context, Kolkata has been successful in adopting PPP in housing in terms of cost and quality because its government focused on appropriate regulations rather than rapid changes (Sengupta, 2004; 2006).

12.3.3 Examples of bad regulations

The following are pointers and examples of "bad" regulations based on the interviewee's responses to the question around the challenges and problems they encountered during PPP implementation.

- **Lack of competitive processes:** The National Housing Corporation (NHC) PPP process was non-competitive, and it depended on the ability of the private partner to submit a quick proposal.
- **Poor PPP contracts and tender documents:** The Tanzanian PPP practitioners reported that some the agreements used in the transactions had no exit clause. Generally, these agreements had determination clauses that provide for the circumstances upon which agreements can be determined. It was observed that some partners were given more projects to add on to the already awarded projects, meaning that there was double allocation of projects without following the necessary procedures. There has been a trend among partners to add more plots on the acquired project on the pretext of expanding the magnitude of the projects.
- **Inadequate legal framework:** This arose in case where the provisions in the agreements are badly crafted to the extent that they contradict each other and distort the whole meaning of their presence in the agreement. This has led to Kavishe *et al.* (2019) suggesting that current PPP policy and guidelines need further improvement
- **Contradictory provisions:** For this reason, the World Bank (2016) provides a detailed summary and a non-exhaustive sample of general legislation such as PPP laws/Concession laws as well as sector-specific legislative provisions such as regulatory framework for PPPs in infrastructure such as roads.
- **Agreements biased in favour of (some) partners:** The majority of developing countries have skill capacity issues. While PPPs overcomes that with the private sector bringing in the desired skills capacity, this potential results in unbalanced contractual agreements (Rothballer and Gerbert, 2016).
- **Non-adherence to the rules and regulations:** The following example is drawn from Kavishe and Chileshe (2019), a study on PPP implementation challenges. Some interviewees stated, "It has been discovered that there are some projects which are run without adhering to

the rules and regulations set by regulatory authorities. In most, the construction site is quite different from the agreed and authorised design. This once proved, might engage the corporation in ordering the private partner contractor to stop, demolish, or be fined."

- **Inadequate PPP policy and legal institutional framework:** This could manifest itself in "poor procedure for determining financial capacity of private partners" (Kavishe and Chileshe, 2019). For instance, previous studies in Tanzania have shown that the lack of an adequate PPP legal framework to guide the implementation of projects has delayed progression of projects (EIU, 2015; URT, 2009). The same conclusions have been drawn in other developing economies such as Thailand (Trangkanont and Charoenngam, 2014).

12.3.4 Best practice, recommendations for good governance

Having identified the need of good governance and examples of bad government, this subsection highlights a few examples of the best practice required for maintaining the status quo of the "good regulations" while negating the bad regulations by providing some remedial solutions. Most are drawn from the studies undertaken by the authors from selected literature such as Osei-Kyei and Chan (2017) and World Bank (2016). The study findings reinforce the need for practical solutions tailored to local or host environment contexts. For instance, based on the advocated solutions, the government could strengthen and enhance the quality of its existing organisations such as the PPP coordinating and financing units as vehicles for improving the regulatory environment[23.] Other best practice includes the following:

- Transparency in procurement process;
- Trust between parties;
- Constant communication and monitoring; partner's compatibility;
- Formation of PPP facilitation funds;
- PPP training;
- Formulation of clear contracts;
- Proper checks and balances;
- Provision of enabling environment through tax holidays for investors and private developers;
- Strong commitment from both parties; and
- Ability of the regulatory and legal frameworks to sufficiently allocate responsibilities across institutions.

12.4 Managing political and regulatory risk

Infrastructure projects operate in varying and occasionally volatile and uncertain environments. As part of the risk management (RM) process, one requirement expected of businesses prior to seeking new project or operating in any environment is to undertake a scan of that environment through a SWOT (strength, weaknesses, opportunities and threats) or PESTLE (political, economic, sociocultural, technological, legal and environmental) analysis (Chileshe *et al.,* 2013). Inherent in that environment will be a number of risks, of which the political and regulatory risks forms part of the PESTLE analysis. According to Cui *et al.* (2018), within the developing economies, the political and regulatory risks still exist during the construction stage. This is in contrast to the developed countries where the risks disappear or are mitigated due to the prevailing sophisticated legal frameworks.

12.4.1 Conceptualisation of risks

Before discussing the necessary management of regulatory and political risks influencing the infrastructure projects, and the identification of the potential strategies for mitigating these risks as well as the desirable CSFs for implementing RM strategies, it is necessary to define the following terminology: "risk" and "risk management (RM)t."

Risk: According to Xiaopeng and Pheng (2013), cited in Jiang *et al.* (2019), political risk in international construction projects is defined as "the risks of adverse consequences arising from political events or government action known as 'intervention,' and the risks of change or discontinuity in the business environment as a result of political change." The Project Management Institute (PMI) (2017) defines project risk as "an uncertain event or condition that, if it occurs, has a positive or a negative effect on a project objective." In the construction context, risk is the likelihood of the occurrence of a definite event/factor or combination of events/factors which occur during the whole process of construction, to the detriment of the project (Wang et al., 2004 cited in Chileshe *et al.*, 2013).

Risk management: According to Last (2001, cited in Thomas *et al.,* 2003), RM involves deciding what an acceptable risk is, how the risk level can be reduced to an acceptable level, and monitoring the reduction in risk after exposure control actions have been taken.

Figure 12.1 illustrates the process of managing risks. As suggested by Chileshe *et al.* (2013), it can also be interpreted from the external and internal environment context. The first step of establishing the infrastructure project context is critical as a number of threats and opportunities exist. While the focus is political and regulatory risks, these need to be understood within the context of the prevailing challenges. For instance, due to the unprecedented level of infrastructure projects in Africa, including mega-projects funded by China, a number of challenges affect the delivery of infrastructure development and investment in Africa. For more detailed explanations around these challenges, Chapter 14 discusses these with examples from developing and developed economies.

These challenges include the regulatory process for infrastructure delivery, of which their quality [of regulations] affects the level of private investment in infrastructure (Moszoro *et al.,* 2016). Other challenges identified in literature include those of political instability, and other economic factors like weak infrastructure in some African countries are among the questions that prey on people's mind when thinking of investing in Africa (Mehta, 2015). According to the World Bank (2016), Tanzania's infrastructure and social services deficits are already massive and projected to increase in the future. However, some authors such as Luiz (2010) have pointed to how the delivery of infrastructure projects in Africa requires the capacity to deliver massive, complex projects in an efficient manner. For instance, only 38 percent of the African population has access to electricity, the penetration rate for internet is less than 10 percent, and only a quarter of Africa's road network is paved (Mayaki, 2014).

There is also the significance and importance of infrastructural development nested within its impact on raising the productivity of humans and physical capital leading to economic growth (Mehta, 2015). For example, infrastructure spending in Africa is estimated to reach $93 billion per year, facilitated by tax revenues and other domestic resources (Mehta, 2015); within sub–Saharan Africa, infrastructure spending is estimated to reach US $180 billion per annum by 2025 (PwC, 2014). On the other hand, many developing countries suffer from poor infrastructure (Sharma, 2012), and in particular, East Africa has met a number of transports infrastructure challenges in the past years (CCE News, 2018). Therefore, in understanding the regulatory and political risks affecting the delivery of infrastructure projects, the following activities, similar to what is

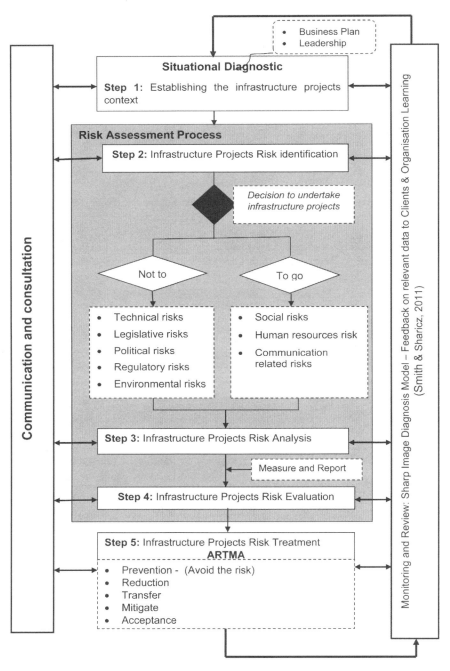

Figure 12.1 Diagnostic infrastructure project risk management
Source: Chileshe *et al.* (2013)

undertaken during the situational diagnostic, would need to be considered (Searcy, 2009). These include understanding the need for infrastructure projects within the specific country (context), surveying the internal environment, and surveying the external environment.

12.4.2 CSFs for managing RM practices for infrastructure projects

The connections between CSFs and RM as a precursor for effective delivery of infrastructure projects through PPPs are beginning to emerge in literature (Cui et al., 2018). Therefore, in this chapter, the diagnostic infrastructure project RM as illustrated in Figure 12.1 is proposed. This framework follows the well-established steps for undertaking RM processes. The importance of CSFs and effective RM is also gaining traction amongst scholars in developing countries. For instance, given the significance of the political and regulatory risks, some studies have been undertaken to identify the CSFs associated with managing infrastructure projects in Africa, as well as coping mechanisms and strategies for dealing with RM practices (Chileshe, 2016). The same study highlighted the synergies between effective implementation of RM and engagement of best practices of infrastructure projects, leading to investment and development. RM is also effective in addressing time and cost overruns associated with infrastructure projects (Van Thuyet et al., 2007). Studies have also shown correlations between the aggregate project risk and cost performance of the projects (Ogbu and Adindu, 2019).

Some coping mechanisms and strategies for dealing with RM practices are also suggested in literature (Chileshe, 2016). The political risk normally associated with government intervention is associated with and mostly prevalent during the prevalent phase of PPP related projects as well as infrastructural projects (Grimsey and Lewis, 2002; Kumaraswamy and Zhang, 2001; Li et al., 2005; Zou et al., 2008). Within the proposed framework, mitigating strategies would be undertaken in Step 5. The following are selected RM related bottlenecks in managing infrastructure projects in Africa identified by (PwC, 2014):

- Lack of skills and RM knowledge;
- Availability of specialist risk consultants;
- Internal capacity to handle major capital projects, political risk and interference; and
- RM implementation costs.

In addition to the above, PwC (2014) further identified the following four obstacles affecting African countries:

- Political instability;
- Policy incoherence;
- Reported corruption; and
- Legal uncertainties.

It should also be pointed out that, as with many developing countries, the RM concepts are relatively new to Tanzania (Chileshe and Kikwasi, 2014a). The political and regulatory risks affecting the infrastructure projects could be managed by the broader RM approaches and techniques as defined by the Project Management Body of Knowledge (PMBOK). However, there is a need to underpin these RM principles and implementation techniques with CSFs. However, while these suggestions are evident in literature, and in practice for traditional projects, caution must be undertaken in implementing these solutions. Studies have shown that the regulatory function is in practice rather complex since it requires balancing a multiplicity of other objectives or goals, which may vary according to specific economic conditions (Carmona, 2010). The following CSFs as identified from a number of studies undertaken in a number of selected sub-Saharan African countries such as Zambia (Manelele and Muya, 2008), Tanzania (Chileshe and Kikwasi, 2014a; 2014b), and Ghana (Agyakwa-Baah and Chileshe, 2010) are suggested as the basis for identifying the coping strategies and lessons learned:

- Management style;
- Awareness of RM processes;
- Co-operative culture;
- Positive human dynamics;
- Customer requirements;
- Goals and (strategic) objectives of the organisation;
- Consideration of external and internal environment;
- Effective usage of methods and tools;
- Teamwork and communication; and
- Availability of specialist RM consultants.

Based on the discussions around the political and regulatory risks, challenges and CSFs, both the government and policymakers could use the findings as the basis for re-examining the existing PPP policies and regulations. They could also reflect on the existing situation with a view to improving the delivery of future soft and hard infrastructure projects. According to Cui *et al.* (2018), proper risk allocation contributes to the enhanced performance of PPP projects. However, these risks such as legal and political risk (direct and indirect) are prevalent throughout the infrastructure's project life cycle (Thomas *et al.*, 2003); this is affected by the inability of practitioners in developing countries to identify and manage project risks, which has hindered the success of PPPs (Zhang, 2005).

12.5 Benchmarks for regulatory governance key standards

This sub-section builds on the regulatory and political risks to identify and mitigate the risks. Benchmarking is identified as one of the mechanisms for assessing the effectiveness of the regulatory systems (Brown *et al.*, 2006). The same study by Brown *et al.* (2006) describes in detail the need for benchmarks and description and justification of the independent regulatory. The need for benchmarking is also quite evident in PPP literature. For instance, according to Caselli *et al.* (2015), governments in developing and developed economies should start by reviewing and benchmarking their PPP policies and processes against a best-practice checklist, as presented within the same study, and identify those areas most in need of change. Therefore, the focus of this sub-section is to build on the widely accepted principles for the independent regulator model of regulatory through the mapping of the descriptors to the Tanzanian case study. According to Brown *et al.* (2006), the following ten principles are important:

- Independence
- Accountability
- Transparency and public participation
- Predictability
- Clarity of roles
- Completeness and clarity in rules
- Proportionality

Regulatory process for infrastructure systems development

- Requisite powers
- Appropriate institutional characteristics
- Integrity

12.5.1 Benchmarking principle 1: Independence

Brown *et al.* (2006) identify independence and accountability of the regulator as one of the two important dimensions of regulation. The World Bank (2016) also attributes institutional and regulatory framework as critical for the success of PPP programs. Therefore, for the effectiveness of the regulations supporting the delivery of infrastructure projects, these should be managed by an independent regulator. Therefore, the World Bank (2016) recommended for Tanzania to establish a strong and capable central PPP unit. However, despite its independence, the impact on PPP projects is questioned[10]. The study also showed that the PPP unit has been ineffective and underutilised. Likewise, the location of the Tanzanian PPP units within government institutions raises questions about their independence and possible conflict of interest (Mourgues and Kingombe, 2017). This suggests that PPP regulations are not enforced, and there was evidence of a number of proposals that were initiated without being submitted to PPP units for assessment and approval due to the fear of government bureaucracy. This finding further suggests that for the regulators to be fully independent, adherence and implementation of approaches to evaluating regulatory effectiveness as described in section 12.2 are paramount.

12.5.2 Benchmarking principle 2: Accountability

In sub-section 12.3, the aspects of identifying "bad" and "good" governance were identified and discussed. Most importantly, the role of the regulator was acknowledged by Brown *et al.* (2006); amongst others was the need to make decisions about the independence and accountability of the regulator and the relationship between the regulator and policymakers. In developing countries, this function is underpinned by the legal and regulatory frameworks. Accountability is also among the main and critical reasons for attracting private investors (Moszoro *et al.*, 2016). Using Tanzania as an example, the PPP unit acts as the regulator and evidence of its accountability was demonstrated in the Kavishe *et al.* (2018) study, which proposed a PPP conceptual model for housing projects comprising the following five phases: preparation, planning, procurement, building and operating. One of the functions within the last phase was monitoring and discharging by the PPP units. The two-tiered monitoring mechanism provided a double check, ensuring effectiveness and accountability. Finally, output specifications can be used to measure project performance.

12.5.3 Benchmarking principle 3: Transparency and public participation

According to Brown *et al.* (2006), the principle of "transparency and public participation" implies that "the entire regulatory process must be fair and impartial and open to extensive and meaningful opportunity for public participation." This principle is very important as studies have shown that about 10–30 percent of real value in infrastructure projects is lost through corruption and non-transparent activities (Rothballer and Gerbert, 2016). In addition, studies such as Githagia *et al.* (2019) have linked security risks and corruption to an increase in the long-term costs of infrastructure projects. Therefore, adherence to the principles enables the mitigation of the pervasiveness of corrupt practices in infrastructure project procurement and execution (Kingsford Owusu and Chan, 2019). Other supporting evidence of this principle can be found in the studies undertaken by Chileshe *et al.* (2020) around the Kenyan infrastructure projects and Kavishe and Chileshe (2019) in Tanzania, which found the CSF of "transparency and equity in the procurement process" among the most important for infrastructure projects. Likewise, in Malaysia, Abdel Aziz (2007) identified process transparency and disclosure as one key principle of PPPs.

12.5.4 Benchmarking principle 4: Predictability

As with other aspects associated with identifying "bad" and "good" governance, the benchmarking principle of "predictability" is one of the important dimensions of regulation[1] The expectation of the regulator is the provision of reasonable, although not absolute, certainty as to the principles and rules (Brown *et al.*, 2006, pg. 60). Predictability is also identified as important for effective regulation and stability of PPPs (Alloisio and Carraro, 2016).

12.5.5 Benchmarking principle 5: Clarify roles

According to the Country Profile on PPPs in Tanzania, the legal and institutional framework, the *General Principles of the Art. 4 of PPP* Act are underpinned by the following:

- Principles of fairness;
- Equity of treatment;
- Transparency; and
- Competitiveness and cost effectiveness of procurement processes.

The clarity of roles for the two Tanzanian entities charged with overseeing the PPP framework is very clear. The PPP Coordination Unit (PPP CU), which was established by the *2010 PPP Act* within the Tanzania Investment Centre (TIC), coordinates and oversees the mainland Tanzanian PPP projects. It's also responsible for providing advice to contracting authority on PPPs, developing guidelines in all PPP matters, and raising PPP awareness in Tanzania. The PPP Financing Unit (PPP FU) is located within the Ministry of Finance and has the duty of assessing and examining all PPP proposals in their financial aspects, including fiscal risks.

12.5.6 Benchmarking principle 6: Completeness and clarity in rules

According to Brown *et al.* (2006, pg. 61), this principle requires the need for the regulatory system, through laws and agency rules, to provide all stakeholders with clear and complete timely advance notice of the principles, guidelines, expectations, responsibilities, consequences of misbehaviour, and objectives that will be pursued in carrying out regulatory activities. However, adherence to this principle appears problematic for some developing countries (Kavishe *et al.*, 2018; Mourgues and Kingombe, 2017). For instance, within the Tanzanian context, Kavishe et al.[10] established that the majority of the stakeholders were not aware of the benefits that arise from undertaking PPPs. Likewise, Mourgues and Kingombe (2017) established that PPPs suffer from low transparency and limited public scrutiny, which undermines democratic accountability. This indicates a greater need for the PPP units to create more awareness on the PPP benefits in order to encourage both private and public sectors to collaborate.

12.5.7 Benchmarking principle 7: Proportionality

Brown *et al.* (2006) suggest the regulatory powers need to have an array of powers and remedies, but this is not the case in the majority of developing countries. Using the case of Tanzania, the main PPP regulatory units, namely the PPP Coordinating Unit (CU) and PPP Financial Unit (FU) have limited powers with no impact on the proposed infrastructure PPP projects despite their existence. This was discussed at length in sub-section 12.2 under the evaluation of the effectiveness of regulatory systems.

12.5.8 Benchmarking principle 8: Requisite powers

This principle calls for the regulatory agencies, under the respective county laws, to possess all powers required to perform their mission (Brown *et al.*, 2006). However, as observed by Mourgues and Kingombe (2017), this principle can only flourish dependent on the size and structure of government, and largely to the extent to which investment decision-making powers are devolved. The existence of two PPP separate units in Tanzania rather limits the functions to that of an advisory role.

12.5.9 Benchmarking principle 9: Appropriate institutional characteristics

This principle is underpinned by the need for education or training opportunities for commissioners and staff that are comparable to what is available at regulated entities (Brown *et al.*, 2006), especially for the developing countries that face a number of challenges around capacity building.

12.5.10 Benchmarking principle 10: Integrity

This principle refers to the rules that decision-makers associated with infrastructure projects should be adhering to and in so doing avoid undertaking improper activities (Brown et al., 2006). However, while most countries put in measures to safeguard against corruptive activities as described in principle #6, PPP contracts and activities by their very nature are quite complex and prone to integrity issues. For example, Stern (2020) points to the number of contractual links (over 1,000) contained within PPPs; each link is dependent on other contracts in the chain, which raises opportunities for bribery in the quest to break the chains within the contractual links.

12.6 Reviewing standards, procedures and tools

One of the major challenges facing developing and developed countries is the harmonisation of laws, procedures and practices across government institutions (EIU, 2015). However, the establishment of a comprehensive and transparent fiscal accounting and reporting standard for PPPs is acknowledged as one of the components of enabling institutional frameworks for PPPs (EU, 2015). The following lists key principles that might help policymakers and practitioners in the management of infrastructure projects (Abdel Aziz, 2007; Stern, 2020). The following examples are from Malaysia and Tanzania:

- Good practices must be maintained throughout the life of the project;
- PPP process transparency and disclosure;
- Standardisation of PPP procedures and contracts;

- Performance specifications and method specifications;
- Availability of a PPP institutional/legal framework;
- Availability of PPP policy and implementation units;
- PPP infrastructure projects should be carefully planned and clearly defined in scope, size and objectives; and
- Competitive, transparent and fair tendering process.

As an illustration for one of the principles, PPP process transparency and disclosure, there should be a thorough, inclusive, and transparent procedure for procurement of PPP projects in transportation infrastructure (Bengesi *et al.*, 2016). The public procurement process is too bureaucratic and opens up an avenue for corrupt transactions by some dishonest public officials who normally inflate costs. There is also no comprehensive and transparent procedure to procure PPP projects.

12.7 Infrastructure contracts and achieving international commitments

According to the World Bank (2016), infrastructure projects and associated PPP contracts often have significant ongoing financial implications for governments. Therefore, these contracts must be assessed and managed appropriately. This suggests that Tanzania has not done enough to invest in infrastructure development possibly due to the traditional approach of relying solely on public funding in infrastructure development and failing to tap the existing potential from the private sector (Bengesi *et al.*, 2016). Among other interventions geared towards fast tracking Tanzania's economic growth includes the PPP funding mechanism, which has been established to particularly complement government resources to invest in the transportation infrastructure (Bengesi *et al.*, 2016).

According to the CEE News (2018), the majority of East Africa relies on external funding for most infrastructure projects. The region's project funding is dominated by China (20.9 percent; down from 25.9 percent in 2019), while governments account for 13.7 percent of total funding. In addition to China, the international and African direct foreign Iivestments (DFIs) also play a significant role in East Africa's project funding, accounting for 13.2 percent and 12.6 percent, respectively. The funding, as well as foreign aid relationships, is valuable to donors as a means of improving development outcomes and influencing recipient country policy (Zeitz, 2020).

Therefore, securing and achieving international commitments is crucial for the successful execution of the infrastructure projects. This situation is not unique to Tanzania and affects the majority of African nations. For example, as recent as 2015, African nations have renewed their efforts to accelerate their priority projects under the Programme for Infrastructure Development, which began in 2012; it needs US \$68 billion by 2020 and an additional US \$300 billion for projects planned to 2040 (Davison, 2015). There is also support from international funding bodies such as the World Bank for continued investment in Africa.

However, demonstrated by previous studies such Chileshe (2016), inadequate funding and revenue problems are amongst the challenges identified as inhibiting the managing of infrastructure in Africa. Other sources of funding for infrastructural projects come from donors such as the World Bank. For instance, to improve the competitiveness of the East African Community (EAC) states, in 2014 the World Bank committed \$1.2 billion to support their infrastructure development (World Bank, 2014). In contrast, in developed countries (e.g., the US), the transportation projects are funded by taxpayers, despite the infrastructure sector contributing approximately 42 percent of the total expenditures on public construction projects (El Adaway *et al.*, 2019). To highlight the significance of funding and the role played by China in funding the majority of mega-infrastructure projects in Africa, Table 12.1 presents a summary of eight of the 10 top

Regulatory process for infrastructure systems development

Table 12.1 Summary of eight of the 10 mega-infrastructure projects in Africa funded by China

No.	Country	Cost (US dollars, in billions)	Description
1	Nigeria	$12	The railway project is 1,402 km in length and upon completion, it will link Lagos, the nation's economic capital, with the eastern city of Calabar, passing through 10 states.
2★	Tanzania	$10	**Coastal railway:** The port is being built in Bagamoyo, a coastal town in Tanzania. It will handle about 20 million containers annually and will be the largest port on the East African coastline, bigger than Mombasa in Kenya. Its construction started in October 2015 but was halted earlier this year due to financial constraints facing the Tanzanian government.
3	South Africa	$10	**Modderfontein new city project:** A Chinese company, Shanghai Zendai, is building the city that will be home to at least 10,000 residents upon completion. The city will have finance and trade facilities, an industrial zone, sports and recreation facilities and an African heritage theme park.
4	Kenya	$3	**Standard gauge railway:** This is the biggest infrastructural project in Kenya since independence. It is worth $3.8 billion. China Exim Bank has funded 85 percent of the project (about $3.1 billion). Construction on the 609 km rail line began in October 2013. The first phase, connecting Mombasa to the capital, Nairobi, was set to be completed by December, 2017. The railway line will ease transport of passengers and cargo between the two cities. It is part of a modern standard gauge railway that will connect Kenya to Uganda, Rwanda and South Sudan.
5	Congo	$6	**Infrastructure for mines barter deals:** The deal was to develop the minefields in Mashamba and Dima basins and Kolwezi. In return for the loan, the Democratic Republic of Congo was to supply copper mines, with approximately 10.6 million tonnes of copper for exploration and mining by Chinese companies.
6	Chad	$5.6	**Chad–Sudan railway:** The 1,344 km railway is being constructed in three phases and will also link the two nations with Cameroon. Its constructions started in October 2014. China Export-Import Bank has funded $2 billion with the rest coming from Chinese loans from government.
7	Nigeria, Ethiopia, Kenya, **Zambia**, Senegal, Mali, Cameroon and Ivory Coast	$4.34	**Dangote PLC cement expansion:** This project will increase cement production by 25 million metric tonnes and boost the overall production to over 70 million metric tonnes annually.
8	**Mozambique**	$3.1	**Mphanda Nkuwa dam and hydroelectric station project:** The project also covers the construction of Moamba-Major dam that will supply drinking water to the residents of Maputo.

Source: Original adapted from (PWC, 2014) The countries in bold (also Angola) registered the fastest growth rates in 2013, exceeding SADC's seven percent growth rate target.

projects in Africa. This is indicative of the massive infrastructure development fuelled by mega projects.

12.8 Chapter summary

As demonstrated in this chapter, with the unprecedented need for the delivery of infrastructure development and investment in both developed and developing economies, there is a need for a better understanding of the regulatory process needed for the delivery of the infrastructure systems development. However, due to the varying conditions and challenges affecting the internal and external environments in which the construction industry operates, some countries have weaker or stronger legal and regulatory systemsm which affect the delivery of the infrastructure projects. This is evidenced through a number of mega and infrastructure projects failing to eventuate or terminating early. At the national level, and using Kenya as an example, it was found that when PPP contracts were used as a vehicle for developing infrastructure, a number of CSFs such as "acceptance and support given by the community," "project feasibility," "the laws, regulations and guidelines put in place," "available financial market" and "having a well organised and committed public agency" could enhance the delivery of the PPPs and subsequent infrastructure projects.

In order to have a better understanding of whether the regulatory mechanisms are achieving the desired results and contributing to the effective delivery of the infrastructural projects, there is a need for monitoring and periodical review of their performance to ensure that they are functional and on target in meeting the government's objectives and aspirations. This can only be achieved by undertaking an evaluation of their regulatory effectiveness. The chapter uses existing approaches of evaluating regulatory effectiveness to highlight how these might have been undertaken in selected cases in developing countries.

Having regulatory agencies for monitoring the existing frameworks is desirable for successful implementation of infrastructure projects. However, this should be accompanied by the establishment of mechanisms for recognising good and bad infrastructure regulations. While effective mechanisms are evident in literature and to some extent well practiced in developed countries, this is nevertheless problematic in developing countries due to a myriad of challenges. These have ranged from lack of capacity building to weaker legal and regulatory systems. This chapter used some examples of previous studies undertaken by the authors to highlight some "bad" and "good" regulations. A detailed description of these PPP implementation challenges is further provided by the same authors in Chapter 14 on "empowering public-private partnership in major infrastructure systems." From the studies and literature review, transparency in the procurement process, trust between parties and training appeared to be among the best practices.

In order to make informed decisions around undertaking the infrastructure projects, there is a need to evaluating the political and regulatory risks. The main purpose of the evaluation is to make decisions, based on outcomes of the risk analysis, about which risks need treatment, whether an activity should be undertaken, and treatment priorities. However, while the process and practise of RM is not new, the level of awareness and implementation is nevertheless still at infancy levels in developing countries, particularly when using PPP contracts in infrastructure projects. In addition to implementing RM practices, the chapter demonstrated the need for articulation of the benchmarks for regulatory governance key standards; reviewing of the standards, procedures and tools; application of robust infrastructure and contracts; establishment of best practices in achieving international commitments; and understanding the policymaking, ownership and operations. The findings from this chapter can be used by practitioners, policymakers and governments, especially those in developing countries, as a basis for re-examining the existing

PPP policy and regulations, and for reflecting on the existing situation with a view to improving the delivery of future infrastructure projects.

12.9 Chapter discussion questions

1. The successful approaches in evaluating regulatory effectiveness vary across the developing countries due to the different prevailing regulations and legal systems. In addition, the following two groups of government, and regulatory bodies, have oversight over the elements that contribute to that effectiveness. Using case examples, discuss these approaches by highlighting some the regulatory indicators as implemented from the two groups.
2. The enabling environments in which PPPs are widely used to delivery infrastructure projects is different in both developed or developing (emerging) economies. How then can the political and regulatory risks affecting the infrastructure projects be identified and assessed effectively? What kind of considerations need to be taken into account when developing coping strategies that can not only be effective but equally most suited for the developed or developing countries?
3. Why is effective project RM an important factor in streamlining the regulatory process for infrastructure systems development?
4. Why is the application of integrity and transparency principles important in establishing mechanisms for recognising good and bad infrastructure regulations? Discuss how the integrity and transparency at each phase of the infrastructure cycle can be applied.
5. The application of benchmarking and lessons learned as a solution for infrastructure project (i.e., PPP) capacity building is acknowledged as being limited in sub-Saharan countries. Identify some prevailing mechanisms for articulation of the benchmarks for regulatory governance key standards currently in place in your country or you are familiar with.
6. As with most developing economies, different standards ensure that the infrastructure project contributes to the development of the economic, social, academic, moral and environmental backgrounds of their respective societies. Discuss the existing compliance measures you are aware of.
7. Inadequate funding and revenue problems are identified among the major challenges affecting the development and management of infrastructure projects in most developing countries. Propose some robust funding and financing frameworks for such projects.
8. Execution of contracts associated with infrastructure projects is often fraught with conflicts, claims and disputes between the recipient of the funding and investors. Suggest effective measures for regulating these contractors particularly in developing countries where legal frameworks are not fully established or robust.

12.10 Case study: Insights into Tanzania's delayed infrastructure project planned, delayed and suspended: Bagamoyo Port project

The coastal railway that was designed to be built in Bagamoyo, a coastal town in Tanzania, was scheduled to start its construction in October 2015, and it was initially planned to take two years to complete, including rail and road links (Muthengi, 2015). However, this was eventually halted in 2019 after protracted delays due to financial constraints facing the Tanzanian government. It would have been able to handle about 20 million containers annually and was intended to be the largest port on the East African coastline, bigger than the Mombasa in Kenya. One reason for the project was to set up a special economic zone to transform the country into a regional business hub. As shown in Table 12.1, initially, the project had a projected cost of about US $11 billion and got most of the funding for construction from the China Merchants Holdings International Company and State Reserve General Fund of Oman (African Business, 2017). The table further shows the influence of China in managing and funding the majority of mega projects in developing countries such as those in Africa.

12.10.1 The China factor

The significance of the "China factor" is evidenced by its involvement within the construction activities in Africa. For instance, according to CEE News (2018), the domination of China in these activities is mostly through funding; it was responsible for building 40 percent of the projects (down from 54.7 percent the previous year). In contrast, both private domestic companies and European Union countries construct only 14.8 percent of projects. Furthermore, according to Mutiso[61], China has also risen to become the single largest trade partner for many African countries. The same report by Mutiso (2016) acknowledges China as a major source of financial support for various development projects being undertaken on the continent. According to the study by PwC (2014), China's involvement in the region spans project funding to direct construction. The underpinning rationale being that infrastructure development is a key pillar of the China–Africa relationship. However, despite the identified benefits of the planned Bagamoyo Port such as the transformation of the economy and putting Tanzania on par with regional business competition Muthengi (2015), a number of issues delayed the project. These were procurement and contract related issues; political risks and investor conditions were reasons put forward for the lack of progress and near suspension of this infrastructure project. These are now discussed within the context of prevailing literature and the case study.

12.10.1.1 Bilateral agreements

Sub-section 12.3.3 identified some examples of "bad regulations" with agreements which favour one partner amongst them. While the failure of this case study was not around that aspect, this was more due to the failure to honour the "bilateral agreement" due to a number of factors. Studies such as Vhumbunu (2019) highlighted how China had been engaging in bilateral agreements and memorandums of understanding (MoUs) with African countries in their individual capacities and constructing cross-sectoral infrastructure projects at the national level mostly through its state-owned enterprises (SOEs) within the construction sector and financial services sector.

12.10.1.2 Procurement procedures for infrastructure project

According to Ng'wanakilala and Merriman (2015), the tendering and procurement procedures for these infrastructure processes were through a consortium of Chinese railway companies led by China Railway Materials (CRM), which was picked to help Tanzania build the railway line. Unlike the PPP projects which use joint ventures (JVs) (Kavishe et al., 2018), the Bagamoyo Port project used the build, own, operate, transfer (BOOT) model of procurement, where the investors had to construct, own and run the project for an agreed period before transferring it to the government (The Citizen, 2019). Such types of contracts are common for infrastructure projects in developing countries, and have been used in Vietnam.[65] However, as further acknowledged by Bengesi et al. (2016), this was also one of the Tanzanian government's attempts to initiate a PPP model in the transportation infrastructure through a build, own, operate and transfer model in which the government engaged a project of building the biggest port in East Africa in the coastal tourist town of Bagamoyo.

However, this appears to be one of the reasons leading to the suspension of the project as the Tanzanian government believed the procurement approach barred the Tanzanian government from making any other new port development plan within its coastal waters for a certain period of time. The view was that proceeding with the project with the prevailing conditions as set by the investor would be tantamount to selling Tanzania to China.[64] The discussed status quo aligns with Swedlund (2017), who questioned whether the growing presence of Chinese financial assistance is decreasing the bargaining power of traditional donors vis-à-vis African governments. However, it's still a departure from Xu *et al.*'s (2019) assertion that Chinese aid projects are positively correlated to Africa's economic development.

12.10.1.3 Contractual issues

The case study highlights some misunderstanding around the interpretations of the contractual arrangement. The main sticking point in this case appeared to be the time period the investors were given to run the facility, as expected of the BOOT procurement contracts, and the overall project terms (Mwangasha, 2019). The second appeared to be that of the transparency associated with the process in terms of how the investors recoup their money, which will have to be agreed upon by both parties (*The Citizen*, 2019). As a result of the identified contractual issues, both sides (the Tanzanian government and Chinese investors) were unable to come to a legally binding agreement (*Business News*, 2019). The findings and lessons from the Bagamoyo Port case are also consistent with literature regarding construction disputes. For instance, studies by Elsayegh *et al.* (2020) undertaken in the United States established that the top two reasons for construction disputes are related directly or indirectly to contract administration issues, and higher transportation infrastructure project costs attributed to what is now commonly referred to as conflicts, claims and disputes (C2D). The case under study further highlights the need for managing the risks such as exit strategy for public default or convenience (Iseki and Houtman, 2012).

Overall, the plausible explanation for the project's delay and suspension confirms the observation proposals by Ramamurti (2003) that one of the three explanations for government reneging includes that "political change" puts new leaders in charge with incentives to renege on old promises. Examples of failing to honour agreements made by previous administrations from a political viewpoint are evident in literature. The EIU (2015) highlights the case of Zambia where the government failed to honour promises. The case further highlights the issues of trust and integrity in infrastructure related projects. For instance, Kavishe and Chileshe (2019) identified "trust and integrity" among the CSFs and key areas for improvement for PPPs in affordable housing schemes (AHS) projects in Tanzania. While the study was related to affordable housing, the issues are also pertinent in infrastructure projects. To highlight and elaborate on this case, the definition of "integrity" as provided by Kanter (1994 cited in Jamali, 2004, pg. 420) as to where "partners behave toward each other in honourable ways that enhance mutual trust without abusing the information they gain, nor undermining each other" bears some resemblance to the issues surrounding the Bagamoyo project. Finally, the take home point is what happens or rather should happen regarding the resuscitation of the Bagamoyo project, as Siemiatycki (2010) also cautions that after a conflict has arisen, unwillingness to assume the cost of compromise may lead to legal battles; public agencies can be left holding substantial risk when disappointed concessionaires decide to exit the agreement.

12.10.1.4 Political risks

The issue of political risks is also at play in the case under consideration as the levels of uncertainties and associated risks involved in transportation infrastructure projects unavoidably make PPP contracts very complex (*Business News*, 2019). There is also sufficient evidence of politically motivated resistance traffic revenue risk factors for BOT road projects in both developing countries such as Nigeria (Babatunde and Perera, 2017) and developed economies such as Europe (Carbonara *et al.*, 2015). The political risks and contractual issues raised could have been managed as part of the broader RM approaches and risk assessment process as illustrated in steps 2, 3 and 4 in Figure 12.1.

12.11 Case discussion questions

1. What key lessons learned can you draw from the case study?
2. Align your lessons learned with countries delivering major infrastructure projects in developing (emerging economies) and developed countries.
3. What capacity building challenges hold developing countries back from delivering infrastructure development?
4. From a policy viewpoint, what lessons or success and failure factors (e.g., suspension of the infrastructure project) can you take from the case study for future infrastructure projects?

References

Abdel Aziz, A.M. (2007). Successful delivery of public-private partnerships for infrastructure development. *Journal of Construction Engineering and Management*, **133**(12), pp. 918–931.

Abdul-Aziz, A.R. and Kassim, P.S.J. (2011). Objectives, success and failure factors of housing public-private partnerships in Malaysia. *Habitat International*, **35**(1), pp. 150–157.

Abdullahi, B.C. and Aziz, W.N. (2011). The role of private sector participation in achieving anticipated outcomes for low-income group: A comparative analysis of housing sector between Malaysia and Nigeria. *African Journal of Business Management*, **5**(16), pp. 6859–6890.

African Business (2017). *Infrastructural projects in Africa*. Available from: https://www.africanbusinessexchange.com/infrastructural-projects-in-africa/ (cited 21st May 2020).

Agyakwa-Baah, A. and Chileshe, N. (2010). Critical success factors for risk assessment and management practices (RAMP) implementation within the Ghanaian construction organisations. In: Wang, Y., Yang, J., Shen, G.Q.P., and Wong, J. (Eds), *Proceedings of the 8th International Conference on Construction and Real Estates Management (ICCREM 2010)*, 1–3 December 2010, Brisbane, Australia, 1, pp. 345–352.

Alloisio, I. and Carraro, C. (2016). Public–private partnerships for energy infrastructure: A focus on MENA region. In: Caselli, S., Corbetta, G., and Vecchi, V. (Eds.), *Public private partnerships for infrastructure and business development: Principles, practices, and perspectives*, Ch. 9, pp. 149–168. Palgrave Macmillan.

Al-Saadi, R. and Abdou, A. (2016). Factors critical for the success of public-private partnerships in UAE infrastructure projects: Experts' perception. *International Journal of Construction Management*, **16**(3), pp. 234–248.

Babatunde, S.O. and Perera, S. (2017). Analysis of traffic revenue risk factors in BOT road projects in developing countries. *Transport Policy*, **56**, pp. 41–49.

Bayat, F., Noorzai, E., and Golabchi, M. (2019). Identifying the most important public–private partnership risks in Afghanistan's infrastructure projects. *Journal of Financial Management of Property and Construction*, **24**(3), pp. 309–337.

Bengesi, K.M.K., Mwesiga, P., Mrema, T. (2016). *Research report on the assessment of the Public-Private Partnership (PPP) in transportation infrastructure in Tanzania: The way PPP is understood, challenges and the way forward*. Dar es Salaam: Economic and Social Research Foundation, ESRF Policy Brief, No. 2.5.

Brown, A.C., Stern, J., Tenenbaum, B., and Gencer, D. (2006). *Handbook for evaluating infrastructure regulatory systems*. Washington, DC: The World Bank.

Regulatory process for infrastructure systems development

Business News (2019). Tanzania's China-backed $10 billion port plan stalls over terms: official. Available from: https://www.reuters.com/article/us-tanzania-port-idUSKCN1ST084 [cited 21 May 2020].

Caselli, S., Corbetta, G., and Vecchi, V. (2015). *Public private partnerships for infrastructure and business development: Principles, practices, and perspectives.* London: Palgrave Macmillan.

Carbonara, N., Costantino, N., Gunnigan, L., and Pellegrino, R. (2015). Risk management in motorway PPP projects: Empirical-based guidelines. *Transport Reviews*, **35**(2), pp. 162–182.

Carmona, M. (2010). The regulatory function in public-private partnerships for the provision of transport infrastructure. *Research in Transportation Economics*, **30**(1), pp. 110–125.

CCE News (2018). Tanzania now catches up with Kenya in infrastructure projects. *Construction and Civil Engineering.* Available from: https://cceonlinenews.com/2020/02/05/tanzania-now-catches-up-with-kenya-in-infrastructure-projects-2/ [cited 1 May 2020].

Chileshe, N. (2016). Critical success factors for managing infrastructure projects in Africa: A critical review and lessons learned. In: Mwanaumo, E.M. et al. (Ed.), *3rd International Conference on Infrastructure Development and Investment Strategies for Africa (DII-2016)*, The Development and Investment in Infrastructure (DII) Conference Series, pp. 82–93.

Chileshe, N. and Kikwasi, G.J. (2014a). Critical success factors for implementation of risk assessment and management practices within the Tanzanian construction industry. *Engineering, Construction and Architectural Management*, **21**(3), pp. 291–319.

Chileshe, N. and Kikwasi, G.J. (2014b). Risk assessment and management practices (RAMP) within the Tanzanian construction industry: Implementation barriers and advocated solutions. *International Journal of Construction Management*, **14**(4), pp. 239–254.

Chileshe, N., Wilson, L.J., Zuo, J., Zillante, G., and Pullen, S. (2013). Strategic risk assessment for pursuing sustainable business in the construction industry: Diagnostic models. In: Wells, G. (Ed.), *Sustainable business: Theory and practice of business under sustainability principles*, pp. 155–177. Cheltenham, UK: Edward Elgar.

Chileshe, N., Njau, C.W., Kibichii, B.K., Macharia, L.N., and Kavishe, N. (2020). Critical success factors for public–private partnership (PPP) infrastructure and housing projects in Kenya. *International Journal of Construction Management*, DOI: 10.1080/15623599.2020.1736835.

Cui, C., Liu, Y., Hope, A., and Wang, J. (2018). Review of studies on the public-private partnerships (PPP) for infrastructure projects. *International Journal of Project Management*, **36**, pp. 773–794.

Davison, W. (2015). *Africa Union seeks investors for infrastructure repairs. Bloomberg:* Available from: http://www.bloomberg.com/news/articles/2015-01-30/african-union-targets-private-investors-to-repair-infrastructure [cited 16 July 2016].

Debela, G.Y. (2019). Critical success factors (CSFs) of public–private partnership (PPP) road projects in Ethiopia. *International Journal of Construction Management*, DOI: 10.1080/15623599.2019.1634667.

EIU (The Economist Intelligence Unit). (2015). Evaluating the environment for public–private partnerships in Africa: The 2015 Infrascope. EIU, London. Available from: http://www.eiu.com/AfricaInfrascope2015 [cited 25 May 2020].

El-Adaway, I.H., Elsayegh, A.S., Abotaleb, I.S., Smith, C., Bootwala, M., Eteifa, S. (2019). Contractual guidelines for contractors working under projects funded by south eastern US dots. Proceedings, *Annual Conference - Canadian Society for Civil Engineering*, June.

Elsayegh, A., El-Adaway, I.H., Assaad, R., Ali, G., Abotaleb, I., Smith, C., Bootwala, M., Eteifa, S. (2020). Contractual guidelines for management of infrastructure transportation projects. *Journal of Legal Affairs and Dispute Resolution in Engineering and Construction*, **12**(3).

Githaiga, N.M., Burimaso, A., Bing, W., Ahmed, S.M. (2019). The belt and road initiative: Opportunities and risks for Africa's connectivity. *China Quarterly of International Strategic Studies*, **5**(1), pp. 117–141.

Grimsey, D. and Lewis, M.K. (2002). Evaluating the risks of public private partnerships for infrastructure projects. *International Journal of Project Management*, **20**(2), pp. 107–118.

Iseki, H. and Houtman, R. (2012). Evaluation of progress in contractual terms: Two case studies of recent DBFO PPP projects in North America. *Research in Transportation Economics*, **30**, pp. 73–84.

Jamali, D. (2004). Success and failure mechanisms of public private partnerships (PPPs) in developing countries: Insights from the Lebanese context. *International Journal of Public Sector Management*, **17**(5), pp. 414–430.

Jiang, W., Martek, I., Hosseini, M.R., and Chen, C. (2019). Political risk management of foreign direct investment in infrastructure projects: Bibliometric-qualitative analyses of research in developing countries. *Engineering, Construction and Architectural Management*, Vol. ahead-of-print No. ahead-of-print. https://doi.org/10.1108/ECAM-05-2019-0270.

Kanter, R.M. (1994), "Collaborative advantage: The art of alliances", *Harvard Business Review*, July–August Issue, 72.

Kavishe, N., Jefferson, I., and Chileshe, N. (2018). An analysis of the delivery challenges influencing public private partnership in housing projects: The case of Tanzania. *Engineering Construction and Architectural Management*, **25**(2), pp. 202–240.

Kavishe, N., Chileshe, N., and Jefferson, I. (2019). Public–private partnerships in Tanzanian affordable housing schemes: Policy and regulatory issues, pitfalls and solutions. *Built Environment Project and Asset Management*, **9**(2), pp. 233–247.

Kavishe, N. and Chileshe, N. (2019). Critical success factors in public-private partnerships (PPPs) on affordable housing schemes delivery in Tanzania: A qualitative study. *Journal of Facilities Management*, **17**(2), pp. 188–207.

Kingsford Owusu, E. and Chan, A.P.C. (2019). Barriers affecting effective application of anticorruption measures in infrastructure projects: Disparities between developed and developing countries. *Journal of Management in Engineering*, **35**(1), art. no. 04018056.

Kumaraswamy, M. and Zhang, X. (2001). Governmental role in BOT-led infrastructure development. *International Journal of Project Management*, **19**(4), pp. 195–205.

Le, P.T., Chileshe, N., Kirytopoulos, K., and Rameezdeen, R. (2020). Investigating the significance of risks in BOT transportation projects in Vietnam. *Engineering, Construction and Architectural Management*, **27**(6), pp. 1401–1425.

Li, B., Akintoye, A., Edwards, P.J., and Hardcastle, C. (2005). Perceptions of positive and negative factors influencing the attractiveness of PPP/PFI procurement for construction projects in the UK: Findings from a questionnaire survey. *Engineering, Construction and Architectural Management*, **12**(2), pp. 125–148.

Luiz, J. (2010). Infrastructure investment and its performance in Africa over the course of the twentieth century. *International Journal of Social Economics*, **37**(7), pp. 512–536.

Mañelele, I. and Muya, M. (2008). Risk identification on community-based construction projects in Zambia. *Journal of Engineering, Design and Technology*, **6**(2), pp. 145–161.

Manelele, I. and Muya, M. (2008), "Risk identification on community-based construction projects in Zambia", *Journal of Construction*, Vol. 12 No. 1, pp. 2–7.

Mayaki, I. (2014). Why infrastructure development in Africa matters. Africa Renewal, Available from: http://www.un.org/africarenewal/web-features/why-infrastructure-development-africa-matters. [cited 16 July 2016].

Mboya, J.R. (2013). "PPP Country Paper-Tanzania," paper submitted to SADC-DFRC 3P NETWORK Public-Private-Partnership Working Group, Limessstrasse 26, 61273 Wehreim, Wehrheim, Available from: www.sadcpppnetwork.org/wp-content/uploads/2015/02/tanzania_27012014.pdf [cited 19 November 2016].

Mehta, A. (2015). Is investing in Africa risky? All Africa Excellence, Available from: http://ahmedmheta.com/is-investing-in-africa-risky/ [cited 16 July 2016].

Moskalyk, A. (2011). *Public-private partnerships in housing and urban development*. Nairobi, Kenya: UN-HABITAT.

Moszoro, M., Araya, G., Ruiz-Nunez, F., and Schwartz, J. (2016). What drives private participation in infrastructure developing countries. In: Caselli, S., Corbetta, G., and Vecchi, V. (Eds.), *Public private partnerships for infrastructure and business development: Principles, practices, and perspectives*, pp. 19–44. Palgrave Macmillan.

Mourgues, T. and Kingombe, C. (2017). How to support African PPPs: The role of the enabling environment. In: Leitão, J., de Morais Sarmento, E., and Aleluia, J. (Eds.), *The emerald handbook of public–private partnerships in developing and emerging economies*, pp. 269–310. Emerald Publishing Limited.

Muthengi, A. (2015). Bagamoyo port: Tanzania begins construction on mega project. Africa Live: BBC News updates. Available from: https://www.bbc.com/news/world-africa-34554524 [cited 20 May 2020].

Mutiso, L. (2016). 10 Mega infrastructure projects in Africa funded by China, AFK Insider. Available from: http://afkinsider.com/121477/10-mega-infrastructural-projects-in-africa-funded-by-china/#sthash.cGpSw3ML.dpuf [cited 16 July 2016].

Mwangasha, J. (2019). Why Magufuli scrapped Bagamoyo Port deal with Chinese investor. Construction Kenya (CK), Available from: https://www.constructionkenya.com/3128/bagamoyo-port-construction/ [cited 21 May 2020].

Ng'wanakilala, F. and Merriman, J. (2015). Tanzania awards $9 bn rail projects to Chinese companies. Available from: https://www.reuters.com/article/tanzania-railways-idUSL5N0YM09H20150531 [cited 21 May 2020].

Regulatory process for infrastructure systems development

Ogbu, C.P. and Adindu, C.C. (2019). Direct risk factors and cost performance of road projects in developing countries: Contractors' perspective. *Journal of Engineering, Design and Technology*, **18**(2), pp. 326–342.

Osei-Kyei, R. and Chan, A.P.C. (2017). Empirical comparison of critical success factors for public–private partnerships in developing and developed countries: A case of Ghana and Hong Kong. *Engineering, Construction and Architectural Management*, **24**(6), pp. 1222–1245.

Project Management Institute (PMI) (2017). *A guide to the project management body of knowledge*. Newtown Square: Project Management Institute.

PwC (2014). *Trends, challenges and future outlook: Capital projects and infrastructure in East Africa, Southern Africa and West Africa*. Available from www.pwc.co.za/infrastructure [cited 17 July 2016].

Ramamurti, R. (2003). Can governments make credible promises? Insights from infrastructure projects in emerging economies. *Journal of International Management*, **9**(3), pp. 253–269.

Rothballer, C. and Gerbert, P. (2016). Preparing and structuring bankable PPP projects. In: Caselli, S., Corbetta, G., and Vecchi, V. (Eds.), *Public private partnerships for infrastructure and business development: Principles, practices, and perspectives*, Ch. 4, pp. 58–80. Palgrave Macmillan.

Searcy, C. (2009). Setting a course in corporate sustainability performance measurement. *Measuring Business Excellence*, **13**(3), pp. 49–57.

Sengupta, U. (2004). Public-private partnerships for housing delivery in Kolkata. *International Conference on "Adequate and Affordable Housing for All*, Toronto from Citeseer, pp. 24–27.

Sengupta, U. (2006). Government intervention and public–private partnerships in housing delivery in Kolkata. *Habitat International*, **30**(3), pp. 448–461.

Sharma, C. (2012). Determinants of PPP in infrastructure in developing economies. *Transforming Government: People, Process and Policy*, **6**(2), pp. 149–166.

Siemiatycki, M. (2010). Delivering transportation infrastructure through public-private partnerships. *Journal of the American Planning Association*, **76**(1), pp. 43–58.

Stern, O. (2020). How to manage corruption risk in African infrastructure projects. Available from: https://www.theafricareport.com/26650/how-to-manage-corruption-risk-in-african-infrastructure-projects/ [cited 28 May 2020].

Swedlund, H.J. (2017). Is China eroding the bargaining power of traditional donors in Africa?. *International Affairs*, **93**(2), pp. 389–408.

Tanzanian Procurement Act (2011), "The united republic of Tanzania act supplement, act no. 9 of 2011, 0856 − 01001X, 30th December 2011, to the gazette of the united republic of Tanzania no. 52 vol. 92 dated 30th December, 2011", available at: www.ppra.go.tz/phocadownload/attachments/Act/Public_Procurement_Act_2011.pdf ISBN (accessed 15 January 2020)

The Citizen (2019). Talks on proposed Bagamoyo port "still going on." Available from: https://www.thecitizen.co.tz/news/-Talks-on-proposed-Bagamoyo-port-still-going-on-/1840340-5113898-14ppj3v/index.html [cited 21 May 2020].

Thomas, A.V., Kalidindi, S.N., and Ananthanarayanan, K. (2003). Risk perception analysis of BOT road project participants in India. *Construction Management and Economics*, **21**(4), pp. 393–407.

Trangkanont, S. and Charoenngam, C. (2014). Critical failure factors of public-private partnership low-cost housing program in Thailand. *Engineering, Construction and Architectural Management*, **21**(4), pp. 421–443.

URT (2009). Prime Minister's Office: National public private partnership (PPP) Policy, "Restricted Circulation." Dar es Salaam, Tanzania.

Van Thuyet, N., Ogunlana, S.O., and Kumar Dey, P. (2007). Risk management in oil and gas construction projects in Vietnam. *International Journal of Energy Sector Management*, **1**(2), pp. 175–194.

Vhumbunu, C.H. (2019), "African Regional Economic Integration in the Era of Globalisation: Reflecting on the Trials, Tribulations, and Triumphs", *International Journal of African Renaissance Studies - Multi-, Inter- and Transdisciplinarity* Vol. 14, No. 1, pages 106–130.

Wang, S., Dulaimi, M. and Aguira, M. (2004), "Risk management framework for construction projects in developing countries", *Construction Management and Economics*, Vol. 22 No. 3, pp. 237–252.

World Bank (2014). World Bank Group to provide $1.2 billion to improve infrastructure and competitiveness of East African Community. Available from: http://www.worldbank.org/en/news/press-release/2014/11/29/world-bank-group-to-provide-12-billion-to-improve-infrastructure-and-competitiveness-of-east-african-community [cited 16 July 2016].

World Bank (2016). Tanzania economic update the road less travelled unleashing public private partnerships in Tanzania. Africa region macroeconomics and fiscal management global practice. Available from: http://www.worldbank.org/tanianla/economlcupdate [cited 28 May 2020].

Xiaopeng, D. and Pheng, L.S. (2013), "Understanding the critical variables affecting the level of political risks in international construction projects", *KSCE Journal of Civil Engineering*, Vol. 17 No. 5, pp. 895–907.

Xu, Z., Zhang, Y., and Sun, Y. (2019). Will foreign aid foster economic development? Grid panel data evidence from China's aid to Africa. *Emerging Markets Finance and Trade*, DOI: 10.1080/1540496X.2019.1696187.

Zhang, X. (2005). Criteria for selecting the private-sector partner in public–private partnerships. *Journal of Construction Engineering and Management*, **131**(6), pp. 631–644.

Zeitz, A.O. (2020). Emulate or differentiate? Chinese development finance, competition, and World Bank infrastructure funding. *Review of International Organizations*, DOI: 10.1007/s11558-020-09377-y.

Zou, P. X., Wang, S., and Fang, D. (2008). A life–cycle risk management framework for PPP infrastructure projects. *Journal of Financial Management of Property and Construction*, **13**(2), pp. 123–142.

13
MANAGING RELATIONSHIP RISKS ON MAJOR INFRASTRUCTURE PROJECTS

David Bryde, Simon Taylor and Roger Joby

13.1 Introduction

Failing to manage relationship risks is a common and significant cause of delay on infrastructure projects. Relationship risks may arise from interactions between stakeholders in an infrastructure project, such as between the sponsor (client) and their supply chains, the sub-contractors on site, the local community, regulatory bodies and governments. This chapter will therefore identify relationship risks that have an impact upon infrastructure projects and propose some possible approaches to management. While recognising the need to deal with the relational aspects arising from the interactions between all these various stakeholders, the chapter focuses on managing the relationships between the different firms, including the sponsor (client), that come together to form the project and deliver the infrastructure. In the next section of the chapter, we set out the aim and objectives.

13.1.1 Chapter aim and objectives

Our overall aim is to outline an operational model for infrastructure project delivery that emphasises behavioural aspects. With such a model is an appropriate response to dealing with the risks associated with the behaviours of the different parties to the project, where undesirable behaviours lead to negative project outcomes. To achieve this aim we derive five high level objectives.

- To understand the need for new operational models for infrastructure project delivery, such as relational-based ones, in the context of poor performance and low productivity for such projects;

- To explore the nature of the relationships between firms involved in major infrastructure projects, focusing on how these relationships require managerial interventions that pay particular attention to relational aspects;
- To explain the key concepts of relational risk, relational governance and relational contracting, as they apply to major infrastructure delivery;
- To introduce frameworks for managing relational risk that incorporate relational indicators/critical success factors, foster an organisational climate conducive to effectively dealing with relational risk and provide processes for relationship management;
- To distinguish, through two cases from industry, between an operational model that emphasises behavioural aspects and a model with more of an emphasis on structure.

We now turn our attention to the first of these objectives and discuss the performance and productivity of major infrastructure projects, shining a light on issues relating to the relationships formed by the different parties involved in project delivery.

13.1.2 *Learning outcomes*

The following learning outcomes have been identified for this chapter. Readers will be able to:

- Describe the nature of relationships on major infrastructure projects;
- Differentiate between relation risk, relational governance and relational contracting;
- Appreciate the value of frameworks for managing relation risk; and
- Comprehend how frameworks for relationship management can be applied to major infrastructure projects.

13.2 Performance and productivity of major infrastructure projects

Over the past decade, government and non-government organisation (NGO)-sponsored reports focused on aspects of poor performance and flat or declining productivity in major infrastructure projects, consistently highlighting delivery failures. In 2010, Infrastructure UK, a collaboration of the government, the Institute of Civil Engineers (ICE) and industry representatives identified £20–£30bn worth of savings over the next 10 years with the development of new business models to improve productivity, achieve greater supply chain integration and promote innovation. These savings would come from the cost of infrastructure, delivering significant benefits in performance and value for money (HM Treasury, 2010). Halfway through this period, the picture of productivity and performance problems had not changed. The UK government's National Audit Office (NAO) assessed in June 2015 that 34 percent of the 149 projects in the government's major projects portfolio, with an estimated whole-life cost of £511bn, were flagged as being in doubt of successful delivery or unachievable without taking some form of corrective action. The NAO identified five main recurring issues at the root of the problem:

- An absence of portfolio management at both the departmental and government level;
- A lack of clear, consistent data with which to measure performance;
- Poor early planning;
- A lack of capacity and capability to undertake a growing number of projects;
- A lack of clear accountability for project leadership (National Audit Office, 2016).

In 2017, the Global Infrastructure Initiative's (GII) "Voices on Infrastructure" series, a collection of articles by industry leaders and McKinsey experts, echoed the need and opportunity for significant improvement in productivity (Global Infrastructure Initiative, 2017). The focus of this GII publication was on project delivery and execution, with attention paid to relational aspects, especially ways to improve contractor-owner (client) relationships. The GII report specifically shone a light on poor procedures adopted by project owners (clients) in their approaches to managing the delivery of infrastructure projects as the major cause for project over-runs and delays. The report urged owners (clients) to learn from best practice, where the best owners (clients) of infrastructure projects view their supply chain within the project coalition as a strategic partnership and, where appropriate, enter into long-term, multi-project relationships. Best practice owners (clients) also pay attention to the interactions between themselves and contractors while undertaking a specific infrastructure project. The importance of people working together – again with an implicit recognition of the need for effective relationships between parties – was a key theme of the UK Infrastructure and Projects Authority's (IPA) priorities for 2020.

Hence, in the next section, we explore the nature of the relationships present in major infrastructure projects, not only between contractor and owner but also between other parties to these projects. It is through the establishment of effective working relationships between firms and individuals that the need "to build strong, flexible, professional and capable teams" to deliver major infrastructure is met (Infrastructure and Projects Authority, 2019). This exploration reveals the complexities of the various relationships and the importance of paying attention to relational aspects when delivering major infrastructure.

13.3 The nature of relationships in major infrastructure projects

It has long been recognised that major infrastructure projects typically involve the creation of a project "coalition" of different firms that come together to deliver the project (Pryke, 2004; Soetanto *et al.,* 2002; Winch, 1989). There are also other key stakeholders to infrastructure projects, including:

- Government/project initiators;
- The general public;
- Pressure groups;
- The media;
- The firms that are involved in delivering the infrastructure (Li, 2012).

These different firms and stakeholders have their own specific expectations relating to the project, including what they might hope to get out of being part of the project coalition and the effects having the infrastructure in place could have on them (Olander, 2007). For stakeholders outside the coalition their interest in an infrastructure project could take the form of concerns about some aspect of the undertaking. That concern might be in the delivery phase, e.g., the possibility of air, water, noise pollution during construction or in the post-delivery phase, e.g., the economic benefits of the new infrastructure to local communities (Li, 2012). These expectations and concerns may often conflict with others' expectations and concerns (Olander, 2007). In addition to their expectations, the different firms that make up the coalition, along with the client, each bring their own values, ways of working, past experiences, etc. that may be very different to those brought by others. Finally, firms may be coming together for the very first time and hence are not aware of other's expectations, values, and ways of working.

The forming of the project coalition to undertake infrastructure projects results in a network of firms; within the network there are numerous relationships within and between these firms. The relationships take various forms. Pryke (2004) identifies three broad types: contractual, performance incentives and information exchange networks. Each firm has dyadic contractual relationships with the client or with firms in the supply chain. Pryke argues that some firms interact with each other in order to incentivise performance. Finally, the exchange of important information in a timely fashion, such as providing accurate data on project progress, both between and within firms, requires the presence of effective internal and external relationships.

The intricacies of the network of firms that make up the project coalition, coupled with the diversity of stakeholder concerns described earlier – create a layer of complexity when it comes to managing infrastructure projects (Mok *et al.*, 2017). For example, while the client will have a direct relationship with the main contractor through the contract, they will have an indirect relationship with sub-contractors, whose performance the main contractor relies on in order to meet their own contractual obligations. Hence, a social network approach, will help effectively manage relationships between stakeholders of infrastructure projects (Mok *et al.*, 2015). This has resulted in a strand of literature focused on applying social network analysis (SNA) techniques to infrastructure projects in order to better understand the forms of networks and the roles of actors within the networks (see, for example, the works of Pryke (2004 and 2005). It is also consistent with seminal project management literature stating that issues of poor performance in projects in general, including infrastructure projects, result from a lack of attention paid to the social dynamics between actors in the network and a failure to recognise the role inadequate inter-firm collaboration plays in project failures (Suprapto, 2015).

For the remainder of this chapter we focus primarily on the key relationship between client and main contractor. However, we acknowledge the need to take a holistic view of the whole network including actors that sit outside the project coalition. We also touch upon specific approaches to managing the relationships between the main contractor and their sub-contractors in the supply chain. In the next section, we introduce the notion of relational risk, which arises from the delivery of infrastructure and needs managing to ensure the achievement of high levels of performance and productivity.

13.4 Relational risk

The multi-faceted and complex nature of the relationships that form to deliver infrastructure means that the environment in which these relationships form and develop is characterised by a high degree of uncertainty. Projects are not only technical systems but also social systems (Böhle *et al.*, 2016) and both are important causes of uncertainty. With such high levels of uncertainty in respect to the social system formed to deliver major infrastructure projects, there exists relational risk in terms of how each party (client, main contractor, sub-contractors and other members of the supply chain) engage and interact with each other (Bryde *et al.*, 2019; Zhang and Qian, 2017). As with well-established conceptions of project risk, which stress the two sides of the risk coin and highlight the danger of a restricted focus on the threats related to specific events that might occur to the neglect of opportunities (Ward and Chapman, 2003), relational risk involves both threats and opportunities (BSI, 2019). Further, the focus of relational risk management is more on dealing with the high level of uncertainty that exists in relation to the workings of the relationships formed by the actors in the project coalition, rather than on the likelihood and consequence of individual events occurring. This focus on the uncertainty around collaborative relationships does not replace traditional risk management; rather, it is an enhancement of the risk management of infrastructure projects.

A useful lens through which to view relational risk is agency theory. Agency theory explains how relationships work in situations where a principal-agent (P-A) relationship exists. A principal engages an agent to undertake a service on their behalf, such as through the letting of a formal contract, and in doing so, delegates a level of authority to the agent to make decisions on the principal's behalf. The project coalition formed to deliver major infrastructure contains a multitude of P-A relationships. Foremost is that between the client (owner) and the firms they directly engage with through a contractual relationship, including the organisation responsible for managing the project (Müller and Turner, 2005). In addition, there are all the P-A relationships that exist between different firms in the supply chain. The presence of a dysfunctional P-A relationship can manifest itself in various ways, to the detriment of the project, including:

- Demonstrating opportunistic behaviour;
- Acting in one's own best interest;
- Disproportionally allocating risks or responsibilities;
- Taking advantage of information asymmetries, i.e.. failing to share crucial information, including the concealment of negative outcomes (Shrestha and Martek, 2015).

Such problems can hinder the forming and functioning of the social system of the project coalition, leading to a lack of trust between the parties and creating a major source of relational risk. There is a wealth of literature on how agency problems adversely affect the delivery of major infrastructure projects; see Flyvberg *et al.* (2009), Cantarelli *et al.* (2010), Hellowell and Vecchi (2015), Shrestha and Martek (2015), Smith *et al.* (2018) and Bryde *et al.* (2019). A focus on understanding the sources of relational risk, such as agency problems present in P-A relationships and then putting in place mitigation strategies to address them, is a key way of ensuring the social systems that exist in the project coalitions delivering major infrastructure function effectively. Hence, in the remainder of this chapter we focus on approaches involving governance and control systems that proactively manage relational risk. As a prelude to this, we begin by defining relational governance.

13.5 Relational governance

Conceptions of project governance identify two broad categories: contractual governance and relational governance (Ul Haq *et al.*, 2019). These are forms of organisational governance, whereby an organisation, in this case the temporary organisation formed to deliver the new infrastructure, aligns its decision-making to achieve its objectives. Contractual governance focuses on mechanisms to ensure compliance, using the formal contract to detail roles, responsibilities, accountabilities, procedures for monitoring and controlling and, typically, penalties for non-compliance rather than incentives for desired outcomes or behaviours. Research on contractual governance shows its effectiveness in managing some aspects of performance in projects, though it is not particularly useful in fostering cooperative relationships between the parties to the project (Ke *et al.*, 2015). Conversely, relational governance is a more informal approach that focuses on building good working relationships amongst the project parties and is characterised by openness and transparency in sharing information, flexible attitudes with a give-and-take on both sides and fostering a sense of solidarity around delivering the project's goals (Bryde *et al.*, 2019). Performance is maximised when both contractual and relational governance mechanisms are utilised (Levitt *et al.*, 2009). One mechanism for undertaking relational governance is through the contract, and this requires a different approach, referred to as relational contracting, which we cover in the next section.

13.6 Relational contracting

The notion of relational contracting brings us back to agency theory and, specifically, the role the contract plays in the theory as a key mechanism addressing agency-related problems. Agency theory identifies two types of contracts (Florical and Lampel, 1998). First, there are outcome-based contracts, in which the principal agrees to pay the agent an amount, typically fixed, for an agreed set of deliverables. In projects, we see these established as fixed price contracts. Second are behaviour-based contracts, where behaviour equates to actions carried out, such as work undertaken. As part of such contracts a fee-for-service, typically person-hours worked, is paid. In projects, these are time and material (T and M) contracts. Pure outcome-based and behaviour-based contracts are at opposite ends of the spectrum of forms of contract – with there being various hybrid forms in between that incorporate elements of both (Kalnins and Mayer, 2004). The key task is to choose a form of contract that is fit for purpose, suits the requirements of the project in hand and, crucially, meets the needs of both client and contractor, a concept that is referred to as "contractual completeness" (Handley and Benton, Jr., 2009). Agency theory helps to explain in what circumstances principals can achieve contractual completeness. For example, early research on agency theory showed that behaviour-based contracts were suitable when a principal and agent had worked together previously, such as partnering arrangements, and hence have a long-standing relationship based on mutual trust (Eisenhardt, 1989). Conversely, in project coalitions where there is no such past relationship and information asymmetry is high, outcome-based forms of contract are more appropriate.

Of course, regardless of whether the contract is outcome-based, behaviour-based, or a hybrid, as highlighted in the discussion of the nature of the relationships in infrastructure projects above, all contracts have a relational aspect to them, with the contractual relationship being one of the three types of relationship present in the network of firms making up the project coalition (Pryke, 2014). What is different about relational contracts, in comparison with traditional contracts, is that relational contracts make the relationship explicit, which is reflected in definitions of relational contracting: "relational contracting is a transaction or contracting mechanism that seeks to give explicit recognition to the commercial 'relationship' between the parties to the contract" (Colledge, 2005: pg. 30).

So here, the emphasis is very much about intent, certainly on the part of the client, and bought into by the other parties in the project coalition, to use the contract not only in its traditional function of commercial governance but also as a mechanism by which relational risk is mitigated and managed during project delivery. It is a philosophy of a value chain, linking together all the interdependent parts of the project, through the numerous relationships, that is a key business objective (Rowlinson and Cheung, 2004). Therefore, the focus of attention in the management of an infrastructure project is on ensuring the relationships function effectively. Relational contracting is a broad concept and, in terms of a project management approach to infrastructure delivery, includes the following characteristics:

- Recognition of mutual benefits to achieve a win–win outcome and the sharing of risks, both threats and opportunities, with an open and transparent model of compensation;
- Greater interdependence between parties leading to collaborative working;
- Bilateral or unified governance structures;
- The development of a learning culture, including the sharing of explicit and tacit knowledge;
- Investment in people including education, training and skills development;
- Rich interactions between the parties;
- A fostering and recognition of the value of innovation, collaborative working, problem solving and creativity (Colledge, 2005; Rowlinson and Cheung, 2004: Sakal, 2005).

The intent and philosophy behind relational contracting is reflected in developments in procurement in the infrastructure project management practitioner communities. At the forefront of these developments is the NEC forms of contract. The first NEC contract came out in 1993, labelled as the "New Engineering Contract." Since then there have been several evolutions, with the latest suite of contracts, NEC4, launched in 2017. Governments and industries worldwide have endorsed the NEC suite of contracts and credited it with contributing to the success of major infrastructure projects, such as the construction of the venues for the 2012 Olympic and Paralympic Games in London. As well as being applicable for a wide variety of work, a key characteristic of NEC contracts is the emphasis on the relationship. As stated by NEC, the contracts "stimulate good management of the relationship between the two parties to the contract and, hence, of the work involved in the contract" (NEC, 2020). The contracts facilitate desirable behaviours, such as cooperation and collaboration, based on mutual trust between the parties. Fundamental to achieving this is having contracts that are clear, simple and easy to understand, written in plain English, with a straightforward and easily understood structure. They seek to foster a collaborative and cooperative spirit amongst the parties in the project coalition, achieved through clear lines of responsibility and governance. The characteristics of relational contracting, described in the previous paragraph, are present in the details of the NEC contract suite. For example, a key principle is to be proactive rather than reactive through the application of foresight, applied through collaborative working, which mitigates problems and reduce threats (Hughes and Waterhouse, 2018).

Documentary evidence testifies to the success of relational contracting in general and the NEC suite of contracts specifically. The Global Infrastructure Initiative (2017) gives examples of how better programme and budget outcomes are achieved through relational contracting, by pooling delivery risk and sharing profits among the owner (client), engineer, and constructor. Specifically, they describe how effective relational contracts often establish separate pools of money to pilot new ideas and incentivise innovation across multiple team members. For example, Sutter Health in Northern California were experiencing unreliable outcomes from a US $7 billion capital plan incorporating more than 255 active projects. This prompted a new relational contracting approach, including the introduction of integrated forms of agreement, integrated project delivery, and lean construction. As a result, the organisation turned things around, delivering 15 capital projects with a value of US $1.5 billion within schedule and budget.

In relation to the effectiveness of NEC contracts, Wright and Fergusson (2009) analysed the case of a NZ $100 million programme to refurbish and upgrade New Zealand's largest hydro power station, which utilised a NEC Engineering and Construction Contract (ECC). Through interviews with members of the project coalition, the authors reported benefits of effective collaborative relationships between project participants, enabled by the relational contracting approach. They specifically highlight a link made by one of the workers on site between the positive and enjoyable work environment and the project's exemplary safety record, with the ECC contract having completed more than 77,000 person-hours of effort in a high-risk environment, with zero lost time due to injuries.

One question to ask is to what extent is relational contracting in place in my organisation and my projects? To answer this question, Harper *et al.* (2016) provide a useful set of statements that encompass the different characteristics of relational contracting. They can act as a checklist for an organisation to benchmark its contracting approach. The authors identify eight relational contracting norms:

- "Role integrity" – organisations strive to overcome their own internal goals for the good of the project;

- "Reciprocity" – an organisation's success is a function of all working together on the project and being successful;
- "Flexibility" – modifications and changes are an accepted part of the project;
- "Propriety of means" – organisations act with a fairness and formality so that their actions have no detrimental effect on other parties of the project;
- "Reliance and expectations" – there is confidence that promises made and commitments given by the project participants will be honoured;
- "Restraint of power" – project parties restrain from exploiting each other when there is an opportunity to do so;
- "Contractual solidarity" – a belief that project success occurs because of the combined efforts of all, rather than by parties competing with each other;
- "Harmonization of conflict" – a spirit of mutual accommodation towards cooperative ends exists amongst all parties in the project. (Harper *et al.,* 2006).

The authors then develop a set of 38 statements across the eight relational norms to assess the extent to which a relational contacting approach is present. This useful tool allows organisations to measure the level of integration of relational contracting norms in their project teams and to develop innovative approaches to increase the emphasis on the relationships between the members of the project coalition delivering infrastructure projects. International standards to establish and improve collaborative relationships in the business world recognise the importance of internal assessment, regardless of the specific environment in which they exist, i.e., project or non-project. For example, ISO 44002:2019[E] describes a process for an organisation to undertake a structured assessment of its capabilities and maturity in order to engage in a successful collaborative initiative (British Standards Institute, 2019). These steps, which are 1) capability and environment for collaboration 2) assessment of strengths and weaknesses 3) assessment of collaborative profile 4) appointment of collaborative leadership, 5) definition of partner selection criteria and 6) implementation of relationship management plan, are applicable to major infrastructure project delivery.

A final topic to consider under the heading of relational contracting is the factors critical to the success of such an approach in practice. Here the importance of education and experience cannot be underestimated. A move from a traditional contracting approach to a relational contracting one can involve significant changes in culture and ways of working, which can be incredibly challenging for some people. This challenge is greater when members of the project coalition are used to certain way of doing things, through long-standing and deeply engrained customs and practices. Education of all parties, both inter- and intra-organisation, becomes crucial (Bryde *et al.,* 2019). For example, if new forms of contract, such as NEC, are being used, not just the client and main contractor, but also sub-contractors, need training on its workings, including the spirit underpinning the use of the contract. Like with other types of change management, there are steps that smooth the transition to a new way of working. Activities to raise awareness of the benefits and re-configuration of a team's way of working – both physically and through virtual methods – enhance communication between members of the project coalition and foster closer working relationships (Wright and Fergusson, 2009). Increased experience of the positive outcomes of the relational contracting approach will overcome obstacles to its use; though where there is no such experience, having a project coalition well educated in the relational contacting approach will mitigate any lack of experience.

Senior management attention to the transformational implications of moving from traditional to relational contracting is especially important in infrastructure projects, where there is history of adversarial relationships between members of the project coalition, reinforced through a traditional contracting approach (Gil, 2009). For example, any good intentions of using a relational contracting approach can be easily undone if the client fails to set the incentives for cooperative working. This failure could take various forms, e.g., a "pain/gain" sharing clause written by the client in which the any gains, such as cost savings through early completion, are shared equally between client and contractor but any pain, such as cost overruns, is borne disproportionally by the contractor. Such clauses are unlikely to drive the desired behaviours that are characteristic of a positive relational approach. To conclude this section, we are back to the earlier definition of relational contracting which stresses the importance of intent. New forms of contracts that incorporate the principles of relational contracting are unlikely to produce the desired outcome if the intent, especially on the part of the client, to adhere to its spirit is not there.

So far, we have discussed the context in which managing relationship risk on major infrastructure projects is important. We have highlighted the need to address performance and productivity shortcomings, discussing how a focus on the relationships of members of the project coalition is one way of meeting this need. We have discussed the inherent challenges that arise from the specific nature of relationships present in major infrastructure projects, using the torch of agency theory to shine a light on the problems inherent in these relationships. We then set out some of the key concepts: relational risk, relational governance and relational contracting. Next, we discuss the different aspects of how organisations effectively manage the relationship risks present in major infrastructure projects.

13.7 Frameworks for managing relational risk

13.7.1 Relational indicators/critical success factors (CSFs)

As is highlighted in the discussion of relational contracting above, frameworks for managing relational risk associated with major infrastructure delivery will be multi-dimensional, encompassing a wide range of aspects of project management. These aspects will incorporate people, systems and organisations. Meng (2012) derives a list of 10 key "relational indicators" that provide a useful checklist of the disparate elements that would be incorporated into a comprehensive framework for managing relational risk. We have already identified some of these elements when discussing the characteristics of relational contracting. The 10 relational indicators are:

1. Mutual objectives
2. Gain and pain sharing
3. Trust
4. No-blame culture
5. Joint working
6. Communication
7. Problem solving
8. Risk allocation
9. Performance measurement
10. Continuous improvement

Having derived the relational indicators, the authors go on to explore granularities in the relationships between indicators and performance. For example, joint working had a particularly positive impact on the reduction of time delays. While communication, risk allocation, no-blame culture, problem solving, and performance measurement had a similar impact on reducing cost overruns, Meng's study reported other nuances relating primarily to time delays and cost overruns but also quality defects, highlighting that areas of poor performance, or where performance improvements/priorities are set, should influence the operationalisation of any framework for managing relational risk. As well as the importance of a no-blame culture, the fourth relational indicator identified by Meng, along with other research into the performance of infrastructure projects, highlights the importance of a shared culture. The presence of such a culture not only provides the environment for effective relationships within the project coalition but it also enables the values of other key external stakeholders to infrastructure projects, such as the general public, to be considered as part of project planning and delivery (van Gestel *et al.*, 2008).

The relational indicators derived by Meng are generic, encompassing the different relationships that exist amongst the various firms that make up the project coalition and, as a rule, the conditions of success for a relationship. That said, as with the nuances relating to the different dimensions of project performance, there is evidence relating to specific relationships. For example, Pal *et al.* (2017) specifically focused on the relationship between contractor and supplier, rather than taking the usual focus on owner (client) and contractor. In doing so, they identified CSFs that main contractors should pay particular attention to in order to ensure their relationships with their own subcontractors and/or suppliers are effective and will deliver positive on-time and within-budget project outcomes. Two of these CSFs are included in the list of 10 relational indicators identified by Meng (2012): continuous improvement and problem solving. Continuous improvement is an "on-going effort to improve product, service, and associated processes incrementally" (Pal *et al.*, 2017: pg. 1227) and problem solving is "developing early warning system to detect potential issues and utilizing regular formal/informal feedbacks to identify opportunities for improving performance further" (Pal *et al.*, 2017: pg. 1227). The common thread here is collaboration and the sense that the best way to deliver improvements and to solve project problems is for the parties to work together with a common purpose. Hence, we conclude that while a generic framework provides a common thread to support relational-focused initiatives across the project coalition, there are contextual factors that can help focus managerial efforts, such as priorities of performance and the specific relationship being targeted, such as the main contractor with their own subcontractors/suppliers that should be taken into consideration.

13.7.2 Organisational climate

In the discussion of relational indicators and CSFs, we highlighted the importance attached by authors to aspects of culture: no-blame and shared. Another related concept that applies to infrastructure projects is organisational climate. Often, the terms culture and climate are interchangeable, even though they are distinct. There is a long-standing strand of literature, going back many decades, that discusses their distinctions. A comprehensive delineation of the concepts of culture and climate is beyond the scope of this chapter but for an informative discussion of the relationship between organisational culture and climate, see the work of Wallace *et al.* (1999). Briefly, culture goes deeper than climate and is about the beliefs, values and assumptions that give an organisation its identity. Members of an organisation recognise and communicate the basic elements of the culture through stories, icons and rituals. Climate describes the shared perceptions of the people in a group or organization, referring to the feel of the environment. Stressing the characteristic of shared perception and feeling, organisational climate in a project context has been succinctly described as "what it feels like to work here" and encompasses the management style at the organisational level in which the project is being undertaken and at the project level

(Gray, 2001: pg.104). The organisational climate is a set of internal characteristics that influence behaviours, which in turn mitigates risk and leads to successful project outcomes (Sharma and Guptab, 2012).

Research into relational management in infrastructure project delivery stresses the need to create an organisational climate within the project coalition that is conducive to collaborative working and which enables the characteristics of relational contracting, for example, to flourish. For example, Hannevik *et al.* (2014) highlight the importance of regular assessment of organisational climate, such as at the start of major phases in the project life cycle. To create a climate conducive to success they stress the need for a climate characterised by communication and cooperation, within the project coalition and with external stakeholders in the network. To do this, they draw upon the competing values framework (CVF) as a useful managerial tool. The CVF helps in organising and understanding a variety of organisational phenomena, such as competences, culture, design and quality, leadership roles, life cycle stage development, financial strategy, information processing and brain functioning (Cameron, 2020). Specifically, Hannevik *et al.* (2014) describe how communication and cooperation amongst the parties to the project coalition can be enabled through management competencies relating to managing teams (facilitating effective and cohesive teams) and managing interpersonal relationships (supportive feedback, listening and solving conflicts). Finally, we draw attention to the crucial role that the project leadership plays in creating a climate where the members of the project coalition work effectively work together. Managerial interventions in the team-working process ensure there is inter-team collaboration; they are crucial in enabling improved project performance where there is a focus on relational aspects (Suprapto *et al.,* 2015).

13.8 Frameworks for relationship management

There are frameworks that provide support to establishing and maintaining effective relationships between collaborating organisations. Some of these are generic to any organisational context and others are specific to projects and infrastructure delivery. A useful generic framework is an eight-stage process for establishing a successful collaborative business relationship called "ISO 44001:2017 Collaborative business relationship management systems" (CBRMS) (British Standards Institute, 2019; International Standards Organisation, 2017). While not being specifically developed for relationships formed in project environments, such as infrastructure delivery, the standard provides a useful framework for embedding a relational approach in the project arena at the operational level. Its focus is on managing relationships between organisations with differing goals, objectives, expectations, cultures and behaviours, which characterises the project coalition formed to deliver infrastructure. This operational level equates to the delivery by the project coalition of new infrastructure.

The eight steps that make up the operational process are:

- Operational awareness
- Knowledge
- Internal assessment
- Partner selection
- Working together
- Value creation
- Staying together
- Exit strategy activation

This process takes place as part of a wider corporate level system, including such areas as leadership, planning, support, performance and improvement. A useful document to link the operational level of the project with the corporate level is a relationship management plan (RMP). The RMP provides for joint governance throughout the life cycle of the project and while the parties are part of the project coalition. It provides a framework for documenting the process for managing the relationship. It also gives the project management function a dynamic record of how relationships evolve and how the project manages relational risk. The eight-step process encompasses activities pre- and post-partner selection but for the purposes of this chapter we highlight some key points from the standard relating to step 5 – "working together" – which covers the period when the project coalition is formed and collaborating together on infrastructure delivery. We qualify this by emphasising that, as described in the standard, there are important steps that need taking at the corporate level, before the forming of the project coalition and beyond "working together," that have a significant influence on the effectiveness of relationships.

Under the heading "working together," ISO 44001 details a 15-step process founded on the key principles of executive sponsorship, operational structure, governance and roles and responsibilities (British Standards Institute, 2019; pg. 44). In terms of executive sponsorship, the need for strong executive support is highlighted, with senior representatives of firms in the project coalition taking responsibility for the successful execution of collaborative relationships, including establishing and rewarding the desired behaviours (sharing of information, innovative thinking and problem-solving).

In terms of structure, governance and roles and responsibilities, the aim and objectives of the project must be validated and agreed by all, and at the same time ensure that while each party in the project coalition might have different aims and objectives, they are complementary and not contrary to those of the project. A project leader needs appointing with the skills to operate outside traditional command and control modes of management; that is, the ability to influence and engage all parties within the project coalition is important. The project management structure should draw upon representatives from the different parties and be focused on delivering the project objectives. A strategy for communicating with all stakeholders, both within and outside the project coalition, needs to be established and to be informed by well-understood processes for stakeholder identification, mapping and needs analysis. This strategy will identify and force agreement on areas for not sharing information, if applicable, and the methods of communication to apply, in a consistent way. This feeds into establishing a knowledge management process. Defining the information requirements and access upfront mitigates for conflict at a later stage, which typically undermines the relationships. Part of effective working together is establishing a joint risk register. The individual risk registers of the project coalition parties, with the sharing of analyses and the adopting of a common approach, inform this. Such a common approach promotes transparency and builds trust.

Throughout the period of working together, there should be an assessment of project management processes and, where there is potential for process improvements, appropriate modifications made. The undertaking of periodic reviews of the collaborative competence of the project coalition links to the ongoing measurement of delivery and performance against agreed objectives. Given an expectation that there are likely to be issues between parties within the project coalition, there should be a dispute resolution process agreed on by everyone. Similarly, at some point parties will leave the coalition and, indeed, the coalition will disband at the end of the project, so there needs to be an exit strategy. Defining the rules of disengagement provides clarity, enabling the focus to be on the period of engagement. Linked to this is the establishment of both a mobilisation and demobilisation strategy that considers the staffing implications on relationships as the project coalition is formed and disbanded. There is a high level of uncertainty at these two points in time and this uncertainty can have negative effects in terms of individual behaviour. Effective working together also encompasses appropriate contractual arrangements and internal agreements, which we discussed earlier in the chapter in the section on relational contracting. It is in this step of the operational processes that the project management function updates the already-created RMP in order to provide a management platform to align with the project objectives. It then provides clarity for all involved in delivering the project by capturing the key elements of the collaborative relationship, including the critical success factors and expectations of the project coalition parties facilitated through collaboration. Having established a clear way of working together, the focus moves onto the next step in the overall process, i.e., establishing approaches to build additional value from the project coalition ("value creation") and then onto the final two steps, "staying together" and "exit strategy" activation. For details of how to undertake these three steps, see British Standards Institute (2019), pp. 60–76.

The UK Infrastructure and Projects Authority (IPA), specifically focusing on organisational design and development (ODD), put a framework in place specifically designed for infrastructure delivery (Infrastructure and Projects Authority, 2014). Having a well-established ODD provides a structured approach that is necessary when collaborative working is required. As with the ISO standard 44001, the IPA predicate the use of the framework upon a recognition of the importance of establishing structures, systems and processes to enable effective relationships upfront. In the case of the IPA, they acknowledge the importance of upfront activity early in the project life cycle by means of a project initiation routemap (PIR). The PIR, as conceptualised by the IPA, is broad in its scope, acting as an aid to strategic decision-making, which supports aligning infrastructure sponsor and client organisations' delivery capability. It also provides an objective and systematic approach to initiating an infrastructure project and diagnostics to address problems associated with capability gaps. An ODD strategy considers the context in which an infrastructure project takes place and aspects relating to the design and development of the organisation put in place to achieve the objectives. Stakeholder workshops typically take place to help define the ODD strategy. There are many considerations when developing a comprehensive strategy. In terms of managing relationships, these include consideration of risk, reward, culture, environment, and working practices; the thread linking them is the drive of behaviours, which are consistent with a relational approach. Two aspects highlighted in the PIR, with examples from infrastructure projects in practice, are behavioural assessments and leadership. A behavioural and cultural assessment by infrastructure clients and owners to select partners places an early emphasis on behaviour and culture, which sets the tone for what follows and enables the implementation of a management model of collaborative working. As has been mentioned elsewhere in this chapter, leadership has a crucial role in supporting behavioural and cultural change by recognising the importance of interpersonal skills and the building of relationships.

Finally, in this section, we highlight our own framework for collaborative working and effective relationship management in a project environment, which is called "The CURED Framework™" (Bryde *et al.*, 2019; Association of Project Management, 2020). Generated from academic study of a number of cases, including two large-scale infrastructure projects, the framework provides mechanisms for resolving agency problems in client-contractor relationships to deliver project success. These mechanisms are categorised in five areas: contract (C), understanding (U), resources (R), education (E) and delegation (D). The framework includes several fundamental principles in each area, operationalised during project delivery through various actions. These principles are, first, the right contract promotes shared goals and expectations, thereby reducing the likelihood of agency-related problems occurring, such as opportunistic behaviour, excessive performance monitoring or concealment of negative outcomes. Next, effective communication promotes shared understanding about goals and risks and reduces the agency-related problem of information asymmetry. Then, having the right resources, in terms of staff, in the right place at the right time is crucial to avoiding agency-related problems and in promoting trust. Next, education through robust training programmes helps to minimise agency-related problems and is conducive to project success. Finally, delegation by the client leads to empowerment of the contractor and the supply chain, leading to flexible, trusting and productive relationships (APM, 2020).

13.9 Chapter summary

We have discussed various management concepts and methods for infrastructure project delivery that emphasise behavioural aspects. We refer to academic and practitioner studies that argue for new approaches to delivery and have such an emphasis as an appropriate response to some of the causes of poor performance and low productivity levels reported as outcomes of major infrastructure projects. A key concept that underpins the emphasis on relationships is that of the project coalition, where different firms come together to form a network in which there are numerous complex inter- and intra-relationships between these firms. These relationships result in certain behaviours on the part of members of the project coalition and there is relational risk that they will work against the project. To counter this, we outline an approach referred to as "relational governance," which focuses on building good working relationships between the parties to the project coalition. Within a relational governance approach, we further highlight the crucial role of the contract, introducing the concept of "relational contracting." To apply concepts and methods that put emphasis on relational aspects requires a structured approach, with management interventions, especially from the client (owner). We highlight elements of such a structured approach including the need to identify key relational indicators/critical success factors and to establish an organisational climate conducive to good working relationships. We finish our theoretical discussion by introducing frameworks to both set up and to maintain effective working relationships between the members of the project coalition. There are useful lessons to take from internationally accredited frameworks that are generic to all collaborative relationships that involve more than one firm and from frameworks specific to projects and to infrastructure projects. As with the concepts and methods, the importance of owner (client) involvement is key, as is the imperative of starting to design and develop the relational approach early in the initiation stage of the project. Lastly, taking two comparable projects tasked with delivering new urban rail infrastructure, we describe how an operational model emphasising relational aspects resulted in a project more adaptable to dealing with uncertainty and better able to maintain effective working relationships, with resultant desirable outcome, than a project which put the emphasis primarily on structural aspects.

13.9.1 Chapter discussion questions

- Review the discussion of performance and productivity of major infrastructure projects. Discuss your experiences of dysfunctional relationships between different parties being a major cause of performance/productivity shortcomings.
- Discuss using concepts of relational governance and relational contracting to mitigate and manage relational risk. Are these concepts realistic management approaches for infrastructure project delivery?
- This chapter presented a number of relational indicators/critical success factors. In your opinion, which are the most important and how would you go about ensuring they are present?

Managing relationship risks on major infrastructure projects

- This chapter argued for the need to create an organisation climate conducive to collaborative team working. What were your reactions to the descriptions of the type of climate that is required?
- Review the discussion of frameworks for relationship management. Discuss the importance of such frameworks in establishing and maintaining effective relationships between the parties involved in infrastructure delivery.

13.10 Case study (for the purpose of confidentiality project location and name withheld)

13.10.1 Introduction

In this section, we look at two complex urban rail programmes as examples of major infrastructure projects. Each programme is comprised of numerous interlinked individual projects. We highlight some of the structural, process and operational decisions they made in respect to dealing with relational aspects to manage progress and adapt to change, specifically highlighting how these decisions affected the behaviours of the project team and other key stakeholders. For reasons of confidentiality, we cannot specifically name the programmes. Still we can provide some key characteristics of each. These characteristics provide insight into the scope, project organisation and the challenges faced and help to explain the contextual factors that influenced the project management approach each adopted.

- Both programmes were significant in size and complexity (programme budgets were greater than $1 billion and they each had an overall baseline duration of more than 10 years).
- Both sought to bring new rail assets and systems into use in an ongoing operational environment of two busy urban centres.
- Both would need to develop, test, assure and deploy safety-critical systems to control transport and to maximise capacity and availability.
- Both would lead to a significant operational change to the running of each railway, meaning that they needed to undertake a substantial amount of operations and maintenance consultation and collaboration.
- Each used a private-public partnership (PPP) funding and delivery model. They had an incentivised output specification based on technical standards and performance. As such, each case adopted a "thin client" approach – the client outsourced the management and delivery and focused its own involvement on just the commercial and compliance aspects, an idea considered best suited to achieving the required outcomes. They coupled this with the establishment of a supply chain, which would be able to innovate and adapt as needed.

Here we focus on the different approaches that both programmes took at a significant point where accountability for delivery changed and how those decisions influenced very different behavioural outcomes. We will also see how these behaviours factored into each programme and dealt with critical challenges. It is important to note that we base our observations on our own opinions that, while informed, are subjective by nature. For this comparison, we will refer to the two infrastructure projects as programme A and programme B.

13.10.2 Overview of the programmes

Both programmes were initially set up as part of a PPP. The client organisations took a primarily contractual and technical compliance role. In both cases, the client had significant technical and operational expertise in the railway industry and would need to second this expertise into the supply chain. It is important to note that this "thin client" approach is an aspiration for many modern clients undertaking complex projects involving public transportation infrastructures. However, it relies on an inherent ability to delegate and trust third parties to deliver in sometimes new and unfamiliar ways. This often involves cultural change for the client, which can be challenging, and as such, there can be ongoing challenges to the project if a client wishes to be a "thin client" but in practice refuses to "let go" and truly delegate authority and decision-making.

As with most programmes, the initial phase of design proved complex and challenging. Both programmes were subject to delays, which is common on such projects given the inconsistent rate at which highly sophisticated designs and scope mature. Also given the size and length of the programmes, there was a concern about potential obsolescence of elements of the infrastructure, requiring an element of continuous design throughout much of the programme's life. At this point, there is little to differentiate between the two approaches; however, during the early construction phase, there was a critical decision to move away from the PPP model. This meant that, while the programmes were in flight, accountability for successfully delivering the programmes would need to revert to the client organisation.

13.10.3 Programme A

As the client took over accountability for the programme, it increased focus on the following key areas:

13.10.3.1 Internal delivery capability

Up until the early construction phase, the client team had primarily acted in a contractual and administrative capacity by ensuring the following of safety and technical standards, ensuring that relevant approvals were undertaken by all parties, and operational and maintenance input was managed. Direct control of the programme would now reside with the client team, which the senior management in the client identified as a capability gap. Hence, there was staff training undertaken and substantial investment in project controls and reporting.

13.10.3.2 Creating a fully integrated team

While the members of the project coalition that formed the delivery team had been co-located since the early part of the project, there had been a natural level of physical separation due to the different roles undertaken and a need for impartiality by the client. Re-forming the project coalition into a new and effective structure meant fully integrating the different parties. This brought with it a level of cultural challenge, as the new structure, including a limited degree of hierarchy, bedded in. To meet this challenge, the project leadership team evaluated each company against its management capability and then they selected the best person for a role in the delivery team, regardless of the organisation they came from. While this created some initial resistance and questioning by some companies, especially when role allocation was not as they might have expected, the process was critical to getting the team working together and delivering efficiently.

13.10.3.3 Implementing a behavioural transformation programme

Given the significant cultural impact of this organisational merger to form a fully functioning delivery team within the project coalition, the primary focus was on developing and ensuring the presence of the right behaviours across the wider team. A behavioural transformation programme was embarked upon which specified what behaviours were expected of everyone and how they could be tangibly applied in each role. Guidance around the application of these behaviours was particularly crucial, as it is sometimes difficult to translate soft skills into action.

The changes outlined above resulted in an operational model that emphasised behavioural aspects. The programme management team was broad instead of deep. This created a notional integrated flat structure, where levels of autonomy and accountability were intrinsic to day-to-day management of the programme. Achieving a fully integrated cross-organisation project team with a standard set of defined behaviours meant that every person felt empowered to overcome challenges as they arrived. When there were issues where parts of the programme had too much control over individual projects, management took action to maintain the integrated approach. Given the complexity and uncertainty of the implementation phase, the programme management team implemented a "top-down" approach to planning and programme controls. They chose several high-level key events and management focus was on productivity and milestones for achieving these deliverables, as opposed to a more detailed level of planning. Project communications continually reinforced the importance of these key events and their enabling activities. There were high levels of inter-project communication, with regular dissemination of every critical issue and decision to the broader team. The result was that everyone involved understood what needed to be done and why. This transparent decision-making method also drove accountability, which was crucial when the programme faced challenges and action was required. Note that in this sense, there are clear parallels with the concept of "agile" project management, although most responsible for infrastructure delivery in the rail industry perceive the concept as the sole domain of software development.

13.10.4 Programme B

As the client took over accountability for the programme, it increased focus on the following key areas, which differed from those described in programme A:

13.10.4.1 Supply chain capability

Given the stage of the programme, and using feedback from lessons learnt, it was decided that, where practical, critical suppliers and products should be re-evaluated to de-risk delivery and realise any opportunities that a new supply chain could bring. This was clearly a significant decision. However, cost pressures and the need to migrate to a new railway in a safe way, without an adverse impact on existing operations, was a driving force for the client. After undertaking this analysis, the programme management team decided to change suppliers in critical areas. While there were clear commercial consequences, it was felt that in the end this would increase the realisation of programme benefits and help negate the inherent dis-benefits of delivery.

13.10.4.2 Commercial capability

The client undertook an innovative procurement approach to identifying and evaluating new products and suppliers to yield significant cost and risk benefits. This approach was a step change for client commercial capability, and they spent considerable time and resources to realise these ambitions. They recognised that to achieve the benefits identified in the new supply chain contract there needed to be ongoing investment in commercial capability and business processes.

There was training of the client team in commercial management and an increased focus on developing and following robust procedures and enforcing commercial compliance by all parties.

13.10.4.3 Programme controls

Given the scale and complexity of not just the technical scope of the programme but of the many organisations involved – some in the project coalition were from different countries – the client invested heavily in programme control systems, resources and processes. The driver was the need for integrated and data-rich client decision-making capability. The client imbedded these requirements into an existing supplier, and they mandated them into the new ones. Client requirements analysis during the procurement process had highlighted the need for supply chain programme controls capability and the programme management team enforced adherence to these requirements as standard during delivery.

Rather than emphasising behavioural aspects, as was the case in programme A, the changes made in programme B resulted in an operational model focused more on structures. As is standard industry practice, the shape of the programme was around asset class and critical areas of the programme that had on-boarded new suppliers ensured that new commercial practices were enforced. The enhanced focus on programme controls capability brought with it new processes and data quality requirements that needed to be developed and implemented on a live programme, which proved challenging. The scale of the organisation and the level of management hierarchy, while necessary at the time, meant that in certain areas where the client did not have direct control, information moved slowly up and down the supply chain and hence between parties to the project coalition. Rigorous enforcement of commercial processes also meant it was necessary to create alternate versions of project progress, one for commercial purposes and the other to manage progress and interfaces. The focus on commercial capability ultimately meant that success hinged on the suitability of the contractual processes and the scope defined within, as well as the client's ability to administer significant levels of change. Ensuring robust impact assessment and tight change control created challenges given the inherently iterative approach to design and development.

13.10.5 What can we learn from these two different approaches?

It is true to say that both programmes took a "best practice" approach to developing their management capability, and at the time, there were distinct environmental factors that informed their decisions. Even so, there were apparent differences in approach: programme A focused on behavioural aspects and programme B on structural changes. It is interesting to see the effect that these different approaches had on the programmes during some of the most challenging times. Programme A proved inherently more adaptable at dealing with uncertainty. It was able to maintain alignment of goals across the project coalition and key individuals with management responsibility helped create a culture where each person not only knew what they needed to do but also why they needed to do it. This meant that everyone saw clearly that each decision made was in the best interests of the programme, allowing teams to flex and change as necessary, which lessened the impact of disruption due to the unknown. On the other hand, programme B's focus on building a highly detailed, hierarchical and structured approach to delivery, while understandable at the time, ultimately resulted in less ability to adapt to the significant changes that development and delivery brought. It might have been able to avoid or mitigate some considerable setbacks it suffered during the delivery of the infrastructure if it had incorporated elements of a behavioural approach.

13.10.6 Case discussions questions

1. What might be the motivations behind an owner (client) designing an operational model emphasising structural issues rather than behavioural ones, and how would you counter them if you were promoting a different, more relational-based, strategy?
2. What role does the owner (client) play in ensuring the success of a relational approach to infrastructure project delivery?
3. What are the arguments for a hybrid approach, incorporating elements of a relational approach and a structural approach, and what contingent factors would influence its design?
4. What lessons are there in terms of the required roles, responsibilities, skills, competences and behaviours from the approach taken on programme A?

References

Association for Project Management (2020). Resolving agency issues in client–contractor relationships [online] Available from: https://www.apm.org.uk/resources/find-a-resource/research-series/resolving-agency-issues-in-client-contractor-relationships/ [cited 15 April 2020].

Böhle, F., Heidling, E., and Schoper, Y. (2016). A new orientation to deal with uncertainty in projects. *International Journal of Project Management*, 34(7), pp. 1384–1392.

British Standards Institute (2019). BSI ISO 44002:2019 Collaborative business relationship management systems – Guidelines on the implementation of ISO 44001 Available from: https://bsol.bsigroup.com/Bibliographic/BibliographicInfoData/000000000030363673 [cited 26 April 2020].

Bryde, D.J., Unterhitzenberger, C., and Joby, R. (2019). Resolving agency issues in client–contractor relationships to deliver project success. *Production Planning & Control* [online], 30(13), pp. 1049–1063.

Cameron, K. (2020) An introduction to the Competing Values Framework [online] Available from: https://www.thercfgroup.com/files/resources/an_introduction_to_the_competing_values_framework.pdf [cited 18 April 2020].

Cantarelli, C.C., Flybjerg, B., Molin, E.J.E., and van Wee, B. (2010). Cost overruns in large-scale transportation infrastructure projects: Explanations and their theoretical embeddedness. *European Journal of Transport and Infrastructure Research*, 10(1), pp. 5–18.

Colledge, B. (2005). Relational contracting – creating value beyond the project. *Lean Construction Journal*, 2(1), pp. 30–45.

Eisenhardt, K.M. (1989). Agency theory: An assessment and review. *Academy of Management Review*, 14(4), pp. 57–74.

Florical, S. and Lampel, J. (1998). Innovative contractual structures for inter-organizational systems. *International Journal of Technology Management*, 16(1), pp. 193–206.

Flyvberg, B., Garbuio, M., and Lovallo, D. (2009). Delusion and deception in large infrastructure projects: Two models for explaining and preventing executive disaster. *California Management Review*, 51(2), pp. 170–194.

Gil, N. (2009). Developing cooperative project client–supplier relationships: How much to expect from relational contracts? *California Management Review*, 51(2), pp. 144–169.

Global Infrastructure Initiative (2017). *Voices on infrastructure: Transforming project delivery* [online] Available from: https://www.mckinsey.com/~/media/McKinsey/Industries/Capital%20Projects%20and%20Infrastructure/Our%20Insights/Voices%20on%20Infrastructure%20Transforming%20project%20delivery/Voices-on-Infrastructure-Transforming-project-delivery.ashx [cited 28 April 2020].

Gray, R.J. (2001). Organisational climate and project success. *International Journal of Project Management* [online], 19(2), pp. 103–109.

Handley, S.M. and Benton, Jr. W.C. (2009). Unlocking the business outsourcing process model. *Journal of Operations Management*, 27(5), pp. 344–361.

Hannevik, M.B., Lone, J.A., Bjørklund, R., Bjørkli, C.A., and Hoff, T. (2014). Organizational climate in large-scale projects in the oil and gas industry: A competing values perspective. *International Journal of Project Management*, 32(4), pp. 687–697.

Harper, C.M., Molenaar, K.R., and Cannon, J.P. (2016). Measuring constructs of relational contracting. *Journal of Construction Engineering Management* [online], 142(10), pp. 04016053 Available from: https://ascelibrary.org/doi/pdf/10.1061/%28ASCE%29CO.1943-7862.0001169 [cited 2 May 2020].

Hellowell, M. and Vecchi, V. (2015). The non-incremental road to disaster? A comparative policy analysis of agency problems in the commissioning of infrastructure projects in the UK and Italy. *Journal of Comparative Policy Analysis: Research and Practice*, 17(5), pp. 519–532.

HM Treasury (2010) *Infrastructure cost review: Main Report* [online] Available from: https://assets.publishing.service.gov.uk/government/uploads/system/uploads/attachment_data/file/192588/cost_review_main211210.pdf [cited 28 April 2020].

Hughes, K. and Waterhouse, P. (2018) *Understanding NEC4: Term service contract* [online], 1st ed. London: Routledge Available from: https://www.routledge.com/Understanding-NEC4-Term-Service-Contract/Hughes-Waterhouse/p/book/9780815348368 [cited 3 May 2020].

Infrastructure and Projects Authority (2019). *People, performance and principles: The IPA's priorities for 2020* [online] Available from: https://ipa.blog.gov.uk/2019/09/24/people-performance-and-principles-the-ipas-priorities-for-2020/ [cited 28 April 2020].

Infrastructure and Projects Authority (2014). *Improving infrastructure delivery: Project initiation routemap – Organisational Design and Development Module* [online] Available from: https://www.gov.uk/government/publications/improving-infrastructure-delivery-project-initiation-routemap [cited 15 April 2020].

International Standards Organisation (2017). *ISO 44001:2017 Collaborative business relationship management systems* [online] Available at: https://ww.iso.org/standard/72798.html [cited 15 April 2020].

Kalnins, A. and Mayer, K.J. (2004). Relationships and hybrid contracts: An analysis of contract choice in information technology. *The Journal of Law, Economics, and Organisation*, 20(1), pp. 207–229.

Ke, H., Cui, Z., Govindan, K., and Zavadskas, E.K. (2015). The Impact of contractual governance and trust on EPC projects in construction supply chain performance. *Inzinerine Ekonomika-Engineering Economics* [online], 26(4), pp. 349–363.

Levitt, R.E., Henisz, W.J., and Settel, D. (2009). Defining and mitigating the governance challenges of infrastructure project development and delivery. *Conference on Leadership and Management of Construction* [online], Fallen Leaf Lake, Lake Tahoe, California, 5–8 November, pp. 2–17 Available from: https://pdfs.semanticscholar.org/f558/6c05af9881baa4b7d29dc2c3012f1c17f223.pdf?_ga=2.33955813.555005262.1588002130-2075666569.1588002130 [cited 27 April 2020].

Li, T., Ng, S., and Skitmore, M. (2012). Conflict or consensus: An investigation of stakeholder concerns during the participation process of major infrastructure and construction projects in Hong Kong. *Habitat International*, 36(2), pp. 333–342.

Meng, X. (2012) The effect of relationship management on project performance in construction. *International Journal of Project Management*, 30(2), pp. 188–198.

Mok, K.Y., Qiping, G.S., Yang, R.J., and Li, C.Z. (2017). Investigating key challenges in major public engineering projects by a network-theory based analysis of stakeholder concerns: A case study. *International Journal of Project Management*, 35(1), pp. 78–94.

Mok, K.Y., Qiping, G.S., and Yang, R.J. (2015). Stakeholder management studies in mega construction projects: A review and future directions. *International Journal of Project Management*[online], 33(2), pp. 446–457.

Müller, R. and Turner, J.R. (2005). The impact of principal–agent relationship and contract type on communication between project owner and manager. *International Journal of Project Management* [online], 23(95), pp. 398–403

National Audit Office (2016). Delivering major projects in government: A briefing for the Committee of Public Accounts [online] Available from: https://www.nao.org.uk/wp-content/uploads/2016/01/Delivering-major-projects-in-government-a-briefing-for-the-Committee-of-Public-Accounts.pdf [cited 28 April 2020].

NEC: (2020) *Procure, Manage, Deliver* [online]. Available from: https://www.neccontract.com/ [cited 3 May 2020].

Olander, S. (2007). Stakeholder impact analysis in construction project management. *Construction Management and Economics*, 25(3), pp. 277–287.

Pal, R., Wang, P., and Liang, X. (2017). The critical factors in managing relationships in international engineering, procurement, and construction (IEPC) projects of Chinese organizations. *International Journal of Project Management*, 35(7), pp. 1225–1237.

Pryke, S.D. (2005). Towards a social network theory of project governance. *Construction Management and Economics*, 23(9), pp. 927–939.

Pryke, S.D. (2004). Analysing construction project coalitions: Exploring the application of social network analysis. *Construction Management and Economics*, 22(8), pp. 787–797.

Rowlinson, S. and Cheung, Y.K.F. (2004). A review of the concepts and definitions of the various forms of relational contracting. In: K. Varghese, and S. Kalidindi (Eds.), *International Symposium of the CIB W92 on Procurement Systems - Project Procurement for Infrastructure Constructions*. India Institute of Technology (CD ROM), India, pp. 227–237. Available from: https://eprints.qut.edu.au/17861/ [cited 2 May 2020].

Sakal, M.W. (2005). Project alliancing: A relational contracting mechanism for dynamic projects. *Lean Construction Journal*, 2(1), pp. 67–79.

Sharma, A. and Guptab, A. (2012). Impact of organisational climate and demographics on project specific risks in context to Indian software industry. *International Journal of Project Management* [online], 30(2), pp. 176–187.

Shrestha, A. and Martek, I. (2015). Principal agent problems evident in Chinese PPP infrastructure projects. Proceedings of the 19th International Symposium on Advancement of Construction Management and Real Estate [online] pp. 759–770 Available from: https://link.springer.com/chapter/10.1007/978-3-662-46994-1_62 [cited 19 April 2020].

Smith, E., Umans, T., and Thomasson, A. (2018). Stages of PPP and principal–agent conflicts: The Swedish water and sewerage sector. *Public Performance and Management Review* [online], 41(1), pp. 100–129.

Soetanto, R., Proverbs, D.G. (2002). A conceptual tool for assessing client performance in the construction project coalition. *Journal of Civil Engineering Science and Application* [online], 4(2), pp. 60–68.

Suprapto, M., Bakker, H.L.M., and Mooi, G. (2015). Relational factors in owner–contractor collaboration: The mediating role of teamworking. *International Journal of Project Management*, 33(6), pp. 1347–1363.

Ul Haq, S., Gu, D., Laing, C., and Abdulla, I. (2019). Project governance mechanisms and the performance of software development projects: Moderating role of requirements risk. *International Journal of Project Management* [online], 37(4), pp. 533–48.

van Gestel, N., Koppenjan, J., Schrijver, I., van de Ven, A., and Veeneman, W. (2008). Managing public values in public-private networks: A comparative study of innovative public infrastructure projects. *Public Money and Management*, 28(3), pp. 139–145.

Wallace, J., Hunt, J., and Richards, C. (1999) The relationship between organisational culture, organisational climate and managerial values. *International Journal of Public Sector Management*, 12, pp. 548–564.

Ward, S. and Chapman, C. (2003) Transforming project risk management into project uncertainty management. *International Journal of Project Management*, 21(2), pp. 97–105.

Winch, G. (1989). The construction firm and the construction project: A transaction cost approach. *Construction Management and Economics*, 7(4), pp. 331–345.

Wright, J.N. and Fergusson, W. (2009). Benefits of the NEC ECC form of contract: A New Zealand case study. *International Journal of Project Management*, 27(3), pp. 243–249.

Zhang, L. and Qian, Q. (2017). How mediated power affects opportunism in owner–contractor relationships: The role of risk perceptions. *International Journal of Project Management*, 35(3), pp. 516–529.

14

EMPOWERING PUBLIC-PRIVATE PARTNERSHIP IN MAJOR INFRASTRUCTURE SYSTEMS

Neema Kavishe and Nicholas Chileshe

14.1 Introduction

14.1.1 Empowering infrastructure public-private partnerships

Public-private partnerships (PPPs) are an effective way of transferring life-cycle costs of infrastructure systems off public-sector budgets and simultaneously create investable assets for the private sector. They have demonstrated their value in infrastructure projects around the world. Private investment in infrastructure, in partnership with the public sector, can motivate accountability in the delivery of critical assets, stretch public capital and help local, state and national governments deliver highways, bridges, ports, airports and other social infrastructure such as schools and housing faster and cheaper, and ensure that they are properly maintained. Infrastructure is a base and also a key driver to any country's economy. However, the development of new infrastructure remains ineffective and insufficient to most sub-Saharan African countries like Tanzania. Global experience clearly shows that well designed PPPs and properly managed contracts can make substantial contributions to the infrastructure delivery and economy as well. The lack of well-prepared PPPs, inadequate feasibility studies, risky policies, unbalanced risk allocations and poor enabling of the environment, to mention a few, all undermine the successful implementation of these projects and also discourages the private sector participation. Therefore, in addressing the challenge to boost the economic and social aspects of a country, there is a need to improve and empower the implementation of infrastructure PPPs through different aspects. For that reason, this chapter focuses on explaining the standard procedures for PPPs' preparation and also identifies the key challenges hindering their successful implementation. Furthermore, inadequate management has been considered among the prevailing challenges; thus, this chapter also describes the key steps to follow to manage the complex PPP process. On the other side, since investors are profit oriented, they always require detailed information to convince them whether the project can service the debt. In that regard the need to conduct bankable feasibility studies is inevitable, hence their aims and processes are detailed herein. With the aim of empowering PPPs the chapter also reviews the different types of risks and how to best allocate the risks to attain value for money. Finally, as explained in chapter 12, the success of any infrastructure project's delivery is heavily dependent on a conducive and enabling environment in which these projects are implemented; therefore, this chapter also emphasises the need for creating a conducive enabling environment to empower PPPs. At the end of the chapter, a Tanzanian case study is included to draw lessons from.

14.1.2 Chapter aim and objectives

The aim of this chapter is to examine how governments can enhance existing PPP models and practices in major infrastructure systems. The main objectives are to:

- Explore the approaches and stages to preparing PPPs;
- Identify the challenges for PPPs in Tanzania;
- Establish the strategies for managing the rigorous PPP process;
- Analyse the process of conducting a bankable feasibility study;
- Describe the process of structuring a balanced risk allocation;
- Examine ways of creating a conducive PPP environment;
- Recommend desirable strategies for enhancing PPP models and practices.

14.1.3 Learning outcomes

The following are the learning outcomes identified for this chapter. Readers will be able to:

- Learn the best strategies to enhance existing PPP models and practices in major infrastructure systems;
- Review the major challenges of implementing PPP projects;
- Identify strategies for managing the rigorous PPP process;
- Understand the process of conducting a bankable feasibility study;
- Comprehend the process of structuring a balanced risk allocation;
- Understand ways of creating a PPP conducive environment.

14.2 Preparing PPPs

PPP projects normally go through various stages or phases of preparation unlike non-PPP (traditionally procured) projects. It is important to note that governments are advised to develop PPP projects which can deliver better value for money, are cost-benefit, defendable and fiscally responsible. But all these criteria can be met once the project is fully designed (World Bank, 2017). Achieving these criteria makes the PPP process complex UNESCAP (2011) and expensive; however, the benefits are significant and the likelihoods of success is high. According to the World Bank (2014) PPP guidelines, the actual preparation is preceded by identification of priority project and then screening the identified priority projects for PPP potential. The projects that pass the screening stage are further developed and appraised. A lot of groundwork is needed before a project is concluded as a PPP project.

The PPP reference guide further advises that, during the preparation stage, it is very important to engage potential users of the proposed facility and the populations likely to be affected by the project. Their engagement allows for testing the quality of the project and provides features for its optimisation (Foo *et al.*, 2011; Umar *et al.*, 2018; World Bank, 2017). In some cases, before a project is to be implemented, there could be persons who are to be relocated as part of preparation; therefore, the PPP reference guide insists on having extensive communication with people to be relocated and it should be planned and publicised and thoroughly communicated to mitigate the environmental impact. Failure to undertake these preparations as advised results in disputes. Examples of previous projects' experiences in Tanzania have demonstrated that disputes are likely to arise when adequate preparations and communications are not done. The National Housing Corporation (NHC) experienced a number of disputes when tenants refused to vacate their properties that were due to be redeveloped. Likewise, 10 projects were stalled for several reasons including pending cases in the courts. Therefore, it is important to note that PPP skills and knowledge are necessary throughout all stages from the identification of priority project through to implementation and contract management. Various studies have identified that lack of PPP skills is among the prominent obstacles hindering adequate preparation of PPP projects in most developing countries (Osei-Kyei and Chan, 2017) as Kavishe *et al.* (2018) further explain in the next section. Similarly, El-Gohary *et al.* (2006) contended that the involvement of stakeholders should be considered early on when preparing and planning for a PPP project. Foo *et al.*'s (2011) study confirms that stakeholder management has the ability to offer different stakeholders a chance to make meaningful contributions to a PPP project.

14.3 The challenges for PPPs in Tanzania

The implementation of PPPs in Tanzania faces many challenges as evidenced in many other developing countries such as India (Babatunde *et al.*, 2015), Nigeria (Ibem and Aduwo, 2012; Sengupta, 2004; 2006; Ukoje and Kanu, 2014), Ghana (Osei-Kyei and Chan, 2017) and Malaysia (Ismail and Azzahra Haris, 2014). These challenges originate from a number of issues related to the cultural, social, economic, political and environmental aspects of a particular country. PPP is not new in Tanzania; since the early 1990s, the public and the private sector have actively collaborated in public service delivery. However, it is still at its early stages primarily because of a lack of direct experience and inadequate new investment. Despite new developments undertaken by the Tanzanian government in enhancing the regulatory frameworks via PPP regulations in 2011, and the Public Procurement Act 2011, as observed by Mboya (2013), a number of challenges still remain, such as the relative infancy of rather complex PPPs and lack of experience across the stakeholder chain. Also based on a study by Kavishe and An (2016), the top six challenges in the implementation of PPP projects in Tanzania include the following:

- Inadequate PPP skills and knowledge;
- Poor PPP contract and tender documents;
- Inadequate project management and monitoring by public sector;
- Inadequate legal framework;
- Misinformation on financial capacity of private partners; and
- Lack of competition.

The following subsections present a brief discussion of individual challenges as listed above.

14.3.1 Inadequate PPP skills and knowledge

PPP is still a new approach to most developing countries like Tanzania and because of its complex nature they require someone with adequate PPP knowledge and skills to identify the right PPP candidate, planning, preparing the implementation and managing. Currently, the majority of PPP practitioners (public and private) have little experience in implementing PPP projects. This is proved through the performance of past PPP projects such as the National Housing Corporation (NHC) PPP housing projects. For example, according to Kavishe (2018), the public partner (NHC) had secured a total of 190 projects worth Tshs.410.11 billion under the public–private partnership strategy. Out of those 190, only 29 projects (15 percent) worth Tshs.41.06 billion were completed. Forty-eight projects (25 percent) worth Tshs 96.64 billion were uncompleted and had exceeded the contract duration because of various challenges and delays. Furthermore, up to 100 projects were under preparation stage for a long time due to PPP incompetence, disputes and refusal from previous tenants to vacate the properties to be redeveloped. The lack of skills and capacity eventually resulted in poor planning for the PPP project, poor risk identification, allocation and management as well as poor PPP project management. Inadequate PPP skills also caused its slow progress and failures (Moskalyk, 2011). The benefits of having well-trained and experienced public sector officials is that they can clearly determine where difficult issues would most likely arise through feasibility studies, and they will possess the ability to select the appropriate PPP candidate projects. The over-reliance of the public officials on the private partners for acquiring the desirable PPP skills is risky and can create inequity among the partners as posited by the equity theory Scheer *et al.* (2003).

14.3.2 Poor PPP contract and tender documents

Poor PPP contracts and tender documents are other key challenges that have been experienced in various projects. The poor performance of PPPs in Tanzania was also attributed to unfavourable contract terms for the contracting authorities. This challenge was also present in other developing countries such as South Africa Fombad (World Bank, 2014) and Thailand (Osei-Kyei and Chan, 2017), including missing important clauses in contracts, contradictory contractual provisions, agreements biased in favour of some partners and non-adherence to rules and regulations. This implied that the local practitioners, both public officials and private investors, in Tanzania had very little experience in preparing PPP contracts and tender documents, which resulted in the absence of robust and clear contractual agreements.

14.3.3 Inadequate project management and monitoring by the public sector

As with most developing countries, the application of project management best practice has always been an issue (Kavishe and Chileshe, 2019). The Tanzanian construction industry is characterised by poor project performance. It is associated with contractors' inadequate project management and human resource skills (Kikwasi and Escalante, 2018) and also a default traditional construction procurement system (TCPS), which provides a poor relationship management system. For example, in the Tanzanian PPP housing projects the work of managing and project monitoring was entirely left to the private partners, giving them a loophole to make alterations on the authorised design, which led to disputes.

14.3.4 Inadequate legal framework

Despite the absence of a legal framework PPPs existed in the 1980s and 1990s, but they have mostly been undertaken through the privatization program and are mostly involved in direct service delivery (URT, 2009). In the last 12 years, they were few new investments in physical infrastructure, with few exceptions in the power and communications sectors; however, there has been little success to these projects. Most Tanzanian PPPs schemes were shorter because most were in the form of concession. For example, the Kilimanjaro International Airport (Bowers and Khorakian, 2014) signed a 25-year concession agreement with Kilimanjaro Airports Development Company (KADCO) in 1998 and the Port of Dar es Salaam awarded a 10-year concession (2000–2010) to the Tanzania International Container Terminal Services (TICTS). Both projects were unsuccessful as they were undertaken in the absence of PPP guidelines. The Tanzania PPP policy was established in 2009, the PPP act and PPP regulations were enacted in 2011; they were amended in 2014 and 2015, respectively.

14.3.5 Misinformation on financial capacity of private partners

In this challenge it was revealed that the contracting authority had incorrect information on the financial capacity of the private partner (Kavishe and An, 2016; Kavishe *et al.,* 2018). For instance, in the PPP housing project in Tanzania it was observed that a number of these projects failed and the majority were not completed as projected because private partners did not have enough financial capacity to undertake the project.

14.3.6 Lack of competition

Lack of competition is also among the PPP challenges identified in Tanzania. Essentially, during their very early PPP housing projects, the NHC selection of private partner was first come, first served (Kavishe, 2010). There was no room for competition; the selection process entirely depended on the ability of the private partner to submit a quick proposal. Other factors such as skills and capacity, experience, viability of the project and integrity, to mention a few, were not prudently considered, which would have assisted the corporation in securing more potential investors. The public sector's desperation resulted from a lack of funds and the fear of loss of some of their plots. Therefore, private partner selection decision was made based on a single proposal.

14.4 Managing the rigorous PPP process

According to World Bank (2017) a typical PPP process involves four main steps:

i. Planning and identification of possible candidates;
ii. Preparation involving screening of the PPP candidate, appraising and preparing the contract
iii. Implementation and procurement involving structuring, tendering and awarding.
iv. Contract management during construction and through operations.

These steps, as shown in Figure 14.1, must be carefully followed when designing and implementing a PPP project. The responsible units are to be involved in each step. Generally, PPPs are complex because of their unique contractual arrangement process, parties involved, financers, suppliers, contractors, engineers, operators and customers (UNESCAP, 2011). The number of players or stakeholders involved demonstrates their complexity. This complexity calls

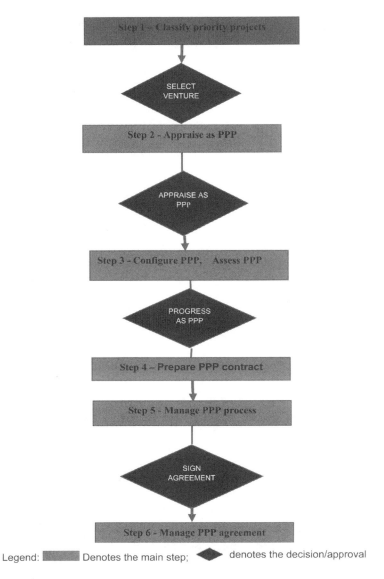

Figure 14.1 Typical PPP process

Source: Original adapted from World Bank (2017)

for commitment from every player and sometimes an increased use of external advisors and expertise. In this regard it is very important that the PPP process is carefully managed from start to finish. Basically, the management of the PPP process becomes easy and successful once the project players/parties are aware of the PPP process and possess the relevant skills and knowledge. According to the World Bank (2014) many governments set out a standardised PPP process to ensure that PPPs are developed in a consistent way with the government's objectives. As simple as the processes may look, a rigorous process is involved that requires PPP skills, knowledge and experience.

Empowering public-private partnership in major infrastructure systems

Basically, the task of monitoring the PPP process usually falls to the contracting authority; however, this is also assessed and approved by the responsible PPP units. The dedicated PPP units, among other roles, are expected to offer technical support in monitoring and implementing the process for efficiency and success. The next section discusses each step as illustrated in Figure 14.1.

14.4.1 Classify priority project

The initial step towards identifying a potential PPP project is through identifying a priority public investment project (World Bank, 2017). Most governments have their own defined processes and methodologies for public investment planning. However, these may vary from place to place. Some governments may set out infrastructure strategies or carry out a thorough feasibility study and cost-benefit analysis or have other ways of doing it. The project identification task is usually undertaken regardless of how it will be procured. Once the priority projects are identified then they must be approved by the responsible PPP unit before developing them further.

14.4.2 Appraise as PPP

Once the projects have been suitably identified, the next step is to screen the priority projects to check for PPP potential. Based on available information, the aim of the screening stage is to evaluate whether the project may provide better value for money if implemented as a PPP. According to the reference guide, governments must consider five key criteria when deciding whether or not to pursue a project as a PPP including assessing the:

i.   Feasibility and economic viability of the project;
ii. Commercial viability (if the project can attract good sponsors and lenders);
iii. Fiscal responsibility (whether the revenue requirements are within the users' capacity);
iv. Value for money; and
v.   Ability to manage the project.

Additionally, in the screening stage government may also decide which potential PPP projects should be developed first depending on the scale of need and project readiness.

14.4.3 Configure as PPP projects

Based on the PPP reference guide, this stage calls for financial structuring, payment mechanisms, allocation of risks and responsibilities to parties in the PPP contract. This is undertaken with the support of information obtained in the detailed feasibility study and economic viability analysis. It is important to have a list of all identified risks associated with the PPP project. Then the allocation of risks is done by deciding which party to the PPP contract can best handle the risk so as to achieve better value for money (Kavishe *et al.*, 2018).

14.4.4 Prepare PPP contract

In this stage the contracting authority drafts the procurement package and a set of prequalification criteria to make sure that they align with project goals. This process requires approval from the responsible PPP unit as part of managing the process and making sure that all important aspects from the previous stage have been fulfilled and the current state permits the drafting of the contract. It is important to note that drafting the PPP contract takes a significant amount of time and resources and sometimes may require PPP experts or advisors depending on the nature of the project and the available skill capacity. The contract is the heart of the partnership, outlining the relationship between the partners, their responsibilities, rights, risk allocation as well as other relevant information. Therefore, this is done before issuing the request for proposal to create clarity and limit ambiguity.

14.4.5 *Manage PPP process*

This stage, also referred to as the procurement stage, is where the public sector selects the private partner to collaborate and implement the project with. The prepared procurement package from the former stage as described in the above section is used in this stage. The main purpose of this stage is to obtain a competent private partner or investor. To manage the process, the government needs to decide on the procurement approach upfront, sell the proposed PPP to the public to interest potential bidders and lenders, qualify the bidders, manage the bidding process and finally attain a financial close. The bidding process might not be similar in every country and some may use a multistage bidding process; for instance, a recent study by Kavishe and Chileshe (2019) recommended a two-stage bidding process in order to secure a strong and sound winning bidder. Stage one shortlists the prequalified bidders and in stage two, a preferred bidder is selected. But this is not cast in stone; it depends on the procurement strategy decided upon by the respective government.

14.4.6 *Manage PPP agreement*

After procuring PPP projects the next step is to monitor and enforce the agreed PPP contract throughout the contract period. Managing PPP contracts is different from managing non-PPP (traditional government) contracts. This is also a very sensitive stage as a number of projects in developing countries have failed due to lack of management skills (Chileshe and Kikwasi, 2014a; Kikwasi and Escalante, 2018). According to MacDonald and Brittan (2007), contract management is defined as the process of managing and administrating contracts from the time they're awarded to the end of the service or operational period. Studies have confirmed that inadequate project management and monitoring by the public sector is among the key challenges that led to poor outcomes and failures to achieve the project goals in some instances (Kavishe and An, 2016; Umar *et al.*, 2018). According to World Bank (2014), the aims for managing PPP contracts are the following:

i. To ensure services are delivered continuously and as per agreed quality.
ii. Risk allocations and management and all other contractual responsibilities are maintained.

MacDonald and Brittan (2007) have identified contract management as comprising the following four main components:

- Putting up the contract management team to determine the management team structure and when to start;
- Managing relationships to develop partnership protocol;
- Managing service performance to define the fundamentals of measuring performance; and
- Contract administration through payment mechanism, dealing with disputes, variations and market testing.

It is important to note that the base for successful contract management should be placed earlier on in the PPP implementation stage (World Bank, 2014). Many aspects of contract management have to be set out in PPP agreements.

14.5 Conducting a bankable feasibility study

A feasibility study is a detailed analysis to determine the viability of a project. It is undertaken to help the public sector determine whether traditional public sector procurement or a PPP is the best option for the proposed project South African National Treasury PPP Unit (SANT PPP unit, 2004). Based on PPIAF (2009) the key objectives of carrying out a feasibility study are to:

- Ensure all the risks associated with a project are identified, allocated and mitigated;
- Provide sufficient information to the government that will help them decide whether to proceed with the project as PPP;
- Provide the bases of negotiations; and
- Lessen the transaction costs of PPPs and evade unnecessary delays.

According to World Bank (2017) "bankable" means that a project can attract both equity finance from its shareholders and the required amount of debt. Therefore, a bankable feasibility report should comprise adequate information that can be used to convince lenders that the project can service the debt. Simply put, the operating cash flows are high enough to cover debt service and an acceptable margin. In this case, if too much risk is allocated to the private partner, lenders will not be comfortable and they will be compelled to reduce the amount they had intended to lend (World Bank, 2014). Therefore, it is the role of the government to ensure that bankable feasibility studies are conducted to guarantee bankability. It is widely known that most African countries are experiencing infrastructure deficit, driven by the growing populations. Despite the massive infrastructure needs, there is a serious shortage of bankable PPP projects for the large number of investors to invest in Rothballer and Gerbert (2015). A feasibility study for PPPs is indispensable and normally has a different focus and requirements from a government procurement project's feasibility study. Therefore, because PPPs are complex, their feasibility studies need expert skills to undertake bankable feasibility studies (Mutambatsere, 2017). According to the South African National Treasury PPP Unit (2004) a feasibility study must comprise the following processes:

- Needs analysis;
- Option analysis;
- Project due diligence;
- Value assessment;
- Economic valuation; and
- Procurement plan.

Several studies (Cheung *et al.*, 2012; Ismail, 2013; Jefferies *et al.*, 2002; Mutambatsere, 2017; URT, 2009) have identified that an inadequate feasibility study is among the challenges hindering the successful delivery of PPPs. For example, a World Bank (2016) report for Tanzania highlighted that there is unequal treatment between solicited proposals and unsolicited proposals. Contracting authority submits the solicited bids to both a partial and a full feasibility study but for the unsolicited bids the task is left in the remit of the private partner, which in most cases results in dishonesty and artificial demands (Kavishe, 2018). A study from Thailand (Trangkanont and Charoenngam, 2014) reported that cases had occurred where the private sector had persuaded the public sector that a PPP project was viable and its demand was high when this was the opposite. Therefore, through a bankable feasibility study, accurate and genuine reports of demand can be obtained through stakeholder engagement (Amadi *et al.*, 2014) to help determine the demand risk, which is important information being looked upon by both lenders and investors in making their decisions.

14.6 Structuring a balanced risk allocation

Structuring a balanced risk allocation is among the first steps toward structuring the PPPs. In the context of a PPP, risk allocation means deciding which party to the PPP contract will handle the risk or bear the cost. However, in making such decisions it is very important to identify all the risks and assess each one in order to find out which party will be in a position to best manage or handle the risk (Grimsey and Lewis, 2002). A balanced risk allocation is one of the key drivers for achieving value for money (EPEC, 2018). The study by Grimsey and Lewis (2004) noted that PPPs have highly increased project risk awareness, something that public procurement failed to achieve previously. This made risk identification, risk allocation and risk management a vital part of PPP processes (Grimsey and Lewis, 2004). Various studies have identified several risks associated with PPPs. Table 14.1 demonstrates the different risks as identified and compiled by Li and Zou (2012) within a project life cycle. Table 14.1 demonstrates that there are several risks throughout the project life cycle, proving that PPPs are complex and highly prone to risks. A number of PPP guidelines such as EPEC (2018) and World Bank (2014, 2017) have indicated that the risk analysis process takes place in five stages:

i. Risk identification and prioritisation;
ii. Risk assessment and valuation;
iii. Risk allocation;
iv. Risk mitigation; and
v. Risk monitoring and review.

The inability to identify, assess, allocate, mitigate or manage project risks has been reported by many countries to have hindered the success of PPPs (Cheung *et al.*, 2012; Jefferies *et al.*, 2002; URT, 2009). For example, Tanzania has experienced many failures (World Bank, 2016) because of poor risk management processes, lack of experience and lack of information (Chileshe and Kikwasi, 2014b). Therefore, structuring a balanced risk allocation is a precondition towards successful PPPs.

14.7 Creating a conducive PPP environment

Worldwide experience has revealed that the successful implementation of PPPs calls for an enabling environment to be put in place so that they can be implemented effectively and achieve maximum benefit for the public sector (PPIAF, 2009), as discussed in chapter 12. The enabling environment for PPPs is the first pre-condition for their adoption (Mboya, 2012). Studies have considered that creating an enabling environment is one among the top critical success factors for implementing successful PPP projects in developing countries (Debela 2019). According to Debela (2019); PPIAF (2009); and Rothballer and Gerbert (2015), in terms of PPPs, the components of an enabling environment entail the following:

- Favourable and legal frameworks;
- Government support;
- Stable political and social environment;
- Rule of law and good governance;

Empowering public-private partnership in major infrastructure systems

Table 14.1 List of identified risks associated with PPPs in a project life cycle

Risk category	Risk factor
Feasibility Study	Political risk
	Land acquisition risks
	Social and public acceptance risks
	Government leadership risks
	Pre-investment risks
	Environment risks
	Market/demand risk
	Poor public decision-making process
Financing	Interest rate fluctuation
	Economic risk (inflation, foreign exchange)
	Legislation change
	High finance cost
	Poor financial market
	Poor financial attraction of project to investors
Procurement and Design	Design deficiency
	Unproven engineering techniques
	Government corruption risks
	Contract risk
	Inadequate competition for tender
	Inability of concessionaire
	Too many design changes
Construction	Construction cost overruns
	Delays
	Technical risks
	Material/labour non-availability
	Too many late design changes
	Geological risks
	Weather risks
	Environmental pollution risks
	Completion risk
	Construction force majeure events
	Poor quality workmanship
	Difficulties in land acquisition
	Infrastructure risks
Operation	Revenues below expectation
	Interest rate volatility
	Inflation rate volatility
	Legislation change
	Market demand change
	Operational risks

Source: Original adapted from Li and Zou (2012)

- Public sector capacity to design, manage, evaluate and monitor PPPs;
- Dedicated team of professionals to oversee the PPP projects; and
- Transparent procurement process.

The creation of such PPP enabling environments will attract competent private partners. Though many developing countries including Tanzania have introduced PPP policies, laws and regulations to guide the implementation of PPP projects, still it is the government's role to make

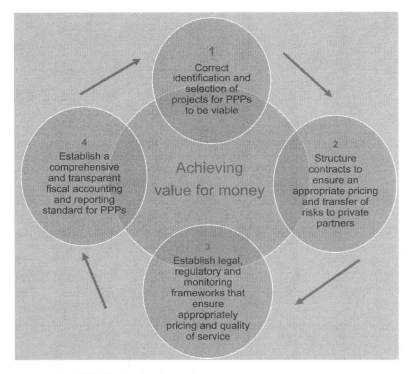

Figure 14.2 Interrelated PPPs' institutional capacity
Source: Original adapted from Jomo et al. (2016)

sure that all other enablers are put in place to achieve a conducive enabling environment. For instance, both public and private sector capacity is still lacking in Tanzania (Kavishe *et al.*, 2018; World Bank, 2016). Jomo *et al.* (2016) have argued that for PPPs to become an effective tool for financing major economic infrastructure projects in developing countries, it is important that these countries instill institutional capacity to create, manage and evaluate PPPs. Figure 14.2 illustrates the four interrelated capacities as suggested by Jomo *et al.* (2016) as significant towards ensuring that PPPs are implemented for the "right reason." It demonstrates that the four institutional capacities are related to the PPP processes/stages within a project life cycle. This implies that every stage within the process requires skills and capacity and every stage is linked or dependent on the other.

14.8 Chapter summary

This chapter demonstrated some key important issues towards empowering PPP collaboration in infrastructure projects. A need to have a better understanding of PPP preparation, managing rigorous PPP processes, providing for an enabling environment, ensuring a bankable feasibility study and structuring a balanced risk allocation have all been highlighted. Likewise, a number of major challenges hindering the successful delivery of PPPs in developing countries have been identified and discussed in detail. It was noted that inadequate PPP capacity, poor PPP contracts and tender documents and misinformation of financial capacity of private partners are the major

Empowering public-private partnership in major infrastructure systems

challenges in developing countries, particularly sub-Saharan Africa. Furthermore, the chapter highlighted inadequate preparation and screening of potential PPPs. It is important to bear in mind that enough justification to important aspects as what will be the benefits of undertaking a particular project; that is, is the project going to address any particular existing problem? All these aspects are centred on the issue of PPP skills and capacities. For example, in order to identify candidate projects, screen them as PPPs, structure them, and manage their transactions and contracts, it is important for the public sector (contracting authority) to possess enough skills to handle the work; otherwise, its deficiency might lead to failures and losses. That reason is why lack of PPP skills is considered the main challenge for PPP infrastructure delivery.

Furthermore, it is worth noting that PPP experiences cannot simply be copied from one country to another; different countries have different characteristics in terms of culture, economy, environment, social aspects and policy. The data presented in this chapter can be used by researchers and practitioners but also governments and policy makers, particularly those in developing countries as a basis for improving and strengthening the implementation of PPPs.

14.9 Chapter discussion questions

1. Inadequate PPP skills and knowledge leading to poor planning and application problem is identified as the major challenge affecting the development and management of PPP infrastructure projects in most developing countries. Propose some robust capacity building strategies to enhance the performance of PPP projects and also mitigate the challenge.
2. Inadequate financial capacity of the private partner and poor PPP contracts and tender documents are identified among the major challenges affecting the implementation and management of PPP infrastructure projects in most developing countries. Propose some robust funding and financing frameworks for such projects and ways of improving the contracts.
3. Do you agree that PPP processes are complex? If yes, what aspects contribute to their complexity?
4. An inadequate feasibility study is identified as one of the major challenges affecting the successful delivery of PPPs; why is that the case?
5. Having learned that the foundation for successful contract management should be placed in a PPP's implementation stage, propose some effective measures to achieve that.

14.10 Case study: Lessons from a failed Dege Eco Village PPP housing project

The Dege Eco Village project was a PPP housing project that started its construction in January 2014; it was being built on 300 acres of land in Kigamboni, Dar es Salaam (Tanzania), 23 kilometres from Magogoni Ferry. This project was to cost TZS 96.64 billion (£34.7 million/USD 653 million) and used a design and build (DB) PPP model. It was a collaboration between the National Social Security Fund (NSSF) as the public partner and Azimio Housing Estate Limited (AHEL) as the private partner (Canadian Council for Public Private Partnerships, 2011). It was an unsolicited proposal, so the partner selection was not competitive. The private partner provided 300 acres of land, 20 percent of the project's value, and equity of up to 35 percent (a total of 55 percent of the project's value) while NSSF was to provide an equity share of 45 percent. The project was being built by M/s Mutluhan Construction Industry Company Limited, a foreign company from Turkey using modern technology known as a tunnel formwork system; the plan included 7,460 apartments for selling. According to an interview with former NSSF Director General, this project was planned to be implemented in three phases to be completed by the end of 2016, 2017 and 2018.

The first phase of the project involved the construction of mixed-use residential apartments catering to high- and medium-income groups. All the residential apartments were for sale, while other social and commercial properties were to be leased or sold. The second phase was planned to cater to a low-income group while the third would construct high end villas, commercial buildings and other social amenities. However, since April 2016 construction activities have stopped and the project has stalled. The main causes are private partner financial difficulties, so project goals have not been achieved to date. It was also revealed that this project did not pass through the PPP units for assessment and approval despite the presence of PPP units in Tanzania; this information was confirmed by the head of the PPP finance unit and the project managers. Furthermore, the government audit report during the construction stage found some irregularities and it was concluded that the project was "not serving the public interest" as claimed. It was revealed that the land for the Dege Eco Village project was priced at more than 30 times its "real" value and unfortunately the auditors could not find evidence to support that the private partner owned the land involved in this project (Policy Forum, 2016).

14.10.1 PPP preparation

As noted earlier in the chapter, there should be sufficient justification for undertaking any PPP project (UNESCAP, 2011). How the project is going to solve an existing problem in the community must be clearly indicated. Public sector due diligence is also a significant aspect of project preparation. This process provides public sector features of a PPP project and determines government capacity level to handle the project. Lessons drawn from this case demonstrate that public sector due diligence, project analysis and justifications were not done or taken into consideration; otherwise, the project would have not proceeded as PPP considering the audit report and its failure.

14.10.2 Regulatory issues

This case study highlights some issues around the PPP regulatory and legal frameworks in Tanzania. As explained in *Chapter 12*, the PPP units are responsible for the assessment, approval and coordination of all PPP projects in Tanzania. These units assisted in the formulation of Tanzanian PPP Policy in 2009, PPP Act 2010 and PPP Regulations 2011. The PPP Coordination Unit was established to coordinate and oversee the mainland Tanzanian PPP projects whereas the PPP Financing Unit has the duty of assessing and examining the financial aspects as well as allocating fiscal risk for all submitted PPP proposals (Mboya, 2012). Despite the existence of these PPP units and regulatory frameworks it is clear from the case study that some of the Tanzanian PPP projects are not assessed and approved by the PPP Financing Unit. Hence, saying they have suitable PPP policies and a regulatory framework is one thing, but making them effective and efficient is another. The audit report undertaken in between construction found numerous irregularities and concluded the PPP project was "not serving the public interest." Because of a lack of a detailed feasibility study, screening and cost-benefit analyses, the site is sadly now a sprawling concrete ghost town (Rosen, 2019).

Empowering public-private partnership in major infrastructure systems

14.10.3 Contractual issues

The major concerns here are the "lack of trust" and "transparency" First, it appears that there was an issue of trust and transparency since the private partner did not reveal enough supporting information regarding the land ownership. Second, the contracting authority (NSSF) did not carefully scrutinise details of the land to obtain enough evidence to support the ownership of the land offered by the private partner; this was very risky as another person could have emerged and claim ownership of the land. The lesson obtained here is that there was a poor PPP contract as identified in Malaysia by Abdul–Aziz and Kassim (2011) and also by other studies (Fomad, 2013; Ngowi, 2006; Trangkanont and Charoenngam, 2014).

14.10.4 Risks

The following two types of risks, financial and political, are discussed below.

14.10.4.1 Financial risk

A private partner's inadequate finances can be a serious risk in a PPP project (Cheung and Chan, 2011), and it can easily happen if adequate scrutiny is not employed (Kavishe, 2018). As evidenced in this case study, the project stopped its construction process in April 2016 because the private partner failed to continue the financing that was laid out in the contract. Based on primary data received from the project team in relation to problems encountered as a result of the private partner's financial difficulties, it was revealed that their inadequate financial capability resulted in other problems: theft at the project site due to workers not being paid their salary, conflicts and disputes between the contractor and workers because of failure to pay them on time, and higher maintenance cost in terms of electricity and security to protect the completed structure. Findings and lessons from the Dege Eco Village case agree with the majority of PPP literature, particularly from the developing countries, regarding private partners' financial difficulties.

14.10.4.2 Political risk

As defined in Chapter 12, political risks are the risks of change or discontinuity in the business environment as a result of political change. The case under study further highlights the need for managing all identified risks, including political risk. Normally, Tanzania carries out its national election every five years. Because this project started in January 2014, it was obvious that the next election would occur in 2015 before the project was completed. Furthermore, the outgoing government had ruled for two terms, which was the maximum period allowed. Therefore, a new government would have been sworn in by December 2015. This should have provided a clear indication of a possibility of political risk resulting from change in administration. The current administration concluded that the project had no value for money, so it did not proceed as a PPP.

14.11 Case discussion questions

1. What key lessons can you draw from the case study?
2. Compare the lessons you drew from the case study with countries delivering major infra-structure projects in developing (emerging economies) and developed countries.

3. What major challenges affect the developing countries from delivering PPP infrastructure projects?
4. From the PPP process viewpoint, what lessons or success and failure factors can you take from the case study for future infrastructure projects?

References

Abdul-Aziz, A.R. and Kassim, P.S.J. (2011). Objectives, success and failure factors of housing public-private partnerships in Malaysia. *Habitat International*, 35(1), pp. 150–157.

Amadi, C., Carrillo, P., and Tuuli, M. (2014). Stakeholder management in public private partnership projects in Nigeria: Towards a research agenda. In: Raiden, A.B. and Aboagye-Nimo, E. (Eds.), *Procs 30th Annual ARCOM Conference*, 1–3 September 2014, Portsmouth, UK: Association of Researchers in Construction Management, pp. 423–432.

Babatunde, S.O., Perera, S., Zhou, L., and Udeaja, C. (2015). Barriers to public private partnership projects in developing countries a case of Nigeria. *Engineering Construction and Architectural Management*, 22(6), pp. 669–691.

Bowers, J. and Khorakian, A. (2014). Integrating risk management in the innovation project. *European Journal of Innovation Management*, 17(1), pp. 25–40.

Canadian Council for Public Private Partnerships. (2011). A guide for municipalities about PPP. Available from: http://www.p3canada.ca/~/media/english/resourceslibrary/files/p3%20guide%20for%20 municipalities.pdf [Cited 15th November 2016].

Chileshe, N. and Kikwasi, G.J. (2014a). Critical success factors for implementation of risk assessment and management practices within the Tanzanian construction industry. *Engineering, Construction and Architectural Management*, 21(3) pp. 291–319.

Chileshe, N. and Kikwasi, G.J. (2014b). Risk assessment and management practices (RAMP) within the Tanzanian construction industry: Implementation barriers and advocated solutions. *International Journal of Construction Management*, 14(4), pp. 239–254.

Cheung, E. and Chan, A.P. (2011). Risk Factors of public-private partnership projects in China: Comparison between the water, power, and transportation sectors. *Journal of Urban Planning and Development*, 137(4), pp. 409–415.

Cheung, E., Chan, A.P., and Kajewski, S. (2012). Factors contributing to successful public private partnership projects: Comparing Hong Kong with Australia and the United Kingdom. *Journal of Facilities Management*, 10(1), pp. 45–58.

Debela, G.Y. (2019). Critical success factors (CSFs) of public–private partnership (PPP) road projects in Ethiopia. *International Journal Construction Management*, doi:10.1080/15623599.2019.1634667.

El-Gohary, N.M., Osman, H., and El-Diraby, T.E. (2006). Stakeholder management for public private partnerships. *International Journal of Project Management*, 24(7), pp. 595–604.

EPEC, (2018). A guide to preparing and procuring a PPP project. Available from: https://www.wbif.eu/ storage/app/media/Library/8.%20Public%20Private%20Partnership/3.%203-PPP-Preparation-and-Procurement-Guide-FINAL-310818.pdf. [cited 3 May 2020.].

Fombad, M. (2013). Accountability challenges in public-private partnerships from a South African perspective. *African Journal of Business Ethics*, 7(1), pp. 11–25.

Foo, L., Asenova, D., Bailey, S., and Hood, J. (2011). Stakeholder engagement and compliance culture: An empirical study of Scottish private finance initiative projects. *Public Management Review*, 13(5), pp. 707–729.

Grimsey, D. and Lewis, M.K. (2002). Evaluating the risks of public private partnerships for infrastructure projects. *International Journal of Project Management*, 20(2), pp. 107–118.

Grimsey, D. and Lewis, M.K. (2004). The governance of contractual relationships in public-private partnerships. *The Journal of Corporate Citizenship*, 15, pp. 91–109.

Ibem, E.O. and Aduwo, B. (2012). Public-private partnerships (PPP) in housing provisions in Ogun state, Nigeria: Opportunities and challenges. In: Laryea, S., Agyepong, S.A., Leiringer, R., and Hughes, W. (Eds.), *Proceedings 4th West Africa Built Environment Research (WABER) Conference*, 24-26 July 2012, Abuja, Nigeria. pp. 653–662.

Ismail, S. and Azzahra Haris, F. (2014). Constraints in implementing public private partnership (PPP) in Malaysia. *Built Environment Project and Asset Management*, 4(3), pp. 238–250.

Ismail, S. (2013). Critical success factors of public private partnership (PPP) implementation in Malaysia. *Asia-Pacific Journal of Business Administration*, 5(1), pp. 6–19.

Kavishe, N. (2018). Improving the delivery of PPP housing projects in developing countries. PhD Thesis. University of Birmingham, UK.

Kavishe, N. (2010). The performance of public private partnership in delivering houses in Tanzania (The case of the National Housing Corporation). Msc. Unpublished thesis. Ardhi University, Dar es Salaam, Tanzania.

Kavishe, N. and An, M. (2016). Challenges for implementing public private partnership in housing projects in Dar es Salaam city, Tanzania. In: Chan, P.W. and Neilson, C.J. (Eds,), *Proceedings of the 32nd Annual ARCOM Conference*, 5–7 September, Manchester, UK. Association of Researchers in Construction Management, 2, pp. 931–940.

Kavishe, N. and Chileshe, N. (2019). Development and validation of public–private partnerships framework for delivering housing projects in developing countries: A case of Tanzania. *International Journal of Construction Management*, DOI:10.1080/15623599.2019.1661065.

Kavishe, N., Jefferson, I., and Chileshe, N. (2018). An analysis of the delivery challenges influencing public private partnership in housing projects: The case of Tanzania. *Engineering, Construction and Architectural Management*, 25(2), pp. 202–240.

Kikwasi, G.J. and Escalante, C. (2018). *Role of the construction sector and key bottlenecks to supply response in Tanzania'*, WIDER Working Paper 2018/131, Prepared for United Nation Universities (UNU-WIDER), Available at: https://www.wider.unu.edu/sites/default/files/Publications/Workingpaper/PDF/wp2018-131.pdf, [cited 28 August 2019].

Jefferies, M., Gameson, R., and Rowlinson, S. (2002). Critical success factors of the BOOT procurement system: Reflections from the Stadium Australia case study. *Engineering Construction and Architectural Management*, 9(4), pp. 352–361.

Jomo, K.S., Chowdhury, A., Sharma, K., and Platz, D. (2016). Public private partnerships and the 2030 agenda for sustainable development. Fit for purpose? UN DESA Working Paper No. 148. ST/ESA/2016/DWP/148

Li, J. and Zou, P. (2012). Risk identification and assessment in PPP infrastructure projects using fuzzy analytical hierarchy process and life-cycle methodology. *Australasian Journal of Construction Economics and Building*, 8(1), pp. 34–48.

Mboya, J.R. (2013). PPP country paper-Tanzania, paper submitted to SADC-DFRC 3P NETWORK public-private-partnership working group, Available at: http://www.sadcpppnetwork.org/wpcontent/uploads/2015/02/tanzania_27012014.pdf [cited 19 November 2016].

Mboya, J. (2012). Tanzania PPP framework: Lessons for enabling environment for PPP pipelines., 6-7 December 2012,. *Regional Conference on PPP, Kampala, Uganda*. 6-7 December 2012.

Moskalyk, A. (2011). *Public-private partnerships in housing and urban development*. Nairobi, Kenya: UN-HABITAT.

MacDonald, M. and Brittan, B. (2007). A guide to contract management for PFI and PPP projects. HM Treasury, UK.

Mourgues, T. and Kingombe, C. (2017). How to support African PPPs: The role of the enabling environment. In: Leitão, J., de Morais Sarmento, E., and Aleluia, J. (Eds.), *The emerald handbook of public–private partnerships in developing and emerging economies*, pp. 269–310. Emerald Publishing Limited.

Mutambatsere, E. (2017). Infrastructure development through public–private partnerships in Africa. In: *The emerald handbook of public–private partnerships in developing and emerging economies*, pp. 45–80. Bingley, UK: Emerald Publishing Limited.

Ngowi, H.P. (2006). Public-private partnerships (PPPs) in the management of municipalities in Tanzania–issues and lessons of experience. *African Journal of Public Administration and Management*, 17(2), pp. 29–31.

Osei-Kyei, R. and Chan, A.P.C. (2017). Implementation constraints in public-private partnership: Empirical comparison between developing and developed countries. *Journal Facility Management*, 15(1), pp. 90–106.

PPIAF, (2009). Tool kit for public-private partnerships in roads and highways: Enabling environment for PPP. Available from: https://ppiaf.org/sites/ppiaf.org/files/documents/toolkits/highwaystoolkit/6/toolkit_files/index.html [cited 14 May 2020].

Policy Forum (2016). 2015/2016 Tanzania governance review: From Kikwete to Magufuli:Break with the past or more of the same? Available from: https://www.policyforum-tz.org/sites/default/files/TGR2015-2016.pdf [cited 30 May 2020].

Rosen, J.W. (2019). This Tanzanian city may soon be one of the world's most populous. Is it ready? Environment the Cities Issue *Available from:* https://www.nationalgeographic.com/environment/2019/04/tanzanian-city-may-soon-be-one-of-the-worlds-most-populous/ [cited 30 May 2020].

Rothballer, C. and Gerbert, P. (2015). Preparing and structuring bankable PPP projects. In: Caselli, S., Corbetta, G., and Vecchi, V. (Eds.), *Public private partnerships for infrastructure and business development*, pp. 57-80. New York USA: Palgrave Macmillan.

Sengupta, U. (2004). Public-private partnerships for housing delivery in Kolkata. *International Conference on "Adequate and Affordable Housing for All"*, 24-27 June 2004, Toronto, Canada.

Sengupta, U. (2006). Government intervention and public–private partnerships in housing delivery in Kolkata. *Habitat International*, 30(3), pp. 448–461.

Scheer, L.K., Kumar, N., and Steenkamp, J.-B.E. (2003). Reactions to perceived inequity in US and Dutch interorganizational relationships. *Academy of Management Journal*, 46(3), pp. 303–316.

South African National Treasury PPP Unit (2004). Public private partnership manual: National treasury PPP practice notes issued in terms of the public finance management act. Available from: https://www.gtac.gov.za/Publications/1160-PPP%20Manual.pdf [cited 15 May 2020].

Trangkanont, S. and Charoenngam, C. (2014). Critical failure factors of public-private partnership low-cost housing program in Thailand. *Engineering, Construction and Architectural Management*, 21(4), pp. 421–443.

Ukoje, J. and Kanu, K. (2014). Implementation of the challenges of the mass housing scheme in Abuja, Nigeria. *American International Journal of Contemporary Research*, 4(4), pp. 209–218.

Umar, A.A., Zawawi, N.A.W., and Abdul-Aziz, A.-R. (2018). A. Exploratory factor analysis of skills requirement for PPP contract governance. *Built Environment Project and Asset Management*, 9(2), pp. 277–290.

UNESCAP (2011). A guidebook on public-private partnership in infrastructure. Issue. *Available from:* https://www.unescap.org/sites/default/files/ppp_guidebook.pdf [cited 28 May 2020].

URT (2009). *Prime Minister's Office: National public private partnership (PPP) policy,'Restricted Circulation'*. Dar es Salaam. World Bank (2017). Public-private partnerships: Reference guide version 3. World Bank, Washington, DC.©WorldBank. Available from: https://openknowledge.worldbank.org/handle/10986/29052 License: CC BY 3.0 IGO [cited 15 May 2020].

URT (2010). Public Private Partnership Act, 2010. Dar es Salaam Government Printer

URT (2011). *Public Procurement Act 2011*. Dar es Salaam Government Printer.

World Bank (2014). Public-private partnerships: Reference guide version 2. World Bank, Washington, DC.©WorldBank. Available from: http://documents.worldbank.org/curated/en/600511468336720455/pdf/903840PPP0Refe0Box385311B000PUBLIC0.pdf [cited 1 May 2020].

World Bank. (2016). Tanzania economic update the road less traveled unleashing public private partnerships in Tanzania, Africa region macroeconomics and fiscal management global practice. Available at: http://www.worldbank.org/tanianla/economlcupdate [cited 10 January 2016].

PART IV

Digitising major infrastructure delivery

15
APPLYING DESIGN THINKING PRINCIPLES ON MAJOR INFRASTRUCTURE PROJECTS

Ximing Ruan and Geraldine Hudson

15.1 Introduction

Major infrastructure projects are generally high profile, incredibly expensive and time-consuming. Mistakes in projects can alienate stakeholders, waste resources and create a lack of confidence in other major projects. Problem-solving and decision-making are key tenets of managing major infrastructure projects. Major infrastructure projects can impact the economy, the environment, benefit society as a whole and contribute to sustainable living. Successful infrastructure project outcomes, therefore, are not limited to the traditional outcomes of a physical end product which is working, functional and well designed; it must provide value and work well for the environment and those that use it. This requires collaboration with stakeholders, including previously siloed departments, to solve what is termed "wicked problems." Wicked problems are at the heart of major infrastructure projects today, as goals of expediency and budget, maximizing performance or efficiency while minimizing costs and meeting a desired level of service. They are replaced by the domain of complexity, where rapidly changing environments and fragmentation of goals require fundamentally new approaches (Chester and Allenby, 2019). This chapter looks at how to address the problems with major infrastructure projects, the multiple stakeholders involved and the complex nature of outputs in terms of building, climate, people and value through two separate yet in some ways aligned processes: design thinking and systems thinking.

Rittel and Weber (1973) introduced the concept of "wicked problems" in order to draw attention to the complexities and challenges of addressing planning and social policy problems. They described a wicked problem this way: one that has innumerable causes, is tough to describe, and does not have a right answer. Common wicked problems in terms of major infrastructure problems could be a lack of open spaces, environmental degradation, and pollution. Because of the very human nature of wicked problems, they cannot be solved in the usual analytical way of conventional project management techniques. They're the opposite of hard but ordinary problems, which people can solve in a finite time period by applying standard techniques. Not only do conventional processes fail to tackle wicked problems, but they may exacerbate situations by generating undesirable consequences (Camillus, 2008). The infrastructures that we rely on today share many of the core design features from when they were initially conceived decades or even a century ago (Chester and Allenby, 2019). However, the processes by which we design, build, operate, manage, rehabilitate, and decommission infrastructures are a wickedly complex problem.

15.1.1 Chapter aim and objectives

The primary aim of this chapter is to examine how infrastructure operators can apply design thinking principles on major infrastructure projects. The main objectives are to:

- Appraise the principles of design thinking and align them with major infrastructure project delivery;
- Examine different types of design thinking models;
- Establish the importance of systems thinking principles;
- Ascertain how systems thinking can be applied on infrastructure projects.

15.1.2 Learning outcomes

The following learning outcomes have been identified for this chapter. Readers will be able to:

- Gain an insight into the use of systems thinking on major infrastructure projects;
- Differentiate the different types of design thinking models;
- Comprehend the importance of systems thinking principles;
- Illustrate how systems thinking can be applied on major infrastructure projects.

15.2 Design thinking

In traditional linear processes, design and construction were separate entities, and it was sufficient and important to recognise their close relationship. Design has become a recognised factor in national infrastructure projects since the 2008 Planning Act (National Infrastructure Commission, 2018). This Act sets out the principles of nationally significant infrastructure projects (NSIPs) for energy, transport, waste and wastewater. Following this, the UK National Planning Policy Framework (NPFF) has urged that major infrastructure projects have a design review to ensure that the value of design support services, particularly relevant in the early stages of a project, ensure good quality outcomes. As specified in Table 15.1, the review is based around 10 guiding principles:

A common theme in the 10 principles of design is not the design aesthetics of the end infrastructure solution but collaboration and a human element to design. These design principles can still be viewed as integrated in a linear way to the project management process. For example, the construction extension to the PMBOK guide (2018) reflects this (see Figure 15.1). It proposes project integration management, stating that the magnitude of stakeholders with varying project expectations (e.g., public taxpayers, regulatory agencies, governments, and environmental or community groups) add significantly to the complexity of infrastructure projects.

Applying design thinking principles on major infrastructure projects

Table 15.1 Top ten design principles

1: Setting the scene	Design thinking should be part of creating the vision and designing the brief for a new project. The project manager and the applicant should think in design terms and define a clear, design-led framework in which the project can develop.
2: Multi-disciplinary teamwork	Collaboration between stakeholders must begin early and be sustained. Stakeholders may include, among others, the client, the design team, technical experts, the community and the local planning authority.
3: The bigger picture	Design does not start and end with the immediate project or site. Holistic thinking is required to ensure that projects are part of an integrated process that fits into bigger strategies such as regional or sub-regional planning.
4: Site master plan	Nationally significant infrastructure projects have far reaching impacts. Good design will do much to integrate the infrastructure project with its environment by creating a facility that responds to its context.
5: Landscape and visual impact assessment	Visual impact assessment should be used as a design tool to inform location, orientation, composition and height. This should take in a large number of viewpoints right from the beginning of the design.
6: Landscape design	Infrastructure projects benefit society as a whole and should be celebrated. Different structures will require different levels of architectural ambition.
7: Design approach	Difference and variety of design approaches in relation to the context can be virtues. Infrastructure projects benefit society as a whole and should be celebrated.
8: Materials and detailing	Design intent for key details should be developed alongside the concept and scheme design stages so that the architectural potential can be understood by approval bodies and consultees.
9: Sustainability	Sustainability must be integral to the design from the very beginning. A successful proposal will cover every aspect of this, including these: traffic movements (e.g., delivery and refuse), social inclusion of workers and visitors and the use of biomass.
10: Visitor centre	Many large infrastructure proposals offer the opportunity to provide a centre where visitors can learn about the plant operation and be introduced to the concepts of sustainability, energy generation, waste management and humanity's impact on the environment in terms of our ecological footprint and the exploitation of natural resources.

Source: Adapted from Design Council (2012) and Cabe, 2012

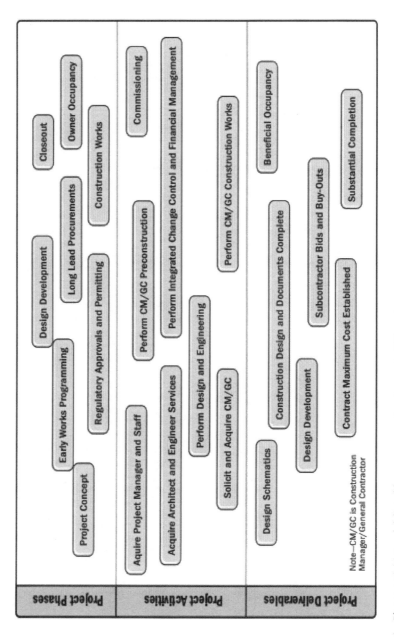

Figure 15.1 Phases, activities and deliverables across a construction project life cycle

Source: *Construction Extension to the PMBOK® Guide (2016) Project Management Institute*

Applying design thinking principles on major infrastructure projects

As design for infrastructure is reviewed and updated, the need for processes that are non-linear and not purely analytical has grown. In February 2020, the UK Design Group published its Design Principles for National Infrastructure (National Infrastructure Commission, 2020). These four principles – climate, people, places and value – aim to guide the planning and delivery of future major infrastructure projects in the UK (see Table 15.2). It seems that these four principles will introduce "wicked problems" from nature that include carbon emissions and, climate change.

Table 15.2 Design principles for national infrastructure

Climate
Mitigate carbon emissions
  and adapt to climate
  change
	The design of our infrastructure must help set the trajectory for the UK to achieve net zero greenhouse gas emissions by 2050 or sooner. This means opportunities must be sought during design and construction to enable the decarbonisation of our society and mitigate and offset residual emissions. Our infrastructure has to support an environmentally sustainable society. It should enable the people and businesses using it to reduce their wider climate impacts, too. The search for these opportunities should not be restricted to the area within the site boundary. And good design incorporates flexibility, allowing the project to adapt over time and build our resilience against climate change.

People **Reflect what society wants** **and share benefits widely**	Infrastructure should be designed for people, not for architects or engineers. It should be to human scale, easy to navigate and instinctive to use, helping to improve the quality of life of everyone who comes in contact with it. This means reliable and inclusive services. It means accessible, enjoyable and safe spaces with clean air that improve health and well-being. The range of views of communities affected by the infrastructure must be taken into account and reflected in the design. While it won't always be possible to please everyone, engagement should be diverse, open and sincere, addressing inevitable tensions in good faith and finding the right balance. And it should not just be designed for people today. Good design will plan for future changes in demographics and population.

Places
Provide a sense of identity
  and improve our
  environment
	Well-designed infrastructure supports the natural and built environment. It gives places a strong sense of identity and, through that, forms part of our national cultural heritage. It makes a positive contribution to local landscapes within and beyond the project boundary. Projects should be inspiring in form and detail, respecting and enhancing local culture and character without being bound by the past. Good design supports local ecology, which is essential to protect and enhance biodiversity. Projects should make active interventions to enrich our ecosystems. They should seek to deliver a net biodiversity gain, contributing to the restoration of wildlife on a large scale while protecting irreplaceable natural assets and habitats.

Value **Achieve multiple benefits** **and solve problems well**	Well-designed infrastructure supports the natural and built environment. It gives places a strong sense of identity and, through that, forms part of our national cultural heritage. It makes a positive contribution to local landscapes within and beyond the project boundary. Projects should be inspiring in form and detail, respecting and enhancing local culture and character without being bound by the past. Good design supports local ecology, which is essential to protect and enhance biodiversity. Projects should make active interventions to enrich our ecosystems. They should seek to deliver a net biodiversity gain, contributing to the restoration of wildlife on a large scale while protecting irreplaceable natural assets and habitats.

Source: Adapted from National Infrastructure Commission (2018)

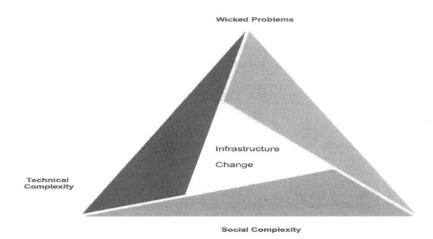

Figure 15.2 Three competing forces
Source: Adapted from Chester and Allenby (2019)

They all meet the definition of being challenging to address and introducing complexities in cases in terms of planning and social processes.

Chester and Allenby (2019) state that wicked complexity is the result of three competing forces that are, given the current approaches for designing and managing infrastructures, inimical to rapid and sustained change in a future marked by acceleration and uncertainty. These three competing forces, shown in Figure 15.2, have separate pulls on our ability to change major infrastructure, resulting in change becoming a wicked, complex process.

Design thinking and systems thinking show how the pulls of technical complexity, social complexity and wicked problems can be addressed. As an academic discipline and a management tool, design thinking has increased in awareness in the last 40 to 50 years. In the 1950s, creative thinking theories were brought to Stanford Engineering by John E. Arnold. The previous linear nature of analytical approaches were not sufficient to solve wicked problems Rittel and Weber (1973): they had to be explored through an experimental approach that explored multiple solutions. Liedtka, et al (2013) concurs that design thinking is a problem-solving process as well as an innovation process. Buchanan (2001) noted that the question of whose values matter and who ought to participate in the design process has changed over time evolving from 1950s' beliefs about the "ability of experts to engineer socially acceptable results" toward a view of audiences as "active participants in reaching conclusions." Recently design thinking has been held at the forefront of organisational thinking by international design and innovation firms such as IDEO. Its founder, David Kelley, stated, "The main tenet of design thinking is empathy for the people you're trying to design for (see Figure 15.3). Leadership is exactly the same thing – building empathy for the people that you're entrusted to help." Also influential, the Institute of Design at Stanford University, firmly connected the value of design thinking to successful innovative organisations. Design thinking could be the shift that is needed from the linear, top-down infrastructure projects and the way to solve the problems which arise from the new design principles of major infrastructures. This shift will be the change necessary to move from projects which proceed with very little input from the people to projects which engage with communities from the outset on what is really required: "How are they actually living and working and moving around

Applying design thinking principles on major infrastructure projects

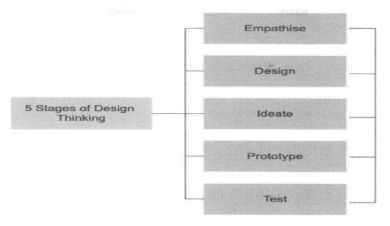

Figure 15.3 Design thinking model
Source: Adapted from Hasso Plattner Institute (2012)

and how do they see that changing in the future and engaging as a source of ideas around which you then build the appropriate infrastructure" Fisher (2016).

15.3 Design thinking models

The three main models of design thinking are shown in Table 15.3.

For the Hasso Plattner Institute, design thinking is considered the ability to combine empathy, creativity, and rationality in analysing and fitting solutions to context (Hasso Plattner, 2012). These stages allow for a "human centred" approach to problem-solving, and it is this process that will lead to eventual product development, through marrying integrative thinking on opposing ideas and constraints.

In the context of major infrastructure projects, the project must follow these steps:

- **Empathise:** Directly getting to know the customer in order to understand user needs and gain user insights, and understand the customer context.
- **Define:** Interaction with the customer and understanding the customer's problem leads to the definition. What is the customer's need? What problem does the customer need solving? That is what the innovation should address.
- **Ideation:** This leads to the third stage (ideation), which frames the options and solutions and possibly reinterprets the problem in the form of brainstorming, mind mapping, story boarding and other group thinking activities.

Table 15.3 Design thinking models

Empathise, Define, Ideate, Prototype, Test Human Centred Design	Hasso Plattner Institute of Design IDEO, Tim Brown	Emphasise, define, ideate, prototype and test Inspiration, ideation, implementation: marrying integrative thinking on opposing ideas and constraints
The 4 Ds or Double Diamond Process Model	British Design Council (2005)	Discover, define, develop and deliver

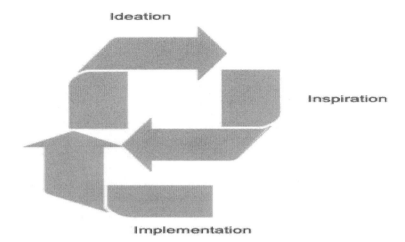

Figure 15.4 Design thinking model
Source: Adapted from IDEO (2012)

- **Prototype:** Ideation becomes tangible in the fourth stage, where a scaled down model of the product solution is built.
- **Test:** Means not necessarily the end of the process; in addition to fault locating, this stage could in fact generate further ideas to be passed back through the process.

The IDEO model of design thinking (see Figure 15.4):

- We live and work in a world of interlocking systems.
- Many of the problems we face are dynamic, multi-faceted and inherently human.
- Design thinking addresses the big questions:

  1. How will we navigate the disruptive forces of the day, including technology and globalisation?
  2. How can we effectively grow and improve in response to rapid change?
  3. How can we effectively support individuals while simultaneously changing big systems?

These questions can be answered through the processes of ideation, inspiration and implementation.

- **Ideation:** The mode of the design process in which you concentrate on idea generation. IDEO describe this in terms of "going wide" in terms of concepts and outcomes. The main aim of the ideation stage is to use creativity and innovation to develop solutions. By expanding the solution space, the design team will be able to look beyond the usual methods of solving problems in order to find better and more elegant and satisfying solutions to problems that affect a user's experience of the product.
- **Inspiration:** The inspiration phase is about framing the problem and its scope, gathering meaningful data from customers and their pains, and then synthesizing and interpreting the collected data for actionable steps in the ideation phase.

- **Implementation:** This third and last phase is where the designer tries to take an idea to reality by testing its possibilities, and fails early to see flaws before putting it in full practice (Brown, Katz 2009).

The Double Diamond process by the Design Council is for use by designers and non-designers (British Design Council, 2005). The two diamonds represent a process of exploring an issue more widely or deeply (divergent thinking) and then taking focused action (convergent thinking).

- **Discover:** The first diamond helps people understand, rather than simply assume, what the problem is. It involves speaking to and spending time with people who are affected by the issues.
- **Define:** The insight gathered from the discovery phase can help you to define the challenge in a different way.
- **Develop:** This diamond encourages people to give different answers to the clearly defined problem, seeking inspiration from elsewhere and co-designing with a range of people.
- **Deliver:** Delivery involves testing out different solutions on a small scale, rejecting those that will not work and improving ones that will.

There is a common theme in all these models: design thinking brings a humanistic element to innovation. The process of design thinking can be integrated into a rational analytical approach in the area of major infrastructure projects. Design thinking can be described in an organisational context as innovation activities with a human collaborative ethos. Design thinking is iterative and collaborative, and the move to design thinking takes the customer from a high priority in the stakeholder influence/power matrix to an integral part of all stages in the process. This shift in focus may require a change in organisational culture for infrastructure project organisations.

15.4 Systems thinking

Infrastructure projects are inherently demanding as they must deal with conflicting aims and requirements from multiple stakeholders with differing motivations; their execution is constrained by political, financial, technological and resource provisions and supports. Traditional "plan and execute" linear methods – brief, design, construct and deliver – are effective only when uncertainties are low. However, the increased uncertainties and complexities in infrastructure projects in the past three decades call for new approaches to create value rather than the traditional linear methods that merely focus on cutting cost and duration with rigid structures like a Gantt chart. It is vital to rethink how to run modern infrastructure projects and one way of doing that is "systems thinking," since a project can be considered as a complex, multiple-loop, non-linear, social system with a strong impact of human actors on decision-making (Aramo-Immonen and Vanharanta, 2009).

The purpose of systems thinking is to

- Integrate the various silos, and hence see the interdependencies between transport and water, between communications and energy, and many others.
- See what is common, and hence understand the relationship between quality and values, functionality and risk, resilience and sustainability.
- Reassess the scope of quality and value, and hence see that these are much deeper and more useful ideas than those in common usage.

- Recognise the need for radical change, and hence see the opportunities to learn and thus add value to yourself, those around you, and your client.
- Deliver customer-focused strategies, and hence understand the value chain with the customer at the head.
- Realise values from processes driven by purpose, and hence realise the need for a strong creative vision and be able to connect it with practicality.
- Integrate people and processes, and hence see that process is the peg on which to hang all attributes.
- Generate simplicity out of complexity, and hence be able to choose the appropriate level of system description.
- Demonstrate practical rigour, and hence understand how practice can be rigorous and not ad hoc.
- Create tools to manage uncertainty, and hence see uncertainty as fuzziness, incompleteness and randomness and use the interval probability theory for evidential support.
- Think in loops and not straight lines, and hence use diagrams of influence to shed light on complexities (Blockley and Godfrey, 2017

Systems thinking is not a completely new technique for project management as the practice of project management had its origin in systems analysis and systems engineering. Systems thinking was not fully recognised until the middle of the 20th century with Ludwig von Bertalanffy's General Systems Theory (von Bertalanffy, 1968). Consequently, systems thinking has never been an explicit part of mainstream education in project management, and the application and understanding of systems thinking benefits were not fully realised (APM, 2018). Systems thinking is meant to complement traditional linear approaches as those approaches are not sufficient to manage modern infrastructure projects; the dynamics between interacting components and an aggregation of optimised subsystems may result in an ineffective overall solution due to the interactions between the subsystems. In the project management process, systems thinking is also reflected in getting the right information (what) to the right people (who) at the right time (when) for the right purpose (why) in the right form (where) and in the right way (how) (Blockley and Godfrey, 2017). Figure 15.5 and 15.6 illustrates the potential impact of systems thinking.

Research on systems thinking and its application to solving management problems began in the 1950s. Systems thinking has been used as an effective frame for understanding problem situations and in guiding day-to-day decision-making (Yeo, 1993). The basic ideas of systems thinking have been used across a far wider spectrum of infrastructure projects (Blockley and Godfrey, 2017). Systems thinking aims to ensure that the whole system performance is designed to take into account all relevant factors when implementing change (Yeo, 1993).

15.5 Systems thinking on infrastructure projects

Since 1960s, the system engineering approach from the US has provided the conceptual basis for the development of the many modern project-management concepts, procedures and techniques, such as work breakdown structure, organisational responsibility matrix, and earned-value methods for progress measurement. Practitioners in project management witnessed a gradual shift in emphasis in systems thinking from the structured, or hard, "systematic" approach in the 1950s and '60s to a softer "systemic" approach in the 1970s and '80s. The shift to the softer systems thinking gives special attention to those involving human behavioural factors related to culture, value systems, attitudes, human perception, meaning, and learning in human activities,

Applying design thinking principles on major infrastructure projects

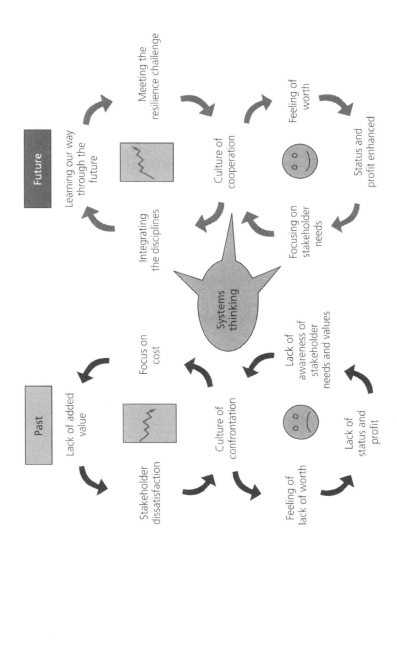

Figure 15.5 The potential impact of systems thinking
Source: Adapted from Blockley and Godfrey (2017, p.202)

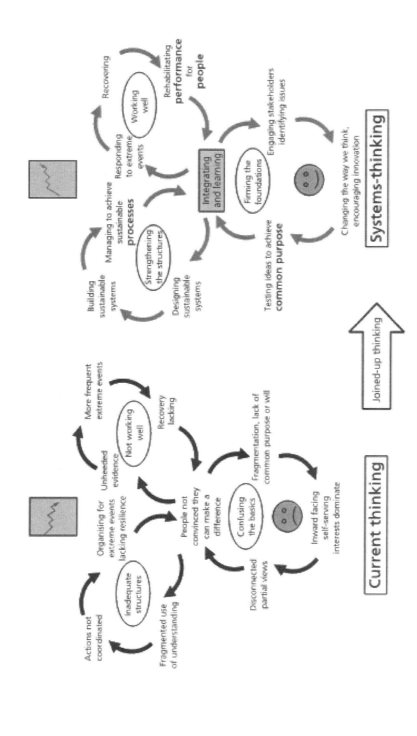

Figure 15.6 Transforming to systems thinking

Source: Blockley and Godfrey (2017, p. 202)

Applying design thinking principles on major infrastructure projects

both at the organisational and individual levels (Yeo, 1993). The aim of systems dynamics in projects is to provide an understanding of the structure of complex systems. Kapsali (2011) points out that project management should be concerned with equipping the team to cope with challenges, which is different from giving top managers a platform to monitor progress; the system to manage projects needs to be unique to those using it and the environment it works in. An open or soft approach allows this adjustment and complexity to be accounted for by acknowledging subjectively what the problem is, and encouraging different viewpoints from multiple stakeholders (van Eck and Ponisio, 2008). Infrastructure systems necessarily involve people of both hard and soft systems. Hard systems are physical systems whose parameters are measurable; soft systems are human and social systems involving people. Soft systems are normally associated with the social sciences, which are rather difficult to pin down, are not very precise and are difficult to test. Measurement of soft systems is difficult, evidence is hard to find, methodology is crucial and often subject to debate, and models tend to be almost always expressed in natural language (Blockley and Godfrey, 2017). Major projects could often benefit from the application of systems thinking to:

- Improve the realism of cost and schedule estimates by understanding that projects are not deterministic. For example, additional tasks are often needed in projects that were not originally expected (such as rework), and this can dramatically slow progress. Systems thinking can help anticipate and manage this effect.
- Improve the integrity and hence value of the product that is delivered by anticipating possible challenges at the interfaces and by anticipating additional enabling tasks and systems beyond the obvious. A close relationship between (systems) engineers and project managers ensures fewer unexpected surprises.
- Improve the understanding of stakeholders' needs throughout the (extended) project life cycle. A systems view encourages broader thinking about how a product or service meets the needs of various stakeholders, and what higher-level goals and constraints exist outside the boundary of the delivered system (APM, 2018).

Complex projects that particularly benefit from systems thinking tend to be characterised by a high number of interactions and a high number of components (Sheffield *et al.,* 2012). Systems thinking supports engineers by:

- Providing a framework for understanding and managing uncertainty and complexity;
- Enabling multiple stakeholders to collaborate and add value;
- Structuring problems to facilitate joined-up thinking; and
- Understanding better interactions between hard and soft systems;
- Providing practical rigour in identifying and agreeing what process holons are necessary and sufficient for success

In reviewing the emerging systems thinking models, Yeo (1993) categorised two methodologies: the hard systems thinking (HST) model and the soft systems thinking (SST) model. HST complies with the principles and presuppositions as below:

- Systems objectively exists in our world and they have good structures and identified goals;
- The parts of the system have the same worldviews, values and interests;
- The system intervener is an outsider of the system and is not influenced by it; and

- Achieving the optimal results is the ultimate goal of problem-solving process (Yan and Yan, 2010).

15.5.1 Hard systems thinking model

The big difference between HST and SST lies in the interpretation of the concept of system. A system is normally defined as an assemblage or combination of things or parts forming a complex or unitary whole. This "whole" can include an ordered and comprehensive set of facts, principles, or doctrines in a particular field of thought or knowledge, or coordinated methods or schemes and procedures. In other words, a system is an organisation of some ideas, things or procedures (Blockley and Godfrey, 2017). HST regards systems as an objective part of our world. This is an ontological definition of system derived from Bertalanffy's General System Theory in 1960s. Based on a mathematical model involving a few (measurable) variables in a linear relationship (Churchman *et al.*, 1957; Checkland, 1981), this approach can be thought of as a form of systems thinking called "hard systems thinking."

Hard systems thinking is incorporated in techniques such as programme evaluation and review techniques (PERT) and critical path analysis, which help to calculate how tasks can be sequenced in a project to minimise time and cost (Jackson, 2003). Systems analysis (SA) and systems engineering (SE) correspond to the two main phases of a project life cycle, with the former dealing with pre-project economic analysis and the latter with project engineering and management. Essentially, SA determines what is to be done, which is often a strategic decision-making process, while SE focuses on how to do it, which is in the realm of operational management. Both approaches strictly follow a hard-systems mental framework and have been adapted and translated into policies and procedures in solving management problems (Yeo, 1993).

Yeo (1993) points out that organisations are viewed as relentless goal-seeking machines from a hard systems thinking perspective. The emphasis is often on the compliance of the systems procedures. Within the project entity, project activities or tasks and manpower are often organized into multi-layered hierarchies, usually in the form of a work-breakdown structure (WSS) and an organisational-breakdown structure. Not surprisingly, the hard systems approach was found to be inadequate in dealing with the many soft, ill structured, and problematic situations in the real world, such as those encountered in the conceptual stage of project definition or those dealing with strategic planning issues, when the definition of clear objectives and formulation of viable alternatives can itself be problematic. The soft and probably "messy" real-world problems often defy precise formulation in the hard sense.

Hard systems thinking, as a bottom-up approach (Pinto and Garvey, 2013), is based on Newtonian science where everything that happens has an identifiable cause and definitive effect. This assumes the engineer can predict the behaviour of any component with certainty if they understand its state at any time. With sufficient knowledge, an engineer can predict the future evolution of the system with a high degree of confidence (Decker, Ciliers, and Hofneyr, 2011). This ability to predict supports the use of risk in risk-resilience methodologies to make optimal decisions.

In contrast, an infrastructure project as a complex system is one where the social component such as human participation or judgement or connectivity is a key component. Traditional approaches such as precise planning are less effective for complex projects due to the way in which unstable systems change (Kopczyński and Brzozowski, 2015). The human aspect introduces relationships between stakeholders as well as complexities not easily represented by hard systems methodologies where reductionist methods can misrepresent the problem domain. These kinds of problems require decision-makers to account for both the technical and social factors to achieve sustainable results (Kirk, 1995).

15.5.2 Soft systems methodology

Since the 1970s there has been significant development in the systems approach, and now a much wider range of project complexity can be dealt with than hard systems thinking was able to deal with alone (Jackson, 2003). Jackson and Keys (1984) developed a framework for classifying systems methodologies called system of systems methodologies (SOSM) (Griffiths, 2017). Based on their more than two decades of research in using the systems approach for solving soft and ill structured problems, Checkland and Scholes (1990) have developed and refined a soft systems methodology (SSM) with seven stages (see Figure 15.7). It is suggested that this sequence need not be followed precisely and, indeed, can be used out of sequence, as dictated by the problem-solving circumstances and the experience of the problem-solver. The SSM model has two parts: the perceived reality and the conceptual model. To start with, there must be at least one person who considers that there is a problem situation that needs to be improved. This is followed by an appreciation and expression of the problem situation in question. The initial stages are the "finding out" stages. The result is a "rich picture" of observed reality or a problem situation which may involve unresolved issues, conflicts, and other problematic and interesting features.

SSM developed because creating clear objectives systematically is not always feasible for large projects. In major projects over-estimates and programme slippage are common occurrences, and the most fundamental cause of the problem is change (Yeo, 1993). Change is inevitable in large projects and it's typical for goals to evolve during the lifespan of mega-projects (Aramo-Immonen and Vanharanta, 2009). Participants should use a systemic methodology to learn what changes are feasible and desirable from the problem context. SSM looks to unfold relationships within projects to enable better decisions to be made (Jackson, 2003). The most accessible component of an SSM study for project management is the "rich picture," which encourages a deep consideration of the problem situation from the perspective of multiple stakeholders, uncovering sympathies and tensions between the various actors. This can form an excellent foundation for the requirements management process (Niu *et al.*, 2011)

15.5.3 Tools for systems thinking

According to Sheffield *et al.* (2012), few project managers employ systems thinking to manage complex projects, even though just a few simple tools could bring unique benefits to problem-solving for these projects. The traditional SE approach is to quantify and seek optimal economic solutions. However, often economic optimisation does not fully capture all the aspects of a decision. The process can inadequately represent or disregard aspects difficult to represent mathematically, such as environmental impacts and social disruptions. Consequently, the public often reject such optimised solutions (Checkland, 2000). As a result, SE is evolving to develop methodologies that try to cope with a problem that is difficult to define. It is expanding its understanding of those aspects of a project that mathematics cannot easily represent. It is recognising the need for satisficing alternatives, that is, those that are not necessarily optimal, but good enough to balance economic, environmental, and social needs (Keating, Calida, Sousa-Poza, and Kovacic, 2010). Complex projects that particularly benefit from systems thinking tend to be characterised by a high number of interactions and a high number of components (Sheffield *et al.*, 2012). Figure 15.8

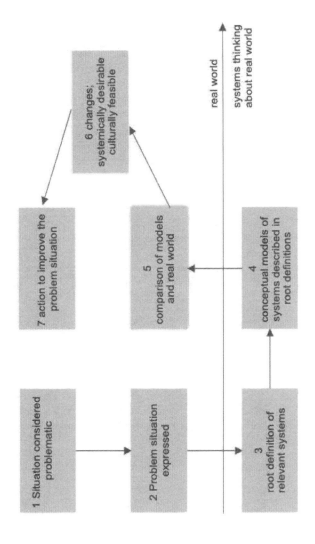

Figure 15.7 Seven-stage model of SSM
Source: Adapted from Checkland (1981)

Applying design thinking principles on major infrastructure projects

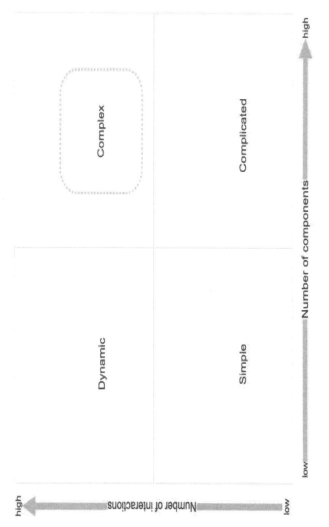

Figure 15.8 Types of systems and projects
Source: Adapted from Sheffield *et al.* (2012)

groups types of systems and projects by the number of components (social and technical) and the number of interactions.

Figure 15.9 shows corresponding types of project management methods and where systems thinking fits in. A low number of interactions and components characterises a simple system managed by a linear "waterfall-style" project management method. A low number of interactions and a high number of components characterises a complicated system managed by compliance to an extensive implementation plan (Slack *et al.*, 2010). A high number of interactions and a low number of components characterises a dynamic system managed by an agile project management method. A high number of interactions and a high number of components characterises a complex system managed by a systems thinking project management method (Leme´ tayer, 2010).

Kopczyński and Brzozowski (2015) identify that the starting point of applying systems thinking in project management is understanding the problem-solving process. Sheffield et al. (2012) recognise a particular technique that can be used for this initial "concept" step, as well as systems thinking techniques from project management that apply to two other phases of the development life cycle (implementation and evaluation) as shown in Figure 15.10.

Maanie and Cavana (2007) outlined five phases, along with relevant systems thinking techniques, that can be used to structure problem-solving and can either be used as a process method or individually, depending on the problem (see Table 15.4).

The International Council on Systems Engineering (INCOSE) and the Association for Project Management (APM) came together to form a Systems Engineering and Project Management (SEPM) Joint Working Group. They produced a paper on systems thinking in which six systems thinking tools were highlighted (APM/INCOSE JWG, 2018). The following techniques and figures are cited from the publication (APM, 2018). How are they used in project management?

1. A fishbone diagram (Figure 15.11) structures thoughts and distinguishes hard and soft variables that affect the problem of interest.
2. A part of soft systems methodology which enables a problem situation to be defined by multiple stakeholders and an initial mental model to be created (see Figure 15.12).
3. Figure 15.13 characterises key organisations and roles that are in, and affected by, the system.
4. Figure 15.14 shows the knowledge concepts of a topic, where the main concept is broken down to show its sub-topics and their relationships.
5. Figure 15.15 details trends that influence the system through collective knowledge of those familiar with the system and its context. Enables activities and events to be visualised to identify potential contextual factors.
6. Figure 15.16 represents the relationships between system elements and identifying reinforcing and balancing processes to explore behaviour over time.

Table 15.4 Systems thinking techniques

Phases	*Systems thinking techniques*
1 Problem structuring	• Affinity diagram hexagon clustering
2 Causal loop modelling	• Affinity diagram hexagon clustering
3 Dynamic modelling	• Rich picture
	• Stock flow diagram
	• Software package (e.g. STELLA)
4 Scenario planning and modelling	• Scenario planning
5 Implementation and organizational learning	• Scenario planning

Source: Adapted from Maani and Cavana (2007)

Applying design thinking principles on major infrastructure projects

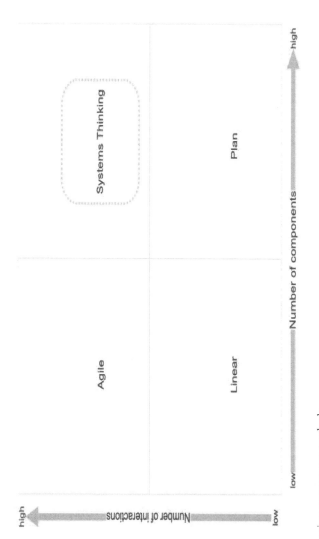

Figure 15.9 Types of project management methods
Source: Adapted from Sheffield et al. (2012)

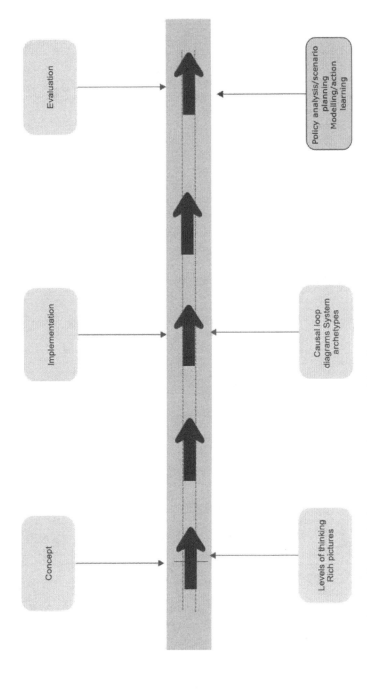

Figure 15.10 Application of systems thinking to system development life cycle

Source: Adapted from Sheffield et al. (2012)

Applying design thinking principles on major infrastructure projects

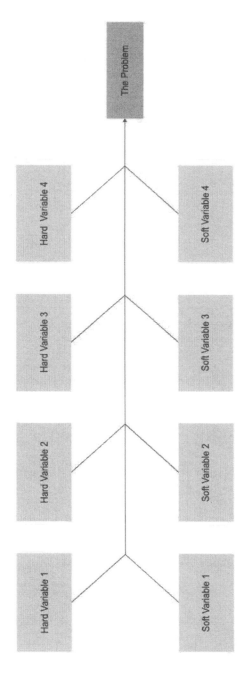

Figure 15.11 Systems thinking fishbone diagram
Source: Adapted from APM/INCOSE JWG (2018)

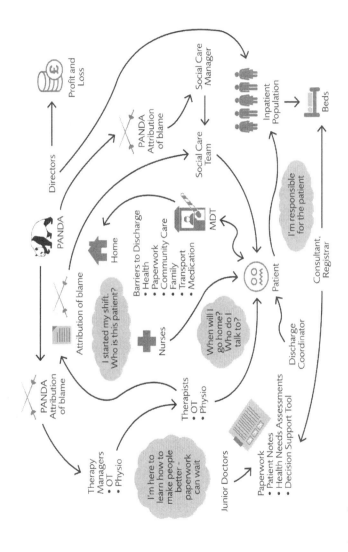

Figure 15.12 Rich picture diagram

Applying design thinking principles on major infrastructure projects

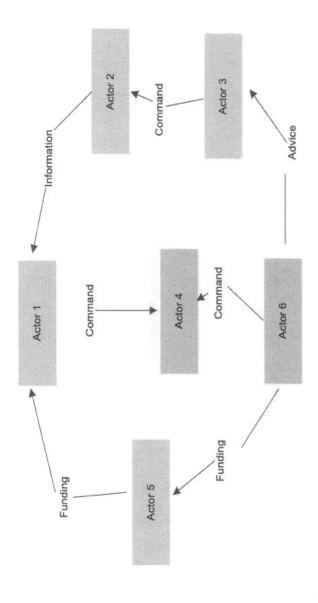

Figure 15.13 Actor map
Source: Adapted from APM/INCOSE JWG (2018)

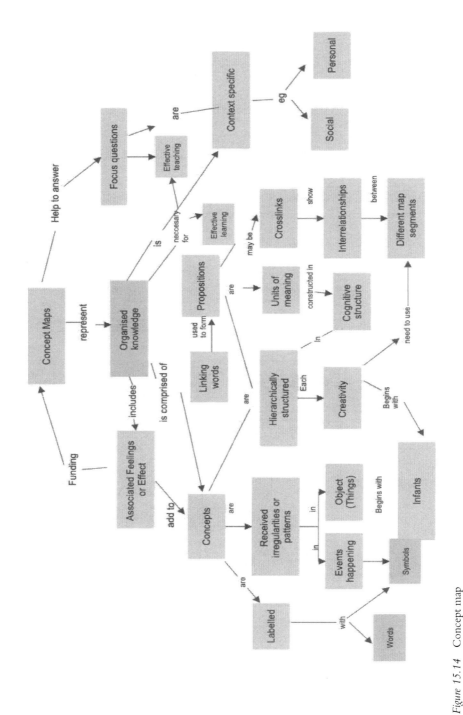

Figure 15.14 Concept map
Source: Adapted from Novak and Canas (2006)

Applying design thinking principles on major infrastructure projects

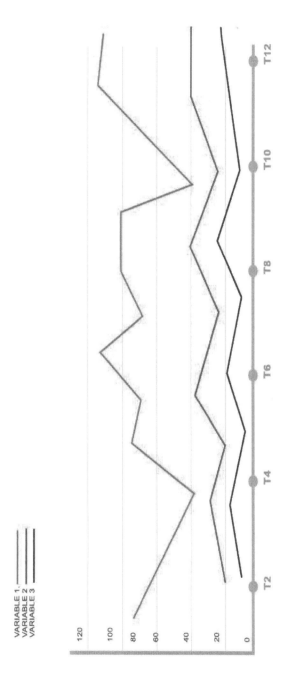

Figure 15.15 Trend map
Source: Adapted from APM/INCOSE (2018)

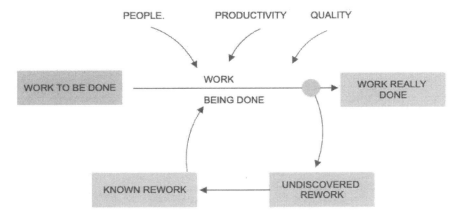

Figure 15.16 Causal loop diagram
Source: Adapted from Cooper (1993)

15.6 Systems thinking models for major infrastructure projects

As the importance of systems thinking is increasingly being recognised in industry, academia and government agencies, a framework for rethinking infrastructure and meeting the resilience challenge must address how we think about the future. Delivering resilience will require innovative systems-thinking skills of reflective practical wisdom that go beyond technique (Blockley and Godfrey, 2017). Blockley and Godfrey (2017) sets out five axioms and fifteen corollaries of system-thinking as below (see Table 15.5 and Table 15.6):

Infrastructure projects face a range of challenges in integrating social and technical aspects to deliver the project outputs to satisfy multiple stakeholder requirements. The challenge in the integration process is these social aspects often override the technical aspects of problems. Mathematical models competently represent the natural and man-made environment of a technical problem, but they can result in an optimal solution that the public do not accept. Municipal engineers are hesitant to invest in such expensive studies that may miss the complexities prevalent in the social aspects of a technical problem. They need a way to think through what a sociotechnical problem is to judge investments in studies that the public will find acceptable (Pezza and Pinto, 2019). As a way forward, Pezza and Pinto (2019) propose an enterprise system model to capture the sociotechnical aspects of society grappling with difficult infrastructure problems. The proposed model combines systems principles applicable to a public works infrastructure with complexity theory to help engineers make decisions based on systems thinking; it offers a framework to help engineers judge a situation and determine which systems approach is appropriate. According to Pinto and Garvey (2013), an enterprise system, as a democratic system, is a way to represent society as a network of interdependent people whose processes and supporting technology are not fully under the control of any single entity (see Figure 15.17). However, using a network to represent an enterprise system is insufficient. The network needs to nest systems and their subsystems in a hierarchical representation as a special case of a network.

Applying design thinking principles on major infrastructure projects

Table 15.5 The five axioms of systems thinking

Axiom A1: Impelling purpose	What: A system discriminated from its surroundings is a set of models of our reality created by people for reasons that give meaning to and determine the purpose of the system. Why: We identify a system because we are curious and want to understand (science), wish to modify the world around us to improve the human condition (engineering and medicine) or to express our emotions (art and religion). Our purpose in identifying the system is our highest goal. That goal provides us with meaning and motivates us to put in effort to add value. Identifying purpose draws on our emotional intelligence to help us reflect on and understand why we think and act, what we value and how we can improve on working together.
Axiom A2: Appropriate layers	What: Systems models are holons; that is, they are both parts and wholes and hence are layered according to levels of detail and abstraction. Why: Thinking about a system in layers helps us cope with size, scale and dimensionality. Models of holons at different levels can be different but still be interdependent.
Axiom A3: Complex interdependency	What: Holons are connected to certain other holons with which they exchange messages. Why: In a complex world everything seems to be interconnected and hence interdependent. Outcomes are often unintended. We can simplify by focusing on local connections in a manner similar to the internet of connected computers.
Axiom A4: Ubiquity of change	What: Systems models change at varying rates but none are permanent and invariable. Some changes may be unforeseen. Some changes may be small but some may be "revolutionary paradigm shifts" involving new ways of thinking (Kuhn, 1962). Why: Most of us think of matter or substance as the permanent stuff of which something is composed, and of form as the way that stuff is put together. Interestingly, Eastern traditions emphasise flow and change as things come into being and cease to be. This view ties in with the spontaneous random fluctuations of energy in a quantum space.
Axiom A5: Evolutionary learning	What: Complex systems often cannot be "solved"; rather they have to be managed to desirable outcomes. Why: Learning reduces uncertainty. Learning is too often seen as rote learning of facts and techniques and of "how to do something." Learning to learn, or "learning power," has much to offer in finding our way through uncertainty.

Source: Blockley and Godfrey, 2017, p. 36

Table 15.6 Corollaries of the five axioms of systems thinking

Corollary C1: Regarding reality	What: Reality is actual but only accessible through human cognition. Why: Engineers are familiar with models but we need to make explicit the notion that models are a representation of reality, not the reality itself.
Corollary C2: Regarding worldviews	What: Systems models are created through the "spectacles" of a set of worldviews. Why: If models are a representation of our reality and not the reality itself, we can think of the models as filters or spectacles through which we must look to see our reality.
Corollary C3: Regarding understanding	What: We can only control effectively what we understand. Why: Perhaps this is self-evident, but each of us understands things in a different way. That is why we need to collaborate in teams to share understanding in different ways.
Corollary C4: Regarding embedment	What: Physical natural or artificial systems are "hard" systems. Both are embedded in "soft" people and social systems. Hard systems are objective, whereas soft systems are subjective and inter-subjective. Why: This aspect of rigour sounds unduly "academic," esoteric and remote from practice. However, it underpins our understanding of uncertainty and the way we learn our way through it, and so is important.
Corollary C5: Regarding functions	What: The purpose of a hard system is a function. The function of an artificial system is decided by us – it is man-made. A model of a natural system helps us understand the behaviour of a part of a reality. As a consequence, we may ascribe a function to it. Why: The function of man-made artefacts is a familiar concept to most of us. However, the function of a natural system can be a source of confusion if we do not explicitly recognise that that function is ascribed by us through our models.
Corollary C6: Regarding fitness for purpose	What: Hard systems are not universally true (i.e., true in all contexts and circumstances). Rather, they are dependable and fit for purpose to a degree and in a context. Dependability corresponds to our common-sense notion of truth or fact. Statements deduced from dependable models correspond to reality in a particular context or situation. Why: Models, by their very nature are partial representations of our reality. Consequently, they are incomplete and are only dependable in the context to which they are relevant.
Corollary C7: Regarding duty of care	What: The dependability of a systems model requires those people involved to exercise a proper duty of care, to test the model to an appropriately dependable level based on evidence, to demonstrate sufficient competence and integrity, and to be transparent about their values. Why: Dependability has to be judged based on the testing of a model. The tests have to be as searching and rigorous as is appropriate for the particular problem. Practical rigour requires diligence and duty of care that leaves no stone unturned, with no sloppy or slip-shod thinking.
Corollary C8: Regarding subsidiarity	What: The principle of subsidiarity (as set out in the Treaty of Lisbon 2007 (EUR-Lex, 2016)) is that systems models should be created at the lowest practical level that is consistent with delivering their purpose. Why: The idea here is that decisions should be as local as possible because that is where the problems are best understood.

Applying design thinking principles on major infrastructure projects

Table 15.6 Corollaries of the five axioms of systems thinking

Corollary C9: Regarding emergence	What: Holons have emergent properties. These are attributes that apply at only one or more layers as a result of interactions between holons at lower levels that do not exhibit these attributes. Why: Emergent properties arise or come forth from interdependencies at more detailed layers. They are more common than many people realise. For example, the pressure of a gas is the result of the buzzing around of gas molecules at a lower level of description. The human ability to walk and talk emerges from the cooperation between our many subsystems.
Corollary C10: Regarding connectivity	What: Connections create relationships and patterns of relationships. Why: In Chinese thought all things are interconnected. The internet is a web of interconnected computers. The brain is a network of highly interconnected neurons. Our infrastructure is an interconnected network of facilities.
Corollary C11: Regarding stakeholder interests	What: There is an increased chance of success if stakeholder interests are aligned. Why: Common sense tells us that we are more likely to be successful if we pull together, as oarsmen do in a boat race. We are more likely to pull together if we have a common purpose.
Corollary C12: Regarding processes	What: Systems models are processes. Why: If we accept that change is ubiquitous, then everything is a process. This is helpful because it shifts our focus and leads to a new understanding of change. It provides us with a means of integrating many ideas and enables us to create simplicity in complexity. Unsurprisingly, perhaps, many people find it hard to think of a table as a process as they cannot reject the idea that it is a thing composed of "stuff" (wood). It may help to think about the life cycle of the table from raw material, through design and making to usage, maintenance and disposal, to see that the table is constantly being and becoming. Everything exists in the process of time.
Corollary C13: Regarding feedback	What: Processes may be loopy, involving feedback and feedforward. Why: Most engineers are familiar with the ideas of feedback and feedforward in hard systems. They apply equally in soft systems, where they are often called "loops of influence."
Corollary C14: Regarding leadership	What: Managing a process to a desirable outcome requires appropriate leadership and collaborative learning. Why: Traditional learning is something we do to acquire knowledge that may be useful to us in some way. We tend to think via a prescribed framework that promotes a strong distinction between the academic and the vocational, which devalues practical wisdom. To change, people need vision. Leadership is about engaging with that vision and then building and coaching teams to achieve it; it applies at all layers.
Corollary C15: Regarding outcomes	What: Unexpected and unintended changes may result in future consequences that may be opportunities to create benefit or hazards that threaten damage. Why: We must protect ourselves from the harmful effects of unintended consequences. That is why we need to be alert to the possibility of "incubating failure" (see Section 8.6). Just as importantly, we must take advantage of possible benefits from unintended consequences; they lead to new opportunities and genuine innovation.

Source: Blockley and Godfrey, 2017, pp. 36–37

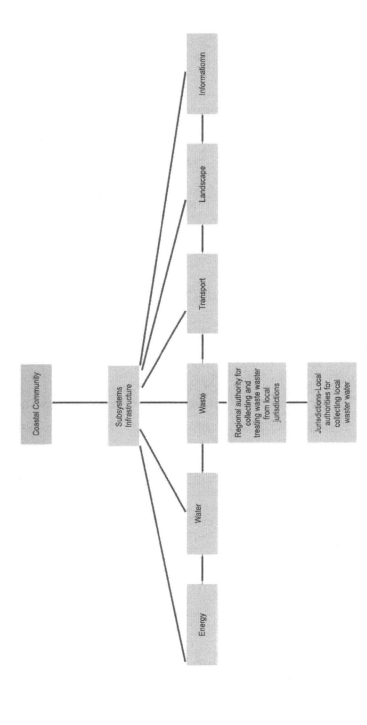

Figure 15.17 An example of an enterprise system
Source: Adapted from Pinto and Garvey (2013)

Applying design thinking principles on major infrastructure projects

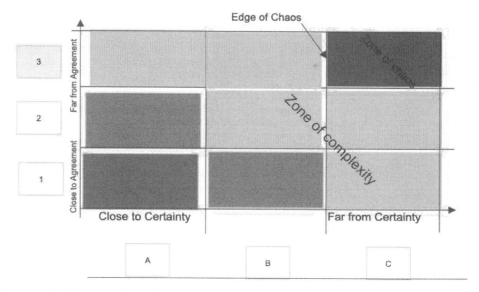

Figure 15.18 The zones of complexity
Source: Adapted from Pezza and Pinto (2018)

According to Pezza and Pinto (2018), a framework for systems thinking should start with identification of the complexity with the following chart (see Figure 15.18).

The zones of complexity support engineers in categorising and communicating an appropriate systems approach (Stacey, 2011). "Far from Agreement" implies a discord among stakeholders surrounding the issue. "Far from Certainty" means the situation is unique and new to the decision-makers, and extrapolation from the past is an insufficient method to project outcomes. The green zone of "Close to Agreement" and "Close to Certainty" implies consensus among stakeholders and the suitability of mathematical models to optimise alternatives. Based on the evaluation of the problem complexity and considerations on a range of elements, the systems thinking approach should be selected as illustrated below (see Table 15.7).

We believe we have shown that systems thinking and design thinking are both valid approaches to wicked problem resolution in major infrastructure projects. Although there are core differences, a balance could be struck between them.

15.7 Chapter summary

As shown in this chapter, design thinking and systems thinking are ways in which the pulls of technical complexity, social complexity and wicked problems can be addressed. The process of design thinking can be integrated into a rational analytical approach in the area of major infrastructure projects. In a project management process, systems thinking is also reflected in getting the right information (what) to the right people (who) at the right time (when) for the right purpose (why) in the right form (where) and in the right way (how). The aim of systems dynamics in projects is to provide an understanding of the structure of complex systems.

Table 15.7 Clarifications of systems

Characteristic	Simple system Apply hard systems thinking	Complex system Apply soft systems thinking
Stacey's zones	Dark green, a hard systems approach. Light green, a hybrid approach to address uncertainties.	Yellow or red, a soft systems approach.
Number of elements	Small	Large
Interactions between elements	Few	Many
Predetermined attributes	Yes	No
Interaction organization	Highly organized	Loosely organized
Laws governing behaviour	Well defined; deterministic or stochastic methods	Undefined; emergence behavior
System evolution over time	Not evolve	Evolves
Subsystems pursue own goals	No	Yes (purposeful)
System affected by behavioral influences	No	Yes
Predominantly closed or open to the environment	Largely closed	Largely open
Predictable	Yes	No
Method of analysis	Risk-resilience-informed decisions	Interactive planning
Type decision	Risk-resilience-informed decisions	Satisficing solutions

15.7.1 Chapter discussion questions

1. Within the context of major infrastructure delivery, identify the key benefits and limitations of design thinking.
2. Critically discuss the value of the different types of design thinking models appraised in this chapter.
3. What are some of the advantages and disadvantages of using systems thinking principles on major infrastructure project delivery?
4. Identify an incomplete infrastructure project of your choice, and critically discuss how you will accelerate the completion of the project by incorporating systems thinking principles.

15.8 Case study: Connecting Bristol strategy (UK)

A SMART city is defined by the British Standards Institute (BSI) as "the effective integration of physical, digital and human systems in the built environment to deliver sustainable, prosperous and inclusive future for its citizens." The UK Department for Business, Innovation and Skills (BIS) considers smart cities a process rather than a static outcome, in which increased citizen engagement, hard infrastructure, social capital and digital technologies make cities more liveable, resilient and better able to respond to challenges. The European Union (2018) expands further: a smart city goes beyond the use of information and communication technologies (ICT) for better resource use and less emissions. It means smarter urban transport networks, an upgraded water supply, better waste disposal facilities and more efficient ways to light and heat buildings. It also means a more interactive and responsive city administration, safer public spaces and meeting the needs of an ageing population. In the UK, there are to date 18 SMART cities. Bristol has been the UK's leading SMART city since 2017. Bristol's Smart City Research and Development network platform is overseen by Bristol City Council and the University of Bristol and operates under the name Bristol is Open.

15.8.1 Background

One of the larger cities in the UK, Bristol has a population of 463,400 (as of May 2020) that is estimated to exceed 500,000 as early as 2027. Historically, Bristol is a port and trading city but today is a high growth centre for media, information technology and creatives. The high tech and digital economy is fuelling growth across the region. Bristol unveiled its first SMART city strategy in September 2019, which brought together existing SMART city functions such as the Bristol is Open Connected Cities Programme and the Bristol Operations Centre. Bristol City Council state that the SMART city strategy takes a holistic approach to refocus their innovation to a challenge-led approach to the problems and issues facing the city. As shown illustrated in Figure 15.19, their new approach identifies six essential pillars.

15.8.2 Highlights of the SMART city strategy

Extracts from the SMART City strategy, adapted with permission from Bristol City Council, are highlighted below.

15.8.3 Introduction

The One City Plan is an ambitious vision for the future of Bristol that is being done decade by decade up to 2050. It is built on six interdependent stories: connectivity, health and well-being, homes and communities, economy, environment, and learning and skills. This strategy sets out how Bristol City Council will support Bristol's SMART city journey. The aim is to strengthen the city's digital foundations so that it becomes well-connected and better placed to deliver the technological innovation needed to keep the city moving, healthy and safe, in a sustainable way. Everything done will focus on ensuring Bristol achieves its exceptional potential.

Figure 15.19 Bristol SMART city essential pillars

Source: Bristol City Council (2019)

- This includes guiding principles about how the city will work collaboratively with partners to co-design solutions, where possible, with a focus on social value. The strategy will be underpinned by a portfolio of initiatives that will deliver challenge-led innovation and build digital foundations.
- The view is that a SMART city is a liveable, sustainable and prosperous city. It is attractive to both people and businesses because of the high quality of life, easy access to jobs, amenities and services, and vibrant culture. The approach is shifting to using civic and social innovation in combination with technology to enhance, not degrade, the messy, human magic of Bristol.
- To be "smart" the city needs to embrace this and focus on blending hard and soft infrastructure with a human-centred approach to:

1. Make Bristol more liveable, workable and sustainable.
2. Manage the city and civic resources as effectively and intelligently as possible.
3. Deliver world-class citizen-centric city services.
4. Underpin a continuous process of reinvention, transformation and creativity.
5. Support economic development and long-term prosperity.

15.8.4 Being smart and sustainable in response to global and future challenges

Some of the challenges that affect Bristol are global in nature, such as the need to respond to climate change and city growth in a sustainable way. In November 2018, the council unanimously set a goal for the city as a whole to be carbon neutral by 2030. This ambitious pledge and a commitment to the United Nations Sustainable Development Goals are written into the 2050 One City Plan. Looking to the future, the so called Fourth Industrial Revolution will create both challenges and opportunities for the city as trends like automation, the internet of things (IoT), artificial intelligence (AI) and disruptive business models influence the city. These forces will change how people work, live, play and move around the city and address energy use challenges in the future. A smart city is one that is able to explore these future scenarios and use those insights to plot a course that allows the city to remain resilient and successful across a range of possible outcomes. Developing a low-carbon and sustainable city will enhance Bristol's ability to remain competitive in the global economy and be more resilient. Smart technologies have an important role to play in reducing carbon emissions and are becoming more commonplace. For instance, smart systems can return excess energy to the grid and supply it at the most efficient time, helping to minimise overall peak demand emissions. City data is increasingly being used to provide people with accurate and timely information for bus, rail and car travel through smart apps and at pre-boarding locations, which can help make sustainable modes of travel more attractive.

15.8.5 World-class connectivity in Bristol

Bristol has smart city capabilities and assets that are already some of the best in the UK – including the Bristol Network (B-NET), Bristol Is Open, the Bristol Operations Centre and the Open Data Platform – but it has an aspiration to become the UK's most connected city with high-quality, secure and reliable digital connectivity. This will ensure that all communities and businesses across Bristol can benefit from a world-class network and communication infrastructure, meet growing demands for digital services and make sure that no one is excluded from the digital economy due to poor connectivity. Effective roll-out of this core infrastructure is critical and it provides the backbone for improved wireless and mobile networks, like ultrafast public Wi-Fi, better 4G and next generation wireless technologies. It will need to be multi-tenant and led by customer requirements. This underlying connectivity will provide a route to securely deploy smart technologies, such as the IoT, a giant network of connected "things" (which also includes people), and unlock the potential of city data and analytics. Importantly it enables the city to deploy smart technologies without affecting the integrity of the council's operational network. Intelligent data is key to a smart Bristol. It needs to harness city data to support decision-making, generate practical insights and take an evidence-based approach to city management. If developed alongside core principles of cyber resilience, open standards, common platforms and interoperability, then the city's world-class connectivity will improve the way the city is managed along with many other aspects of everyday life. Collaborating with partners across the region to deliver network and data connectivity at scale will further unlock this area of England's full potential.

15.8.6 City-wide innovation ecosystem

The city wants to nurture a city-wide approach to developing a smart Bristol which will encourage digital innovation and city-led initiatives across a diverse and inclusive network of institutions, communities, and individual entrepreneurs. It will ensure that smart city innovation is done by the city, for the city. Firstly, the collective creativity of the city will be harnessed to help co-design solutions to complex city challenges. Across Bristol there are rich, diverse, energised communities of innovators and entrepreneurs, all passionate about making Bristol a great place to work, visit and live. Together their influence on the digital fabric of this city could be transformational. Another goal is to foster a more inclusive, aspirational approach to smart cities that is more in line with our reputation as a learning city. Linking the development of a smart Bristol with skills and education engagement, for people of any age, will sustain access to employment and job creation in key areas of the regional economy. Facilities and activities across the city, from well-connected community hubs, living labs and incubator spaces, to hackathons and innovation events, will be augmented by a virtual open innovation infrastructure. This will enable the city to empower community-led innovation and attract the most ambitious and innovative partners from around the world.

15.8.7 Responsible innovation

A future smart Bristol might be enabled by technology and data, but it will only be sustained by trust. Cities are increasingly important in the emerging data economy, representing a major source of data that enables modern tech firms and start-ups to prosper. While Bristol wants to become well-connected and data-enabled, there is a risk that this transformation undermines the social benefits that the city seeks to deliver. It wants to lead on the ethics and governance side of smart city and digital technologies, and ensure that solutions comply with privacy and data protection regulations of personal data, to ensure it does not create even wider imbalances of socioeconomic power. It is also important to remain mindful of environmental impacts of smart solutions, such as energy intensive ICT practices. Smart cities need smart citizens. The plan is that a people-centred approach will lead to a more inclusive city-wide transformation. However, we need a broader ethos of responsible innovation to ensure that we are alive to the ethical, societal and regulatory challenges of smart cities. A smart Bristol should be about the use of technology and data for public good.

15.8.8 Public service innovation

Rising demands, diminishing budgets and big ambitions means that the council specifically needs to explore different, better ways to manage the city and deliver public services. The challenge to do more with less, especially for health and social care, means that innovation and reinvention is a necessity. Bristol City Council is undergoing its own digital transformation to make change possible. New technologies, such as cloud infrastructure, can kick-start new conversations and new ways of working that fuel change. But how can the city better use its collective talents, data and technology to improve citizen engagement and make public services simpler, stronger and more effective? Embedding a bold ethos of entrepreneurial civic innovation at the heart of service delivery will enable us to deliver frictionless, well-designed, effective services and an infrastructure that delivers the step changes outlined in this strategy and supports the aspirations of the city and the region. As the city explores how to build better public services, it will look to established centres of excellence and reach beyond the public sector for good practice. User-centred design thinking and agile development methods will ensure that citizen experience informs service design. The goal is to transform analogue, paper-based, or legacy systems used to interact with residents into open, citizen-centric and accessible systems. Cost savings will be utilised elsewhere to deliver core services that citizens continue to need support with and the inclusive-by-design approach is addressing the digital divide.

15.8.9 Explore, enable and lead

In order to implement their SMART city strategy, Bristol City Council will be inclusively innovative by not always leading on the development of solutions. They will work with partners, stakeholders and communities across the city in one of three ways:

- **Explore:** We will investigate options to implement this action;
- **Enable:** We will work with partners to implement this action; and
- **Lead:** We will be directly responsible for implementing this action.

15.8.10 Bristol design principles

Bristol City Council has derived a set of 10 design principles to work on this strategy. These principles will guide what we do and how we do it:

- Transparent, inclusive and ethical;
- Open by default and interoperable;
- Start with challenges and user needs;
- Engage, collaborate and co-design;
- Technology should make things easier;
- Create positive experiences and outcomes; and
- Do the hard work to make the complex simple.

The council states that these principles are meaningless if people are not experiencing them, seeing hem and using them every day. The strategy, they state, is a starting point and they wish to develop a shared set of Bristol design principles that are co-developed and adopted by the city.

15.8.11 Case discussion questions

1. Bristol City is already the UK's number one SMART city; as can be seen in the case, it has big ambitions to integrate all of the SMART city elements into one entity: Connecting Bristol. As a project manager, what problems do you foresee arising with this integration?

2. Can you identify the "wicked" problems identified that the Connecting Bristol project may address and look to solve?
3. Is design thinking an appropriate process for this project? Why?
4. How can systems thinking help you to conceptualise and understand the project, its stakeholders and its linkages?

References

Ahiaga-Dagbui, D.D., Love, P.E.D., Smith, S.D. and Ackermann, F. (2017). Toward a systemic view to cost overrun causation in infrastructure projects: A review and implications for research. *Project Management Journal*. 48 (2), pp. 88–98.

Andrew, T.N. and Petkov, D. (2003). The need for a systems thinking approach to the planning of rural telecommunications infrastructure. *Telecommunications Policy*. 27 (1), pp. 75–93.

Aramo-Immonen, H. and Vanharanta, H. (2009) Project management: The task of holistic systems thinking. *Human Factors and Ergonomics in Manufacturing & Service Industries*. 19 (6), pp.582–600.

Blockley, D. and Godfrey, P. (2017). *Doing It Differently, Second Edition* [online]. London: ICE Publishing.

Bristol Council. (2019). *Connecting Bristol: Laying the foundations for a smart, well connected future*. Available from: https//www.connectingbristol.org/wpcontent/uploads/2019/Connecting_Bristol_300819_WEB.pdf [accessed 24 July 2020].

Brown, T. and Katz, B. (2009). *Change by Design: How Design Thinking Transforms Organizations and Inspires Innovation* [online]. New York: Harper Business.

Buchanan, R. (2001) **Design Research and the New Learning**. *Design Issues; JSTOR*. pp.3–23.

Camillus, J.C. 2008, "Strategy as a wicked problem", *Harvard business review*, [Online], vol. 86, no. 5, pp. 98

Checkland, P. (1999). *Soft Systems Methodology: A 30-Year Retrospective*. Chichester: Wiley.

Checkland, B. (1981) *Systems Research and Behavioral Science*. 17(S1), pp. S77–S78, Chichester: John Wiley

Checkland, P. (2000), Soft systems methodology: a thirty year retrospective. *Syst. Res.*, 17: S11–S58. doi:10.1002/1099-1743(200011)17:1+<::AID-SRES374>3.0.CO;2-O

Chester, M.V. and Allenby, B. (2019). Infrastructure as a wicked complex process. *Elementa (Washington, D.C.)*. 7 (1), p. 21.

Design Council (2005) **what is the Framework for Innovation? Design Council's Evolved Double Diamond**. Available from: https://www.designcouncil.org.uk/news-opinion/what-framework-innovation-design-councils-evolved-double-diamond [Accessed May 2020].

Eck, Pascal van and Ponisio, Maria Laura, "IT Project Management from a Systems Thinking Perspective: A Position Paper" (2008). *International Research Workshop on IT Project Management 2008*. 2. https://aisel.aisnet.org/irwitpm2008/2

Emes, M. (2018). Systems thinking: How is it used in project management? *Association for Project Management: Research Fund Series*.

European Commission. (2019). Smart cites. Available from: https//ec/Europa.eu/regional-and-urban-development/topics/cities-and-urban-development/city-initiatives/smart-cities.en [accessed 24 July 2020].

Humphries, M. (2018). *National Infrastructure Planning Handbook 2018* [online]. Third ed. Haywards Heath: Bloomsbury Professional.

Institute, P.M. (2016a). *Construction Extension to the PMBOK® Guide* [online]. 1st ed. Newtown Square, PA: Project Management Institute.

Liedtka, J., King, A. and Bennett, K., eds. (2013) *Solving Problems with Design Thinking: Ten Stories of what Works*. Columbia, USA: Columbia Business School Printing.

Lowe, B. (2010). Design driven innovation: Changing the rules of competition by radically innovating what things mean. *The Journal of Consumer Marketing*. 27 (7), pp. 647–648.

Maanie, E.; Cavana, R. *A Methodological Framework for Systems Thinking and Modelling (ST&M) Interventions*. Geelong. Australia, 8 - 10 November 2000. Researchgate.

Mahmoud-Jouini, S.B., Midler, C. and Silberzahn, P. (2016). Contributions of design thinking to project management in an Innovation context. *Project Management Journal*. 47 (2), pp. 144–156.

National Infrastructure Commission (2018). *Developing design principles for national infrastructure*. Available: at https://www/nic.org.uk/content/uploads/Developing-Design-Principles-for-national-infrastructure-projects [accessed 21 July 2020].

Novak, J.D.; Cañas, A.J. (2006) The Origins of the Concept Mapping Tool and the Continuing Evolution of the Tool. *Information Visualization*. 5(3):175–184. doi:10.1057/palgrave.ivs.9500126

Pereira, J.C. and Russo, Rosaria de F.S.M. (2018). Design thinking integrated in agile software development: A systematic literature review. *Procedia Computer Science*. 138, pp. 775–782.

Pezza, D.A. and Pinto, C.A. (2019). Applying systems thinking to coastal infrastructure systems. *Public Works Management and Policy*. 24 (1), pp. 71–87.

Pinto, C.A.; Garvey, P. R., "Advanced Risk Analysis in Engineering Enterprise Systems" (2013). *Engineering Management & Systems Engineering Faculty Books*. 7. https://digitalcommons.odu.edu/emse_books/7

Plattner, H., Meinel, C. and Leifer, L., eds. (2012) *Design Thinking Research, Measuring Performance in Context*. New York: Springer Heidelberg.

Rittel, H. W. J., & Webber, M. M. (1973). Dilemmas in a general theory of planning. *Policy Sciences*, 4, 155–169.

Rose, J. (1997). Soft systems methodology as a social science research tool. *Systems Research and Behavioral Science*. 14 (4), pp. 249–258.

Rubenstein-Montano, B., Liebowitz, J., Buchwalter, J., McCaw, D., Newman, B. and Rebeck, K. (2001). A systems thinking framework for knowledge management. *Decision Support Systems*. 31 (1), pp. 5–16.

Sheffield, S. Haslett, T (2012), "Systems thinking: taming complexity in project management", *On the Horizon*, Vol. 20 Iss: 2 pp. 126–136

Slack, N. (2010) *Operations and Process Management: Principles and Practice for Strategic Impact*. Harlow: Financial Times Prentice Hall.

Tabish, S.Z.S. and Jha, K.N. (2012). Success traits for a construction project. *Journal of Construction Engineering and Management*. 138 (10), pp. 1131–1138.

Treasury, HM. (2020) *National Infrastructure Comission Framework Document*. Available from: www.gov.uk.

Union, E. (2018) *Digital Single Market Smart Cities*. Available from: https://ec.europa.eu/digital-single-market/en/smart-cities [Accessed May 2020] [

Yeo, K.T. (1993). Systems thinking and project management—time to reunite. *International Journal of Project Management*. 11 (2), pp. 111–117.

16

DIGITAL TRANSFORMATION AND THE CYBERSECURITY OF INFRASTRUCTURE SYSTEMS IN THE OIL AND GAS SECTOR

Sulafa Badi and Huwida Said

16.1 Introduction

Digitisation has rapidly increased within the oil and gas industry. The sector is embracing technologies such as the "digital oilfield" and internet of things (IoT) solutions to increase the efficiency of asset monitoring, the speed of problem identification and resolution, and effective decision-making (Rosner *et al.,* 2017). Supervisory control and data acquisition (SCADA) systems, industrial control systems (ICS), and programmable logic controllers (PLCs) are also increasingly integrated into operational technology (OT) networks to automate physical processes through valves, pumps, motors, sensors, and controllers, allowing operational systems to be centrally accessed, measured, monitored, and controlled. With the twenty-first century called the golden age of digital transformation in the oil and gas sector, the increased connectivity of cyber-physical systems (CPSs) will generate new opportunities for organisations such as greater safety, reliability, and efficiency.

While digital transformation allows for tremendous operational improvements, it also generates a plethora of cybersecurity threats (Mihelič and Vrhovec, 2018). The increased connectivity of information technology (IT) and OT systems through the deployment of remote sensor-based field equipment increases the vulnerability of oil and gas infrastructure to cyberattacks. Anonymous cyber attackers often exploit network vulnerabilities and inflict physical damage to these critical infrastructures. Although the risks associated with cyberattacks are ubiquitous across all industries, oil and gas infrastructures are more exposed given their criticality and the devastating impacts that occur once a risk materialises. Price Waterhouse Cooper (PwC)'s global state of information security survey in 2016 (PwC, 2016) showed that cybersecurity incidents increased by 38 percent across industries in 2015. Among oil and gas organisations, incidents escalated by an overwhelming 93 percent, and losses due to data breaches amounted to an average of US $1.5 million per company.

An example of a cyberattack on oil and gas infrastructure is the Baku-Tbilisi-Ceyhan pipeline cyberattack (see the chapter's case study). The case study underlines the significant cyber risks that critical oil and gas infrastructures are experiencing (Speake, 2015). The attack outlined in the case study is not unprecedented; indeed, oil and gas organisations are increasingly subject to cybersecurity breaches, which are growing in both number and sophistication. An infamous incident is the loss of 30,000 virus-stricken computers by Aramco, one of the world's largest oil and gas organisations, in 2012 (Bronk and Tikk-Ringas, 2013; Leyden, 2012). The 2013 cyber-espionage campaign across Europe, Asia, and North America also infected thousands of organisations throughout the energy supply chain, including pipeline operators, electricity generation firms, energy grid operators, and industrial equipment suppliers. The operational functioning of critical infrastructure systems and the reputation of major oil and gas organisations are increasingly threatened by website sabotages, production distributions, and spear-phishing plots with sinister motives such as maliciously acquiring information about drilling technologies, programs, bids, and refinery engineering data. This situation emphasises the need for adequate consideration of the cybersecurity of oil and gas infrastructure systems.

16.1.1 Aim and objectives

This chapter aimsto explore the key cybersecurity issues and challenges facing oil and gas infrastructure systems as they become more digitised. The objectives are as follows:

- Explore the digitisation trends in oil and gas infrastructure systems;
- Assess the major cyber vulnerabilities generated by the increased digitisation of oil and gas infrastructure;
- Underline the technological, organisational, and environmental determinants of cybersecurity in oil and gas infrastructure systems.

The premise of this chapter is that cybersecurity should be viewed from the perspective of an organisation's sphere of influence. An oil and gas organisation is considered a key stakeholder and decision-maker that occupies an important position in the oil and gas upstream, midstream, or downstream supply network. Hence, the oil and gas organisation is responsible for the cybersecurity of its operations and those of the major oil and gas infrastructure systemd within which it is embedded. This chapter explores three key determinants of effective cybersecurity in oil and gas organisations: technological, organisational, and environmental. First, the technological determinants concern increasing the cybersecurity of technologies utilised within the firm, across the supply chain, and in the oil and gas operations. Second, the organisational factors concern the cybersecurity characteristics of the organisation itself such as its cybersecurity culture, capabilities, responsibility, and resources. Third, the environmental determinants of cybersecurity include the governmental and regulatory frameworks that govern cybersecurity. The identified determinants of cybersecurity are valuable to a wide range of oil and gas stakeholders (including engineering managers, ICS defenders, chief information officers [CIOs], chief security officers [CSOs], and CEOs, among others). They provide oil and gas organisations with a decision-making framework to guide their resource allocation priorities and make informed decisions to mitigate against the ever-expanding cybersecurity threats to critical oil and gas infrastructure systems.

Digital transformation and the cybersecurity of infrastructure systems

16.1.2 Learning outcomes

The following learning outcomes have been identified for this chapter. Readers will be able to:

1. Identify major digitisation efforts in oil and gas infrastructure systems.
2. Evaluate the cybersecurity vulnerabilities in oil and gas infrastructure systems.
3. Evaluate the role of technological, organisational, and environmental determinants in the future-proofing of cybersecurity in oil and gas infrastructure systems.

16.2 The oil and gas sector

The oil and gas sector is a fundamental part of the global economy and assumes considerable influence in international politics and trade. This sector employs a significant amount of the global labour force, with the United States' oil and gas sector alone supporting over 10 million jobs (Muspratt, 2019). Large multinational corporations dominate the sector, with almost 200 operating worldwide. The largest and most profitable organisations are state-owned such as the Chinese firm Sinopec ($377 billion USD revenue in 2018), Saudi Aramco ($355.9 billion USD revenue in 2018), and China National Petroleum ($324 billion USD revenue in 2018). Certain organisations are publicly owned, such as Shell ($322 billion USD revenue in 2018), BP ($303.7 billion USD in 2018), and Total ($156 billion USD revenue in 2018), while others such as ExxonMobil ($241 billion USD revenue in 2018) are privately owned (Muspratt, 2019). The oil and gas operation streams are divided into three main segments: upstream, midstream, and downstream. The upstream supply chain is concerned with the exploration, development, and production of crude oil and natural gas; the midstream includes long-distance pipeline and ship transportation activities; and the downstream involves refinery operations, petrochemicals, gasification, and sales of oil and gas products (Lu *et al.*, 2019).

The oil and gas sector has been experiencing substantial disruption in its supply and demand patterns in recent years (WEF, 2017). On the supply side, renewable energy is gaining popularity and becoming more cost effective. New resources such as light oil (shale oil) are on the rise due to technological innovations such as horizontal drilling and hydraulic fracturing. Indeed, forecasts predict that 12 percent of the global energy supply will be provided by non-traditional sources by 2025. Demand patterns are also changing. Sustainable development efforts worldwide and the endorsement of the COP21 Paris Agreement by numerous countries (United Nations Framework Convention on Climate Change, 2015) have led to strict emission regulations. It is also predicted that the growth of electric vehicles will reduce the demand for oil products by 1.5 million barrels per day by 2025 (WEF, 2017). Demand for renewable energy is also increasing, facilitated by the development of greater battery storage capacity and technological innovations (WEF, 2017). These disruptive supply and demand forces have been coupled with persistent uncertainty, market volatility, and depressed oil prices, which fell by as much as 70 percent between 2014 to 2019 (Depersio, 2019). Production has also been reduced by three to five percent during this period with a significant impact on oil and gas organisations (Lu *et al.*, 2019). Moreover, market volatility has driven many organisations to downsize their workforce, with total job losses reaching 440,000 globally in 2017 (Jones, 2017). Some organisations have been acquired by other firms (e.g. ConocoPhillips' $35.6 billion acquisition of Burlington Resources) or have merged with other companies (e.g. Marathon Petroleum and Endeavor Logistics had a $23 billion merger) (Vara, 2019). Organisations unable to cope with this turbulent environment have gone out of business, with at least 100 US oilfield service organisations going bankrupt during 2015 and 2016 (Offshore Technology, 2017). Investors' fears are also increasing, particularly as the oil and gas sector's total return to shareholders (TRS) is much lower than others such as technology, healthcare, and retail (WEF, 2017).

These recent major disruptions have driven many oil and gas organisations to pursue the digitisation of their operations. The rapid development of Industry 4.0 technologies has attracted the attention of oil and gas organisations seeking efficiency gains in these highly uncertain economic and socio-political environments. The main objectives of digitisation are reducing costs, enhancing safety, and increasing workforce productivity. The digital transformation in the oil and gas sector will be discussed in the next section.

16.3 Digital transformation in the oil and gas sector

The oil and gas sector is not foreign to technological development: it has progressively adopted innovative solutions to increase efficiency, productivity, and safety. Figure 16.1 illustrates the sector's technological development trajectory.

In recent years, Industry 4.0[1] initiated what has been termed "oil and gas 4.0," with innovative digital technologies such as big data, IoT, and robotics promising to revolutionise the entire oil and gas infrastructure system (Lu *et al.*, 2019). A report by the WEF (2017) argued that digitisation in the oil and gas sector can realise around $1.6 trillion of value to the sector, its customers, and wider society over the next 10 years (2016–2025). Benefits to the environment will include a 1,300 million tonne reduction of CO_2-equivalent water preservation, approximately 800 million gallons, and minimisation of oil spillage by roughly 230,000 barrels. Table 16.1 lists the most prominent digital technologies and their applications in oil and gas infrastructure systems.

Digital transformation could span the following four key areas in oil and gas (depicted in Figure 16.2): (1) asset life cycle digital management, (2) "beyond the barrel,", (3) circular collaborative ecosystems and (4) "energising new energies" (WEF, (2017). Asset life cycle digital management will rely on automation, robotics, artificial intelligence, and wearable devices to analyse, control, and optimise asset utilisation and maintenance action. These objectives will be achieved through robotics, remotely controlled operation centres, and predictive maintenance. Beyond the barrel refers to digital marketing and customer services, while circular collaborative ecosystems seek to build strong relationships among supply chain actors by securing transactions through blockchain/smart contracts. Energising new energies refers to the use of technologies to establish new energy choices and platforms (WEF, 2017). Despite significant operational gains, digital technologies are subject to multiple cybersecurity risks, as will be explained in the next section.

16.4 Cybersecurity risks in the oil and gas sector

16.4.1 What is cybersecurity risk?

A computer network consists of various components. Some are considered essential elements that assist in system operation and the transmission of data among different sections of an organisation, supply chain, or ecosystem. Network components include routers, switches, transmission media, and other operating system devices. In everyday life, the organisation takes known measures to secure networks, applications, or systems. Yet, with advancements in technology and the growth of information explosion generating increasing amounts of data in electronic forms, these measures can be exposed and lead to potential attacks. The revolution of network technology is proceeding rapidly in ways that provide further opportunities for attackers to launch attacks on vulnerable networks, increasing cybersecurity risks (Anchugam and Thangadurai,

Digital transformation and the cybersecurity of infrastructure systems

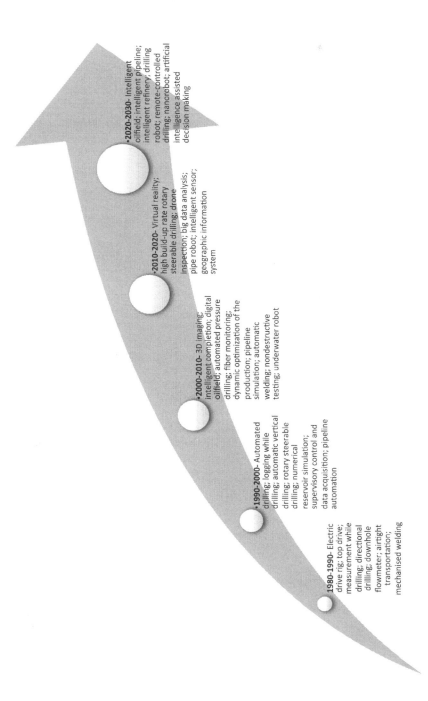

Figure 16.1 Oil and gas sector technological development trajectory

Source: Original based on information from Lu et al. (2019)

Table 16.1 Main digital technologies and their applications in oil and gas infrastructure systems

Technology	Definition	Advantages in oil and gas ecosystems	Existing and future applications
SCADA (supervisory control and data acquisition) systems	Technologies and customisable software applications used in the monitoring, control, and operational coordination of physical assets and processes across the oil and gas supply chain.	SCADA remote monitoring and control system capabilities enable the safe, effective, and efficient operations of a large number of manufacturing, automation, and process systems.	A SCADA system was first introduced in 1912 in the Chicago power industry: a basic system of telephone and voice communication was implemented, allowing the status and functioning of remote power stations to be communicated and controlled by a centrally located control room supervisor (Henrie, 2013). This early use of SCADA systems increased the effectiveness and efficiency of the energy power grid's operations, generating superior business opportunities (Henrie, 2013). A SCADA system in the form of an industrial control computer was introduced in 1959 in the Texaco Port Arthur refinery, and SCADA's technological capabilities have continued to expand, with significant developments occurring in the 1960s utility industry, shaping modern-day SCADAs which are highly complex network-based environments (Henrie, 2013).
Big Data	The term "big data" refers to the use of sophisticated technologies in the capture, processing, analysis, visualization, and storage of data sets that are difficult to handle by traditional data-processing technologies due to their size and complexity. Among these technologies are cloud computing and machine learning. Big data also works in conjunction with IoT technologies.	Big data facilitates the capture, analysis, and visualisation of the complex and considerably large amount of information that can support effectiveness in decision-making, reduction of costs, increased productivity, and superior management of risk and uncertainty. Big data can be used in reservoir monitoring by the deployment of advanced technologies to improve the field's depletion planning and reduce deep exploration drilling. Coupled with IoT technologies, big data can be utilised to increase the quality of predictive maintenance (Rommetveit et al. 2008; Mayani et al., 2018).	By 2026, the value of the oil and gas big data global market is predicted to be $10.935 billion (Lu *et al.*, 2019) An example is the p-Frame platform developed by Sinopec (Zhao, 2016), a massive software platform for seismic exploration that can handle 200 GB of seismic data loads per minute.

Technology	Definition	Advantages in oil and gas ecosystems	Existing and future applications
Internet of Things (IoT)	The IoT is a 'group of infrastructures interconnecting connected objects and allowing their management, data mining, and access to the data they generate' (Dorsemaine *et al. 2015*, p. 73). It incorporates sensors and technologies for mobile communication, data and signal processing, discovery and search engines, and security and privacy.	IoT technologies can be used to support swift decision-making, reduce system failure, increase safety, and enhance the efficiency of oil and gas operations.	IoT can be used in oil and gas supply chain optimisation (Lu *et al.,* 2019).
Blockchain	Blockchain is a decentralised, immutable ledger technology, widely known as the technology underpinning cryptocurrency in the financial sector, such as bitcoin (a peer-to-peer [P2P] payment network with no central authority) (Miraz and Ali, 2018). It is "a ledger of digital transactions, which is not under the control of any singular individual group or company" (Kinnaird and Geipel, 2017). The key attribute of this transactional database is that it can consist only of "blocks," data which cannot be changed or removed that is added or linked to other blocks to create a chain of blocks (Kinnaird and Geipel, 2017). Depending on the scope of use, there are three types of blockchain: private, public, or consortium.	Blockchain is a transaction-cost-lowering technology, as it supports the ability to conduct contracts without the need for a "trusted third-party intermediary" or "a controlled transactional database" (Beck and Müller-Bloch, 2017). It also reduces transaction time and ensures reliable payment if utilised in the design of a self-enforcing smart contract. Blockchain technology is secure with low cybersecurity risks.	A survey by Deloitte Consulting (2016) has shown that 55 percent of oil and gas companies consider blockchain as crucial to increasing their competitive advantage, while 45% view this technology as possessing potential to transform the industry. Blockchain application in oil and gas is increasing. For example, in 2017 Natixis, a global bank operating across the energy and natural resources value chain, adopted blockchain technology to digitise its U.S. crude oil trade transactions. BP, Royal Dutch, and Shell, as part of a consortium of energy companies, are also developing a blockchain-based energy trading platform (Sarrakh *et al.,* 2019). Likewise, HP Billiton is working to implement a blockchain-based tracking application for wellbore samples (Sarrakh *et al.,* 2019).

408

Table 16.1 Main digital technologies and their applications in oil and gas infrastructure systems (Continued)

Technology	Definition	Advantages in oil and gas ecosystems	Existing and future applications
Digital Oilfield	The concept of the "digital oilfield" is a combination of digital technologies that serve to automate workflows and processes such as downhole sensors, "digital twins," decision support centres (DSCs), complex data capture solutions, production optimisation, reservoir management technologies and satellite communication systems. Digital oilfields are espoused as a solution to increase reliability, enhance productivity through standardisation and automation, improve collaboration, reduce costs, enhance surveillance, and minimise safety and environmental risks in operational and management processes across the oil and gas value chain (Sankaran *et al.,* 2009; Baker Hughes, 2017; Mayani et al. 2018).	Smart field technologies offer operators detailed, accurate, and on-time information and knowledge about the field's operations, supporting effective control processes. Safety is enhanced and workload is reduced. Smart technology also increases operational data transparency and effectiveness. Reservoir life cycle modelling and optimisation technology provide a more proactive approach to production planning and problem identification. Intelligent oilfields can increase production by 2–8% and recovery by 2–6 % (Yang *et al.,* 2016).	The digital oilfield is gaining momentum in the oil and gas industry as an idea capable of increasing the industry's competitiveness and sustainability, with companies such as BP, Royal Dutch Shell, Chevron, Equinor, and Anadarko Petroleum being the major players in developing this digitisation concept. Oilfield service providers such as Halliburton, Schlumberger, Weatherford International and Baker Hughes are also key players. The first 'smart well' concept was introduced in 1997 by Saga Petroleum in their North Sea Snorre Field. Other companies, as exemplified by ConocoPhillips and Chevron's I-field program, followed suit in the mid-2000s. Additionally, the Kuwait Oil Company (KOC) launched the Kuwait Intelligent Digital Field (KwIDF) program in 2010, a highly ambitious project utilising cutting-edge digital technologies (Carvajal *et al.,* 2017; Sarrakh *et al.,* 2019).
Artificial intelligence	The term "artificial intelligence" (AI) refers to computer systems that perform cognitive tasks such as learning and problem-solving in ways that imitate human capacity.	AI can be used for swift and speedy decision-making, predictive maintenance, prediction, and allocation of oil and gas resources.	In collaboration with Microsoft, Shell pioneered Shell Geodesic™, an intelligent drilling solution that uses drilling data and processes algorithms for swift real-time decision-making and better prediction of their outcomes (Microsoft, 2018). It reduced risk and uncertainty, minimised well misplacement, and improved safety.

Sources: Kinnaird and Geipel (2017); Beck and Müller-Bloch, (2017); Sarrakh *et al.* (2019)

Digital transformation and the cybersecurity of infrastructure systems

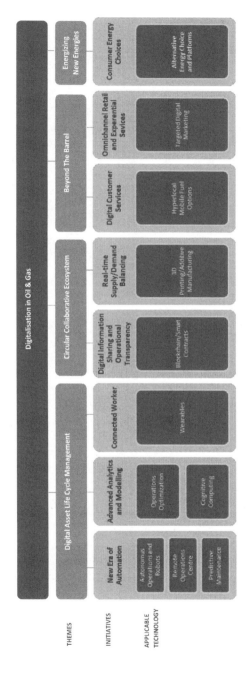

Figure 16.2 Digital transformation in the oil and gas sector
Source: Original based on information from the WEF (2017)

2016). Cybersecurity risk is defined as the risk of intentional disruption of the proper functioning of the processing and information exchange space created by the IT system. As illustrated in Figure 16.3, the cybersecurity risk associated with a critical infrastructure system is often seen as a function of three main interdependent components: (1) the likelihood of the cyberattack threat, (2) the vulnerability of the system (the likelihood that the cyberattack will be successful) and (3) the scale of the consequences (Petit *et al.*, 2015).

In addition, cybersecurity threats are often classified into three types according to the intentions of the initiator: (1) an "unintentional threat" occurs when the initiator accidentally causes the cyberattack but with no premeditated intentions; (2) a "financial gain threat" aims to exploit an opportunity for monetary benefit; and (3) a "malicious threat" involves a devious attempt to inflict damage by infecting the company's computer resources (Henrie, 2013). The threat initiator, or agent, could be internal or external to the company. In an "internal agent" attack, also known as an insider attack (Ruppert, 2009) or internal penetration (Diaz-Gomez *et al.*, 2011), the culprit is a trusted company employee or contractor who has been granted privileged access to the company's computer system and who possesses extensive knowledge concerning the company's internal processes, its computer systems, and its security weaknesses (Ruppert, 2009). An internal agent attack could occur, for example, when an employee makes an unintentional mistake, attempting to accomplish a required task by accessing unauthorised data or by using the system in an inappropriate manner. Internal attacks also include the agent working to uncover the system's weaknesses or vulnerabilities with the intent to report errors or to act maliciously to cause damage, prompted by greed, disloyalty, or other sinister motives (Ruppert, 2009). Studies have found that while internal agent threats are more frequent, particularly due to unintentional mistakes, they rarely cause significant financial damage to the organisation (Richardson, 2011). Insider threats for financial gain do occur, but infrequently, such as in the 2011 case of an Australian wastewater ex-employee who attempted to coerce his water treatment company to re-hire him to fix the computer system they intentionally damaged (Henrie, 2013). An internal agent could also have damaging motives such as espionage. An infamous case is that of the young US soldier behind the leak of nearly 750,000 classified and sensitive US Government documents published by WikiLeaks in 2010.

An "external agent" attack, in contrast, is inflicted by an initiator who has not been authorised to access the computer and network systems of an organisation (Diaz-Gomez *et al.*, 2011). External agents may seek to take control of an IT system to monitor its activities or to control it remotely to inflict system damage or cause significant disruption to its functionality (Mateski *et al.*, 2012). External agent threats are more often driven by financial motives or malicious intentions (Henrie, 2013; Quigley *et al.*, 2015; Stohl, 2007). The loss of 30,000 virus-stricken computers in 2012 by Saudi Aramco, the world's largest oil producer, and the crippling of its operations for almost two weeks resulted from a cyberattack motivated by the desire to damage the country's petroleum production (Bronk and Tikk-Ringas, 2013; Leyden, 2012). A global survey conducted by PwC in 2015 has indicated that executives in oil and gas organisations attributed 22 percent of detected incidents to competitors. This is an alarming figure that highlights the fierce competitiveness of the sector and how the pursuit of survival and profitability may drive some organisations to unethical and illegal practices.

Returning to the risk equation in Figure 16.3, the *threat* variable is widely considered to be extremely difficult to control by oil and gas organisations. Indeed, cybersecurity threats could be described as uncertain risks (Renn, 2008) given the inherent unreliability of the empirical data that can be employed in objective estimations of the probability of specific cyber risks. Thus, as Quigley *et al.* (2015) argue, "fuzzy" or subjective methods of risk estimation are often

Digital transformation and the cybersecurity of infrastructure systems

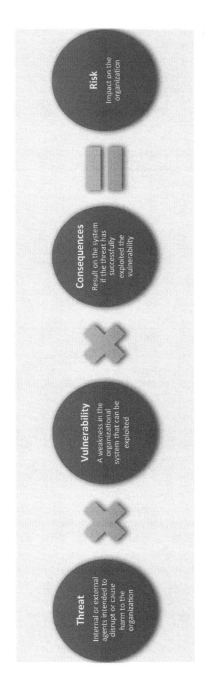

Figure 16.3 Cybersecurity risk
Source: Original based on Henrie (2013)

used in cybersecurity risk management processes. Hence, given the limited control over the threat variable, organisations often focus on the *vulnerability* and *consequences* variables (Henrie, 2013). Vulnerabilities are weaknesses in a company's communication infrastructure, network, or individual computers that can be exploited. These security holes can be defined, identified, and classified through a vulnerability analysis process or vulnerability assessment. Such evaluations include a series of systematic measures that are used to review and prioritise security weaknesses. The assessments help organisations to determine their security status and level of exposure to threats (Yaqoob *et al.*, 2017). The definition of a vulnerability assessment is often confused with penetration testing (pen test). Both methods report the vulnerabilities in a system or network, but the difference is that penetration testing verifies the existence of vulnerabilities by attempting exploitation (Yaqoob *et al.*, 2017). A penetration test is an authorised attack simulation on systems and networks. The purpose of compromising systems during penetration testing is to make an organisation aware of the security weaknesses in their systems and processes and subsequently to address these vulnerabilities to mitigate a real attack scenario (Denis *et al.*, 2016). A system's vulnerability can be reduced through enhanced cybersecurity.

16.4.2 Cybersecurity risks in oil and gas infrastructure systems

As mentioned earlier, the oil and gas sector runs its operational processes through three main streams: the downstream, midstream, and upstream. According to Rick and Iyer (2017), the segment most vulnerable to cyberattacks is the upstream supply chain. As Figure 16.4 indicates, the upstream segment of the oil and gas supply chain encompasses many complicated operations and depends on other streams to operate. This segment manages the drilling control system, remote data acquisition, remote gateways and data conversion and analysis. Additionally, the midstream entails cybersecurity risks in transportation and logistics operations, whereas the downstream has other cybersecurity issues associated with sales and refining processes (Rick and Iyer, 2017).

The oil and gas operation streams are at risk due to network vulnerability (Aljubran et al., 2018). These vulnerabilities could be the entry point for many attacks to the critical infrastructure system. Table 16.2 highlights the points of entry for different cyberattacks and the potential vulnerabilities in the oil and gas operation streams (Rick and Iyer, 2017).

Several widely used technologies in oil and gas are known to be prone to technological vulnerabilities. One such example are SCADA systems, which are increasingly utilised to control physical processes in oil and gas operational networks. The vulnerability within these technologies is often exploited by cyber attackers to compromise cyber-physical systems with threats including viruses, malicious files, Trojan horses, and zombie attacks (Henrie, 2013). A notorious illustration of this problem is the 2010 Stuxnet cyberattack on the Iranian uranium enrichment facility at Natanz, in which attackers were able to gain access to the SCADA system and cause substantial damage to the uranium enrichment centrifuges (Farwell and Rohozinski, 2011; Langner, 2011). The prevalence of SCADA systems and the ability of attackers to easily acquire detailed information on their functioning have resulted in such systems being frequently targeted by cybercriminals across all sectors (Keeney *et al.*, 2005; Miller and Rowe, 2012). Worldwide incidents include the cyberattacks that shut down a nuclear power plant in the United States (Krebs, 2008), a wastewater treatment facility in Australia (Slay and Miller, 2007), and a steel mill in Germany (Lee *et al.*, 2014).

A major vulnerability in the oil and gas operation streams is the remote access control of the data. Losing control over the remote access authorisation for the individual (or devices) to

Digital transformation and the cybersecurity of infrastructure systems

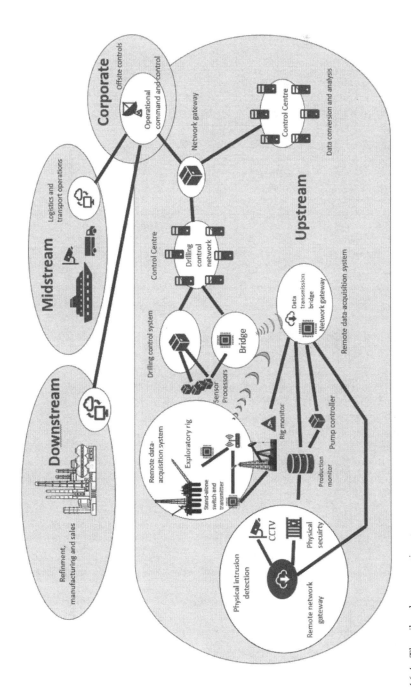

Figure 16.4 The oil and gas operation streams
Source: Original based on Rick and Iyer (2017)

Sulafa Badi and Huwida Said

Table 16.2 Points of entry for different cyberattacks and vulnerabilities across the oil and gas operation streams

Possible attacks	Data theft and physical attacks	DoS, MIMA, data theft, DNS and ARP poisoning	Data theft, physical attacks	DoS, MIMA, DNS poisoning
Data level	Corporate	Upstream	Midstream	Downstream
Vulnerability 1	Production information	Exploratory information	Supply chain logistics	Refinery information
Vulnerability 2	Potential partners and acquisition targets	Field production and sensor information	Distribution network	Industrial plants and manufacturing
Vulnerability 3	Details of drilling sites	Operational information, e.g., drilling sites, reservation equipment	Storage information	End-user distribution
Vulnerability 4	Trading positions	–	Pipeline and transportation	Retail data
Vulnerability 5	Financial and organisation reports	–	–	Consumer data
Vulnerability 6	Capital projects	–	–	–

Source: Rick and Iyer (2017)

critical components leads to malware injections, the introduction of malicious software intended to damage a computer, server, or the whole IT network. There are different types of malware that have caused massive destruction for IT network systems and which seek to breach network systems in many ways. Consequently, data compromise, theft, deeper intrusion, and diffusion can readily occur. Hackers monitor the malware spread on a system using command and control (C and C) servers. Malware is a threat actor that is continuously present on a user's machine and reports to the hacker's C&C machine. This communication process (between the Malware file and C and C) enables the hacker to update and upgrade the malware remotely. Multiple types of malware files might threaten user machines, concealing their existence and spread. These files are small, written in Active Server Pages (ASP), Javascript (JS), or Hypertext Preprocessor (PHP) code, and can directly communicate with the attacker's C and C server and download files on a breached victim server. oil and gas organisations also reserve the data regarding drilling techniques, oil reserves, and other chemical products. Such valuable data can be targeted by hackers, who use different approaches to exploit the corporate network and facilitate data theft activities (Chen *et al.,* 2016). The data theft can be implemented through domain name server (DNS) session hijacking, compromising the webmail server, and other virtual private network (VPN) and database server breaches.

Since the oil and gas sector has embraced automation and network connectivity, it is becoming more reliant on cloud services technology. Automation is utilised in monitoring the crucial functions that measure the onsite equipment pressure, temperature, chemical reactions, and oil leaks (Drias *et al.,* 2015). The safety and instrument systems (SIS) emergency system termination is also dynamically monitored and controlled remotely. However, such systems experience considerable threats of compromise by attackers. For example, in man-in-the-middle (MITM) attacks an attacker positions themself between two parties' conversation channels and obtains sensitive information these parties are exchanging on a private line, unaware that an attacker is monitoring their communication. Such attacks can lead to data theft, corruption of information, and obstruction of communication, which consequently result in loss of confidentiality and potential loss of control over equipment, devices, and operation systems (Conti *et al.*, 2016; Rob *et al.*, 2014). The different types of MITM attacks are distinguished by the methods used or the vulnerabilities exploited to execute the attack. One example is targeting the address resolution protocol (ARP), which is responsible for resolving an IP address into a corresponding media access control (MAC) address (Tripathi and Mehtre, 2014). It is vulnerable to attacks because it is unauthenticated. In the ARP poisoning attack mechanism, the attacker sends ARP spoofed messages to two communicating hosts, tricking them by claiming that it is the other host, which leads to both hosts updating their ARP cache with the attacker's MAC address. This deception results in these hosts forwarding all the communications to the attacker.

An example of how hackers can exploit encrypted communication is the Secure Socket Layer (SSL) strip. This method allows the attacker to capture usernames and passwords on secure hypertext transfer protocols (HTTPS) websites. In short, the SSL strip is an MITM attack that focuses on the victim's browser communications and changes HTTPS to hypertext transfer protocol (HTTP). In this case, some users might notice that the URL has changed to HTTP and will not fall victim to this attack, but others might not recognise or understand the change. Furthermore, numerous open source tools can be used to intercept the secure communication (HTTPAS, 2016). Therefore, oil and gas organisations need to observe that the SSL/TLS protocol can also be compromised without altering any security features on the website. The hacker can even bypass security measures using open source tools such as Wi-Fi Pineapple and Kali Linux tools for hackers (Oh *et al.*, 2017). Oh *et al.* (2017) found in their experiment that over 45 percent of popular websites are susceptible to one of the attacks that can exploit vulnerabilities in the SSL/TLS protocol. Due to these vulnerabilities, the communication between two hosts will not be secure, and the confidentiality and integrity of the conversation will be compromised. Consequently, an attacker can conduct an MITM attack on the system.

The oil and gas operation streams have many different points of entry for attacks, as shown in Figure 16.4. Some methods target the sector's infrastructure (damaged cables, bomb shelling, etc.) or seek to disable the automation systems by launching denial-of-service (DoS) attacks. This type of attack involves attempting to shut down a computer network temporarily or permanently by overloading the targeted machines with a flood of requests that exceed their capacity. More significantly, most DoS attacks are employed to conceal other criminal activities such as network infiltration or data theft. While the target is distracted by frantic attempts to address the DoS attack, the attacker seeks to infiltrate the network through malware. According to a study by the Washington, DC (2009) policy think tank Center for Strategic and International Studies (CSIS), the oil and gas industry is faced with the highest rate of DoS attacks across all industries. Seventy-one percent of respondents in the oil and gas sector reported experiencing network infiltrations, compared to only 54 percent of respondents in other sectors. DoS attacks were found to be particularly severe, often targeting computer-based operational control systems such as SCADA. According to the survey respondents, the cost of 24 hours of network inactivity following a major incidence, would exceed $8 million per day in the oil and gas sector (Baker *et al.*, 2009).

Another type of attack targets the network boundary (or nodes). Here vulnerability and attacks occur when there are multimedia and smartphone device interfaces integrated into the network. Another factor is that the network could run weak defences such as firewalls and intrusion detection/prevention systems (IDS/IPS). IDS/IPS are controls that analyse and monitor network traffic for abnormal behaviour and deny any traffic based on a security profile (Radmand *et al.*, 2010). Moreover, phishing attacks where an attacker attempts to fraudulently acquire sensitive information from a victim by impersonating a trustworthy third party are also increasing (Jagatic *et al.*, 2007, p. 94). Phishing typically involves a malicious email campaign that encourages the victim to click a link, sign into a spoof website, or provide confidential financial information. Therefore, oil and gas organisations need to focus their effort on counteracting such attacks (Chen *et al.*, 2016).

16.5 Cyber-proofing oil and gas infrastructure systems

Shin *et al.* (2017) define cybersecurity as "actions required to preclude unauthorised use of, denial of service to, modification to, disclosure of, loss of revenue from, or destruction of critical system or informational assets." In this section, the determinants of cybersecurity in oil and gas infrastructure systems are discussed from technological, organisational, and environmental perspectives.

16.5.1 Technological determinants of cybersecurity

Technology is the first line of defence to cyber-proof critical oil and gas infrastructure systems. Oil and gas organisations suffer from information interruptions by hackers, which frequently occur due to vulnerabilities in the company communication, port, and service channels (Khurana, 2011). As noted earlier, there are several types of attacks on a network, for example, DoS and MITM attacks. To increase the cybersecurity of systems, the interruption of systems can be avoided by using Transport Layer Security (TLS). TLS is a cryptographic protocol that prevents eavesdropping and tampering and enables client-server applications to communicate safely across a network (Oh *et al.*, 2017). Companies' websites ensure sensitive communication that runs over the secure HTTPS channel uses the TLS protocol. The HTTPS protocol is designed to increase privacy on the Internet when communicating with websites and sending sensitive data. This protocol has made MITM attacks increasingly difficult as the data is transmitted in an encrypted manner. To implement this measure, the owner must purchase an Secure Socket Layer (SSL) certificate, which acts as an online identification card and encrypts any data that passes through the HTTPS protocol. Now, when a user requests data from the server, it examines the SSL certificate, which provides authorised confirmation of the website's identity. Subsequently, an encryption method is decided and used to encrypt and decrypt the data (Aloul, 2010). If an MITM attack occurs, this data is encrypted and unreadable. However, it should be noted that accessing an HTTPS website does not mean that the system is secured from other types of attacks.

Within any network, there are several types of firewalls. Duan and Al-Shaer (2013) define a firewall as "a network element that controls the traversal of packets (data that transmitted between network nodes) across the boundaries of a secured network based on a specific security policy." Furthermore, there are two kinds of packet filtering: stateful and stateless (Turnbull, 2005). The significant difference between stateful and stateless packet filtering is the part of the packets examined. More specifically, stateless packet filtering inspects only the header of the packet, whereas stateful packet filtering examines entire packets, including the status of the connection that the packet is part of (Turnbull, 2005). This process helps to determine if the packet is entering or leaving through a vulnerable system, and accordingly, how the firewall acts on the packet, that is, whether the firewall accepts, drops, or rejects it. Once a system starts to run several services and applications, different types of packets are transmitted and received through the hosts over the network. The traffic of the packets is exposed to considerable threats, especially if they are connected to the Internet. If the firewall is configured, it can prevent these persistent attacks but cannot avert zero-day attacks (i.e., a flaw in a software that is unknown to the patching

Digital transformation and the cybersecurity of infrastructure systems

developer). Indeed, firewalls are considered the primary approach to prevent malicious attacks against the network, but having one firewall can be viewed as a single point of failure, since all of the network traffic is flowing through it. The massive number of packets passing through it will overload the network's firewall and introduce latencies to the network. DoS attacks are a major threat to any organisation's computer systems and are easily achievable by malicious users. DoS attacks can be executed by sending a large number of packets to the target system, which causes its resources and services to crash, thereby preventing legitimate users from accessing the system (Chand and Mathivanan, 2016). A distributed firewall will provide a network with the advantage of minimising the workload on the centralised firewall. Furthermore, implementing firewall policies throughout the clients/servers prevents internal attacks (Kaur *et al.,* 2016).

Blockchain promises to strengthen the cybersecurity of oil and gas organisations. It is a decentralised, immutable ledger technology, widely known as the technology underpinning cryptocurrency in the financial sector such as bitcoin, a peer-to-peer (P2P) payment network with no central authority (Miraz and Ali, 2018). Blockchain is "a ledger of digital transactions, which is not under the control of any singular individual group or company" (Kinnaird and Geipel, 2017). The key attribute of this transactional database is that it can have only "blocks," data which cannot be changed or removed that is added or linked to other blocks to create a chain of blocks (Kinnaird and Geipel, 2017). Blockchain is a transaction-cost-lowering technology because it supports the ability to conduct contracts without the need for a trusted third-party intermediary or a controlled transactional database (Beck and Müller-Bloch, 2017). It also reduces transaction time and ensures reliable payment if used in the design of a self-enforcing smart contract. A survey by Deloitte Consulting (2016) has found that 55 percent of oil and gas organisations consider blockchain crucial for increasing their competitive advantage, while 45 percent view it as having potential to transform the sector. Blockchain application in oil and gas is increasing. For example, in 2017, Natixis, a global bank that operates across the energy and natural resources value chain, adopted blockchain technology to digitise its US crude oil trade transactions. BP, Royal Dutch, and Shell, as part of a consortium of energy companies, are also developing a blockchain-based energy trading platform (Sarrakh *et al.,* 2019). Likewise, HP Billiton is working to implement a blockchain-based tracking application for wellbore samples (Sarrakh *et al.,* 2019). Blockchain offers a promising solution to secure the oil and gas supply chain and its distributed nature of operations. Arutyunov (2018) proposed a blockchain approach to connect remote assets across oil and gas operations including SCADA systems. Through the blockchain ledger and its built-in identity and access policies, all devices and applications are securely connected, ensuring their continuous collaboration. The system is also foolproof, as any malicious device or malware attempting to access the network and attack an industrial control network will be intercepted by consensus protocols established between devices that can quickly determine the unauthenticity of the malicious device or application and subsequently isolate it from the network. Arutyunov (2018) refers to this blockchain-based cybersecurity system as "self-healing, without human intervention," mitigating against system-wide cybersecurity risks and ensuring the proper functioning of the oil and gas operations.

A rapidly growing technology that promises to increase the cybersecurity of oil and gas infrastructure is artificial intelligence (AI). Rosner *et al.* (2017), for example, developed a cognitive analytic approach which is proposed as an AI-based solution for monitoring and protecting both the IT infrastructure and the operational technology network. This technology helps analysts ascertain the precise location of a potential threat and assess the validity and severity of the attack. More research is required within the area of AI and its cybersecurity applications in the oil and gas sector.

16.5.2 Organisational determinants of cybersecurity

Technology alone cannot fully protect an oil and gas organisation and the critical oil and gas infrastructure system within which the organisation is embedded. In the pursuit of greater cybersecurity, technology-centred measures are often viewed as ideal solutions to mitigate cybersecurity risks. However, in recent years, scholars have contended that cybersecurity requires a more nuanced approach than technology can offer, proposing instead a people-centred approach that favours employee awareness and a strong cybersecurity culture. Indeed, due to the high connectivity and digitisation of contemporary organisations, employees continuously interact with cyberspace. Hence, employees' utilisation of cyberspace directly impacts the integrity, privacy, confidentiality, and quality of an organisation's information resources (Da Veiga, 2016). The educated, responsible, and ethical use of cyberspace will ensure greater cybersecurity, whereas ignorance, negligence, and malicious intentions could introduce risks with dire consequences not only within organisations, but also at the national and even international levels (Da Veiga, 2016). Internal agent threats in oil and gas form a staggering 60 percent of the overall incident profiles (Howard and Wallaert, 2015). Employee-related data leaks, whether intentional or accidental, are found to be the most significant cybersecurity risk in organisations (McAfee and SAIC, 2011), mostly due to employees' ignorance and lack of responsible and ethical use of cyberspace (Da Veiga, 2016). The significance of internal agent threats was emphasised in a roadmap published by United States Homeland Security (2009) as a key area for future research to strengthen national cybersecurity. The success of organisational cybersecurity depends to a significant degree on the behaviour of employees and what they choose or choose not to do. Consequently, the human element could be considered the Achilles heel of cybersecurity, and it requires careful attention to future-proof critical infrastructure systems against cyber threats (Da Veiga, 2016). Due to the multifaceted nature of organisational cybersecurity, we will discuss its key determinants under three headings: cybersecurity culture, cybersecurity capability, and responsibilities and resources.

16.5.2.1 Cybersecurity culture

Organisational cybersecurity is a people-centred phenomenon that encompasses, among other important issues, a cybersecurity culture (Gcaza et al., 2017; Knapp et al., 2006; Panguluri et al., 2017). Cybersecurity culture has been defined as "the attitudes, assumptions, beliefs, values, and knowledge that employees/stakeholders use to interact with the organisation's systems and procedures at any point in time" (Da Veiga, 2016; p. 1008). It is "the way things are done when interacting in cyberspace" in a particular organisation (Da Veiga, 2016). Da Veiga and Eloff (2010) developed a framework and assessment instrument for cybersecurity culture. It is posited that quantifying the cybersecurity culture of an organisation is largely beneficial as the assessment can illuminate the security risk exposure of the organisation. This information could then be integrated into security risk frameworks (Shameli-Sendi et al., 2016) and incident response decision frameworks (He et al., 2015), as well as utilised in the calculation of insurance premiums against cyber risk (Gordon et al., 2003). A strong cybersecurity culture is one that "1) promotes good cybersecurity practices; 2) security is traditionally considered an important organisational value, 3) employees value the importance of security, (4) practicing good security is the accepted way of doing business, (5) the overall environment fosters security-minded thinking, and (6) information security is a key norm shared by organisational members" (Knapp et al., 2006, p. 32). Cybersecurity culture was found to change over time through strategies and programs that assist in promoting safe cyberspace through security measures, ethical behaviour, and privacy codes of conduct (Da Veiga, 2016).

16.5.2.2 Cybersecurity capability

An organisation's cybersecurity capability is an important element that determines its resilience against cyberattacks. Cybersecurity is a team effort, and responsibility for cybersecurity should be based on a top-down approach with accountability cascading from the board of directors and diffusing downwards throughout the organisation (Shoemaker *et al.*, 2018). An organisation's top management should have a clear understanding of the cybersecurity risks facing their firm, be aware of the measures being taken to mitigate these risks, and periodically assess whether these strategies are fit for purpose (Evans *et al.*, 2016). It is also important for an oil and gas organisation to objectively and systematically assess its cybersecurity capability. Cybersecurity capability maturity models (Caralli *et al.*, 2010; DoE, 2014; NIST, 2014) can be utilised to evaluate the organisation's cybersecurity performance and benchmark this performance against other firms to identify improvement opportunities. Examples of such maturity models include the National Institute of Science and Technology's (NIST) framework for improving critical infrastructure cybersecurity (NIST, 2014), the United States Department of Energy's (DoE, 2014) Cybersecurity Capability Maturity Model (C2M2), and the Carnegie Mellon Software Engineering Institute's CERT Resilience Management Model (CERT RMM) (Caralli *et al.*, 2010). Payette *et al.* (2015) also proposed a cybersecurity project management maturity model for the IT project component of critical infrastructure. Since these IT systems are implemented through projects, the maturity models integrate project management maturity models to assess the organisation's capability and "design in" cybersecurity within the existing IT systems rather than retrofitting it later.

16.5.2.3 Responsibilities and resources

It is paramount that cybersecurity be considered the main priority of oil and gas organisations. Cybersecurity should be handled as an organisation-wide responsibility and should be promoted collaboratively starting at the helm of the organisation. Chief executive officers and board members should foster a cybersecurity culture and ensure a robust cybersecurity strategy and an adequate budget. In 2015, the average information security budget within the oil and gas sector was $6.3 million, according to a global survey conducted by PwC that involved CIOs and CSOs from 103 oil and gas organisations and 29 nations. While the reported average budget has shown that organisations are willing to invest in their cybersecurity, these amounts remain small as they represent only 16 percent of IT spending within the oil and gas sector, compared to the 19 percent average in other sectors. The survey also found that only 44 percent of oil and gas organisations have allocated funds to finance security event correlation technologies. A subsequent report by McKinsey (2019) blamed the sharp drop of oil prices in 2014 on the considerable reduction in IT budgets across the sector, amounting to a significant 30–50 percent decrease in IT capital expenditures.

Focusing on cyber resilience, the responsibility for cybersecurity should cascade down in a collaborative network that includes security officers, internal auditors, risk management committees, legal counsel, and compliance officers working cooperatively toward a risk-based approach to cybersecurity (Shoemaker *et al.*, 2018). The PwC (2018) Global State of Information Security Survey 2018 underlined the critical role of CIOs in the digital transformation of oil and gas organisations. The chief Information security officer was seen as the main point of responsibility for cybersecurity. However, the report found that CIOs were excluded from business planning and decision-making. CIOs also have limited control of end-to-end data flows as data from operational technology are often managed by the asset operating team and not shared enterprise-wide.

This disconnect between central IT and OT made it difficult for CIOs to utilise operational data within the organisation (McKinsey, 2019). For example, while a single drilling rig in an oilfield can generate about 0.3 GB of data/well/day (Lu *et al.*, 2019), only a minute fraction of this information is used in decision-making (WEF, 2017). Therefore, it is recommended that oil and gas organisations reinforce the role of chief information security officers by stronger involvement in business planning and decision-making and greater access to operational cybersecurity data which could be effectively utilised within the organisation.

16.5.3 Environmental determinants of cybersecurity

The environment within the oil and gas organisation is situated to have a considerable impact on its cybersecurity. Indeed, from a cybersecurity perspective, an examination of the organisational environment must consider both the national and the international context in terms of worldwide internet connectivity. A national cybersecurity environment that promotes safety, confidentiality, and privacy can significantly enhance the security and resilience of the country's critical infrastructure systems (NIST, 2014). Several regulatory frameworks have been introduced by governments worldwide to strengthen the security of national critical infrastructure systems against cyber threats. Table 16.3 lists several cybersecurity regulatory frameworks used in Australia, Canada, Japan, Switzerland, the United Kingdom, and the United States.

Table 16.3 Cybersecurity regulatory frameworks

Country	Regulation/framework	Year introduced	Description
Australia	Cyber Security Strategy	2009	The Australian government outlined its approach to cybersecurity to protect the privacy of its citizens and their cyberspace. The strategy is built on improving public awareness of risks and confidence. It prioritises threat awareness, cultural change, business–government partnerships, international engagement, legal and law enforcement, knowledge, skills, and innovation.
	Australian Signals Directorate Strategies to Mitigate Cyber Security Incidents	2010	This document provided mitigation strategies to prevent malware attacks, limit cybersecurity incidents, detect and respond to cyber events, facilitate data recovery and system availability, and avert insider attacks.
	CERT Australia	2010	The Computer Emergency Response Team (CERT) is the national body for handling security incidents. It operates as a trusted source for cybersecurity information.
Canada	Action Plan 2010–2015 for Canada's Cyber Security Strategy document	2010	Canada's strategic cybersecurity plan is to secure the welfare of citizens and governments from cyber threats and create a cybersecurity alliance with other countries.

(continued)

Digital transformation and the cybersecurity of infrastructure systems

Table 16.3 Cybersecurity regulatory frameworks (Continued)

Country	Regulation/framework	Year introduced	Description
Japan	JPCERT	1996	The Japan Computer Emergency Response Team Coordination Centre (JPCERT/CC) is a team responsible for responding to cybersecurity incidents.
	The First National Strategy on Information Security	2006	The Japanese government outlined its approach to cybersecurity by implementing plans to respond to threats and by promoting awareness to ensure the government and citizens are taking security precautions.
Switzerland	International Organisation for Standardisation (ISO): ISO/IEC 27032:2012	2012	This standard provides guidelines for improving cybersecurity and addressing common issues.
	International Organisation for Standardisation (ISO): ISO/IEC 27002:2013	2013	This standard offers guidelines on the best security management practices and both the implementation of and information about security standards.
United Kingdom	The UK Cyber Security Strategy	2009	The United Kingdom's strategic cyber plan objectives are to create the most secure and safe cyberspace possible for business and to defend against cyberattacks.
	CERT-UK	2014	The Computer Emergency Response Team (CERT) is the national body responsible for responding to, handling, and reporting on cyber breaches and incidents.
United States	The National Strategy to Secure Cyberspace	2003	The United States' strategic cybersecurity plan is to secure cyberspace by maintaining a response system, a vulnerability reduction program, and an awareness and training program.
	National Institute of Standards and Technology: SP 800-53 Rev. 4	2013	This standard provided guidelines for the security and privacy of the government, including on the selection, specification, and standards for security controls.
	The North Atlantic Treaty Organisation (NATO): National Cybersecurity framework manual	2012	This manual proposed a national security strategy that consists of three approaches, five mandates, three dimensions, and five dilemmas.

Sources: Australia Government (2011); Australian Signals Directorate and Australian Cybersecurity Centre (2017); Information Security Policy Council (2006); International Organisation for Standardisation (ISO) (2012, 2013); Joint Task Force Transformation Initiative (2013); Klimburg, (2012); Luiijf and Besseling (2013); Shafqat and Masood (2016); Sabillon *et al.* (2016); United Kingdom Government (2014); Government of Canada (2010).

Cybersecurity regulations and frameworks must be understood as a proliferating scheme; as the rules in the selected countries not only increased, newer versions of the same standards were implemented to reap tangible benefits for the security infrastructure. Generally, the aims and characteristics of each framework may differ because individual nations seek to achieve specific objectives (Luiijf and Besseling, 2013). For instance, Australia's national strategy emphasises the importance of protecting the privacy of its citizens in cyberspace by outlining the government's approach to identifying risks and organized crimes. Similarly, the United States' strategy is directed towards securing cyberspace by establishing a response system and vulnerability reduction, awareness, and training programs. Conversely, Canada's strategy aims to protect the country's citizens, businesses, and government from threats. Defending the government system, cyberspace, and establishing alliances with other countries form the three pillars of the strategy (Sabillon *et al.*, 2016). Japan's security strategy comprises a threat response plan and awareness programs to ensure security precautions are taken by both the government and the citizens (Information Security Policy Council, 2006). Lastly, the United Kingdom's security strategy objectives are to secure cyberspace by addressing cybercrimes (Sabillon *et al.*, 2016). Different governments' efforts to tackle cybersecurity issues are not limited to implementing a national security strategy but extend to appointing a team of experts known as a Computer Emergency Response Team (CERT) to respond to, deal with, and prepare for national cyber breaches and cybercrimes. CERT Australia, CERT Japan, and CERT-UK are examples of the national teams which operate in different countries (Shafqat and Masood, 2016).

Governments play a crucial role in securing national critical infrastructure. Cybersecurity regulations, policies, and strategies may serve as the foundational elements, but other initiatives could also build a holistic approach to cybersecurity – for instance, the government guiding the private sector to develop cybersecurity innovative technologies, working with the private sector to safeguard critical supply chains, developing with the private sector a shared repository of cybersecurity actionable information in collaboration with law enforcement and intelligence agencies, and encouraging new generations of young citizens to pursue careers in cybersecurity (Coldebella and White, 2007). Indeed, as argued by Harknett and Stever (2009), the responsibility for cybersecurity should be placed on a balanced triad of governmental collaborations, private sector involvement, and engaged cyber citizenship. This triadic model will cultivate the resiliency and sustainability required in an increasingly challenging cybersecurity environment.

16.6 Chapter summary

The challenge in the digital economy is that no chain is stronger than its weakest link.
    (C. Wernberg-Tougaard, Oracle Corporation)

The premise of this chapter is that cybersecurity is the responsibility of every organisation within the oil and gas infrastructure system. An oil and gas organisation is viewed as a key stakeholder and decision-maker that occupies an important position in the oil and gas upstream, midstream, or downstream supply network. The oil and gas organisation should ensure the cybersecurity of its operations and those of the major oil and gas infrastructure system within which it is embedded. In this chapter, we explored the key cybersecurity challenges facing oil and gas infrastructure systems as they become more digitally transformed. The main digitisation trends in the oil and gas sector were examined and the potential benefits of integrating digital technologies, such as SCADA, digital oilfields, big data, and IoT, among others, within the oil and gas operational streams were assessed. While these technologies were implemented to minimise operating costs and enhance business agility, they also bring to the fore significant cybersecurity threats. In addition, the authors assessed the major cyber vulnerabilities and risks that allow critical infrastructure systems to become subject to major cyberattacks such as DoS and MITM attacks, to name a couple. The technological, organisational, and environmental determinants of cybersecurity that may work to cyber-proof oil and gas infrastructure systems were then underlined. First, the technological determinants are concerned with increasing the cybersecurity of technologies utilised within the firm, across the supply chain, and in oil and gas operations. Second, the organisational factors are concerned with the cybersecurity characteristics of the organisation itself, such as its cybersecurity culture, capabilities, responsibilities, and resources.

Digital transformation and the cybersecurity of infrastructure systems

Third, the environmental determinants of cybersecurity include the governmental and regulatory frameworks that govern cybersecurity. All three key determinants need to work in unison to safeguard critical oil and gas infrastructure against cyber threats. It is recommended that key oil and gas stakeholders (including CIOs, CSOs, and CEOs) integrate the identified determinants into their decision-making frameworks to guide their resource allocation priorities and make informed decisions to cyber-proof critical oil and gas infrastructure systems.

16.7 Chapter discussion questions

1. Discuss major digital transformation trends in the oil and gas sector and assess the potential uses of Industry 4.0 technologies across the oil and gas operation streams.
2. Develop a strategy to increase the cybersecurity of an organisation of your choice, taking into account its technological vulnerabilities as well as cybersecurity determinants at the technological, organisational, and environmental levels.
3. Assess the cybersecurity culture of an organisation of your choice and recommend how the culture could be strengthened to increase cybersecurity.

16.8 Case study: The Baku-Tbilisi-Ceyhan pipeline cyberattack

In August 2008, the Baku-Tbilisi-Ceyhan pipeline, a 1768-kilometre oil pipeline that runs from the Caspian Sea to the Mediterranean, exploded close to the town of Refahiye, Turkey (Speake, 2015). The pipeline is strategically important as it is used by major oil and gas corporations to transport crude oil across Europe. The investigation following the incident has revealed that the pipeline was deliberately over-pressurised by cyber attackers who penetrated the network and gained access to the pipeline control system. Although the entire pipeline is continuously monitored by a network of sensors and camera systems, the severely damaging explosion did not trigger a single alarm or distress signal. The pipeline operators only became aware of the attack 40 minutes after it occurred because the culprits suppressed the alarm systems that would normally send alerts to the control room when pressure readings become abnormal. The pipeline was physically ruptured and the escaped oil ignited, resulting in a fiery explosion that caused the pipeline to be out of commission for almost a month. The identity of the attackers remains unknown, but geopolitical instability in the region, including the possibility of an imminent Russian attack on Georgia (through which the pipeline runs), may point to state-sponsored motivations to disrupt the pipeline operations.

Investigations following the attack point to a suspicious team of two individuals with laptops who were seen near the pipeline area. According to Lee *et al.* (2014), the first target of the attack was the communication network, which was accessed through a vulnerability in the new networked IP-based camera system that is used for surveillance all along the pipeline. It is claimed that these IP-based camera systems may have formed the first point of attack since they are frequently misconfigured to communicate openly with the Internet. Through these camera systems, the attackers may have been able to blind pipeline operators to physical intrusions while they gained access to control components at remote valve or compression stations. These include remote terminal units (RTUs) and PLCs. Subsequently, the attackers changed pressure settings by tampering with the automated pressure reliefs. They also tampered with the leak detection system, alarm servers, input traffic from field devices, and satellite communication links which send information to the control room about the abnormal functioning of the pipeline. This interference with the communication system meant operators were oblivious to the rupture and explosion of the pipeline for almost 40 minutes.

According to Lee *et al.* (2014, p.5), the incident affected the following systems:

- Camera system and communication network;
- Leak detection system;
- Automated pressure reliefs;
- Alarm server or input traffic from field devices;
- Pipeline field devices found in valve or compression stations (e.g., RTUs and PLCs);
- Satellite terminals or the actual transmission of signals.

The following functions were affected by the attack: loss of view (LoV) or spoofed view, possible loss of control (LoC), suppression of alarms, direct writing or command of the control element, and camera system monitoring and storage of recordings. The losses to oil and gas companies, including British Petroleum, amounted to millions of dollars, while environmental damage was caused as the oil spilled into a water reserve near the pipeline.

16.8.1 *Case discussion questions*

1. What cybersecurity vulnerabilities and risks may have led to the attack?
2. What technological, organisational, and environmental determinants could have increased the cybersecurity of the pipeline?
3. From a cybersecurity viewpoint, what lessons can you draw from the case study to future-proof oil and gas infrastructure systems?

References

Aljubran, M., Al-Ghazal, M., and Vedpathak, V. (2018). Integrated cybersecurity for modern information control models in oil and gas operations. SPE International Conference and Exhibition on Health, Safety, Security, Environment, and Social Responsibility, April, Abu Dhabi, UAE, pp. 16–18.

Aloul, F.A. (2010). Information security awareness in UAE: A survey paper, International Conference for Internet Technology and Secured Transactions, London, United Kingdom.

Anchugam, C. and Thangadurai, K. (2016). Classification of network attacks and countermeasures of different attacks, In: Kumar, D., Kumar Singh, M., Jayanthi, M. K. (eds.) *Network security attacks and countermeasures*, pp. 115–156, IGI Global, Hershey, USA.

Arutyunov, R. (2018). The next generation of cybersecurity in oil and gas. *Pipeline and Gas Journal*, **245**(6). Available from: https://pgjonline.com/magazine/2018/june-2018-vol-245-no-6/features/the-next-generation-of-cybersecurity-in-oil-and-gas [accessed 6 March 2020].

Australia Government. (2011). Australia Government cyber security strategy. Available from: http://mddb. apec.org/documents/2011/tel/tel43-spsg-wksp/11_tel43_spsg_wksp_005.pdf [accessed 12 May 2020].

Australian Signals Directorate and Australian Cybersecurity Centre (2017). Strategies to mitigate cyber security incidents. Available from: https://www.cyber.gov.au/sites/default/files/2020-04/PROTECT%20-%20Strategies%20to%20Mitigate%20Cyber%20Security%20Incidents%20%28February%202017%29.pdf [accessed 12 May 2020].

Baker Hughes (2017). *APM: Driving value with the Digital Twin*. Available from: https://www.bhge.com/system/files/2017-10/D2%20S2%20APM%20Driving%20Value%20with%20the%20Digital%20Twin.pdf [accessed 20 March 2020].

Digital transformation and the cybersecurity of infrastructure systems

Baker, S. A., Waterman, S., and Ivanov, G. (2009). *In the crossfire: Critical infrastructure in the age of cyber war.* McAfee, Incorporated, Santa Clara, CA.

Beck, R. and Müller-Bloch, C. (2017). Blockchain as radical innovation: A framework for engaging with distributed ledgers. 50th Hawaii International Conference on System Sciences, January 2017, Hawaii, pp. 5390–5399.

Bronk, C. and Tikk-Ringas, E. (2013). The cyber-attack on Saudi Aramco. *Survival,* **55**(2), pp. 81–96.

Carvajal, G., Maucec, M., and Cullick, S. (2017). *Intelligent digital oil and gas fields: Concepts, collaboration, and right-time decisions.* Gulf Professional Publishing, Cambridge, MA.

Caralli, R. A., Allen, J. H., and White, D. W. (2010). *CERT resilience management model: A maturity model for managing operational resilience (CERT-RMM Version 1.1).* Addison-Wesley Professional, Boston, MA.

Chand, N. and Mathivanan, S. (2016). A survey on resource inflated denial of service attack defense mechanisms, 2016 Online International Conference on Green Engineering and Technologies (IC-GET), Tamil NADU, India.

Chen, X., Zhou, Y., Zhou, H., Wan, C., Zhu, Q., Li, W., and Hu, S. (2016). Analysis of production data manipulation attacks in petroleum cyber-physical systems, ICCAD '16: Proceedings of the 35th International Conference on Computer-Aided Design, November 2016, Austin, TX, pp. 1–7.

Coldebella, G. P. and White, B. M. (2007). Foundational questions regarding the federal role in cybersecurity. *Journal of National Security Law and Policy,* **4**(1), pp. 233–245.

Conti, M., Dragoni, N., and Lesyk, V. (2016). A survey of man in the middle attacks. *IEEE Communications Surveys and Tutorials,* **18**(3), pp. 2027–2051.

Da Veiga, A. (2016). A cybersecurity culture research philosophy and approach to develop a valid and reliable measuring instrument. SAI Computing Conference (SAI), IEEE, London, UK, pp. 1006–1015.

Da Veiga, A. and Eloff, J. H. (2010). A framework and assessment instrument for information security culture. *Computers and Security,* **29**(2), pp. 196–207.

Deloitte Consulting (2016). *Blockchain: Enigma, paradox. Opportunity.* Deloitte LLP, United Kingdom.

Denis, M., Zena, C., and Hayajneh, T. (2016). Penetration testing: Concepts, attack methods, and defense strategies. IEEE Long Island Systems, Applications and Technology Conference (LISAT), Farmingdale, NY, pp. 1–6.

Depersio, G. (2019). Why did oil prices drop so much in 2014? Available from: https://www.investopedia.com/ask/answers/030315/why-did-oil-prices-drop-so-much-2014.asp [accessed 21 December 2019].

Dorsemaine, B., Gaulier, J. P., Wary, J. P., Kheir, N., and Urien, P. (2015). Internet of things: A definition and taxonomy. 9th International Conference on Next Generation Mobile Applications, Services and Technologies, IEEE, Cambridge, MA, pp. 72–77.

Drias, Z., Serrhouchni, A., and Vogel, O. (2015). Analysis of cybersecurity for industrial control systems. International Conference on Cyber Security of Smart Cities, Industrial Control System and Communications (SSIC), Shanghai, China, pp. 1–8.

Duan, Q. and Al-Shaer, E. (2013). Traffic-aware dynamic firewall policy management: Techniques and applications. *IEEE Communications Magazine,* **51**(7), pp. 73–79.

Evans, M., Maglaras, L. A., He, Y., and Janicke, H. (2016). Human behaviour as an aspect of cybersecurity assurance. *Security and Communication Networks,* **9**(17), pp. 4667–4679.

Diaz-Gomez, P. A., ValleCarcamo, G., and Jones, D. (2011). Internal vs. external penetrations: A computer security dilemma. Proceedings of the International Conference on Security and Management (SAM) (p. 1). The Steering Committee of The World Congress in Computer Science, Computer Engineering and Applied Computing (WorldComp). 18–21 July, Las Vegas, NV.

Gcaza, N., von Solms, R., Grobler, M. M., and van Vuuren, J. J. (2017). A general morphological analysis: Delineating a cyber-security culture. *Information and Computer Security,* **25**(3), pp. 259–278.

Gordon, L. A., Loeb, M. P., and Sohail, T. (2003). A framework for using insurance for cyber-risk management. *Communications of the ACM,* **46**(3), pp. 81–85.

Government of Canada (2010), Action Plan 2010-2015 for Canada's Cyber Security Strategy, Available from: https://www.publicsafety.gc.ca/cnt/rsrcs/pblctns/ctn-pln-cbr-scrt/index-en.aspx [accessed 10 March 2020]

Farwell, J. and Rohozinski, R. (2011). Stuxnet and the future of cyber war. *Survival,* **53**(1), pp. 23–40.

Harknett, R. J. and Stever, J. A. (2009). The cybersecurity triad: Government, private sector partners, and the engaged cybersecurity citizen. *Journal of Homeland Security and Emergency Management,* **6**(1). DOI: https://doi.org/10.2202/1547-7355.1649.

He, Y., Maglaras, L. A., Janicke, H., and Jones, K. (2015). An industrial control systems incident response decision framework. IEEE Conference on Communications and Network Security (CNS 2015), Florence, Italy.

Henrie, M. (2013). Cyber security risk management in the SCADA critical infrastructure environment. *Engineering Management Journal*, **25**(2), pp. 38–45.

Homeland Security (2009), A roadmap for cyber security research. Available from: http://www.dhs.gov/sites/default/files/publications/CSD-DHSCybersecurity-Roadmap.pdf [accessed 12 May 2020].

Howard, S. and Wallaert, T. (2015). Improving cybersecurity defenses in oil and gas applications. *Pipeline and Gas Journal*, **242**(2). Available from:http://www.pgjonline.com/magazine/2015/february-2015-vol-242-no-2/features/improving-cybersecurity-defenses-in-oil-and-gas-applications [accessed 14 February 2020].

Information Security Policy Council. (2006). The first national strategy on information security. Available from: https://www.nisc.go.jp/eng/pdf/national_strategy_001_eng.pdf [accessed 12 May 2020].

International Organisation for Standardisation (ISO) (2012). ISO/IEC 27032:2012. Available from: https://www.iso.org/standard/44375.html [accessed 12 May 2020].

International Organisation for Standardisation (ISO) (2013). ISO/IEC 27002:2013. Available from: https://www.iso.org/standard/54533.html [accessed 12 May 2020].

Jagatic, T. N., Johnson, N. A., Jakobsson, M., and Menczer, F. (2007). Social phishing. *Communications of the ACM*, **50**(10), pp. 94–100.

Joint Task Force Transformation Initiative (2013). SP 800-53 Rev. 4: Security and privacy controls for federal information systems and organizations. Available from: https://csrc.nist.gov/publications/detail/sp/800-53/rev-4/final [accessed 12 May 2020].

Jones, V. (2017). More than 440,000 global oil, gas jobs lost during downturn, Rigzone. Available from: https://www.rigzone.com/news/oil_gas/a/148548/more_than_440000_global_oil_gas_jobs_lost_during_downturn/ [accessed 21 December 2019].

Kaur, S., Kaur, K., and Gupta, V. (2016). Implementing OpenFlow based distributed firewall. International Conference on Information Technology (InCITe) - The Next Generation IT Summit on the Theme – Internet of Things: Connect your Worlds, 6–7 October, Noida, India, pp. 172–175.

Keeney, M., Kowalski, E., Cappelli, D., Moore, A., Shimeall, T., Rogers, S. (2005). Insider threat study: Computer system sabotage in critical infrastructure sectors. Available from: https://resources.sei.cmu.edu/asset_files/SpecialReport/2005_003_001_51946.pdf [accessed 12 May 2020].

Khurana, H. (2011). Moving beyond defense-in-depth to strategic resilience for critical control systems, IEEE Power and Energy Society General Meeting, Detroit, MI, 2011, pp. 1–3.

Kinnaird, C. and Geipel, M. (2017). Blockchain technology: How the inventions behind Bitcoin are enabling a network of trust for the built environment, Arup Blockchain Technology Report. Arup, London, UK. Available from: file:///C:/Users/Sulafa%20Badi/Downloads/Arup%20%20Blockchain%20Technology%20Report_comp%20(1).pdf [accessed 10 July 2020].

Klimburg, A. (2012). National cyber security framework manual. Available from: https://ccdcoe.org/uploads/2018/10/NCSFM_0.pdf [accessed 12 May 2020].

Knapp, K. J., Marshall, T. E., Kelly Rainer, R., and Nelson Ford, F. (2006). Information security: Management's effect on culture and policy. *Information Management and Computer Security*, **14**(1), pp. 24–36.

Krebs, B. (2008). Cyber incident blamed for nuclear power plant shutdown. Washington Post. Available from: http://www.washingtonpost.com/wp-dyn/content/article/2008/06/05/AR2008060501958.html [accessed 12 May 2020].

Langner, R. (2011). Stuxnet: Dissecting a cyberwarfare weapon. *IEEE Security and Privacy*, **9**(3), pp. 49–51.

Lasi, H., Fettke, P., Kemper, H. G., Feld, T., and Hoffmann, M. (2014). Industry 4.0. *Business and Information Systems Engineering*, **6**(4), pp. 239–242.

Lee, R., Assante, M., and Connway, T. (2014). ICS CP/PE (Cyber-to-Physical or Process Effects) case study paper – Media report of the Baku-Tbilisi-Ceyhan (BTC) pipeline Cyber Attack. Available from: https://ics.sans.org/media/Media-report-of-the-BTC-pipeline-Cyber-Attack.pdf [accessed 12 May 2020].

Leyden, J. (2012). Hack on Saudi Aramco hit 30,000 workstations, oil firm admits. Available from: https://rb.gy/6iffgc [accessed 12 May 2020].

Li, L. (2018). China's manufacturing locus in 2025: With a comparison of made-in-China 2025" and "industry 4.0. *Technological Forecasting and Social Change*, **135**, pp. 66–74.

Lu, H., Guo, L., Azimi, M., and Huang, K. (2019). Oil and gas 4.0 era: A systematic review and outlook. *Computers in Industry*, **111**, pp. 68–90.

Luiijf, E. and Besseling, K (2013). Nineteen national cybersecurity strategies, *International Journal of Critical Infrastructure*, **9** (1.2), pp. 3–31.

Mateski, M., Trevino, C., Veitch, C., Michalski, J., Harris, M., Mauuoka, S., and Frye, J. (2012). Cyber Threat Metrics, Sandia Report. Available from: http://www.fas.org/irp/eprint/metrics.pdf [accessed 12 January 2020].

Mayani, M.G., Svendsen, M., and Oedegaard, S.I. (2018). Drilling digital twin success stories the last 10 years. SPE Norway One Day Seminar. 18 April 2018, Society of Petroleum Engineers, Bergen, Norway.

McAfee and SAIC (2011). Underground economies. Intellectual capital and sensitive corporate data now the latest cybercrime currency. Available from: http://freepdfs.net/rp-underground-economiesnational-defense-industrialassociation/dd8111e5875c655df5def936e6110948/ [accessed 12 May 2020].

McKinsey (2019). A New mandate for the oil and gas Chief Information Officer. Available from: https://www.mckinsey.com [accessed 9 December 2019].

Microsoft (2018). Shell expands strategic collaboration with Microsoft to drive industry transformation and innovation. Available from: https://news.microsoft.com/2018/09/20/shell-expands-strategic-collaboration-withmicrosoft-to-drive-industry-transformation-and-innovation/ [accessed 25 May 2019].

Mihelič, A., and Vrhovec, S. (2018). Obligation to defend the critical infrastructure. Offensive cybersecurity measures. *Journal of Universal Computer Science*, **24**(5), pp. 646–661.

Miller, B. and Rowe, D. (2012). A survey SCADA of and critical infrastructure incidents. In Proceedings of the 1st Annual Conference on Research in Information Technology, New York, NY: ACM, pp. 51–56.

Miraz, M. and Ali, M. (2018). Applications of blockchain technology beyond cryptocurrency. *Annals of Emerging Technologies in Computing*, **2**(1), pp. 1–7.

Muspratt, A. (2019). The top 10 oil and gas companies in the world: 2019, Oil and Gas IQ. Available from: https://www.oilandgasiq.com/strategy-management-and-information/articles/oil-and-gas-companies [accessed 14 April 2020].

Oh S., Kim E., Kim H. (2017) Empirical analysis of SSL/TLS weaknesses in real websites: Who cares? In: Choi, D., Guilley, S. (eds.) International Workshop on Information Security Applications, pp. 174–185. Springer, Cham.

Offshore technology (2017). Size matters: inside the General Electric-Baker Hughes merger. Available from: https://www.offshore-technology.com/features/featuresize-matters-inside-the-general-electric-baker-hughes-merger-5711079/ [accessed 21 December 2019].

Panguluri, S., Nelson, T. D., and Wyman, R. P. (2017). Creating a cybersecurity culture for your water/wastewater utility. In: Clark, R. and Hakim, S. (eds.) *Cyber-physical security*, pp. 133–159, Springer International Publishing, Switzerland.

Payette, J., Anegbe, E., Caceres, E., and Muegge, S. (2015). Secure by design: Cybersecurity extensions to project management maturity models for critical infrastructure projects. *Technology Innovation Management Review*, **5**(6), pp. 26.

Petit, F., Verner, D., Brannegan, D., Buehring, W., Dickinson, D., Guziel, K., and Peerenboom, J. (2015). *Analysis of critical infrastructure dependencies and interdependencies, Argonne National Lab. (ANL)*, Argonne, IL. Available from: https://publications.anl.gov/anlpubs/2015/06/111906.pdf [accessed 14 March 2020].

PwC (2016), The Global State of Information Security® Survey 2016. Available from: https://www.pwc.ru/en/publications/gsiss-2016.html [accessed 15 January 2020].

PwC (2018), The Global State of Information Security® Survey 2018. Available from: https://www.pwc.com/us/en/services/consulting/cybersecurity/library/information-security-survey.html [accessed 21 December 2019].

Radmand, P., Talevski, A., Petersen, S., and Carlsen, S. (2010). Taxonomy of wireless sensor network cybersecurity attacks in the oil and gas industries. 24th IEEE International Conference on Advanced Information Networking and Applications. 20–23 April, Perth, Western Australia, pp. 949–957.

Renn, O. (2008). White paper on risk governance: Toward an integrative framework. In: Renn, O. and Walker, K. (eds.) *Global risk governance*, pp. 3–73, Springer, Dordrecht.

Richardson, R. (2011). 2010/2011 CSI Computer Crime and Security Survey, Computer Security Institute, New York, NY.

Rick, K. and Iyer, K. (2017). Countering the threat of cyberattacks in oil and gas, BCG perspectives by Boston consulting group. Available from: https://image-src.bcg.com/Images/BCG-Countering-the-Threat-of-Cyberattacks-in-Oil-and-Gas-Mar-2016_tcm52-186245.pdf [accessed 12 May 2020].

Rob, R., Tural, T., McLaren, G., Sheikh, A., and Hassan, A. (2014). Addressing cybersecurity for the oil, gas and energy sector, 2014 North American Power Symposium (NAPS), 24 November 2014, Pullman, WA, USA, pp. 1–8.

Rommetveit, R., Bjorkevoll, K.S., Odegaard, S.I., Herbert, M.C., Halsey, G.W., Kluge, R. (2008). eDrilling used on Ekofisk for Real-Time Drilling Supervision, Simulation, 3D Visualization and Diagnosis, Intelligent Energy Conference and Exhibition, 25–27 February, Amsterdam, The Netherlands.

Rosner, M., Herve, P., Moore, K. (2017). Using a cognitive analytic approach to enhance cybersecurity on oil and gas OT systems, Offshore Technology Conference, 1–4 May, Houston, TX, pp. 371–372.

Ruppert, B. (2009). Protecting against insider attacks, SANS Institute InfoSec Reading Room. Available from: http://www.sans.org/reading_room/whitepapers/incident/ protecting-insider-attacks_33168 [accessed 17 May 2020].

Sabillon, R., Cavaller, V., and Cano, J. (2016). National cyber security strategies: Global trends in cyberspace. *International Journal of Computer Science and Software Engineering*, **5**(5), pp. 67–81.

Shoemaker, D., Kohnke, A., and Sigler, K. (2018). *How to build a cyber-resilient organization*. CRC Press, Boca Raton, UAE.

Slay, J., Miller, M. (2007). Lessons learned from the Maroochy water breach. In: Goetz, E., Shenoi, S. (eds.) *ICCIP 2007: International Conference on Critical Infrastructure Protection* (Vol. 253, pp. 73–82). Springer, Boston, MA.

Quigley, K., Burns, C., and Stallard, K. (2015). 'Cyber Gurus': A rhetorical analysis of the language of cybersecurity specialists and the implications for security policy and critical infrastructure protection. *Government Information Quarterly*, **32**(2), pp. 108–117.

Sankaran, S., Lugo, J. T., Awasthi, A., and Mijares, G. (2009). The promise and challenges of digital oilfield solutions: Lessons learned from global implementations and future directions. SPE Digital Energy Conference and Exhibition, 7-8 April, Houston, TX.

Sarrakh, R., Suresh, R., Suresh, S., and Al Nabt, S. (2019). Smart solutions in the oil and gas industry: A review. *Journal of Clean Energy Technologies*, **7**(5), pp. 72–76.

Shafqat, N. and Masood, A. (2016). Comparative analysis of various national cyber security strategies. *International Journal of Computer Science and Information Security*, **14**(1), pp. 129–136.

Shameli-Sendi, A., Aghababaei-Barzegar, R., and Cheriet, M. (2016). Taxonomy of information security risk assessment (ISRA). *Computer Security*, **57**, pp. 14–30.

Shin J., Son H., and Heo, G. (2017). Cybersecurity risk evaluation of a nuclear Iand using BN and ET. *Nuclear Engineering and Technology*, **49**(3), pp. 517–524.

Speake, G. (2015). Cybersecurity 2015: Connected pipelines and proliferation of threats to infrastructure. 242(5). Available from: http://www.pgjonline.com/magazine/2015/may-2015-vol-242-no-5/features/cybersecurity-2015-connected-pipelines-and-proliferation-of-threats-to-infrastructure [accessed 8 November 2019].

Stohl, M. (2007). Cyber terrorism: A clear and present danger, the sum of all fears, breaking point or patriot games? *Crime, Law and Social Change*, **46**(4-5), pp. 223–238.

The White House, (2018). Strategy for American leadership in advanced manufacturing. Available from: https://www.whitehouse.gov/wp-content/uploads/2018/10/Advanced-Manufacturing-Strategic-Plan-2018.pdf [accessed 19 December 2019].

Tripathi, N. and Mehtre, B. M. (2014). Analysis of various ARP poisoning mitigation techniques: A comparison. International Conference on Control, Instrumentation, Communication and Computational Technologies (ICCICCT), IEEE, Kanyakumari, India, pp. 125–132.

Turnbull, J. (2005). *Hardening Linux*. APress, New York.

United Nations Framework Convention on Climate Change (2015). Report of the Conference of the Parties on its twenty-first session, held in Paris from 30 November to 13 December 2015. Available from: https://unfccc.int/resource/docs/2015/cop21/eng/10.pdf [accessed 20 December 2019].

United Kingdom Government. (2014). UK launches first national CERT. Available from: https://www.gov.uk/government/news/uk-launches-first-national-cert [accessed 10 May 2020].

Vara, V. (2019). The biggest ever mergers and acquisitions in the oil and gas industry, Offshore technologies. Available from: https://www.offshore-technology.com/features/mergers-and-acquisitions-oil-gas-industry/ [accessed 12 May 2020].

World Economic Forum (WEF) (2017). Digital transformation initiative: Oil and gas industry. Available from: http://reports.weforum.org/digital-transformation/wp-content/blogs.dir/94/mp/files/pages/files/dti-oil-and-gas-industry-white-paper.pdf [accessed 16 March 2020].

Yang, J. H., Qiu, M. X., Hao, H. N., Zhao, X., and Guo, X. X. (2016). Intelligence– oil and gas industrial development trend. *Oil Forum*, **35**, pp. 36–42.

Yaqoob, I., Hussain, S. A., Mamoon, S., Naseer, N., Akram, J., and Ur Rehman, A. (2017). Penetration testing and vulnerability assessment. *Journal of Network Communications and Emerging Technologies*, **7**(8), pp. 10–18.

Zhao, G. (2016), Keynote: Seismic Processing and Interpretation Platform in the Big Data Era, 2016 Workshop: Workshop High Performance Computing. Society of Exploration Geophysicists, Beijing, pp. 1-4.

Notes

1 Industry 1.0 refers to the steam engine era; Industry 2.0 to the electrification era, Industry 3.0 to the information era, and Industry 4.0 to the intelligence era; 4.0 promotes industrial change through information technology and is underpinned by nine main technologies including cloud computing, 3D printing, cybersecurity, IoT, big data, digital twin, augmented reality, autonomous robots and system integration (Lu et al., 2019). National efforts to formulate Industry 4.0 strategic plans began in Germany with Germany's 4.0 plan (Lasi et al, 2014) and other countries followed suit such as China's "Made in China 2025" (Li, 2018), and the United States' "Strategy for American Leadership in Advanced Manufacturing" (The White House, 2018), among others.

17
INFRASTRUCTURE MEGAPROJECTS AS ENABLERS OF DIGITAL INNOVATION TRANSITIONS

Eleni Papadonikolaki and Bethan Morgan

17.1 Introduction

Innovation refers to the development of a new product, service or process (Abernathy and Clark, 1985). Novelty and innovations are often observed in projects (Shenhar and Dvir, 2007). Innovations are highly context-dependent and rely on good projects (Shenhar and Dvir, 2007). Infrastructure megaprojects have a particularly close relationship with innovation (*Davies et al.,* 2009). This relation between innovation and its context, such as infrastructure megaprojects. is important for understanding innovation. Digital innovation is differentiated from other innovations as it is highly pervasive and systemic (Egyedi and Sherif, 2008). This chapter increases our understanding of digital innovation and its relation to institutional contexts, especially infrastructure megaprojects. Focusing on the relation between digital innovation and infrastructure megaprojects, such as London Heathrow and Crossrail (Davies *et al.,* 2009; Dodgson *et al.,* 2015), shows how these infrastructure megaprojects influenced institutions through standardisation to promote digital innovation.

The chapter draws upon the work by Davies *et al.* (2009) on how innovation and particularly digital innovation was used in megaprojects to achieve government-led standardisation outcomes. Megaprojects, due to their embeddedness (Blomquist and Packendorff, 1998), longevity and pervasiveness, offer a rich research setting to understand the interplay of institutions and agency. Any organisational unit, actor or agent shapes and is shaped by its environment or structure, also called embeddedness (Giddens, 1984). As projects are inseparable and essentially embedded into their issue, organisational and institutional contexts are quintessential for understanding and managing projects (Blomquist and Packendorff, 1998). Not only should projects' relational context be continuously managed, but their wider institutional environment also merits attention (Blomquist and Packendorff, 1998). Söderlund (2004) acknowledges that whereas project management discipline has its "intellectual roots" in process planning and a Taylorist approach to workflows, it has developed into a hybrid field that incorporates many strands of social science.

The relation between projects and innovation is well-documented in scholarship. Davies (2014) recognised two contrasting models of project-based innovation; one optimal, emphasising planning and formal processes and another, adaptive, governed by uncertainty and adaptation. In the adaptive model, individual agency, informal processes, tacit knowledge and context shapes projects through innovation. This chapter focuses on the adaptive model of innovation (Davies, 2014) as is suitable for studying the highly dynamic context of digital innovations. Using the concept of multi-level perspective (MLP) by Geels (2004) as an analytical lens, this chapter investigates the relation between infrastructure megaprojects and their institutional setting to understand how the UK construction sector is currently undergoing a transition from traditional to digital. The authors begin by appraising relevant literature on innovation, highlighting studies relating to infrastructure megaprojects. Then, a longitudinal data relating to digital innovations in UK infrastructure megaprojects and its analysis is presented. Finally, a summary of the case and managerial implications of the study are discussed.

17.1.1 Chapter aim and objectives

This chapter aims to perform the following:

- Introduce the concept of digital innovation and differentiate it from innovation;
- Explain how actors, institutions and infrastructure megaprojects influence digital innovation;
- Present how six infrastructure megaprojects in the UK supported digital innovation;
- Discuss the socio-technical system view of multi-level perspective (MLP);
- Describe ways that the public sector can influence digital innovation; and
- Illustrate key transitions in digital innovation in the UK construction sector.

17.1.2 Learning outcomes

The following learning outcomes have been identified for this chapter. Readers will be able to:

- Articulate the concept of digital innovation and be able to differentiate it from innovation;
- Analyse how actors, institutions and infrastructure megaprojects influence digital innovation;
- Evaluate how infrastructure megaprojects support digital innovation;
- Appreciate the socio-technical system view of MLP;
- Synthesise ways that the public sector can influence digital innovation; and
- Identify key transitions in digital innovation in the UK construction sector.

17.2 Theoretical background and knowledge gap

17.2.1 Innovation footprint of infrastructure megaprojects

Megaprojects are projects with investment of more than US $1 billion and are often found in infrastructure projects in transport, energy, and communications (Sanderson, 2012). Because megaprojects are large complex systems, systems integration has been proposed as a strategy to deal with complexity (Davies and Mackenzie, 2014; Davies *et al.*, 2014). Systems integration is also applied to facilitate innovation and improve megaproject performance (Davies *et al.*, 2009). Geels (2004) defines socio-technical systems (STS) in an abstract sense as the linkages among elements necessary to fulfil societal functions, such as transport, communication, nutrition, etc. Technology is a crucial element for fulfilling those functions and STS encompass the production, diffusion and use of technology, as technology is a sub-function for these major societal functions. Hence, STS consist of artefacts, knowledge, capital, labour, cultural meaning and so forth. To this end, megaprojects are socio-technical niches.

There is a strong relation between projects and innovation (Shenhar and Dvir, 2007) and especially infrastructure megaprojects, due to their longevity, multi-stakeholder engagement and pervasiveness in the institutional setting, are ideal organisational vessels to study innovation. Megaprojects and project-based organisations (Hobday, 2000) are closely linked as the latter are vehicles to deliver the former. Megaprojects are projects of massive, significant scale with long delivery phases that span years or even decades. In particular infrastructure megaprojects carry societal value due to their social functions. Apart from societal impact, megaprojects are usually notorious for poor delivery performance (Flyvbjerg, 2014). Among others, scholars usually emphasise their front-end management, the promoter's role (Gil and Pinto, 2016), their embeddedness into their context (Blomquist and Packendorff, 1998) and involvement of numerous external stakeholders. Megaprojects have a bespoke nature and high uncertainty.

Megaprojects are undoubtedly long-standing, behave as organisations and depart from the traditional notion of project temporality and uniqueness. Sydow *et al.* (2004) explained that despite the fact that organisations usually outlive their projects, the two have similar learning mechanisms. Whereas there is a general notion of temporality of megaprojects, *Brookes et al.* (2017) questioned the "dichotomy of durability between a longer lasting organisational milieu and an ephemeral project." Second, project typologies, such as those of transportation and oil and gas sectors, allow for a degree of repetition. Repetitiveness may account for less uncertainty and more predictability, even in unique, long-standing, and complex projects (Davies and Brady, 2000), due to the "economics of repetition." Therefore, despite being unique, infrastructure megaprojects have mechanisms for transferring knowledge across other projects and their context and supporting innovation (Davies *et al.*, 2009; Davies *et al.*, 2014; Davies and Mackenzie, 2014) in the sector.

17.2.2 *Contextual transitions from innovation to digital innovation*

Historical advancements in hardware and software gave new IT capabilities to infrastructure megaprojects (Whyte and Levitt, 2011). These megaprojects have a bilateral influencing relation with their context. Giddens' structuration theory (1984) suggests that social systems, such as projects and megaprojects, shape and are shaped by their environment by having a mutually constitutive relationship and being embedded in a wider context. This insight calls for understanding megaprojects and innovations as not only being capable of shaping their environment, but also being shaped by it, according to the duality of structure and agency in structuration (Giddens, 1984). Innovation depends a lot on its context. For Rogers *et al.* (2005), contextual heterogeneity is central to his diffusion of innovations theory, and acknowledging the influence of heterogeneous institutional contexts offers a grounded grasp of innovation.

However, digital innovation is inherently different than innovation, as it is unbounded – that is, it relates to more than one discipline or multiple agents; it is related to distributed agency and the innovation processes and outcomes are interrelated (Nambisan *et al.*, 2017). Digital innovation is different from innovation as digital technologies are used during the innovation process (Nambisan *et al.*, 2017). The transition from innovation to digital innovation creates new challenges and transformations in individuals, organisations and more widely, in society (Nambisan *et al.*, 2017). Digital innovation as a socio-technical phenomenon challenges key assumptions of innovation management. Hence, digital innovation due to its particular features is better understood within its context and at the same time, relates to many different institutions and actors outside the traditional supply and demand chain.

From a neo-institutional and STS view, Geels (2004) and Geels *et al.* (2017) provided a MLP for looking at innovation as STS where actors and institutions interact through different rules and induce transitions in the innovation landscape. In this macro-level view of innovation, Geels (2004) distinguished among (1) actors and social groups, (2) rules and institutions, and (3) technologies and STS that interact in a dynamic manner. These actors and social groups are embedded in the STS and use the rules of the institutions, for example, to introduce new technologies, make investment decisions about technology or develop new regulations and standards. These actors have different resources, such as money, knowledge, tools and opportunities to realise their decisions and influence social rules (Geels *et al.*, 2016). Through various interactions between actors and social groups with rules and institutions, various events can be mapped that eventually shape the transition of the innovation landscape. These transitions shift between the following pathways: "substitution," "transformation," which leads to "reconfiguration" and is possibly followed by "re-alignment and 'disruption"(Geels *et al.*, 2016).

Negotiating and setting rules and standards are part of the innovation process through a neo-institutional and STS view. Standards can be defined as consensus reached by various actors on how to do specific activities according to agreed-upon rules (Farrell and Saloner, 1992). Standardisation, that is, the development of standards, supports also the legitimisation of the technology and these standards (Narayanan and Chen, 2012). Among the various institutions involved, the government is recognised as one of the most important standardization actors (Gao *et al.*, 2014). This is because the government typically promotes technology development, implementation and diffusion by research and development, technology investment and forming state-led standard-setting consortia (Funk and Methe, 2001). The government is not only a regulative but also cognitive and normative institution for supporting standards and innovation (Gao *et al.*, 2014). Key actors in standardisation are firms that build capabilities (Xie *et al.*, 2016) and are organised in megaprojects (Davies and Brady, 2000).

17.2.3 Research setting of innovation in infrastructure megaprojects

Innovation has been traditionally considered "incremental" (evolutionary) by involving gradual minor changes and "radical" by engaging in completely new approaches (Abernathy and Clark, 1985; Burns and Stalker, 1961). In construction, which is largely project-based (Morris, 2004), innovation is considered to have a slow uptake and undergoes slow transitions. In the last decade, the construction industry has been transformed by "wakes" of innovation in project networks (Boland, Jr *et al.*, 2007). From digital three-dimensional (3D) representations of built assets until automated design and construction processes started using Building Information Modelling (BIM) – a three-dimensional data modelling approach – and various realities (Whyte *et al.*, 2000), the construction sector has witnessed changes in technologies, work practices and knowledge across multiple communities (Boland, Jr *et al.*, 2007). Presently, BIM is considered the most representative digital technology and information aggregator in construction globally. Following similar trends in other sectors, the advancement of construction IT has evolved within the context of the digital economy. Accordingly, various digital artefacts and functionalities alter the way construction megaprojects are designed and delivered (Whyte and Lobo, 2010). Lobo and

Whyte (2017) studied UK infrastructure megaprojects and how the project setting affects digital delivery and discussed how the complex institutional forces affected the project setting of these megaprojects. This chapter navigates across the infrastructure megaproject setting focusing on the impact of institutions and agency upon digital innovation.

17.2.4 Knowledge gap

By viewing innovation as a social phenomenon, studies show that it is deeply embedded in its historical and institutional context. Thus, by exploring its embeddedness, insights are generated into how it works as an STS. The emphasis on infrastructure megaprojects as an organisational lens provides a novel view of innovation as an STS and helps frame the phenomenon of digital innovation. Additionally, individual actors and firm-centric agencies may facilitate wakes of innovation (Boland, Jr et al., 2007) and use formal and dynamic approaches to influence their networks. By mapping the relationships among actors, institutions, megaprojects and digital technologies, one can infer their role in transitions of digital innovation. Rather than focusing on the organisational view of developing innovations (Hobday, 1998), this chapter focuses on the institutional structure (hierarchical or networked), agency and processes that influence innovation. It does so by using Geels' (2004) MLP theoretical lens of innovation through the interplay of (1) rules and institutions, (2) actors and social groups, and (3) digital technologies and STS. Figure 17.1 shows the theoretical framework of the chapter as a loosely coupled system of MLP within the area of problematisation: UK construction.

17.3 Research method and data

17.3.1 Methodological rationale

The study setting looks at digital innovation in infrastructure megaprojects. Through the research setting of institutional intervention and megaprojects in the UK, the mutually constitutive relationship between digital innovation and its institutional context is unpacked. The UK has adopted an institutional interventionist mode in relation to innovations in the built environment (Papadonikolaki, 2017). Digital innovations are catalysed by both responsive and anticipatory mechanisms (Morgan, 2019). Data for the case was collected using narrative literature reviews, anecdotal data from industry leaders, policy makers and practitioners who were directly involved in the megaprojects, standardisation and publicly available archival data. By combining retrospective and contemporary data in this way, a longitudinal study was generated (Pettigrew, 1990). The data was then analysed using comparative, synthetic strategies. In this longitudinal case study (Leonard-Barton, 1990) of digital innovation in infrastructure megaprojects, the chronological description serves as a structure to frame the theorization process. Through the theorising

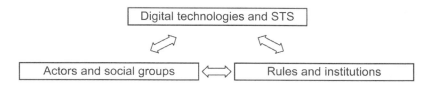

Figure 17.1 Theoretical setting of the chapter framed around MLP: (1) actors and social groups, (2) rules and institutions, (3) digital technologies and STS (infrastructure megaprojects) based on Geels (2004)

mechanism of induction, the researchers formulated propositions on various institutional roles in digital innovation in UK infrastructure megaprojects through the interplay of discursive analysis and empirical data to develop the study narrative.

17.3.2 Methods for data collection and analysis

For this chapter, data was collected on selected megaprojects in the UK, spanning from 1985 to contemporary, ongoing projects, thus covering a substantial time period in the process of digital innovation in UK construction industry. The UK construction sector was the main case study and featured six case sub-units (Yin, 1984) of infrastructure megaprojects. Four completed and two current infrastructure megaprojects or "breakthrough projects" (Shenhar and Dvir, 2007; Wheelwright and Clark, 1992) were studied, namely High Speed 1 (HS1) or the Channel Tunnel Rail Link, Heathrow Terminal 5, the London Olympics, Crossrail, Thames Tideway and High Speed 2 (HS2).

Data were collected through desk research, including government reports, industry reports, standards and mandates. Additionally, publications in academic journals provided useful insights into structuring the narrative. The data collected using desk research (Petticrew and Roberts, 2008) provided an unbiased and replicable account of the existing substantial body of literature relating to the institutional setting, the role of individual agency, and digital innovations in these megaprojects. Due to the emphasis on an institutional lens, both grey and scientific literature were reviewed, consistent with a networked view of innovation in the context of construction. The data were analysed using Langley's (1999) recommendations for using synthetic strategies to analyse process data, as is appropriate for a longitudinal embedded case study comprising multiple sub-unit (Yin, 1984) megaprojects, institutional actors and key events. The predictive potential of such analysis (Langley, 1999) increased the potential value of the findings.

17.4 Data presentation and findings

17.4.1 Framing digital innovation in UK infrastructure megaprojects

The chronology of events was divided into several distinct stages by drawing upon findings by Morgan (2019), who studied how the UK industry improvement agenda influenced organisational change in a consulting firm through a longitudinal study. Critical events in each segment were identified and used to explore the role of key stakeholders in digital innovation and its standardisation (Leonard-Barton, 1990), following a presentation of key events and the interplay among (1) actors and social groups, (2) rules and institutions, and (3) digital technologies in megaprojects as STS according to MLP (Geels, 2004). In chronological order, the six projects identified for detailed study were The Channel Tunnel Rail Link (CTRL or HS1) which ran from 1985 to 1994); Heathrow Terminal 5 (1999–2008); the London Olympics 2012 (2005–2012); Crossrail (2008–2018); Thames Tideway (2012–2023) and High Speed Two (or HS2) (2017–2026). The last two projects – Thames Tideway and HS2 – are ongoing at the time of writing; the other projects are completed.

In particular, the key events that provided temporal breakpoints were used to signpost the transitions of digital innovation through the pathways of reconfiguration, transformation, reconfiguration and re-alignment as outlined and analysed through an MLP lens in the next sub-sections. The ensuing analysis is organised per transitions and (1) actors and social groups, (2) rules and institutions, and (3) digital technologies and STS through the setting of megaprojects. However, as the MLP framework targets the interplay among these factors, the respective subsections make

Infrastructure megaprojects as enablers of digital innovation transitions

Table 17.1 Transition pathways of digital innovation in UK construction along the three categories (institutions, actors and technologies), based on framework by Geels (2004)

Transition pathways	Rules and institutions	Actors and social groups	Digital technologies and STS
Substitution (–1998)	Limited institutional change from government-sponsored industry reports calling for change No rules developed	Firms become sporadically aware of the need for change No new entrants	In complex megaprojects, the need for technological change is accentuated Digital tools complement and substitute existing tools
Transformation (1998–2011)	Institutional change by sponsoring an institutional and collaborative project to inspire change Project outputs become basis of British standards	Incumbents reorient incrementally by adjusting their processes New government-sponsored groups and communities to drive change emerge	Incremental progress in existing technologies Incorporation of symbiotic niche innovations and add-ons that change the processes
Reconfiguration (2011–2016)	Substantial institutional involvement and legitimization via mandates for new technology in public procurement Issuing of suites of standards to lead change	Incumbent firms reorient substantially to new technology and business model New alliances between incumbents and new entrants	From initial add-ons to new hybrids of new and existing technologies Partial or full technical substitution of tools that brings new processes
Re-alignment (2016–)	Institutions are disrupted from the impact of new technology mandates Need to change so as to internalize new processes	Firms have fully reoriented Incumbents collapse because of landscape pressure Opportunities for new entrants	Decline of old technologies create space for competing innovations Processes are disrupted

linkages to other of the three categories (institutions, actors and technologies) where appropriate as the dynamics of the UK construction are interconnected. Table 17.1 summarises the transition pathways as a "nuanced analytical apparatus to analyse unfolding transition processes" (Geels *et al.,* 2016) and specific findings are further presented in the ensuing sections.

17.4.2 Pathway 1: Calling for cultural and technological change – Substitution (up to 1998)

17.4.2.1 Rules and institutions during substitution

The improvement agenda in UK construction sector has included visions of partnering, supply chain management and a lean philosophy, all of which were imported from other sectors such as aerospace and manufacturing (Bresnen and Marshall, 2001) and tested in infrastructure megaprojects. This confirms research that finds that the construction industry has a tradition of importing and not producing technological innovations (Pavitt, 1984). Scholars challenged the extent to which such innovations are indeed applied and effective in construction (Fernie and Tennant, 2013), accusing industry strategists of uncritically adopting "management fashions" (Green, 2011) and defending business as usual.

17.4.2.2 Actors and social groups during substitution

Through an institutional lens, two inter-connected reports were instrumental in influencing the institutional context of innovation in the UK construction sector. The Latham (1994) and Egan (1998) Reports called for increased integration and collaboration among the supply chain to improve industry performance. Improved performance and increased efficiency were recurring themes of both reports. The Latham (1994) Report, "Constructing the Team," criticised the industry for being adversarial, ineffective, fragmented, and with low value for money for the client and proposed the adoption of partnering to increase teamwork and collaboration (Latham, 1994). The Egan Report (1998), "Rethinking Construction," followed the same logic and apart from cultural change was very keen to use digital technologies to improve construction performance.

17.4.2.3 Socio-technical systems during substitution

The Egan Report (1998) was an outcome of the Channel Tunnel project – the first seminal infrastructure megaproject in the UK construction sector – and strongly emphasised using digital technologies to improve processes. The technical complexity of the Channel Tunnel project (also referred to as High Speed 1 or HS1) pushed the boundaries of technological developments in the UK construction sector through information representation and modelling solutions and 3D design tools (Pöttler, 1992). Computer-based tools to deal with interoperability problems among different systems, such as cost management and procurement systems, were the main innovations of the project (Kelsey, 2019). The technological developments in this infrastructure megaproject echoed developments in automotive and aeronautical engineering in the mid-1980s and initiatives in the US for "building product model" definitions to exchange building information amongst Computer-Aided Design (CAD) applications (Eastman, 1999) that replaced error-prone human interventions.

17.4.3 Pathway 2: Reorienting towards new digital technologies – Transformation (1998–2011)

17.4.3.1 Rules and institutions during transformation

Simultaneously, as well as developing policy interventions, the UK government stimulated and facilitated innovation development and market diffusion of digital technologies through institutional projects (Holm, 1995). In the context of the construction sector, such a collaborative or institutional project (Holm, 1995) was the Avanti project (2001–2005), whose objective was to enable effective collaboration among teams through digital technology (Morgan, 2017). Avanti was an inter-institutional collaboration among the Department of Trade and Industry with the support of most of the largest UK construction firms, the International Alliance for Interoperability (IAI, now called BuildingSMART), universities and R&D departments. Such collaborative projects can implement improvement agendas' policies and share the vision when firms lack the confidence and means to invest in their own R&D. Around 2010 in the Victoria Station upgrade project, the team started using the Avanti standard to control information and increase people's trust in data and collaboration. Avanti's focus was on creating collaborative culture and providing processes and digital tools to enable collaboration in teams (Morgan, 2017) through the use of two-dimensional digital design. Avanti became the basis of the BIM British Standard BS1192 issued by British Standards Institution (BSI).

17.4.3.2 Actors and social groups during transformation

One of the first institutional projects to explore collaborative digital working in the UK, the Avanti project marks the beginning of the journey along the BIM maturity trajectory (Bew and Richards, 2008). Funded by the UK government in 2001 under its "Partners in Innovation" scheme, Avanti drew several major industry partners together to work on developing rudimentary shared work practices amongst project teams. Such practices were essential in enabling the collaboration required by digital technologies such as BIM. The project underlined the significant challenges and need for such practices to be developed. Such was the industry appetite for this that Avanti continued to be supported after government funding expired.

In 2006, the Avanti brand ownership was transferred to Constructing Excellence (CE), an industry body that was formed to implement the "Improvement Agenda" as laid out in the Latham and Egan Reports and started as the aggregator of several other industry bodies. It was concluded that despite these coordinated efforts from the two reports, the proposed improvements or innovation had not been readily adopted in construction. Almost two decades after the Egan Report (1998), the Wolstenholme *et al.* (2009) Report, 'Never Waste a Good Crisis," after collaboration with CE, reviewed the success of the 1994 and 1998 reports. Radical change came from the UK government mandated the use of the so-called Level 2 BIM on all public sector projects starting after April 2016 (GCCG, 2011). The Level 2 BIM is a concept developed by Bew and Richards (2008), which is relatively vague and specifies the use of 3D models in a collaborative way.

17.4.3.3 Socio-technical systems during transformation

Digital technologies also evolved along the years. The CAD tools used in the Channel Tunnel project were predecessors of BIM. Building product modelling advancements followed the long-standing debate on the computerisation and digitalisation of construction (Eastman, 1999). Contrary to widespread belief, BIM is not new; it is, in part, an outcome of evolving efforts by industry consortia, such as BuildingSMART, to standardise building information (East and Smith, 2016). In particular, megaprojects have a strong relationship with digital innovations and require collective action from organisations involved to "to overcome considerable resistance to new ways of working" (Davies *et al.,* 2009). Information directors and personnel moved from the Channel Tunnel project to the Heathrow Terminal 5 megaproject and informed the evolution of digital innovations there.

Throughout these megaprojects the language about digital innovation changed. For example, at Heathrow Terminal 5 the digital innovation that now is called BIM was referred to by Davies *et al.* (2009) as a "single-model environment," Heathrow Terminal 5 was the research setting of innovation studies in megaprojects by Davies *et al.* (2009) and their influential model of systems integration. This was developed in the later study by Davies and Mackenzie (2014), drawing on systems integration in complex projects, which are conceived as a "system of systems." Another study drawing on the London Olympics megaproject found that the trajectories of learning had a legacy beyond the built assets created: on the individuals and professions involved (Grabher and Thiel, 2015).

Heathrow Terminal 5 was also a research setting for Harty (2005), who drew on the digital practices used there to find the "unbounded nature" of digital technologies, and direct attention to the important area of interorganisational working that continues to challenge construction practitioners and researchers today. In later work, Harty and Whyte (2010) drew on the same megaproject to observe the "hybrid practices" being employed by practitioners – a theme that persists in contemporary practice with digital innovations.

17.4.4 Pathway 3: Legitimising digital innovation – Reconfiguration (2011–2016)

17.4.4.1 Rules and institutions during reconfiguration

The first construction strategy to specifically ask for change in digital innovation, namely by adopting BIM, was the 2011 Government Construction Strategy (HMSO, 2011). This strategy defined the objective that the government "will require fully collaborative 3D BIM (with all project and asset information, documentation and data being electronic) as a minimum by 2016" (Office,2011). Following the announcement of the mandate in 2010, the British Standards Institute (BSI) issued the suite of Publicly Available Specifications (PAS) number 1192. The PAS1192 contained six parts in total, published between 2013 and 2015. It specified the use of a Common Data Environment (CDE) – an online data sharing platform – how Asset Information Model (AIM) should be created and operated, facility management, safety and security implications of built assets (preparing the ground for BIM Level 3 mandate) and health and safety.

In 2013, the government issued "Construction 2025: Industry Strategy," reaffirming the strong position with regards to BIM and digital way of working in the built environment and emphasising a joint commitment to the BIM vision and programme through partnership between government and industry and close collaboration of these two institutions. The visions further explained the firm stand in ensuring all centrally (governmentally) procured projects would be delivered through a BIM-based approach, eventually leading to a wider offsite manufacturing strategy. In 2016, the 2016–2020 GSC was issued by the cabinet Office and the Infrastructure and Projects Authority (IPA), which built upon the 2011 strategy, emphasising on BIM and Digital Construction as "an important part of the strategy and is helping to increase productivity and collaboration through technology" (Office, 2016), as well as the reliance on infrastructure megaprojects for digitally transforming the sector.

17.4.4.2 Actors and social groups during reconfiguration

In support of the mandate, the UK government created the UK BIM Task Group, a government-funded group managed by the Cabinet Office, in 2011. The BIM Task Group included practitioners seconded by their employers to support the success of the UK BIM Level 2 mandate. The UK BIM Task Group was funded until 2016 and later disbanded to form another Task Group to work on the BIM Level 3 mandate of "Digital Built Britain" (HMG, 2015). After it was disbanded, some of its members formed the UK BIM Alliance, which was publicly funded until 2017 to continue the efforts for increased adoption and implementation of BIM.

17.4.4.3 Socio-technical systems during reconfiguration

The Wolstenholme *et al.* (2009) Report directly influenced the Crossrail project, as Andrew Wolstenholme was the chief executive of this infrastructure megaproject. The innovation strategy

Infrastructure megaprojects as enablers of digital innovation transitions

that followed at Crossrail has been the subject of considerable scholarly and practitioner attention (DeBarro *et al.*, 2015). The longevity and institutional pluralism of Crossrail (2008–2018) made it an ideal vehicle to study the interrelations between megaprojects, institutions and agency. Senior managers from London Olympics (2005–2012) worked in Crossrail and transferred the ideas of digital change.

Crossrail originally specified the need for two-dimensional deliverables but evolved to develop 3D digital deliverables throughout its duration. Whereas it started well before the UK BIM mandates, it is one of the first UK projects to become PAS1192-compliant and use BIM as a digital platform for other innovations, such as laser scanning, using unmanned aerial vehicles (UAVs), drones, virtual reality (VR) and augmented reality (AR). The movement of ideas and leadership around digital innovation that took place was clearly evident. Andrew Wolstenholme, Chief Executive of Crossrail, explained about the inception of a BIM Academy in partnership with Bentley software (Munsi, 2012): "The Academy will support the Government Construction Strategy by increasing the use of BIM in the construction industry and creating a lasting legacy of best practice in innovation. The training received at the Academy will also help contractors use the knowledge and skill gained here on other major projects such as HS2."

17.4.5 Pathway 4: Renewing megaprojects via digital innovation – Re-alignment (2016–ongoing)

17.4.5.1 Rules and institutions during re-alignment

Following the enforcement of the 2016 mandate, the government stepped back to enable the industry to take the lead. However, the industry has not progressed fast enough to follow up on the mandate. In 2018, the UK PAS1192 was adapted at a European level and translated into ISO 19650 which shows that the UK leads standardisation of digital innovations in construction at an international level.

Recently, the Farmer (2016) Review "Modernise or Die," commissioned by the Construction Leadership Council at the request of the UK government, resonated with Wolstenholme *et al.* (2009) regarding (1) productivity losses and (2) lack of collaboration, additionally highlighting (3) lack of innovation and (4) skills shortage as persistent issues of construction. The Farmer (2016) Review called for urgent action in light of the newly-announced megaprojects pipeline in London and the southeast of England. The UK government initialised the standardisation process and further organised innovation activities for digital technology development and diffusion.

17.4.5.2 Actors and social groups during re-alignment

Although the institutions have ceased rule-setting activities since the mandate of 2016, organizations started slowly to fully de-align and re-align. Due to the new mandates for digital delivery, Lobo and Whyte (2017) found that contractors involved in megaprojects aligned the project set-up with their existing capabilities and reconciled differing agendas and capabilities in collaborating firms across the project ecology. Nevertheless, the five years from 2011 to 2016 was a short amount of time to change the industry mindset and more leadership was needed. Despite the leadership from the central UK government, the local government did not effectively drive the mandate, as there were less influential publicly procured projects, such as infrastructure megaprojects to drive digital innovation outside London and the southeast of England.

17.4.5.3 Socio-technical systems during re-alignment

The digital shift required collective action from companies across the supply chain and collaborating would need to be more substantial. For example, in pilot projects organised by the UK government's Ministry of Justice, such as Cookham Wood Prison, a two-stage open book procurement model BIM-based delivery, the supply chain – from suppliers to operators – was involved in dialogue very early on in the design stage to be able to co-create the project and collaborate effectively thereafter.

Senior managers from London Olympics (2005–2012), after working in Crossrail (2008–2018), were moved to the HS2 project (2017–2026). Both these past projects had a strong emphasis on passing legacy to newer megaprojects. Building upon developments from previous projects, the HS2 project aims to gather knowledge and mobilise it to construct the digital twin of the asset, that is a digital asset mirroring the physical infrastructure and able to simulate scenarios. It is expected that with the green light for HS2 in early 2020, the digital legacy of infrastructure megaprojects will support higher collaboration among government departments, with other major infrastructure programmes and professional institutions, to develop standards and make sure that UK government requirements can be met by the supply chain.

The infrastructure megaprojects studied in this chapter are marked not only but their innovative use of digital technologies within the project but by a growing awareness of the digital legacy they create. For example, the HS2 project is looking to redefine digital innovation in the UK construction sector by looking to develop competences not just limited to BIM but about data management in a broader sense. However it is also pioneering solutions that benefit the entire industry and are aimed to create lasting change such as its BIM upskilling platform (HS2, 2019). Similarly, Thames Tideway – a major infrastructure project that will replace London's ageing sewage structure with 25 km of new tunnel and the creation of new public spaces along the Thames – is heavily involved in institutional initiatives to create a digital learning legacy for the construction sector.

17.4.6 Mapping multi-level transitions of digital innovation in the UK construction sector

Digital innovation in the UK construction sector has been associated with various evolving technologies such as CAD and BIM, described previously. BIM is not only a domain of digital artefacts but has historical roots in the long process of structuring and standardising building information across the construction sector (Laakso and Kiviniemi, 2012). As seen through the chronology of the MLP transitions in the previous sections, BIM is an evolving concept and scholars and practitioners move towards more broad descriptions of BIM, such as "Building Information Management" (Becerik-Gerber and Kensek, 2009), "digitally-enabled working" (Dainty et al., 2017), digitisation (Morgan, 2017) and digital innovation, to capture numerous associated innovations.

Drawing on the above data, indicative findings about the transitions of digital innovation across UK construction are mapped below in Figure 17.2, which plots (1) rules and institutions, (2) actors and social groups and (3) digital technologies and STS through the lens of megaprojects, against the timeline of digital innovation in the construction sector following on the theoretical lens of Figure 17.1. This graphic visualises the interrelations among policy (rules and institutions), champions of change (actors and social groups) and digital innovations used in infrastructure megaprojects (digital technologies and STS).

Infrastructure megaprojects as enablers of digital innovation transitions

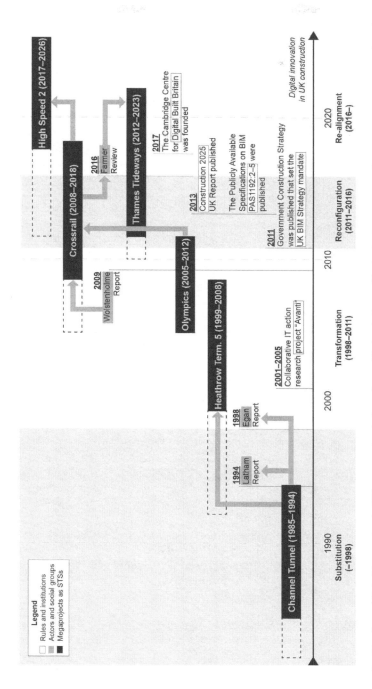

Figure 17.2 Timeline of digital innovation in UK construction influencing and being influenced by institutions, actors and megaprojects as socio-technical systems (STS) of digital innovation

17.5 Discussion

17.5.1 *Megaprojects as niches for digital innovation*

The authors drew on descriptions of megaprojects and institutional theory to identify six megaprojects for this chapter. First, Lundin and Söderholm (1995) described megaprojects as having temporal character but also of institutionalised termination, and as being both fluid and strategic. Hence, from the data it can be seen that megaprojects, with a range of organisations and institutions involved, can also be described as institutional projects that influence transition pathways of digital innovation. At the same time, megaprojects, due to their empirical richness, were ideal instances of socio-technical systems (STS) to show the evolving digital technologies through Geels' (2004) multi-level perspective (MLP) analytical lens. Holm (1995) defined institutional projects as political projects that engage various institutions by necessitating collective action to generate new institutions of political actors. The data revealed the "digital legacy" that megaprojects' leaders were keen to leave, as shown by Wolstenholme (Munsi, 2012) and sources about HS2's digital legacy.

The findings reveal the relation among institutional regulation, digital innovation and UK infrastructure megaprojects. First, it was revealed that the institutional push for digital innovation was detached from relevant technological emergence and therefore the reactive standards issued and the responsive mechanisms (Morgan, 2019) were not fully aligned with the continuously evolving status of digital innovation in UK infrastructure megaprojects. This can be seen through the scarcity of pilot projects to marshal digital innovations and the reactive compliance to digital innovation standards for Crossrail project – albeit ultimately successful. This chapter challenges the timing of regulation, standardisation and its stability, which differs among anticipatory, enabling and responsive standards (Morgan, 2019), in aligning the production system with the new technologies in the UK construction sector. Second, the chapter highlights the importance of influential social actors moving across infrastructure megaprojects and institutions and influencing digital innovation. The boundary-spanning capabilities of such social actors who transfer knowledge across domains (Koskinen, 2008; Levina and Vaast, 2005) supported digital innovation in infrastructure megaprojects.

Nevertheless, the industry still struggles with innovation efficiency (see Table 17.1). The authors suggest that the UK construction industry needs to set up an open innovation paradigm towards the creation of anticipatory standards open and flexible enough as to realise the promise of digital innovation. Some evidence of relevant change can be found in the recent effort called "Project 13," an industry-led collaborative endeavour, seeking to develop a new business model, based on an enterprise, not on traditional transactional arrangements, to improve the UK construction sector (Group, 2017). This open innovation initiative is bringing changes not only to product-oriented sectors but also services (Chesbrough, 2011). Among the ambitions of the Infrastructure Client Group (ICG) (Group, 2017) is to "establish standards that all ICG members will sign up to and enforce across all of their projects."

17.5.2 *Identifying transition pathways of digital innovation in the UK construction sector*

This chapter set out to explore the relation between megaprojects and their institutional setting to understand how their evolution shapes transition pathways of digital innovation. Addressing this research aim, the authors framed the chapter around Geels' (2004) MLP: (1) actors and social groups, (2) rules and institutions, and (3) digital technologies and STS and context: UK

Infrastructure megaprojects as enablers of digital innovation transitions

infrastructure megaprojects. Although MLP has been used in the past to define the transition pathways where different countries belong, e.g., in the case of comparing Germany and the UK as to low-carbon electricity transitions (Geels *et al.*, 2016), this chapter due to its longitudinal character revealed the evolving interplay of institutions, actors and technologies in UK infrastructure megaprojects.

Drawing on Langley's (1999) temporal bracketing strategy, this chapter identified four distinct transition pathways that mark the evolution of digital innovation in the UK construction sector: substitution, transformation, reconfiguration and re-alignment. As shown in the findings each of these phases are distinguished by distinct events and interactions at multiple levels. By undertaking a longitudinal process study of the industry, a long-term view of digital change is afforded. The digital evolution of the industry is particularly important in its current transition through "re-alignment." This chapter suggests that infrastructure megaprojects play a significant role in affording digital evolution, and that its effects influence and are influenced by events at multiple levels: social groups, rules and institutions, digital technologies and STS. Recent industry initiatives illustrate the convergence among these – for example, in initiatives such as Project 13; in HS2's BIM upskilling platform – and harness the significant influence that megaprojects have in driving digital innovation. However, the rapid pace and extent of digital transformation in the construction industry seen in recent years suggests that such initiatives need to become more prevalent and timelier.

17.5.3 Contribution to theory and knowledge

The contribution of this chapter is at two levels. First, at a middle-range theory level, the chapter added to the knowledge base of digital innovation in construction by structuring and synthesising an alternative view of existing and new empirical data on digital innovation in infrastructure megaprojects. Through this longitudinal study of over 30 years, the institutionalisation of digital innovation is a central finding that calls for rethinking and re-organizing institutions and infrastructure megaprojects. From a multi-level perspective (Geels *et al.*, 2016), four discrete transition pathways (substitution, transformation, reconfiguration and re-alignment) were defined through three major (standardisation or regulatory) events. Among these milestones, standardisation and regulation abounded, specifically through the Avanti project (2001–2005) and the suite of PAS1192 standards (2013–2015) and accompanied digital innovation in infrastructure megaprojects as STS.

Second, at a general management theory level, the chapter added to our understanding of digital innovation and infrastructure megaprojects in an analytic way that revealed the interdependences among institutions through the lens of MLP: institutions, actors and technologies as "analytical apparatus to analyse unfolding transition processes" (Geels *et al.*, 2016). Additionally, through MLP, although the areas of (1) rules and institutions and (2) actors and social groups are typically self-explanatory in MLP studies (Geels *et al.*, 2016; Geels, 2004) but technologies and socio-technical systems (STS) are less so. To this end, this chapter extended the MLP analytical lens by presenting infrastructure megaprojects as STS and niches where actors and technology interact in joint endeavours. Eventually, understanding the politics at a macro level of networked innovation (Swan and Scarbrough, 2005) can support the understanding of organisational innovation at a micro level.

17.5.4 Managerial implications

The chapter has several implications for practitioners. Table 17.1 shows two main categories of implicated parties: institutions, such as government bodies, and actors, such as practitioners. First, with regards to institutions, the chapter outlined how the different regulatory and standardisation efforts impact the sector, as seen by such events acting as defining breakpoints of four discrete transition pathways (substitution, transformation, reconfiguration and re-alignment). Institutional projects, such as Avanti, that bring the sector together in an open innovation paradigm are useful vessels to marshal collective expertise and action to influence change and socio-technical transitions. Recently, infrastructure megaprojects also form institutional vessels that broker change.

Second, at a practitioners' level, the chapter has implications for managers of incumbents and new entrants in the market (see Table 17.1). Managers of incumbent firms need to reorient substantially by adopting digital technologies and new business models, e.g., "from transactions to enterprises." In addition, they should be open to form alliances, such as Project 13, between incumbents and new entrants to co-develop standards and respond to digital transition. As the sector is under disruption, managers of new entrants and reoriented incumbents with novel business models stand to gain more from digital innovation.

17.6 Chapter summary

The use of the multi-level view on the process of digital innovation in the UK construction system offers an alternative view on the process. Digital innovations are produced and shaped by the interplay of institutional and organisational factors. The chapter discusses the role of actors who moved across infrastructure megaprojects and institutions and influenced digital innovation in socio-technical niches. The chapter contributes to our understanding of the importance of infrastructure megaprojects as potential niches of digital innovation. Another emergent finding relates to the role of institutions and actors in leading digital innovation through these infrastructure megaprojects. From the data presented, four discrete transition pathways of digital innovation (substitution, transformation, reconfiguration and re-alignment) emerged the last few decades due to varying interactions among institutions, actors and infrastructure megaprojects. Digital innovation became legitimised through standards emerging from infrastructure megaprojects.

The practical implications of this chapter are to reveal the transition mechanisms that lead to digital innovation. Understanding the inter-relationships among infrastructure megaprojects, institutions, actors and how they influenced digital innovation will help prepare for and identify patterns and opportunities for managing the unprecedented pace of emerging digital technologies that influence the construction industry. Additionally, the chapter showed that infrastructure megaprojects due to the range of organisations and institutions involved could also be described as niches. Apart from the construction sector, these findings are valuable for other sectors, because the built environment allows us to study this relatively slow transformation over three decades and identify mechanisms and inter-relations that are hardly noticeable in other sectors, where the pace of innovation is more accelerated.

17.6.1 Chapter discussion questions

After studying this chapter, discuss the following questions:

1. What is digital innovation, and how it is different from innovation?
2. How do actors, institutions and infrastructure megaprojects influence digital innovation?
3. What is the extent to which infrastructure megaprojects support digital innovation?
4. What is a socio-technical system view and multi-level perspective?
5. How can the public sector support digital innovation?
6. What key transitions can be distinguished in digital innovation in UK construction?

Infrastructure megaprojects as enablers of digital innovation transitions

17.7 Case study: Tideway's digital evolution: how digitalisation drives productivity

One of the current megaprojects considered in this chapter, Thames Tideway Tunnel, supplements London's existing Victorian sewage network. Running 25 kilometres through central and greater London, 24 construction sites exist across the capital spanning from Acton in the west to Stratford in the east. The tunnel will enhance the lives of many Londoners, as well as the natural ecosystems of the Thames itself, by ensuring sewage is not pumped directly into the river. Like earlier infrastructure megaprojects, it played a significant role in the development of institutional policies and standards in the UK construction industry. Further, many of its key actors had gained experience at Crossrail and other preceding megaprojects.

What is distinct about Thames Tideway is the rate at which digital change is occurring during its design and construction. In the five years since construction of Thames Tideway began, the potential of digital technologies to transform almost all aspects of professional and personal life has accelerated radically. It is no surprise then that the approach to digital technologies in Tideway has also evolved. Not only has this enabled the contractors to meet their contractual obligations to the client and supply digital records of the built asset upon handover, it had also given rise to many other digital innovations. With the project's complex supply chain and its numerous stakeholders it had to liaise with, this coordination and communication was invaluable. For instance, 3D models were being used extensively to improve the quality and quantity of communication and coordination with a diverse set of stakeholders. Regular design reviews used a walk-through of the model and enabled the team to visualise new issues that demanded attention. Models enabled communication with local residents, who could easily see planned and completed work. 3D models made available for consultation by all contractors at five construction sites had been generated from real time data. Increasingly models had been developed to include 4D elements (incorporating time into 3D) to help with challenges around logistics and construction sequencing.

As well as modelling, numerous other digital innovations were being used to help management and to solve engineering challenges. For example, drones were being used to survey the existing sewers, thus saving staff from working in an extremely unsafe and unknown environment. These types of digital innovations were likely to become increasingly important as the project entered its peak delivery phase. Much had also been achieved through relatively small interventions and investments: for example, the falling cost of mobile technology meant that this could be made available to a larger number of on-site staff. The recent introduction of Flowforma was already being used to digitise the "starters and leavers" process across the joint venture of Costain, VINCI, and Bachy Soletanche (CVB), thus enabling a previously problematic business process to be performed more accurately and quickly.

However, although there was good evidence that digital technologies were having numerous benefits across the project, the challenge remained of how to prove the value of digital innovation was becoming more pressing. With an increasing number of digital technologies available, how could their value be measured and managed?

17.7.1 Case discussion questions

1. How can digital innovations improve productivity across the Tideway Alliance?
2. How can we measure how digital technologies drive productivity gains?
3. Can you think of the advantages and disadvantages of doing so?
4. What measures would you take to encourage digital innovation in a megaproject?

References

Abernathy, W.J. and Clark, K.B. (1985). Innovation: Mapping the winds of creative destruction. *Research policy*, 14, pp. 3–22.

Becerik-Gerber, B. and Kensek, K. (2009). Building information modeling in architecture, engineering, and construction: Emerging research directions and trends. *Journal of Professional Issues in Engineering Education and Practice*, 136(3), pp. 139–147.

Bew, M. and Richards, M. (2008). BIM maturity model. Construct IT Autumn 2008 Members' Meeting. Brighton, UK.

Blomquist, T. and Packendorff, J. (1998). Learning from renewal projects: Content, context and embeddedness. Lundin, R.A. and Midler, C. (Eds.), *Projects as arenas for renewal and learning process*. Boston, MA; Dordrecht; London: Kluwer Academic Publishers.

Boland, Jr, R.J., Lyytinen, K., and Yoo, Y. (2007). Wakes of innovation in project networks: The case of digital 3-D representations in architecture, engineering, and construction. *Organisation Science*, 18, pp. 631–647.

Bresnen, M. and Marshall, N. (2001). Understanding the diffusion and application of new management ideas in construction. *Engineering Construction and Architectural Management*, 8(4), pp. 335–345.

Brookes, N., Sage, D., Dainty, A., Locatelli, G., and Whyte, J. (2017). An island of constancy in a sea of change: Rethinking project temporalities with long-term megaprojects. *International Journal of Project Management*, 35(7), pp. 1213–1224.

Burns, T.E. and Stalker, G.M. (1961). *The management of innovation*. London: Tavistock.

Chesbrough, H.W. (2011). Bringing open innovation to services. *MIT Sloan Management Review*, 52, p. 85. Available from: https://sloanreview.mit.edu/article/bringing-open-innovation-to-services/?gclid=E AIaIQobChMIlMeu0aCQ6gIVAcqyCh3HTwzhEAAYASAAEgKtePD_BwE

Dainty, A., Leiringer, R., Fernie, S., and Harty, C. (2017). BIM and the small construction firm: A critical perspective. *Building Research and Information*, 45(6), pp. 1–14.

Davies, A. (2014). Innovation and project management. In: Dodgson, M., Gann, D.M. and Phillips, N. (Eds.), *The Oxford handbook of innovation management*. Oxford, UK: Oxford University Press.

Davies, A. and Brady, T. (2000). Organisational capabilities and learning in complex product systems: Towards repeatable solutions. *Research Policy*, 29(7–8), pp. 931–953.

Davies, A., Gann, D., and Douglas, T. (2009). Innovation in megaprojects: Systems integration at London Heathrow Terminal 5. *California Management Review*, 51(2), pp. 101–125.

Davies, A., Macaulay, S., Debarro, T., and Thurston, M. (2014). Making innovation happen in a megaproject: London's crossrail suburban railway system. *Project Management Journal*, 45(6), pp. 25–37.

Davies, A. and Mackenzie, I. (2014). Project complexity and systems integration: Constructing the London 2012 Olympics and Paralympics Games. *International Journal of Project Management*, 32 (5), pp. 773–790.

Debarro, T., Macaulay, S., Davies, A., Wolstenholme, A., Gann, D., and Pelton, J. (2015). Mantra to method: Lessons from managing innovation on Crossrail, UK. *Proceedings of the Institution of Civil Engineers-Civil Engineering*, pp. 171–178.

Dodgson, M., Gann, D., Macaulay, S., and Davies, A. (2015). Innovation strategy in new transportation systems: The case of Crossrail. *Transportation Research Part A: Policy and Practice*, 77, pp. 261–275.

East, B. and Smith, D. (2016). The United States National Building Information Modeling Standard: The First Decade. *33rd CIB W78 Information Technology for Construction Conference (CIB W78 2016)*. Brisbane, Australia.

Eastman, C. (1999). *Building product models: Computer environments, supporting design and construction*. Boca Raton, FL: CRC Press.

Egan, J. (1998). *Rethinking construction: Report of the construction task force*. London, UK: HMSO.

Egyedi, T.M. and Sherif, M.H. (2008). Standards' dynamics through an innovation lens: Next generation ethernet networks. Innovations in NGN: Future Network and Services, 2008. K-INGN 2008. First ITU-T Kaleidoscope Academic Conference, IEEE, pp. 127–134.

Farmer, M. (2016). *Modernise or die: The framer review of the UK construction labour market*. London: Construction Leadership Council.

Farrell, J. and Saloner, G. (1992). Converters, compatibility, and the control of interfaces. *The Journal of Industrial Economics*, 40(1), pp. 9–33.

Fernie, S. and Tennant, S. (2013). The non-adoption of supply chain management. *Construction Management and Economics*, 31(10), pp. 1038–1058.

Flyvbjerg, B. (2014). What you should know about megaprojects and why: An overview. *Project Management Journal*, 45(2), pp. 6–19.

Funk, J.L. and Methe, D.T. (2001). Market-and committee-based mechanisms in the creation and diffusion of global industry standards: The case of mobile communication. *Research Policy*, 30(4), pp. 589–610.

Gao, P., Yu, J., and Lyytinen, K. (2014). Government in standardization in the catching-up context: Case of China's mobile system. *Telecommunications Policy*, 38(2), pp. 200–209.

Gccg (2011). Government construction client group: BIM working party strategy paper. Available from: https://www.cdbb.cam.ac.uk/system/files/documents/BISBIMstrategyReport.pdf [cited 8 May 2020].

Geels, F.W. (2004). From sectoral systems of innovation to socio-technical systems: Insights about dynamics and change from sociology and institutional theory. *Research Policy*, 33(6–7), pp. 897–920.

Geels, F.W., Kern, F., Fuchs, G., Hinderer, N., Kungl, G., Mylan, J., Neukirch, M., and Wassermann, S. (2016). The enactment of socio-technical transition pathways: A reformulated typology and a comparative multi-level analysis of the German and UK low-carbon electricity transitions (1990–2014). *Research Policy*, 45(4), pp. 896–913.

Geels, F.W., Sovacool, B.K., Schwanen, T., and Sorrell, S. (2017). Sociotechnical transitions for deep decarbonisation. *Science*, 357(6357), pp. 1242–1244.

Giddens, A. (1984). *The constitution of society: An outline of the theory of structuration*. Cambridge, MA: Polity Press.

Gil, N.A. and Pinto, J. (2016) Collective action at the complex systems project front-end: Governance and performance implications. *Academy of Management Proceedings. Academy of Management*, p. 17194.

Grabher, G. and Thiel, J. (2015). Projects, people, professions: Trajectories of learning through a mega-event (the London 2012 case). *Geoforum*, 65, pp. 328–337.

Green, S.D. (2011). *Making sense of construction improvement*. Oxford, UK: John Wiley and Sons.

Group, I.C. (2017). From transactions to enterprises: A new approach to delivering high performing infrastructure. Institution of Civil Engineers (ICE). Available from: https://www.ice.org.uk/ICEDevelopmentWebPortal/media/Disciplines-Resources/Briefing%20Sheet/from-transactions-to-enterprises.pdf [cited 10 May 2020].

Harty, C. (2005). Innovation in construction: A sociology of technology approach. *Building Research and Information*, 33(6), pp. 512–522.

Harty, C. and Whyte, J. (2010). Emerging hybrid practices in construction design work: Role of mixed media. *Journal of Construction Engineering and Management*, 136(4), pp. 468–476.

HMG. (2015). *Digital Built Britain, Level 3 BIM Strategic Plan*. HM Government. Available from: https://www.cdbb.cam.ac.uk/news/2015DBBStrategy [cited 15 May 2020].

Hobday, M. (1998). Product complexity, innovation and industrial organisation. *Research Policy*, 26 (6), pp. 689–710.

Hobday, M. (2000). The project-based organisation: An ideal form for managing complex products and systems? *Research Policy*, 29(7-8), pp. 871–893.

Holm, P. (1995). The dynamics of institutionalisation: Transformation processes in Norwegian fisheries. *Administrative Science Quarterly*, 40(3), pp. 398–422.

Hs2. (2019). BIM Upskilling Platform. Available from: https://www.bimupskilling.com [cited 20 May 2020].

Koskinen, K.U. (2008). Boundary brokering as a promoting factor in competence sharing in a project work context. *International Journal of Project Organisation and Management*, 1(1), pp. 119–132.

Laakso, M. and Kiviniemi, A. (2012). The IFC standard: A review of history, development, and standardisation. *Journal of Information Technology in Construction*, 17, pp. 134–161.

Langley, A. (1999). Strategies for theorising from process data. *Academy of Management Review*, 24(4), pp. 691–710.

Latham, S.M. (1994). *Constructing the team*. HM Stationery Office London. Available from: https://constructingexcellence.org.uk/wp-content/uploads/2014/10/Constructing-the-team-The-Latham-Report.pdf [cited 19 May 2020].

Leonard-Barton, D. (1990). A dual methodology for case studies: Synergistic use of a longitudinal single site with replicated multiple sites. *Organisation Science*, 1(3), pp. 248–266.

Levina, N. and Vaast, E. (2005). The emergence of boundary spanning competence in practice: Implications for implementation and use of information systems. *MIS Quarterly*, 29(2), pp. 335–363.

Lobo, S. and Whyte, J. (2017). Aligning and reconciling: Building project capabilities for digital delivery. *Research Policy*, 46(1), pp. 93–107.

Lundin, R.A. and Söderholm, A. (1995). A theory of the temporary organisation. *Scandinavian Journal of Management*, 11(4), pp. 437–455.

Morgan, B. (2017). Organising for digitisation in firms: A multiple level perspective. In: Chan, P.W. and Neilson, C.J. (Eds.), *Proceedings of the 33RD Annual Association of Researchers in Construction Management Conference (ARCOM 2017)*, 4–6 September 2017 Cambridge, UK: Association of Researchers in Construction Management.

Morgan, B. (2019). Organising for digitalisation through mutual constitution: The case of a design firm. *Construction Management and Economics*. 37(7), pp. 400–417.

Morris, P.W.G. (2004). Project management in the construction industry. In: Morris, P.W.G. and Pinto, J.K. (Eds.), *The Wiley guide to managing projects*. Hoboken, NJ: John Wiley and Sons.

Munsi, A. (2012). *Crossrail and Bentley Systems launch UK's first dedicated Building Information Modelling Academy*. Available from: http://www.crossrail.co.uk/news/articles/crossrail-bentley-systems-launch-uks-first-dedicated-building-information-modelling-academy# [cited 14 February 2018].

Nambisan, S., Lyytinen, K., Majchrzak, A., and Song, M. (2017). Digital innovation management: Reinventing innovation management research in a digital world. *MIS Quarterly*, 41(1), pp. 223–238.

Narayanan, V.K. and Chen, T. (2012). Research on technology standards: Accomplishment and challenges. *Research Policy*, 41(8), pp. 1375–1406.

Office, C. (2011). Government Construction Strategy. HM Government. Available from: https://www.gov.uk/government/uploads/system/uploads/attachment_data/file/61152/Government-Construction-Strategy_0.pdf [cited 20 May 2020].

Office, C. (2016). *Government construction strategy: 2016 - 2020*. London, UK: Cabinet Office and Infrastructure and Projects Authority Available from: https://assets.publishing.service.gov.uk/government/uploads/system/uploads/attachment_data/file/510354/Government_Construction_Strategy_2016-20.pdf [cited 20 May 2020].

Papadonikolaki, E. (2017). Unravelling project ecologies of innovation: A review of BIM policy and diffusion. *IRNOP (International Research Network on Organizing by Projects)*. Boston, MA: IRNOP (International Research Network on Organizing by Projects).

Pavitt, K. (1984). Sectoral patterns of technical change: Towards a taxonomy and a theory. *Research policy*, 13(6), pp. 343–373.

Petticrew, M. and Roberts, H. (2008). *Systematic reviews in the social sciences: A practical guide*. Oxford, UK: John Wiley and Sons.

Pettigrew, A.M. (1990). Longitudinal field research on change: Theory and practice. *Organization Science*, 1(3), pp. 267–292.

Pöttler, R. (1992). Three-dimensional modelling of junctions at the channel tunnel project. *International Journal for Numerical and Analytical Methods in Geomechanics*, 16(9), pp. 683–695.

Rogers, E.M., Medina, U.E., Rivera, M.A., and Wiley, C.J. (2005). Complex adaptive systems and the diffusion of innovations. *The Innovation Journal: The Public Sector Innovation Journal*, 10(3), pp. 1–26.

Sanderson, J. (2012). Risk, uncertainty and governance in megaprojects: A critical discussion of alternative explanations. *International Journal of Project Management*, 30(4), pp. 432–443.

Shenhar, A.J. and Dvir, D. (2007). Reinventing project management: The diamond approach to successful growth and innovation, Harvard Business Review Press. Available from: https://www.reinventingprojectmanagement.com/material/other/030_HBS.pdf [cited 20 May 2020].

Söderlund, J. (2004). Building theories of project management: Past research, questions for the future. *International Journal of Project management*, 22(3), pp. 183–191.

Swan, J. and Scarbrough, H. (2005). The politics of networked innovation. *Human relations*, 58(7), pp. 913–943.

Sydow, J., Lindkvist, L., and Defillippi, R. (2004). Project-based organisations, embeddedness and repositories of knowledge. *Organisation Studies*, 25(9), pp. 1467–1489.

Wheelwright, S.C. and Clark, K.B. (1992). *Revolutionising product development: Quantum leaps in speed, efficiency, and quality*. New York: The Free Press.

Whyte, J., Bouchlaghem, N., Thorpe, A., and Mccaffer, R. (2000). From CAD to virtual reality: Modelling approaches, data exchange and interactive 3D building design tools. *Automation in Construction*, 10(43–55), pp. 43–55.

Whyte, J. and Levitt, R. (2011). Information management and the management of projects. *The Oxford Handbook of Project Management*. Oxford, UK: Oxford University Press.

Whyte, J. and Lobo, S. (2010). Coordination and control in project-based work: Digital objects and infrastructures for delivery. *Construction Management and Economics*, 28(6), pp. 557–567.

Wolstenholme, A., Austin, S.A., Bairstow, M., Blumenthal, A., Lorimer, J., Mcguckin, S., Rhys Jones, S., Ward, D., Whysall, D., and Le Grand, Z. (2009). Never waste a good crisis: A review of progress since Rethinking Construction and thoughts for our future. Available from: https://pdfs.semanticscholar.org/3391/82688341f5a0da716883ce11bc42abaf5f95.pdf [cited 22 May 2020].

Xie, Z., Hall, J., Mccarthy, I.P., Skitmore, M., and Shen, L. (2016). Standardisation efforts: The relationship between knowledge dimensions, search processes and innovation outcomes. *Technovation*, 48, pp. 69–78.

Yin, R.K. (1984). *Case study research: Design and methods*. Beverly Hills, CA: Sage Publications.

INDEX

Note: *Italicized* page numbers refer to figures, **bold** page numbers refer to tables

25 de Abril Bridge (Lisbon, Portugal) 49
2030 Agenda 73

Aaltonen, K. 32
Abdel Aziz, A.M. 308
Abdul-Aziz, A.R. 357
Abu Bakar, A.H. 85
accountability: benchmarking 307; of regulators 307
Active Server Pages (ASP) 414
adaptability 49
Addis Ababa Action Agenda (AAAA) 82
address resolution protocol (ARP) 415
ADM Capital 193
affinity diagrams 140
Africa: infrastructure delivery in 303; infrastructure funding in 310; infrastructure investments in 6–7, 310; infrastructure spending in 303; resilient infrastructure developments in 57
Africa50 76
African Climate Resilient Infrastructure Summit 48
agency theory 326
Ahren, T. 88
AI (artificial intelligence) **408**, 417
AIIB (Asian Infrastructure Investment Bank) 78, 83
AIM (asset information model) 440
Al Maktoum International Airport 125
Al Sammarae 111
Al Tayer, H.E. Mattar 121
Ali, S.F. 76
Allenby, B. 368
alliances 208–209
Almahmoud, E.S. 138, 139
Al-Shaer, E. 416
American Society of Civil Engineers (ASCE) 227, 246
Amnesty International 65
An, M. 345

Anderson, K. 88
Andres, L. 176–177
Ansar, A 239
APM (Association for Project Management) 380
appraisal 179–181, 270–283; context description 270; economic analysis 276–278; environmental sustainability 271; financial analysis 272; financial profitability 272; financial sustainability **275**, 275–276; in Germany 192; project identification 271; project objectives 270; return on investment 273, **274**; return on national capital 273–275, **274**; risk assessment 278–283; in Sweden 192; technical feasibility 271; through cost-benefit analysis 270–283; in United States 192–193
Aramco 402, 403, 410
Arequipa, Peru, resilient infrastructure developments in 56
Arnold, John E. 369
ARP (address resolution protocol) 415
artificial intelligence (AI) **408**, 417
Arup 12, 202
Arutyunov, R. 417
ASCE (American Society of Civil Engineers) 227, 246
ASDSO (Association of State Dam Safety Officials) 58
Ashurst, C. 28
Asia: infrastructure investments in 6; resilient infrastructure developments in 56
Asian Development Bank 48
Asian Infrastructure Investment Bank (AIIB) 78, 83
Asoka, G.W. 57
ASP (Active Server Pages) 414
asset based community development 139
asset information model (AIM) 440
asset management: model 229–230; technologies 226
Association for Project Management (APM) 380

451

Index

Association of Project Management (APM) Benefits Management SIG 30
Association of State Dam Safety Officials (ASDSO) 58
augmented reality (AR) 441
Australia: cybersecurity regulatory framework in **420**, 421; Infrastructure Australia Act 2008 82; infrastructure systems in 247; sustainable infrastructure in 82
Avanti 438–439, 439, 445
Azimio Housing Estate Limited (AHEL) 355

Bachy Soletanche 447
bad regulations 301–302
Badam, R. 108
Bagamoyo Port project (Tanzania) 313–316; bilateral agreements 314; China factor 314; contractual issues 315; overview 313–314; political risks 316; procurement procedures 314–315
Bahadorestani, A. 140
Baker, E. 140
Baker, R. 85
Baku-Tbilisi-Ceyhan pipeline cyberattack 402, 423–424
Balfour Beatty PLC 86
Baloi, O. 82
bank lending **259**
Barito Pacific Group 193
Barney, J.B. 28
Bartle, J.R. 3–4
BCRs (benefit-cost ratios) 192
behavioural theory **84**
Belt and Road Initiative (BRI) 77, 83
benchmark indicative asset cost (BIAC) 90
benchmark indicative non-asset cost (BINAC) 90
benchmarking 83–84, 88–89; accountability 307; appropriate institutional characteristics 309; clarity of roles 308; completeness and clarity in rules 308; independence 307; integrity 309; predictability 308; principles 307–309; proportionality 309; public participation 307–308; regulatory governance key standards 306–310; requisite powers 309; sustainability 88–89; sustainable infrastructure 89–91; transparency 307–308
benefit-cost ratios (BCRs) 192
benefits 21
benefits realisation: capabilities 28–29; competencies 28; realisation 28–29; resource-based 28; and stakeholder perceptions 30–34
Bengesi, K.M.K. 314
Benin 8
Benjamin, C.E. 257
Bennon, M. 61
Berssaneti, F.T. 254
Bertalanffy, Ludwig von 372, 376

best practices 302
best-cost country sourcing **234**
Bhanot, N. 89
Bhattacharya, A 74, 91, 92, 93
BIAC (benchmark indicative asset cost) 90
BIG (Bjarke Ingels Group) 111
big data **406**
big rooms 157
Biller, D. 176–177
BIM Academy 441
BIM Task Group 440
bin Rashid, H,H, Sheikh Mohammed 121
BINAC (benchmark indicative non-asset cost) 90
Biorck, J. 9
bitcoin 417
Bivens, J. 11
Bjarke Ingels Group (BIG) 111
Blinder, A. 11
blockchain **407**, 417
Blockley, D. 380–390
Bloomberg New Energy Finance (BNEF) 76
BNP Paribas 193
bond insurance **259**
bonds, project 289–291; consent and intercreditor issues 289–291; construction risk 291; credit rating requirements 289; operating period risk 291; principal stages *290*; regulatory requirements 289
Boom, W. 104, 107
BOOT (build, own, operate, transfer) model 314
Bordat, C. 189
Borrmann, A 210–211
Bourdeau, I. 113
Bourke, K. 198
Bourne, L. 32
Bourne, R. 11
Boyd, R. 201, 202
Bratislava Ring Road 33
Brazil: growth rate 10; infrastructure financing in 292–293
BRI (Belt and Road Initiative) 77, 83
Bristol Is Open 396
Bristol Network (B-NET) 396
Bristol Operations Centre 396
Bristol SMART city (UK) 394–399; background 395; design principles 398–399; essential pillars *395*; explore, enable and lead 398; goals of 396; innovation ecosystem in 397; One City Plan 395–396; overview 394–395; public service innovation in 398; responsible innovation in 397–398; smart technologies and sustainability in 396; world-class connectivity 396–397
British Airways 30
British Petroleum (BP) 403, 417
British Standard 198

Index

Brown, A.C. 298, 299, 307, 308, 309
Brunaeu, M. 50
Brundtland Commission 73
Buchanan, R. 369
build, own, operate, transfer (BOOT) model 314
building information modelling (BIM) 9, 210–214, 434; confidentiality 212; design responsibility 212; Dubai Expo 2020 116; intellectual property rights 212; regularisation manifesto *213*; requirements 212; security 212; standard of care 212; transition to 439, 440, 442; *see also* life-cycle performance
BuildingSMART 438
bundling **234**
Burlington Resources 403
business community 138–139
business mindset 86
Butler, D. 59

Cambridge Systematics 244
Cameroon, China's mega-infrastructure projects in **311**
Canada 8; cybersecurity regulatory framework in **420**, 422; infrastructure systems in 247; sustainable infrastructure in 81–82
Cantarelli, C.C. 325
capabilities 28–29
capital expenditure 201
Caracol Industrial Park (Haiti) 61
carbon emissions 13, 108–109
carbon pricing 13
Carnegie Mellon Software Engineering Institute 419
Carvalho, M.M. 254
case studies: Bagamoyo Port project (Tanzania) 313–316; Dege Eco Village (Tanzania) 355–357; E39 Coastal Highway Route (Norway) 217–218; Eko Atlantic city project (Lagos) 94–96; Haitian 2010 earthquake 65; High Speed 2 (United Kingdom) 33–41; infrastructure financing in Brazil 292–293; Standard Gauge Railway (Kenya) 169–170; Tropical Landscapes Finance Facility 193–194
cash flow, projections 269
Castro, V. 223
CAV (connected and autonomous vehicles) 59
Cavana, R. 380
CDE (common data environment) 440
Center for Strategic and International Studies (CSIS) 415
Cerro Verde 56
CERT (Computer Emergency Response Team) 422
certification event sustainability (CES) 110
CES (certification event sustainability) 110

Chad, China's mega-infrastructure projects in **311**
chain of benefits model 24, *25*
Chan, C. 208, 254, 255, 302
Channel Tunnel Rail Link 436
Chen, C. 3–4
Chester, M.V. 368
chief information officers (CIOs) 419–420
Chileshe, N. 301, 303, 308, 310, 315, 350
China: Belt and Road Initiative (BRI) 83; growth rate 10; infrastructure investments in Africa 310, **311**
China Merchants Holdings International Company 314
China National Petroleum 403
China Road and Bridges Company 169
CIM (civil information modelling) 211
circular economy 201–203
cities, resilient infrastructure in 55–57
citizen deliberative councils 139
Civil Engineering Environmental Quality Assessment and Awards (CEEQUAL) 61
civil information modelling (CIM) 211
Cleland, D.I. 31
Clift, M. 198
climate change 12
climate-resilient infrastructure 12, 52–55; adaptation measures 53; defined 53; designing 54–55; planning 54–55
closure of projects 190–191
Coates, D. 109
collaboration skills 86
collaborations, multi-partner 155–168; aligned efforts in **160**, 161; benefits of 165; case study 169–170; challenges/barriers to 161–166; cohesion in **160**, 161; co-locations 157; communication in 160, **160**, 164–165; communication standards/tools 167–168; coordination in **160**, 161; efficiency in 166; innovations in 165–166; knowledge integration capacity 158–159; management approaches 161; mediators 157–158; mutual support in **160**, 160–161; and organisational culture 161–162; overview 155; planning for 156–157; project collaboration quality 159–160; relational challenges in 162–164; stakeholder integration model 156–157; standardisation of best practices in 166–168
collaborative contracting 203–204; delivery partner model **206**; engineering, procurement and construction management (EPCM) **206**; integrated project delivery model **205**; managing contractor model **205**; partnering model **205**; *see also* life-cycle performance
collaborative procurement 203–207
Collinge, W. 140

Index

co-locations 157
command and control (C and C) servers 414
commercial mindset 86
common data environment (CDE) 440
communication: in multi-partner collaborations **160**, 160, 164–165; standards/tools 167–168
Compact With Africa ecosystem 9
competencies 28
competing values framework (CVF) 331
competitive alliance 209
completion 190
Computer Emergency Response Team (CERT) 422
computer network 404
Congo, China's mega-infrastructure projects in **311**
Congo-DRC 78
connected and autonomous vehicles (CAV) 59
ConocoPhillips 403
construction: mega-events 107; project life cycle *366*; sustainable 107, 115–117
Construction Excellence 203–204
Construction Leadership Council 441
Construction Research and Innovation Strategy Panel 198
construction risk **257**; project bonds 291
construction sector, Covid-19's impact on 9–10
consumption, versus satisfaction 22–23
contracting, collaborative 203–204; delivery partner model **206**; engineering, procurement and construction management (EPCM) **206**; integrated project delivery model **205**; managing contractor model **205**; partnering model **205**
contractors, early involvement 207–210
contracts 310–311; Engineering and Construction Contract (ECC). 327; international commitments 310–311; New Engineering Contract (NEC) 327; public-private partnerships (PPPs) 346, 349–350; relational 326–329; time and material (T and M) 326–329
contractual completeness 326
contractual governance 325
contractual solidarity, in relational contracting 328
conversion determinants 277
Cookham Wood Prison 442
Cornelissen, S. 109
Corporate Administrative Support Services (CASS) 123
corporate social responsibility (CSR) 140
Corporate Technology Support Services (CTSS) 123
corrective maintenance 224
cost estimates 271
cost overruns 188–190
Costain 447
cost-benefit analysis (CBA) 87, 177–179, *178*; general principles 268–270; incremental approach 268–270; infrastructure project appraisal 270–283; macroeconomic approach 268

Cote D'Ivoire 8
Courtice, P. 86, 87
COVID-19 9–10, 48, 72
Crawford slip method 140
credit ratings: project bonds 289; use of **259**
critical infrastructure 4
critical path analysis 376
critical success factors 305–306, 329–330
cross-border public-private partnerships 285–288
Cross-Ministerial Strategic Innovation Promotion Program 226
Crossrail 54, 74, 436, 440–441, 442, 444
cryptocurrency 417
Cui. C. 306
CURED Framework 333
CVF (competing values framework) 331
cyberattacks 401–402, 413–416
cybersecurity 416–422; capability 419; culture 418; environmental determinants of 420–422; maturity models 419; organisational determinants of 418; overview 402; regulatory frameworks **420–421**; responsibilities and resources 419–420; technological determinants of 416–417
Cybersecurity Capability Maturity Model (C2M2) 419
cybersecurity risks: case study 423–424; components of 410, *411*; defined 410; external agent attacks 410; financial gain threats 410; insider attacks 410; internal agent attacks 410; malicious threats 410; in oil and gas infrastructure systems 412–416; in oil and gas sector 404–412, 404–416; unintentional threats 410
cyclones 47

Da Veiga, A. 418
Dabla-Norris, E. 267
Dappe, M.H. 176–177
data envelopment analysis (DEA) 88–89
Davies, A. 439
Davis, K. 34
DC (degree centrality) 145
DEA (data envelopment analysis) 88–89
Debela, G.Y. 352
debt sustainability assessments (DSA) 263
debt-investment-growth (DIG) 263
deep-water container terminal (DCT) (Gdansk, Poland) 53
Dege Eco Village (Tanzania) 355–357; contractual issues 357; financial risk 357; political risk 357; preparation 356; regulatory issues 356
degree centrality (DC) 145
delays 188–190
deliberative democracy 139

Index

delivery of infrastructure projects 175–191; appraisal of situation and government needs 179–181; case studies 192–194; cost-benefit analysis 177–179, *178*; delays and cost overruns 188–190; finalising plans 186–188; governance 185; institutional capabilities 185; moving from planning to 188; overview 175; perceptions of benefits and loss 182; prioritisation 176–177; process 185; public involvement 185–186; risk management 186; vision and goals for future assets 181–184

delivery partner model **206**; *see also* collaborative contracting

demand management **234**

demand risk **257**

Demirkesen, S. 156

Deng, Y. 104, 107

denial-of-service (DoS) attacks 415, 417

Design Council 371

design principles: for national infrastructure **367**; top ten **365**

design thinking 363–394; Double Diamond process 371; IDEO model 370, *370*; models *368*, **369**, 369–371, *370*; overview 364; principles **365**; stages of *368*; and systems thinking 371–380

design-build (DB) contracts 208

Deulgaonkar, P. 111

developed countries: sustainable infrastructure governance in 262–266; sustainable infrastructure in 266–267

developing countries: public-private partnerships 297; sustainable infrastructure governance in 262–266; sustainable infrastructure in 266–267

Dewar, J. 289

Dey, P.K. 88

Di Maddaloni, F. 34

DIG (debt-investment-growth) 263

digital collaboration tools 9

digital innovation 431–446; case study 447; framing 436–437; knowledge gap 435; managerial implications 445–446; mapping multi-level transitions 442; in megaprojects 444; overview 431–432; re-alignment 441–442; reconfiguration 440–441; as a socio-technical phenomenon 433; substitution 437–438; theory and knowledge 445; timeline of *443*; transformation 438–440; transition pathways **437**, 444–445; transitions from innovation to 433–434

digital oilfield **408**

District 2020 114–115

documentation 190

Dodouras, S. 111, 112

Doherty, N.F. 28

Doloi, H. 137, 138, 139, 142, 145

domain name server (DNS) session hijacking 414

Donaldson, D. 77

Dooms, M. 32

Double Diamond process 371; *see also* design thinking

drones 441

DSA (debt sustainability assessments) 263

Duan, Q. 416

Dubai Airport 120, 125

Dubai Exhibition Centre (DEC) 127

Dubai International 125

Dubai Metro 108

Dubai Ports World 120, 124

Dubai South 108

Dubai Strategic Vision 2015 108–109, 115

Dubai Supreme Council of Energy 120

Dubai Taxi Corporation (DTC) 123

Dubai Water and Electricity Authority (DEWA) 117–119, 129

Dubai Water Canal 121

Dubai World Central (DWC) 125

E39 Coastal Highway Route (Norway) 217–218

early contractor involvement (ECI) 207–210; availability of right resources 209–210; competitive alliance 209; cost and time versus overall value 209; expertise of resources 210; loss of competitive tension 210; management of 210; procurement process 209; pure alliance 208; *see also* life-cycle performance

earthquakes **60**

East Africa, infrastructure funding in 310

East African Community (EAC) 310

ECC (Engineering and Construction Contract) 327

eco-friendly 111

ecology 111

economic analysis: assessment of non-market impacts and corrections for externalities 277; conversion from market to shadow prices 277; evaluation of greenhouse gas emissions 277–279; fiscal corrections 276; infrastructure project appraisal 276–278; utilisation of conversion determinants to project inputs 277

economic and financial sustainability **75**

Economic Community of West African States (ECOWAS) 184

economic growth 10

economic infrastructure systems 4; benefits of 4–6; classification of *5*; environmental issues 6

economic rate of return (ERR) 268

economics of repetition 433

Edwards, P.N. 184

Egan, J. 85, 438

Egan Report 438

Egypt 8

EIAs (environmental impact assessments) 87

eigenvector centrality (EV) 145

Eko Atlantic city project (Lagos) 94–96

Index

El-Gohary, N.M. 345
Elkington, J. 33, 112, 135
Ellen McArthur Foundation 201
Elsayegh, A. 315
emerging economies, sustainable infrastructure governance in 262–266
Emirates 125
Emirates Central Cooling System Corporation (EMPOWER) 118
Endeavor Logistics 403
end-user community 139
Energy Transitions Commission (ETC) 76
engineering, procurement and construction management (EPCM) **206**; *see also* collaborative contracting
Engineering and Construction Contract (ECC) 327
ENPV (expected net present value) 278–279, 281, *282*
enterprise system 388, *389*
environmental, social and corporate governance (ESG) 84
environmental and social impact assessments (ESIAs) 87
environmental impact assessments (EIAs) 87
environmental sustainability **75**; infrastructure project appraisal 271
environmentally friendly 111
Envision 83
Equator Principles and World Bank EHS Guidelines 61
ERR (economic rate of return) 268
ESG (environmental, social and corporate governance) 84
ESIAs (environmental and social impact assessments) 87
ETC (Energy Transitions Commission) 76
Ethiopia: China's mega-infrastructure projects in **311**; infrastructure investments in 8
Etihad ESCO 118
Eti-Osa Lekki Toll Road Concession Project (Nigeria) 188
Europe, resilient infrastructure development s in 57
European Bank 253
European Bank for Reconstruction and Development (EBRD) 53
European Union 8
European Union (EU) 184
EV (eigenvector centrality) 145
event life cycle (ELC) 130
Exim Bank of China 169
expected net present value (ENPV) 278–279, 281, *282*
Expo 2015 (Milan) 110–111
Expo 2020 (Dubai) 108, 111, 114–115, 114–130; accessibility and indoor environmental quality 116; ecology and planning 116; impacts of 125–126; leadership emphasis on sustainability 130; leadership engagement on environmental

impact 128–129; legacy planning 130; management 116; multi-stakeholder engagement 128; planning 126–127; pollution management 116; recommendations 126–130; sustainable construction 115–117; sustainable design and planning 129; sustainable mobility 120–125; sustainable utilities 117–120; waste management 117
externalities 277
ExxonMobil 403

Fairholm, M.R. 85
Fay, M. 8
Fergusson, W. 327
financial analysis, infrastructure project appraisal 272
financial mindset 86
financial net present value of capital (FNPV(K)) 275
financial net present value of investment (FNPV(C)) 273
financial profitability, infrastructure project appraisal 272
financial rate of return of the investment (FRR(C)) 273
financial risk 357
financial sustainability, infrastructure project appraisal **275**, 275–276
financing: alternative 291; assessment options 283–291; cross-border public-private partnerships 285–288; defined 291; innovative 291; new financial arrangements 291; new funding sources 291; new mechanisms 291; private activity bonds 284; project bonds 289–291; public-private partnerships 284–285, *286*; regional infrastructure funds 288–289; risk **257**
financing vehicles 255; market exposure 255; risk management 255; transaction costs 255
firefighting 20
firewalls 415, 416
fiscal policy 10–12
5D simulation 9
Fletcher, P. 255
flexibility 49; in relational contracting 328
Flowforma 447
Fly Dubai 125
Flyvbjerg, B. 20, 325
force majeure clause 9
foreshore development project 148–152
fossil fuel subsidies 13
Foster and Partners 111
4D simulation 9–10
Fourth Industrial Revolution 396
France, infrastructure systems in 247
Frangopol, D.M. 222
Freeman, R.E. 29
Fricke, E. 49
function infrastructure 4

Index

funding 291

G7 Ise-Shima Principles for Promoting Quality Infrastructure Investment 83–84
Galderisi, A. 57
Garvey, P.B. 388
GCC (Gulf Cooperation Council) 7, 116
GDP (gross domestic product) 8, 238
Geels, F.W. 444
General Service Administration (GSA) 211
general systems theory 372, 376
Gerbert, P. 351, 353
Germany: infrastructure systems in 247; project appraisals in 192
Gesner, G.A. 49
Getz, D. 103
Ghana 8, 306
GHGs (greenhouse gases) 76, 239–241, 277–279
GIB (Global Infrastructure Basel) 76
Gibson, G. 239, 241
GICA (Global Infrastructure Connectivity Alliance) 75
Giddens, Anthony 433
GII (Global Infrastructure Initiative) 323, 327
Gilrein, E.J. 49
Githaiga, N.M. 298, 308
Global Infrastructure Basel (GIB) 76
Global Infrastructure Connectivity Alliance (GICA) 75
Global Infrastructure Hub 8, 75
Global Infrastructure Initiative (GII) 323, 327
Global State of Information Security Survey 2018 419
Godfrey, P. 380–390
good governance 302
good regulations 300–301
Goodpasture, J.C. 26
governance: benchmarking 306–310; contractual 325; in delivery of infrastructure projects 185; good 302; regulatory 299, 306–310; relational 325; of sustainable infrastructure 262–266; transparent and effective 185
governance community 139
government 179–181
Grand Inga Dam (Congo-DRC) 78
Green Alliance 202
green buildings 12
Greenhouse Gas (GHG) Protocol Accounting and Reporting Standard 61
greenhouse gases (GHGs) 76, 239–241, 277–279
Griffiths, S. 56
Grimsey, D. 352
Grimshaw Architects 111
gross domestic product (GDP) 8, 238
growth rate 10
GSA (General Service Administration) 211

Gubic, I. 82
Guinea 8
Gulf Cooperation Council (GCC) 7, 116
Gurara, D. 267

Haas, R. 242
hackers 414–415
Haiti Reconstruction Fund (HRF) 61
Haitian 2010 earthquake 65
Hallegatte, S. 52, 61
Hamel, G. 28
Hannevik, M.B. 331
hard systems thinking (HST) 376–377; *see also* design thinking; systems thinking
harmonization of conflict, in relational contracting 328
Harper, C.M. 327
Hartono, E. 28
Harty, C. 439–440
Harvey (hurricane) 47
Hasso Plattner Institute 369
Heathrow Terminal Five 30, 436, 439–440
heating, ventilation, and air conditions system (HVAC) 116
Hellmuth, Obata + Kassabaum (HOK) 111
Hellowell, M. 325
Henderson, S. 111
Hepburn, G. 258
HFA (Hyogo Framework for Action) 51–52
High Speed 2 (United Kingdom) 33–41, 436, 442; actual costs 36, *38*; available funding *39*; delivery of expected benefits 36–40; forecasted costs 36, *39*; key stakeholders 40–41, **41**; overview 33–34; proposed route *37*; timescale and cost overruns 36, **40**
High-level Political Forum on Sustainable Development (2016) 73
high-net-worth individuals (HNWIs) 127
highways **60**
Hiller, H. 106
HNWIs (high-net-worth individuals) 127
hold-to-maturity framework 266
Holling, C.S. 50
Hong Kong 32, 184
Hornbeck, R. 77
Hotchkiss, J. 109
HP Billiton 417
HST (hard systems thinking) 376–377; *see also* systems thinking
HTTPS (hypertext transfer protocols) 415, 416
Huemann, M. 33
Humphrey, B. 109
Humplick, E. 242
hurricanes 47, 58
Hyogo Framework for Action (HFA) 51–52
Hypertext Preprocessor (PHP) 414
hypertext transfer protocols (HTTPS) 415, 416

457

ICG (Infrastructure Client Group) 444
ICIF (International Centre for Infrastructure
  Futures) 89
IDA (International Development Association)
  61
Idai (cyclone) 47
IDB (Inter-American Development Bank) 65, 74, 76
ideation, design thinking 370
IDEO model 369, 370; *see also* design thinking
Ika, L.A 24
IMF (International Monetary Fund) 262
implementation: design thinking 371; timing 271
INCOSE (International Council on Systems
  Engineering) 380
independence, of regulators 307
Independent Power Producer (IPP) model 119
India: growth rate 10; public-private partnerships in
  301; resilient infrastructure developments in 56
industrial control systems (ICS) 396–397
industry community 138–139
influencing skills 86
information and communication technologies
  (ICT) 226
Infrascope analytical framework 299
Infrastructure and Projects Authority (IPA) 89–91,
  185, 323, 333, 440
Infrastructure Australia 82
Infrastructure Canada 81
Infrastructure Client Group (ICG) 444
infrastructure investments: efficiency considerations
  256; global demand for 253–254; long-term
  viewpoinrt 268; opportunity cost 268; risks **257**;
  by sector **283**
infrastructure project appraisal 270–283; context
  description 270; economic analysis 276–278;
  environmental sustainability 271; financial
  analysis 272; financial profitability 272;
  financial sustainability **275**, 275–276; project
  identification 271; project objectives 270; return
  on investment 273, **274**; return on national
  capital 273–275, **274**; risk assessment 278–283;
  technical feasibility 271
infrastructure projects: financing and development
  of 253–292; leadership 84–87; relationships in
  323–324; stakeholder engagement in 135–148;
  systems thinking on 372–376, 380–390
infrastructure spending: alignment with economic
  growth 238–239; Covid-19's impact on 9–10;
  global trends in 7–8; impact of fiscal/monetary
  policies on 10–12; measuring 239–245
infrastructure systems: critical 4; defined 3,
  3–4; economic 4; functional 4; global outlook
  6–7; non-critical 4; social 4; soft 4; strategic 4;
  sustainability 12–14; types of 4; in United States
  246–247
*Infrastructure to 2030 Mapping Policy for Electricity,
  Water and Transport* 75

innovation: defined 431; and megaprojects 432–
  433; and projects 433; research setting of 434–
  435; transitions to digital innovation 433–434
innovative infrastructure financing 291
inspiration, design thinking 371
Institute of Design 369
institutional sustainability **75**
integrated project delivery (IPD) **205**, 207; *see also*
  collaborative contracting
integrity, of regulators 309
intellectual property rights 212
Intelligent Traffic Solutions (ITS) 121
intended nationally determined contributions
  (INDCs) 76
Inter-American Development Bank (IDB) 65, 74, 76
Intergovernmental Panel on Climate Change
  (IPCC) 52
internal delivery, capability 336
International Alliance for Interoperability (IAI)
  438–439
International Centre for Infrastructure Futures
  (ICIF) 89
international commitments 310–311
International Committee of the Red Cross
  (ICRC) 47
International Conferences on Financing for
  Development 71
International Council on Systems Engineering
  (INCOSE) 380
International Development Association (IDA) 61
International Finance Corporation (IFC)
  Performance Standards 61
International Institute for Sustainable Development
  (IISD) 93
International Journal of Project Management
  24
International Monetary Fund (IMF) 262
International Organisation for Standardisation
  (ISO) 228; ISO 44001 331–332; ISO
  44002:2019[E] 328
Internet of Things (IoT) 10, 396, **407**
intrusion detection/prevention systems (IDS/IPS)
  415
investment financing methods: assessment
  options 283–291; cross-border public-private
  partnerships 285–288; innovative 291; private
  activity bonds 284; project bonds 289–291;
  public-private partnerships 284–285, *286*;
  regional infrastructure funds 288–289
investment risks **257**
IoT (Internet of Things) 10, 396, **407**
Iraldo, F. 111
Irma (hurricane) 47
ISO (International Organisation for
  Standardisation) 228; ISO 44001 331–332; ISO
  44002:2019[E] 328
ISO 44001 331–332

Index

ISO 44002:2019[E] 328
Ivory Coast, China's mega-infrastructure projects in **311**
Iyer, K. 412

Jamali, D. 315
James, P. 111, 112
Japan: cybersecurity regulatory framework in **421**, 422; infrastructure systems in 247; resilient infrastructure systems in 13
Jardim, J. 49
Javascript 414
Jebel Ali Port 120
Jebel Ali Power and Desalination Complex 119
Jiang, W. 302–305
Jomo, K.S. 354

Kali Linux 415
Kanter, R.M. 315
Karlaftis, M. 241
Kassim, P.S.J. 357
Kavishe, N. 299, 301, 307, 308, 315, 345, 346, 350
Keeys, L.A. 33
Kelley, David 369
Kenya: China's mega-infrastructure projects in **311**; Standard Gauge Railway project 169–170; sustainable infrastructure in 77
Kepapsoglou, K. 241
Khalfan, M.M.A. 86
Kilimanjaro Airports Development Company (KADCO) 347
Kilimanjaro International Airport 347
Kingombe, C. 300, 308
knowledge gap 435
knowledge integration 158–159
Kolkata 301
Kong, J. S. 222
KPMG 84
Kunc, M. 23, 24

LA (Licensing Agency) 123
land use: plan 107; planning assessments on 245; and urban pattern 245
landslide **60**
Langley, A. 436, 445
Latham, S.M. 85, 438
Latham Report 438
Latin America, resilient infrastructure developments in 56
Laursen, M. 20
LCCA (life-cycle cost analysis) 199–201, *200*
leader–member exchange (LMX) **84**
Leadership in Energy and Environmental Design (LEED) Gold Standard 108, 128
Lee, G. 210–211
Lee, R 423–424
Lekki Concession Company Limited 188

Lekki Toll Road (Nigeria) 188
Leung, M–Y 140
Lewis, M.K. 352
Li, J. 352
Li, T.H.Y. 32
Liao, H. 110
Licensing Agency (LA) 123
Liedtka, J. 369
life-cycle cost analysis (LCCA) 199–201, *200*
life-cycle performance 197–216; building information modelling 210–214; case study 217–218; circular economy principles 201–203; collaborative procurement models 203–207; early contractor involvement 207–210; life cycle cost analysis 199–201, *200*; overview 197; pre-construction phase 214–216; preliminary design phase 214–216; whole life costing 198–199
liquefaction **60**
Lisboa-Porto High Speed Line 33
listening circles 139
Liu, M. 222
Lobo, S. 434–435, 441
London Olympics 2012 436, 440–441, 442
Lu, W. 80
Luiz, J. 303

Ma, S. 110
Maanie, E. 380
MAC (media access control) address 415
Mackenzie, I. 439
maintenance *see* operations and maintenance
Mair, J. 110
major infrastructure projects *see* megaprojects
major programmes *see* megaprojects
major projects *see* megaprojects
make-or-buy **234**
Malawi 47
Malaysia, public-private partnerships in 301, 308
Maldives 12
Malhado, A. 106
Mali, China's mega-infrastructure projects in **311**
malware 414
managing contractor model **205**; *see also* collaborative contracting
man-in-the-middle (MITM) attacks 414–415, 416
Marathon Petroleum 403
Maria (hurricane) 47
market exposure 255
Martek, I. 325
Masdar City, United Arab Emirates, resilient infrastructure developments in 56
Mason, M. 109
Matchmaker 76
May, A. 106
Mboya, J.R. 345
McAdam, R. 88

Index

McBride, J. 246
McElroy, B. 32
McKinsey 419, 419–420
Medellín, Colombia, resilient infrastructure developments in 56
media access control (MAC) address 415
mediators, collaboration 157–158
medium term debt management strategy (MTDS) 263–266
mega-events 103–130, 104, 128–129; carbon emissions 108–109; case study 114–130; construction 107; design stage 110–111; green infrastructure development 108; green principles in awarding system 110; impacts of 125–126; leadership emphasis on sustainability 130; legacy planning 109–110, 130; multi-stakeholder engagement 128; overview 103–104; planning 126–127; recommendations 126–130; sustainability 111–112; sustainable design and planning 129; transportation 105–106; utilities 106
megaprojects 20, 31, 410; delivery of 175–191; design thinking principles in 363–394; and digital innovations 431–446; financing and development of 253–292; innovation footprint of 432–433; life-cycle performance 197–216; multi-partner collaborations 155–168; prioritisation of 176–177; relationship management in 321–334; resilient infrastructure 47–64; stakeholder engagement in 135–148; sustainable infrastructure 71–93
Meng, X. 329–330
Merriman, J. 314
Metrocable 56
Mexico 8; growth rate 10
Michelin 193
Middle East, resilient infrastructure developments in 56
Middle East and North Africa (MENA), infrastructure investments in 7
Milan, Italy, resilient infrastructure developments in 57
Mills, L.D. 32
mindset 86
MI-ROG 202–203
Mitchell, R.K. 31–32
MITM (man-in-the-middle) attacks 414–415, 416
Mittelstaedt, R.F. 88
Miyamoto 59
MLP (multi-level perspective) 435, 444, 445
Mohammed, H,H, Sheikh 108
Mohammed bin Rashid Al Maktoum Solar Park 118
monetary politcy 10–12
monolines 259
Monte Carlo simulation 281

Morocco 8
Mourgues, T. 308, 309
Mozambique 47; China's mega-infrastructure projects in 311
M/s Mutluhan Construction Industry Company Limited 355–356
multi-effect desalination process (MED) 119
multilateral development banks (MDBs) 75
multi-level perspective (MLP) 435, 444, 445
multi-partner collaborations 155–168; aligned efforts in 160, 161; benefits of 165; case study 169–170; challenges/barriers to 161–166; cohesion in 160, 161; co-locations 157; communication in 160, 160, 164–165; communication standards/tools 167–168; coordination in 160, 161; efficiency in 166; innovations in 165–166; knowledge integration capacity 158–159; management approaches 161; mediators 157–158; mutual support in 160, 160–161; and organisational culture 161–162; overview 155; planning for 156–157; project collaboration quality 159–160; relational challenges in 162–164; stakeholder integration model 156–157; standardisation of best practices in 166–168
Myanmar 77, 78
Myingyan, Myanmar 78

Nairobi, Kenya, resilient infrastructure developments in 57
Nakamura, H. 224, 227–228, 238
National Audit Office (NAO) 322
National Bank for Economic and Social Development (BNDES) 292–293
National Housing Corporation (NHC) 300, 301, 345, 346
National Infrastructure Assessment (NIA) 185
National Infrastructure Commission (NIC) 185
National Institute of Science and Technology (NIST) 419
National Planning Policy Framework (NPFF) 364
National Social Security Fund (NSSF) 300, 355
Natixis 417
neighbourhood community 139
net present value (NPV) 30, 201, 268, 278
Netherlands 12
New Climate Economy (NCE) 76
New Engineering Contract (NEC) 327
new financial arrangements 291
new financing mechanisms 291
new funding sources 291
Newtonian science 376
Nguyen, N.H. 32
Ng'wanakilala, F. 314
Nigeria: China's mega-infrastructure projects in 311; growth rate 10

Index

nominal group technique 140
non-critical ventures 4
non-value-adding processes 21
Nordic Investment Bank (NIB) 53
North/South Metro line (Netherlands) 33

Ochieng, E. 3, 73, 184, 229, 270
off-site construction 10
Ofori, G. 85
Oh, S. 416
oil and gas infrastructure systems: cybersecurity 416–422; cybersecurity risks in 412–416; operation streams *413*; points of entry for cyberattacks **414**
oil and gas sector 403–404, 413–414; cybersecurity risks in 404–416; digital technologies and applications in **406–408**; digital transformation in 404, *409*; disruption in supply and demand 403; downsizing 403; revenues 403; technological development trajectory *405*
Olander, S. 32
Olanrewaju, L.A. 199
Olympic Games 2012 110
Olympic Park 107
Onyenechere, E.C. 57
Open Data Platform 396
open dialogue 139
open forums 139
open space technology 139
operating period risk, project bonds 291
operational expenditures 199
operational risk **257**
operational technology (OT) 401
operations and maintenance 221–248; alignment of infrastructure spending with economic growth 238–239; asset management model 229–230; asset management technologies 226; asset utilisation 231, *232*; best practices 227–230; challenges 245; corrective maintenance 224; cost reduction 233; definition of 223; information and communication technologies 226; inspection, monitoring and diagnosis technologies 226; measurement of infrastructure spending 239–245; overview 221–222; plan-do-check-act (PDCA) 228–229; predictive maintenance 223; preventive maintenance 223; proactive maintenance 223; quality enhancement 231–238; reactive maintenance 224; research 225–226; robotics technologies 226; structural materials, deterioration mechanisms, repairs and reinforcement technologies: 226; Tokaido Shinkansen train 224–225
Opoku, A. 86, 198
Opportunity Pavilion 111
O'Reilly, N. 103
Øresund Crossing (Denmark) 33

Organisation for Economic Co-operation and Development (OECD) 54
organisational climate 330–331
organisational culture 161–162
organisational design and development (ODD) 333
originate-to-distribute framework 266
Oroville Dam 58
Osei-Kyei 302
outsourcing 233
overruns 188–190
Ozorhon, B. 156

packet filtering 416
Paggiaro, A. 109
Parida, A. 88
Paris Agreement 91, 403
parish mapping 139
participatory learning and action approaches (PLA) 139
partnering model **205**; *see also* collaborative contracting
PAS1192 standards 445
Paterson, W.D.O. 242
peer-to-peer (P2P) payment 417
Pelhan, E. 104, 111
Pendleton, A. 255
people, process and technology (PPT) model 210
performance, of major infrastructure projects 322–323
performance indicators: key determinants **244**; for transportation infrastructures 242–244
performance-tuning design 107
Petrie, M. 267
Pezza, D.A. 388
Pfeffer, J. 29
Pheng, L.S. 302–305
phishing 415–416
photo voice 139
PHP (Hypertext Preprocessor) 414
Pinto, C.A. 388
Pisu, M. 89
Pitts, A. 110
plan-do-check-act (PDCA) 228–229
Planing Act 2008 364
political risk 302–305, 316, 357
pollution management 117
Pontines, V. 77
Poon, S.W. 104
population growth 245
Port of Dar es Salaam 347
post-execution phase 190–191
Potomac Electric Power Company 58
power infrastructure, resilient 57–58
power-interest grid/matrix 32, 140

PPP fiscal risk assessment model (PFRAM 2.0) 263
Prahalad, C.K. 28
pre-construction phase 214–216
predictability, of regulators 308
predictive maintenance 223
preliminary design phase 214–216
Presidential Policy Directive 21 – Critical Infrastructure Security and Resilience 58
Preuss, H. 113
preventive maintenance 223
price distortions 13
Price Waterhouse Cooper (PwC) 401, 419
primary stakeholders 29, 32
principal-agent (PA) relationship 325
private activity bonds 284
proactive maintenance 223
probability analysis 281
process optimisation **234**
procurement 233, **234**, 314; collaborative 203–207
production plan 271
productivity, of major infrastructure projects 322–323
profitability 272
programmable logic controllers (PLCs) 401, 423
programme evaluation and review techniques (PERT) 376
Programme for Infrastructure Development 310
Project 13 444
project bonds 289–291; consent and intercreditor issues 289–291; construction risk 291; credit rating requirements 289; operating period risk 291; principal stages *290*; regulatory requirements 289
project identification: analytical framework 260–262, *261*; capacity 254–256; infrastructure project appraisal 271; market 259–260; risks 257–258; transaction costs 258–259
project initiation routemap (PIR) 333
project management: success 21; terminology 21
Project Preparation Facilities (PPF) 93
project review 191
project stakeholder typology model (PSTM) 140
project system closure 190
project value *23*; defined 21
project value management 19–34; case study 35–41; chain of benefits model 24, *25*; elements of 23–24; propositional elements 26–28, *27*; and public infrastructure projects 20
proportionality, of regulators 309
propriety of means, in relational contracting 328
Pryke, S.D. 324
public infrastructure projects 20
public investment efficiency 262
public investment management assessment (PIMA) 263, *264*
public participation 307–308

Public Procurement Act 2011 300
Public Transport Agency (PTA) 123
Publicly Available Specifications (PAS) 440
public-private partnerships (PPPs) 7, 188, 236, **259**, 284–285, 343–355, 352–354; agreements 350; appraisal 350; and bad regulations 301–302; case study 355–357; challenges 345–347; competition 347; configuration 350; contract preparation 349–350; contracts 346; contractual issues 357; cost of *287*; cross-border 285–288; in developing countries 297; feasibility study 351–352; financial capacity of private partners 347; financial risk 357; institutional capacity *354*; legal framework 347; mode *286*; political risk 357; preparation 356; preparing 344–345; priority project classification 349; process management 347–350, *348*; project management and monitoring 346; regulatory issues 356; risk allocation 352, **353**; skills and knowledge 346; in Tanzania 299, 345–347; tender documents 346
Puerto 215
pure alliance 208
Puttock, C. 289

qualitative risk analysis 280–281
quality enhancement 231–238; asset life 235–236; end-to-user experience 231; externalities 233; life cycle view 236; operational and maintenance cost reduction 233; smart technologies 233; sustainability plans 233; user-centric operating framework 231
Quigley 410
Qureshi, Z. 13, 81

Rail Agency (RA) 123
railways **60**
Rajasekar, U. 56
Ramachandran, G. 199, 201
Ray, S. 283
reactive maintenance 224
re-alignment: actors and social groups during 441; digital innovation **437**; rules and institutions during 441; socio-technical systems during 442
reciprocity, in relational contracting 328
reconfiguration: actors and social groups during 440; digital innovation **437**; rules and institutions during 440; socio-technical systems during 440–441
redesign **234**
Reed, James Edward 217
regional infrastructure funds 288–289
regulatory effectiveness: bad regulations 301–302; best practices 302; components of 299–300; good governance 302; good regulations 300–301; international commitments 310–311; overview 298–299; political risk 302–305;

Index

regulatory risk 302–305; standards, procedures and tools 309–310; in Tanzania 299, 300
regulatory governance 299; benchmarking 306–310
regulatory risk **257**, 302–305
regulatory substance 299
relational contracting 326–329
relational governance 325
relational indicators 329–330
relational risks 324–325
relationship management 321–334; case studies 335–338; critical success factors 329–330; frameworks for 331–334; and organisational climate 330–331; overview 323–324; plan (RMP) 332; relational contracting 326–329; relational governance 325; relational indicators 329–330; relational risks 324–325
remote terminal units (RTUs) 423
renewable energy 403
Rentschler, J. 57
repetitiveness 433
resilience 50; aspects of **50**; definitions of 50–51, **51**; global perspective on 51–52; key concepts 49; success-measuring tools 61
resilient infrastructure 47–64; adaptability 49; adaptation measures 53; advancing aspirations for *63*, 63–64; case study 65; cities and urban developments 55–57; climate-resilient infrastructure 52–55; defined 53; designing 54–55; and extreme weather events 47–48; flexibility 49; key concepts 49–50; planning 54–55; power 57–58; robustness 49; strategies for 55–61; tools for measuring success of 61; transportation 58–59, **60**; water 59–61, *62*
Resources and Waste Strategy: At a Glance 77
restraint of power, in relational contracting 328
Rethinking Timber Buildings 80
return on investment 273, **274**
return on national capital 273–275, **274**
reverse osmosis desalination (RO) 119
Rick, K. 412
rightsizing 233
Rio Earth Summit (1992) 71
risk: construction **257**; cybersecurity 404–412; defined 303; demand **257**; financial 357; financing **257**; investment **257**; operational **257**; political 302–305, 316, 357; regulatory **257**; relational 324–325; sovereign **257**; technological **257**; vulnerability *237*
risk allocation 352
risk assessment 276–278; mitigation and defence strategies 283; probability analysis 281; qualitative risk analysis 280–281; sensitive analysis 278–279
risk management 186, 255, 303–305; critical success factors 305–306; domestic infrastructure project *304*
Rittel, H, W.J. 363

Road and Transport Authority (RTA) 120–124
roads, secondary **60**
robotics technologies 226
robustness 49
Roche, M. 109
Rogers, E.M. 432
role integrity, in relational contracting 328
Rosner, M. 417
Rothballer, C. 351, 353
Royal Dutch 417
Royal Lestari Utama (RLU) 193
Russia 8; growth rate 10
Rwanda: infrastructure investments in 8; resilient infrastructure developments in 57; sustainable infrastructure in 82

safety and instrument systems (SIS) 414
Salanick, G.R. 29
salience model 31, 140
Sandy (hurricane) 47, 58
Santhanaman, R. 28
satisfaction, versus consumption 22–23
Saudi Arabia 76; growth rate 10
Saudi Aramco 403, 410
SCADA (supervisory control and data acquisition) systems 401, **406**, 412
Scheer, L.K. 346
Schulz, A. 49
Schutte, I.G. 241
Scottish Future Trust 214
secondary stakeholders 32
Secure Socket Layer (SSL) 415
securitisation 266
Sembcorp Myingyan IPP (Myanmar) 78
Sendai Framework for Disaster Risk Reduction 2015–2030 52
Senegal: China's mega-infrastructure projects in **311**; infrastructure investments in 8
sensitive analysis 278–279
Serra, C.E.M. 23, 24
Shand, W. 82
Sharma, R, 61
Shaw, T 241
Sheffield 377
Shell 403, 417
Sheppard, R. 267
Shi, Q. 107
Shin, J. 416
Shinkansen test train 224–225
Shrestha, A. 325
Siemiatycki, M. 315
Singapore 12, 184
Singh, H. 89
Sinopec 403
situational (contingency) theory **84**
Skidelsky and Miller 238
smart buildings 12

SMART city (Bristol, UK) 394–399; background 395; design principles 398–399; essential pillars *395*; explore, enable and lead 398; goals of 396; innovation ecosystem in 397; One City Plan 395–396; overview 394–395; public service innovation in 398; responsible innovation in 397–398; smart technologies and sustainability in 396; world-class connectivity 396–397

Smith, E. 325

Snowvember (hurricane) 47

social discount rate 277

social distance 9

social infrastructure systems 4

social network analysis (SNA) 142, 324; case study 148–152; foreshore development project 148–152; and network characteristics 144–147; social value performance 145–147

social performance: evaluation of 142; index 146–147; indicators 147; thresholds 143–144, *144*

social sustainability **75**

social value performance 145–147

Society of American Value Engineers (SAVE) 21

socio-technical systems (STS) 433, 434, 435, 438, 439–440, 442, 444, 445

soft infrastructure systems 4; classification of *5*

soft systems methodology (SSM) 377, *378*

soft systems thinking 377; *see also* design thinking; systems thinking

SOURCE 93

South Africa 8, 184; China's mega-infrastructure projects in **311**; growth rate 10

South African National Treasury PPP Unit 351

Southern Tagalog arterial road (STAR) (Philippines) 77

Sovacool, B.K. 56

sovereign risk **257**

Spain, infrastructure systems in 247

Spendolini, M.J. 88

stake 139

stakeholder circle methodology 32

stakeholder engagement 135–153; case study 148–152; Dubai Expo 2020 128; in multi-partner collaborations 166–167; overview 137–138; phases 137–138; schematic diagram *143*; strategies in 141–142; in sustainable infrastructure 80–81; tools and techniques for 139–141

stakeholder(s): as advocate or adversary 32; and benefits realisation 30–34; categories 31, 138–139, *140*; defined 137; end-user community as 139; expectations 142; governance community as 139; identification 137–138, 141; impact assessment 143–144; impact index 32; impacts 142; industry/business community as 138–139; interests 137–138, 142; management 86; management of vs. management for 29–30; mapping with project issues 141–142;

neighbourhood community as 139; primary 29, 32; satisfaction 191; secondary 32; and sustainable infrastructure 80–81

Standard for Sustainable and Resilient Infrastructure (SuRe), 61

Standard Gauge Railway (Kenya) 169–170

standardisation **234**, 434

standards 434

Stanford Engineering 369

Stanford University 369

State Reserve General Fund of Oman 314

stateful packet filtering 416

stateless packer filtering 416

Stip, C. 61

Strategic Infrastructure Planner Framework. 179–180

strategic infrastructure systems: critical infrastructure 4; defined 4; non-critical ventures 4; overview 3; scope of 3–4

Strategic Management: A Stakeholder Approach (Freeman) 29

strategic thinking 86

Strategy and Corporate Governance (SCG) 123

structural materials 226

structuration theory 433

Stuart, Spencer 86

Stuxnet cyberattack 412

submerged floating tube bridge (SFTB) 217

sub-Saharan Africa, infrastructure spending in 303

substitution: actors and socials during 438; digital innovation **437**, 437–438; rules and institutions during 437–438; socio-technical systems during 438

Sultana, M. 241

supervisory cont, digital technologies and applications in **406–408**

supervisory control and data acquisition (SCADA) systems 401, **406**, 412

supplier management **234**

supply chains, capability 337

SuRe 83

sustainability: achieving 135–136; benchmarking 88–89; environmental 271; financial 275–276; guiding principles for dimensions of **75**; leadership 85–87; triple bottom lines of 33

Sustainability Pavilion 111

sustainable construction 107

sustainable development 73–74

sustainable development goals (SDGs) 48, 61, 72, 73, 93, 181–182, *183*

Sustainable Development Investment Partnership (SDIP) 76

Sustainable Development Triple Bottom Line (SDTBL), 112

sustainable energy 109

sustainable environment 109

Index

sustainable infrastructure 12–14, 71–93; benchmarking 83–84, 89–91; case study 94–96; defined 74; in developed and developing countries 266–267; financing 82–83; four dimensions of 74; framework for delivery of 92; global initiatives 75–79; global trends in 77; governance 262–266; influencers 76; innovation/technology integration 79–80; mobilizers 76; overview 73–74; policies and national strategies 81–82; post-COVID-19 development 91–93; prerequisites 79, 79–84; project leadership 84–87; resource efficiency 80; stakeholder management 80–81; systems and tools for **87**, 87–91; tool providers 76; waste mitigation and management 80
sustainable mobility 120–125
sustainable transport 109
sustainable urban development 109
Sutter Health 327
Svejvig, P. 20
Sweden, project appraisals in 192
Swedlund, H.J. 315
Switzerland, cybersecurity regulatory framework in **421**
Sydow, J. 433
syndication 266
system development life cycle 382
systems analysis 376
systems engineering 376, 377
Systems Engineering and Project Management (SEPM) Joint Working Group 380
systems thinking 371–380; actor map 385; benefits of 375; case study 394–399; causal loop diagram 388; causal loop diagram' 388; clarification of systems **394**; concept map 386; corollaries of five axioms of **392–393**; enterprise system 388, 389; fishbone diagram 383; five axioms of **391**; general systems theory 372; hard 376–377; on infrastructure projects 372–376, 380–390; potential impact of 373; project management methods 381; purpose of 371–372; rich picture diagram 384; soft 377; support to engineers 375; in system development life cycle 382; techniques of 390; tools for 377–380; transforming to 374; trend map 387; types of systems and projects 379; zones of complexity 388, 390; see also design thinking

Tabassi, A.A. 85
Tanzania 299, 300, 306; Bagamoyo Port project 313–316; China's mega-infrastructure projects in **311**; public-private partnerships in 345–347
Tanzania International Container Terminal Services (TICTS) 347
Tanzania Investment Centre (TIC) 299
target cost contracting (TCC) 208–209
Taxa de Juros de Longo Prazo (TJPL) 293

Taxa de Longo Prazo (TLP) 293
TCC (target cost contracting) 208–209
team management 191
techical design 271
technical feasibility 271
technological risk **257**
tender documents, in public-private partnerships (PPPs) 301, 346
Thacker, S. 184
Thailand, public-private partnerships in 351
Thames Barrier 30
Thames Estuary 2100 Project 54–55
Thames Tideway 436, 447
Third Riverfront Development 104
time and material (T and M) contracts. 326–329
Tokaido Shinkansen train 224–225
Toor, S.R. 85
town-hall meetings 139
traditional construction procurement system (TCPS) 346
Traffic and Roads Agency (TRA) 123
traits theory **84**
transaction costs 255, 258–259
transformation: actors and social groups during 439; digital innovation **437**; rules and institutions during 438–439; socio-technical systems during 439–440
Transforming Infrastructure Performance (TIP) programme 89
transparency, of regulators 307–308
Transport Layer Security (TLS) 416
transportation infrastructures: highways **60**; life cycle energy framework 240; for mega-events 105–106; performance indicators 242–244; railways **60**; resilient 58–59, **60**; secondary urban roads **60**
travel restrictions 9
triple bottom line (TBL) 135–136
Trojan horses 412
Tropical Landscapes Finance Facility (TLFF) 193–194
Tuas Mega Port (Singapore) 79
Tunde, M.A. 57
Tunisia 8
tunnel formwork system 356
Turkey 8
Turner, J.R. 85
Tyson, J.E. 266

UK BIM Alliance 440
UK BIM Task Group 440
UK Design Group 367
United Arab Emirates 104
United Kingdom: cybersecurity regulatory framework in **421**, 421–422, 422; infrastructure investments in 54; infrastructure systems in 247; sustainable infrastructure in 76–77

Index

United Nations: Climate Change Conference (COP21) 71; Framework Convention on Climate Chang 76; Framework Convention on Climate Change 52, 91; Hyogo Framework for Action 51–52; International Strategy for Disaster Reduction 51; Office for Disaster Risk Reduction 52; Sendai Framework for Disaster Risk Reduction 2015–2030 52; Sustainable Development Goals 48, 61, 396; Sustainable Development Summit (2015) 71, 73; World Conference on Disaster Reduction 51–52

United States 8; cybersecurity regulatory framework in **421**; infrastructure systems in 246–247; project appraisals in 192–193

United States Department of Energy 419

United States Homeland Security 418

United States Navy Bureau of Yards and Docks 21

unmanned aerial vehicles (UAVs) 441

urban development, resilient infrastructure in 55–57

urban roads, secondary **60**

US Army Corps Engineers (USACE) 211

user community 139

Usman, B.A. 57

utilities: mega-events 106; sustainable 117–120

value 21; analysis 21; creation 21; planning 22; satisfaction versus consumption 22–23

value engineering: defined 21; phases 22

value management (VM) 19–34; aim of 23; case study 35–41; chain of benefits model 24, 25; described 22; elements of 23–24; propositional elements 26–28, 27; and public infrastructure projects 20; timeline of 23

Vecchi, V. 325

Vegvesen, S. 217

Vhumbunu, C.H. 314

VINCI Construction Grand Projects 447

virtual private network (VPN) 414

virtual reality (VR) 441

vision, infrastructure: drafting 184; SDG-based 181–182, 183; tensions of 182

Visser, W. 86, 87

Vives, A. 262

volatile organic compounds (VOCs) 116

Von Bertalanffy, Ludwig 372, 376

vulnerability risks 237

Walker, D. 32

Wallace, J. 330

Wang, X. 58

Warner, J. 78

waste management 80, 117

Wastenizer waste-to-energy plants 117

water infrastructures, resilient 59–61, 62

"waterfall-style" project management 380

web-based communication 139

Weber, M.M. 363

Weissman, M. 12

West Coast Main Line (UK) 33

Whitehead, C. 86

Whitford, M. 110

whole life costing 198–199; circular economy principles in 201–203

Whyte, J. 434–435, 440, 441

wicked problems 363–364

Wi-Fi Pineapple 415

Wikileaks 410

willingness-to-pay (WTP) 277

wind **60**

Wolstenholme, Andrew 440–441, 444

Wondimu, P.A. 207, 208, 209

work-breakdown structure (WSS) 190, 376

World Bank 61, 65, 82, 253

world café 139

World Economic Forum (WEF) 83, 179, 239

World Summit on Sustainable Development (WSSD) 110

Worldwide Fund for Nature (WWF) 77

Wright, J.N. 327

Xiaopeng, D 302–305

Xu, Z. 315

Yang, J. 32

Yellow Doctor 224–225

Yeo, K.T. 376

Yoshino, N. 77

Yu, J. 140

Yukl, G. 84, 87

Zambia 306; China's mega-infrastructure projects in **311**

Zietlow, G. 241

Zimbabwe 47

zombie attacks 412

zones of complexity 388, 390

Zou, P. 352

Zuofa, T. 73